Pietro Giuseppe Fré, Alexander Fedotov

Groups and Manifolds

Also of Interest

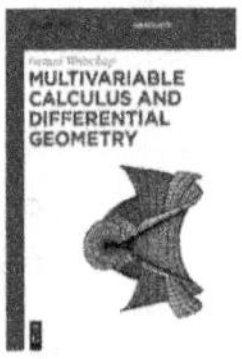

Multivariable Calculus and Differential Geometry
Gerard Walschap, 2015
ISBN 978-3-11-036949-6, e-ISBN 978-3-11-036954-0

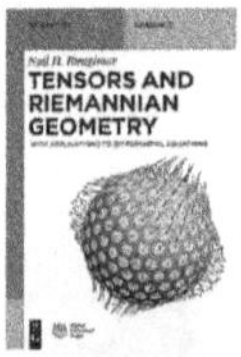

Tensors and Riemannian Geometry
With Applications to Differential Equations
Nail H. Ibragimov, 2015
ISBN 978-3-11-037949-5, e-ISBN 978-3-11-037950-1

Symmetry
Through the Eyes of Old Masters
Emil Makovicky, 2016
ISBN 978-3-11-041705-0, e-ISBN: 978-3-11-041714-2

Series: De Gruyter Studies in Mathematical Physics
M. Efroimsky, L. Gamberg, D. Gitman, A. Lazarian, B. M. Smirnov
ISSN 2194-3532
Published titles in this series:
Vol. 42: Javier Roa: Regularization in Orbital Mechanics (2017)
Vol. 39: Vladimir K. Dobrev: Invariant Differential Operators 2 (2017)
Vol. 35: Vladimir K. Dobrev: Invariant Differential Operators 1 (2016)

Pietro Giuseppe Fré, Alexander Fedotov

Groups and Manifolds

Lectures for Physicists with Examples in Mathematica

DE GRUYTER

Authors
Prof. Pietro Giuseppe Fré
Università degli Studi di Torino
Via P. Giuria 1
10125 Torino
Italy
pietro.fre@esteri.it

Dr Alexander Fedotov
Department of Theoretical Physics
National Research Nuclear University MEPhI
Kashirskoe Shosse 31
Moscow 115409
Russia
AMFedotov@MEPhI.ru

ISBN 978-3-11-055119-8
e-ISBN (PDF) 978-3-11-055120-4
e-ISBN (EPUB) 978-3-11-055133-4

Library of Congress Cataloging-in-Publication Data
A CIP catalog record for this book has been applied for at the Library of Congress.

Bibliographic information published by the Deutsche Nationalbibliothek
The Deutsche Nationalbibliothek lists this publication in the Deutsche Nationalbibliografie; detailed bibliographic data are available on the Internet at http://dnb.dnb.de.

© 2018 Walter de Gruyter GmbH, Berlin/Boston
Typesetting: VTeX UAB, Lithuania
Printing and binding: CPI books GmbH, Leck
Cover image: Science Photo Library / Laguna Design
♾ Printed on acid-free paper
Printed in Germany

www.degruyter.com

This book is dedicated from the part of Pietro Frè to his beloved daughter Laura, to his darling wife Olga and to his younger son Vladimir. On the part of Alexander Fedotov it is dedicated to his darling wife Evgenya and to his beloved daughters Ekaterina and Anna.

Preface

The present book has its initial basis in the course on *Lie Algebras and Lie Groups* that one of us (P.F.) used to give at SISSA when he was Professor of Theoretical Physics there in the years 1990–1996. The stimulus to revise and update those lecture notes to the point of writing a self-contained book on such topics came more recently when it was agreed that P.F., working several years in Moscow as scientific attaché of the Italian Embassy, would give a course on *Groups and Symmetries* at the National Research Nuclear University MEPhI of Moscow. Further momentum the project gained from the association of the first with the second author (A.F.) who conducted weekly sessions of exercises and problems, attached to the Course.

A further motivation to write these introductory *Lectures on Groups and Manifolds for Physicists* came by the imminent publication of other two more advanced books written by the first author for Springer Verlag, namely:

1. *Advances in Lie Algebras and Geometry from Supergravity.*
2. *A Conceptual History of Symmetry from Plato to Special Geometries.*

Symmetry is the keyword underpinning our modern understanding of the Fundamental Laws of Nature and it is also the central focus of most mathematical disciplines, of which Geometry, in this respect, is at the top of the priority list.

By the wording *The Mathematics of Symmetry*, every contemporary Mathematician and Physicist understands *Lie Algebras, Lie Groups, Coset Manifolds* and also *Discrete and Finite Groups*. There is no doubt that such topics are classical and constitute the backbone in the education of students in theoretical physics so that many excellent textbooks exist on the market that have been written both by mathematicians and by physicists. Yet, notwithstanding their established classical status, research continues in these mathematical fields and many developments have taken place in the last twenty years, several of them stimulated and directed by developments in supersymmetric field theories, supergravity and superstrings. As the first author advocates in the above quoted two books, the mathematical needs of the *super-world* have promoted an extensive reframing and a deeper understanding of many chapters in the *Mathematics of Symmetry*, leading to new conceptions that usually are not reflected in the existing textbook literature. As their title suggest, the above quoted two books deal with the task of presenting in a modern unified way the Advances in Geometry and Lie Algebra Theory that are most relevant to contemporary field theories, the supersymmetric ones having top priority. Parallel aim in the writing was that of presenting the conceptual evolution of the *episteme* from ancient times to the contemporary vision. Indeed, trying to summarize the implications for the *episteme* of the last thirty-three years of experimental physics, the first author arrived at the following conclusion.

Leaving apart the issue of **quantization**, that we can generically identify with the **functional path integral over classical configurations**, we have, within our con-

https://doi.org/10.1515/9783110551204-201

temporary physical paradigm, a rather simple and universal scheme of interpretation of the Fundamental Interactions and of the Fundamental Constituents of Matter based on the following few principles:

A) *The categorical reference frame is provided by Field Theory defined by some action* $\mathscr{A} = \int_{\mathcal{M}} \mathscr{L}(\Phi, \partial\Phi)$ *where* $\mathscr{L}(\Phi, \partial\Phi)$ *denotes some Lagrangian depending on a set of fields* $\Phi(x)$.

B) *All fundamental interactions are described by connections* $\mathbf{A}$ *on principal fiber-bundles* $P(G, \mathcal{M})$ *where* G *is a Lie group and the base manifold* $\mathcal{M}$ *is some space-time in* $d = 4$ *or in higher dimensions.*

C) *All the fields* Φ *describing fundamental constituents are sections of vector bundles* $B(G, V, \mathcal{M})$, *associated with the principal one* $P(G, \mathcal{M})$ *and determined by the choice of suitable linear representations* $D(G) : V \to V$ *of the structural group* G.

D) *The spin zero particles, described by scalar fields* ϕ^I, *have the additional feature of admitting non-linear interactions encoded in a scalar potential* $\mathscr{V}(\phi)$ *for whose choice general principles, supported by experimental confirmation, have not yet been determined.*

E) *Gravitational interactions are special among the others and universal since they deal with the tangent bundle* $T\mathcal{M} \to \mathcal{M}$ *to space-time. The relevant connection is in this case the Levi-Civita connection (or some of its generalizations with torsion) which is determined by a metric* g *on* $\mathcal{M}$.

In this framework that is certainly provisional, yet encompasses the whole of Fundamental Physics as we currently understand it, the notions of Lie Group, Lie Algebra, Fiber-Bundle and Connections are the central building blocks.

It follows that contemporary educational programmes in theoretical physics should place the learning of these topics in an equally central position. Furthermore, the approach to such a learning should be tuned to the modern advanced visions discussed by the first author in the quoted two books.

The present textbook, which is introductory, streams from the motivations illustrated above and encodes the following features that make it significantly different from other existing textbooks:

1) The covered topics include both finite groups, Lie groups, Lie algebras and as much as it is necessary of differential geometry to properly understand the notions of manifold, fiber-bundle, coset manifold and connection on a principal fiber-bundle extended to its associated bundles.

2) The exposition is pedagogical and oriented to physical applications. Proofs are presented only for few basic theorems and the preference is given to those that are constructive, leading to useful and applicable algorithms.

3) An extended collection of exercises is presented and solved that helps the students in their process of progressive understanding of primary concepts. The exercises are mainly focused on basic notions of algebra and finite group theory. For Lie groups and manifolds we rather rely on a collection of explicitly worked out examples of increasing complexity.

4) Special emphasis is given to the classification of simple Lie algebras in terms of root systems and Dynkin diagrams, stressing the relation of the ADE classification of simply laced Lie algebras with the ADE classification of finite rotation groups.
5) As anticipated above, the root and weight lattices for Lie algebras are discussed and utilized in several explicit examples of construction of linear representations, illustrating the relation between the Dynkin language and that of tensors, customary in physical theories.
6) A special attention is given to exceptional Lie algebras and to the explicit construction of their fundamental and adjoint representations. Exceptional Lie algebras, considered for long time as *Old Curiosity Shop* items proved instead to be of fundamental relevance in modern theories of the world.
7) In the case of finite group theory, just as in the case of Lie group theory, the teaching approach is based on in-depth case studies where all the advocated conceptions are illustrated in explicit constructions. It goes without saying that we chose cases corresponding to items of special relevance in contemporary theoretical physics. For finite groups we utilize in particular the crystallographic octahedral group and the second smallest simple group with 168 elements that is crystallographic in seven dimensions.
8) One distinctive feature of our presentation concerns the use of MATHEMATICA codes for group-theoretical constructions. We illustrate how all such constructions can be effectively dealt with by means of suitable computer codes and a collection of dedicated MATHEMATICA NoteBooks is illustrated in Appendix B. The actual NoteBooks can be downloaded by the reader of this book from the open source site:

   http://www.degruyter.com/books/978-3-11-055119-8

9) We provide a short but self-contained introduction to the basic notions of differential geometry, namely, connections and metrics that are the basic building blocks for General Relativity and Gauge Theories. Some simple examples are worked out in detail.
10) The last chapter, encoding a self-contained introduction to the geometry of coset manifolds, constitutes the bridge from the basic algebraic lore encoded in this book and more advanced geometrical constructions ubiquitous in all directions of contemporary theoretical physics, from supergravity to statistical mechanics and generic gauge theories.

We think that it is time to reshape and update our teaching of fundamental physics in view of the impressive, although provisional, advances made by the *episteme* in the last thirty years, which one of us tried to summarize in points A)–E). This involves reshaping our teaching of mathematics and in particular of mathematics for physics majors, which is sometimes extremely obsolete and inadequate. In theoretical physics students can proceed to conceptual advances only if their knowledge of the basic mathematical techniques is adequate and computational skills have been properly developed and trained. The present book aims at that goal.

Moscow,
July 2017

Pietro Fré
University of Torino
presently
Scientific Counselor of the Italian Embassy in Moscow

Moscow,
July 2017

Alexander Fedotov
National Research Nuclear University MEPhI

Acknowledgement

With great pleasure one of us (P.G. Fré) would like to thank his former student, collaborator and good friend Marco Billó, whose lecture notes have provided an important basis for the construction of several sections of the initial chapters on finite group theory.

https://doi.org/10.1515/9783110551204-202

Contents

Preface —— VII

Acknowledgement —— XI

Some remarks about notation —— XXI

1 Fundamental notions of algebra —— 1
1.1 Introduction —— 1
1.2 Groups —— 1
1.2.1 Some examples —— 2
1.2.2 Abelian groups —— 2
1.2.3 The group commutator —— 3
1.2.4 Conjugate elements —— 3
1.2.5 Order and dimension of a group —— 3
1.2.6 Order of an element —— 3
1.2.7 The multiplication table of a finite group —— 4
1.2.8 Homomorphisms, isomorphisms, automorphisms —— 4
1.2.9 Rank, generators, relations (a first bird's-eye view) —— 6
1.2.10 Subgroups —— 7
1.3 Rings —— 7
1.4 Fields —— 8
1.5 Vector spaces —— 9
1.5.1 Dual vector spaces —— 10
1.5.2 Inner products —— 10
1.6 Algebras —— 11
1.7 Lie algebras —— 12
1.7.1 Homomorphism —— 13
1.7.2 Subalgebras and ideals —— 13
1.7.3 Representations —— 13
1.7.4 Isomorphism —— 13
1.7.5 Adjoint representation —— 13
1.8 Moduli and representations —— 14
1.9 Bibliographical note —— 15

2 Groups: a bestiary for beginners —— 17
2.1 Survey of the contents of this chapter —— 17
2.2 Groups of matrices —— 17
2.2.1 General linear groups —— 17

2.2.2 Special linear groups — 17
2.2.3 Unitary groups — 18
2.2.4 Special unitary groups — 18
2.2.5 Orthogonal groups — 18
2.2.6 Special orthogonal groups — 19
2.2.7 Symplectic groups — 19
2.2.8 Groups of transformations — 19
2.3 Some examples of finite groups — 20
2.3.1 The permutation groups S_n — 20
2.4 Groups as the invariances of a given geometry — 22
2.4.1 The Euclidean groups — 22
2.4.2 Projective geometry — 24
2.5 Matrix groups and vector spaces — 25
2.5.1 General linear groups and basis changes in vector spaces — 25
2.5.2 Tensor product spaces — 25
2.5.3 (Anti)-symmetrized product spaces — 26
2.5.4 Exterior forms — 27
2.5.5 Special linear groups as volume-preserving transformations — 29
2.5.6 Metric-preserving changes of basis — 29
2.5.7 Isometries — 32
2.6 Bibliographical note — 33

3 Basic elements of finite group theory — 35
3.1 Introduction — 35
3.2 Basic notions and structural theorems for finite groups — 36
3.2.1 Cayley's theorem — 36
3.2.2 Left and right cosets — 37
3.2.3 Lagrange's theorem — 38
3.2.4 Conjugacy classes — 39
3.2.5 Conjugate subgroups — 41
3.2.6 Invariant subgroups — 41
3.2.7 Factor groups — 41
3.2.8 Center, centralizers, normalizers — 42
3.2.9 The derived group — 43
3.2.10 Simple, semisimple, solvable groups — 43
3.2.11 Examples of simple groups — 44
3.2.12 Homomorphism theorems — 44
3.2.13 Direct products — 45
3.2.14 Action of a group on a set — 46
3.2.15 Semi-direct products — 48
3.2.16 Extensions of a group — 50
3.2.17 Abelian groups — 51

3.3 The linear representations of finite groups — 51
3.3.1 Schur's lemmas — 53
3.3.2 Characters — 56
3.3.3 Decomposition of a representation into irreducible representations — 57
3.3.4 The regular representation — 58
3.4 Strategy to construct the irreducible representations of a solvable group — 59
3.4.1 The inductive algorithm for irreps — 60
3.4.2 A) Uplifting of self-conjugate representations — 61
3.4.2 B) Uplifting of not self-conjugate representations — 61
3.4.3 The octahedral group $O_{24} \sim S_4$ and its irreps — 62
3.5 Bibliographical note — 67

4 Finite subgroups of SO(3) and crystallographic groups — 69
4.1 Introduction — 69
4.2 The ADE classification of the finite subgroups of SU(2) — 70
4.2.1 The argument leading to the Diophantine equation — 71
4.2.2 Case $r = 2$: the infinite series of cyclic groups $\mathbb{A}_n$ — 75
4.2.3 Case $r = 3$ and its solutions — 76
4.2.4 Summary of the ADE classification of finite rotation groups — 79
4.3 Lattices and crystallographic groups — 81
4.3.1 Lattices — 81
4.3.2 The n-torus T^n — 82
4.3.3 Crystallographic groups and the Bravais lattices for $n = 3$ — 82
4.3.4 The proper Point Groups — 84
4.3.5 The cubic lattice and the octahedral Point Group — 84
4.3.6 Irreducible representations of the octahedral group — 85
4.4 A simple crystallographic point-group in seven dimensions — 86
4.4.1 The simple group L_{168} — 88
4.4.2 Structure of the simple group $L_{168} = PSL(2, \mathbb{Z}_7)$ — 89
4.4.3 The 7-dimensional irreducible representation — 90
4.4.4 The 3-dimensional complex representations — 93
4.4.5 The 6-dimensional representation — 94
4.4.6 The 8-dimensional representation — 95
4.4.7 The proper subgroups of L_{168} — 95
4.5 Bibliographical note — 102

5 Manifolds and Lie groups — 103
5.1 Introduction — 103
5.2 Differentiable manifolds — 104
5.2.1 Homeomorphisms and the definition of manifolds — 106

5.2.2 Functions on manifolds — 110
5.2.3 Germs of smooth functions — 112
5.3 Tangent and cotangent spaces — 113
5.3.1 Tangent vectors at a point $p \in \mathscr{M}$ — 114
5.3.2 Differential forms at a point $p \in \mathscr{M}$ — 118
5.4 About the concept of fiber-bundle — 120
5.5 The notion of Lie group — 121
5.6 Developing the notion of fiber-bundle — 122
5.7 Tangent and cotangent bundles — 128
5.7.1 Sections of a bundle — 130
5.7.2 The Lie algebra of vector fields — 132
5.7.3 The cotangent bundle and differential forms — 134
5.7.4 Differential k-forms — 136
5.8 Lie groups and Lie algebras — 138
5.8.1 The Lie algebra of a Lie group — 138
5.8.2 Maurer–Cartan forms on Lie group manifolds — 145
5.8.3 Maurer Cartan equations — 147
5.9 Matrix Lie groups — 148
5.9.1 Some properties of matrices — 148
5.9.2 Linear Lie groups — 150
5.10 Bibliographical note — 150

6 Structure of Lie algebras — 151
6.1 Introduction — 151
6.2 Homotopy, homology and cohomology — 151
6.2.1 Homotopy — 153
6.2.2 Homology — 156
6.2.3 Homology and cohomology groups: general construction — 162
6.2.4 Relation between homotopy and homology — 164
6.3 Linear algebra preliminaries — 165
6.4 Types of Lie algebras and Levi's decomposition — 167
6.4.1 Solvable Lie algebras — 168
6.4.2 Semisimple Lie algebras — 171
6.4.3 Levi's decomposition of Lie algebras — 171
6.4.4 An illustrative example: the Galilean group — 176
6.5 The adjoint representation and Cartan's criteria — 177
6.5.1 Cartan's criteria — 178
6.6 Bibliographical note — 180

7 Root systems and their classification — 181
7.1 Cartan subalgebras — 181
7.2 Root systems — 184

7.2.1 Final form of the semisimple Lie algebra — **187**
7.2.2 Properties of root systems — **187**
7.3 Simple roots, the Weyl group and the Cartan matrix — **190**
7.4 Classification of the irreducible root systems — **192**
7.4.1 Dynkin diagrams — **193**
7.4.2 The classification theorem — **194**
7.5 Identification of the classical Lie algebras — **200**
7.5.1 The $\mathfrak{a}_\ell$ root system and the corresponding Lie algebra — **200**
7.5.2 The $\mathfrak{d}_\ell$ root system and the corresponding Lie algebra — **204**
7.5.3 The $\mathfrak{b}_\ell$ root system and the corresponding Lie algebra — **206**
7.5.4 The $\mathfrak{c}_\ell$ root system and the corresponding Lie algebra — **207**
7.5.5 The exceptional Lie algebras — **209**
7.6 Bibliographical note — **209**

8 Lie algebra representation theory — 211
8.1 Linear representations of a Lie algebra — **211**
8.1.1 Weights of a representation and the weight lattice — **212**
8.2 Discussion of tensor products and examples — **220**
8.2.1 Tensor products and irreps — **221**
8.2.2 The Lie algebra $\mathfrak{a}_2$, its Weyl group and examples of its representations — **225**
8.2.3 The Lie algebra $\mathfrak{sp}(4,\mathbb{R}) \simeq \mathfrak{so}(2,3)$, its fundamental representation and its Weyl group — **235**
8.3 Conclusions for this chapter — **245**
8.4 Bibliographical note — **245**

9 Exceptional Lie algebras — 247
9.1 The exceptional Lie algebra $\mathfrak{g}_2$ — **247**
9.1.1 A golden splitting for quaternionic algebras — **251**
9.1.2 The golden splitting of the quaternionic algebra $\mathfrak{g}_2$ — **254**
9.1.3 Chevalley–Serre basis — **257**
9.2 The Lie algebra $\mathfrak{f}_4$ and its fundamental representation — **258**
9.2.1 Explicit construction of the fundamental and adjoint representation of $\mathfrak{f}_4$ — **262**
9.3 The exceptional Lie algebra $\mathfrak{e}_7$ — **268**
9.3.1 The matrices of the fundamental 56 representation — **274**
9.3.2 The $\mathfrak{sl}(8,\mathbb{R})$ subalgebra — **276**
9.4 The exceptional Lie algebra $\mathfrak{e}_8$ — **277**
9.4.1 Construction of the adjoint representation — **279**
9.4.2 Final comments on the $\mathfrak{e}_8$ root systems — **283**
9.5 Bibliographical note — **284**

10 A primary on the theory of connections and metrics — 285
10.1 Introduction — 285
10.2 Connections on principal bundles: the mathematical definition — 285
10.2.1 Ehresmann connections on a principal fiber-bundle — 286
10.3 Connections on a vector bundle — 293
10.4 An illustrative example of fiber-bundle and connection — 297
10.4.1 The magnetic monopole and the Hopf fibration of S^3 — 297
10.5 Riemannian and pseudo-Riemannian metrics: the mathematical definition — 302
10.5.1 Signatures — 303
10.6 The Levi-Civita connection — 305
10.6.1 Affine connections — 305
10.6.2 Curvature and torsion of an affine connection — 306
10.7 Geodesics — 309
10.8 Geodesics in Lorentzian and Riemannian manifolds: two simple examples — 311
10.8.1 The Lorentzian example of dS_2 — 311
10.8.2 The Riemannian example of the Lobachevsky–Poincaré plane — 317
10.8.3 Another Riemannian example: the catenoid — 319
10.9 Bibliographical note — 322

11 Isometries and the geometry of coset manifolds — 323
11.1 Conceptual and historical introduction — 323
11.1.1 Symmetric spaces and Élie Cartan — 324
11.1.2 Where and how do coset manifolds come into play? — 325
11.1.3 The deep insight of supersymmetry — 326
11.2 Isometries and Killing vector fields — 326
11.3 Coset manifolds — 327
11.3.1 The geometry of coset manifolds — 332
11.4 The real sections of a complex Lie algebra and symmetric spaces — 343
11.5 The solvable group representation of non-compact coset manifolds — 346
11.5.1 The Tits–Satake projection: just a flash — 349
11.6 Bibliographical note — 349

12 An anthology of group theory physical applications — 351
12.1 Introduction — 351
12.2 The full tetrahedral group and the vibrations of XY_4 molecules — 352
12.2.1 Group theory of the full tetrahedral group — 353
12.3 The hidden symmetry of the hydrogen atom spectrum — 359
12.3.1 Classical Kepler problem — 361

12.3.2 Hydrogen atom — 364
12.4 The AdS/CFT correspondence and (super) Lie algebra theory — 366
12.4.1 Particles in anti de Sitter space and induced representations — 369
12.4.2 Unitary irreducible representations of SO(2, 3) — 376
12.4.3 The Osp($\mathcal{N}$|4) superalgebra: definition, properties and notation — 380
12.4.4 Structure of the Osp(2|4) multiplets — 391
12.5 Bibliographical note — 397

A Exercises — 399
A.1 Exercises for Chapter 1 — 400
A.2 Exercises for Chapter 2 — 401
A.3 Exercises for Chapter 3 — 402
A.4 Solutions and answers to selected problems — 408

B MATHEMATICA NoteBooks — 441
B.1 Introduction — 441
B.2 The NoteBook e88construziaNew — 441
B.3 The NoteBook g2construzia — 442
B.4 The NoteBook e77construzia — 442
B.4.1 Instructions for the user — 442
B.5 The NoteBook F44construzia.nb — 444
B.5.1 The $\mathfrak{f}_4$ Lie algebra — 444
B.5.2 Maximal compact subalgebra $\mathfrak{su}(2) \times \mathfrak{sp}(6, \mathbb{R}) \subset F_{(4,4)}$ — 445
B.5.3 The $\mathfrak{sl}(2, \mathbb{R}) \times \mathfrak{sp}(6, \mathbb{R})$ subalgebra and the **W**-representation — 446
B.6 The NoteBook L168Group — 447
B.6.1 Description of the generated objects — 448
B.6.2 The basic commands of the present package — 451
B.7 The NoteBook octagroup2 — 456
B.7.1 Objects and commands available to the user — 457
B.7.2 Auxiliary group theoretical routines used by the package but available also to the user — 458
B.8 The NoteBook metricgrav — 459
B.8.1 Description of the programme — 460
B.8.2 Initialization and inputs to be supplied — 460
B.8.3 Produced outputs — 460
B.9 The NoteBook Vielbgrav — 461
B.9.1 Instructions for the user — 461
B.10 Roots of $\mathfrak{e}_8$ — 461
B.10.1 Listing of the $\mathfrak{e}_8$ roots — 462

Bibliography — 469

Some remarks about notation

Concerning cyclic groups we use alternatively the notation $\mathbb{Z}_k$ or C_k. In particular we give preference to the second notation when we look at the cyclic group as an abstract group, while the first notation is preferably used when we think of it as given by integer numbers *mod k*.

In general, when discussing Lie groups we utilize the notation G for the abstract Lie group and $\mathbb{G}$ for the corresponding Lie algebra. In several cases for specific Lie algebras, like the exceptional ones, we also use the notation $\mathfrak{e}_{6,7,8}, \mathfrak{g}_2, \mathfrak{f}_4$ or similarly $\mathfrak{a}_\ell, \mathfrak{b}_\ell, \mathfrak{c}_\ell, \mathfrak{d}_\ell$ for the classical Lie algebras.

According to a quite standard mathematical notation by $\mathrm{Hom}(V, V)$ we denote the group of linear homomorphisms of a vector space V into itself.

We use throughout the book the standard convention that repeated indices are summed over. Depending on the case and on graphical convenience we utilize either the 8 to 2 or the 10 to 4 convention, sometimes also the repetition at the same level of the same index.

The *Euclidean Group* in n dimensions includes rotations, translations and reflections and it is denoted Eucl_n. It includes the roto-translation group $\mathrm{ISO}(n)$ as a proper subgroup.

In relation with the number of elements contained in a set $\mathcal{S}$ we use alternatively either the notation $|\mathcal{S}|$, typically reserved to case when $\mathcal{S}$ is a finite group G or the notation $\mathrm{card}\mathcal{S}$ when $\mathcal{S}$ is some type of set, for instance the set of all roots of a simple Lie algebra. When we consider specific type of objects we also utilize the often employed notation # of *objects such and such*.

Let us clarify notation for elements of a Lie algebra. In Chapter 6 and earlier in Chapter 1, while discussing general theorems on Lie algebras, we have denoted elements $\mathbf{X} \in \mathbb{G}$ of a generic Lie algebra $\mathbb{G}$ by boldface letters to emphasize their nature of vectors in a vector space and to facilitate the distinction with respect to groups, algebras and subalgebras. In Chapters 7, 8 and 9 where we develop and extensively use the formalism of roots and weights we have suppressed the boldface notation for Lie algebra elements in order to come in touch with the standard notation utilized for generators in the Cartan–Weyl basis and also to avoid the excess of boldface symbols that would have resulted.

In the chapters devoted to differential geometry and group manifolds, namely Chapters 5 and 10, vector fields are typically denoted by boldface capital letters: in particular we use such a notation for the left- or right-invariant vector fields that generate the Lie algebra of a Lie group.

Concerning tensor product of vector spaces, when it is convenient for us, we utilize abbreviation $V_{m|n}$, which denotes the tensor product of m copies of a vector space V_n of dimension n.

https://doi.org/10.1515/9783110551204-203

1 Fundamental notions of algebra

Quando chel cubo con le cose appresso
Se agguaglia a qualche numero discreto
Trouan dui altri differenti in esso.
Niccoló Tartaglia

1.1 Introduction

The present chapter summarizes the formal definitions of all the algebraic structures we are going to use in the sequel. These algebraic structures are presented in their natural hierarchical order of increasing complexity:

1 GROUPS
2 RINGS
3 FIELDS
4 VECTOR SPACES
5 ALGEBRAS
6 LIE ALGEBRAS
7 MODULES and REPRESENTATIONS

The purpose of this chapter is to establish the current mathematical language which will be our reference point in all subsequent discussions.

Groups correspond to the simplest in the above list of algebraic structures since they involve just one internal binary operation while all the other structures have more than one, typically a *sum* and a *product*, or also a multiplication by *scalar coefficients*, as is the case of vector spaces and modules. Yet groups constitute the most important algebraic structure because of their basic interpretation as a *set of transformations* acting on some *space* composed of *objects* which can have the most different nature. However, in developing Group Theory, in particular in developing the Theory of Group Representations, all the other algebraic structures of the above list come into play and for this reason we need to review them here. Furthermore, in order to deal with *continuous groups* and their relation with the differential geometry of *homogeneous spaces*, we have to introduce the notions of *manifolds*, *fiber-bundles* and *differential forms*. This will be done in later chapters.

1.2 Groups

In this section we introduce the algebraic definition of a group and we illustrate it with several different examples. Before moving to the next algebraic structure, namely to *Rings*, we fix the basic concepts of order of a group, homomorphisms, isomorphisms,

https://doi.org/10.1515/9783110551204-001

automorphisms, generators, and relations. The analysis of the general algebraic properties of the group structure and the related fundamental theorems are postponed to Chapter 3 where they are considered in full generality and in depth.

Definition 1.2.1. A group G is a set equipped with a binary operation $\cdot$ such that:
1. $\forall a, b, c \in G$, $a \cdot (b \cdot c) = (a \cdot b) \cdot c$, i.e., *associativity* holds;
2. $\exists e \in G$ such that $a \cdot e = e \cdot a = a\ \forall a \in G$, i.e., there is an *identity* element e;
3. $\forall a \in G\ \exists b \in G$ such that $a \cdot b = b \cdot a = e$, i.e., each element a admits an *inverse* b which is usually denoted as a^{-1}.

In the following, we will often indicate the group product simply as ab, or we may indicate it with different symbols, e.g., as $a + b$, when the product law is in fact the usual addition. To stress a specific choice of the product law, we may also indicate the group as, for instance, $(G, +)$.

1.2.1 Some examples

- Consider the set $\{0, 1\}$ with the group product being the usual addition defined mod 2; this is a group, usually denoted as $\mathbb{Z}_2$.
- The set $\{1, -1\}$ equipped with the usual multiplication is a group. This group is isomorphic, i.e., it has the same abstract structure as the group $\mathbb{Z}_2$ considered in the previous example.
- The set of real numbers $\mathbb{R}$, with the group law being the addition, is a group.
- The set $\mathrm{U}(1) \equiv \{e^{i\theta}; \theta \in [0, 2\pi]\}$ with the usual multiplication is a group.
- The groups $\{e, a, a^2 \equiv a \cdot a, \ldots, a^{k-1}\}$ containing all the powers of a single generator a with respect to the group product, with the single extra relation $a^k = e$, are named *cyclic groups*, and are denoted as $\mathbb{Z}_k$.
- The set of permutations of three numbered objects 1, 2, 3 forms a group called S_3, the product law being the composition of the permutations. This group has order 6, containing: the identical permutation, three exchanges: $p_{12} = (1 \leftrightarrow 2)$, $p_{13} = (1 \leftrightarrow 3)$, $p_{23} = (2 \leftrightarrow 3)$, and two cyclic permutations: $p_{123} = (1 \to 2 \to 3 \to 1)$ and $p_{132} = (1 \to 3 \to 2 \to 1)$.

1.2.2 Abelian groups

The group product is *not* required to be commutative. When the product *is* commutative, the group is called an *abelian group*:

$$G \text{ abelian}: \quad ab = ba \quad \forall a, b \in G \tag{1.1}$$

Abelian groups are of course the simplest types of groups. All the groups of the previous examples are in fact abelian, except the permutation group S_3.

1.2.3 The group commutator

Two elements g, h of a group commute if, with the group product, $gh = hg$, i.e., $ghg^{-1}h^{-1} = e$, with e being the identity. Then the *group commutator* of g and h, defined as

$$ghg^{-1}h^{-1} \tag{1.2}$$

indicates if (and how) the two elements fail to commute.

1.2.4 Conjugate elements

Two group elements h and h' are said to be *conjugate* to each other if

$$\exists g \in G \quad \text{such that } h' = g^{-1}hg \tag{1.3}$$

If the group is abelian, each element is conjugate only to itself.

1.2.5 Order and dimension of a group

The number of elements of a group G can be finite, infinite but denumerable, or continuously infinite.

In the first two cases, the number of elements of G is named the *order* of the group, and denoted as $|G|$ (in the second case, $|G| = \infty$). A group of finite order is called a *finite group*.

1.2.5.1 Examples

- The cyclic group $\mathbb{Z}_k$ is a finite group of order k.
- The integers $\mathbb{Z}$ with the group product being the addition form a group of infinite order.
- The set of real numbers (with the zero excluded) $\mathbb{R} \setminus \{0\}$, equipped with the ordinary product, is a continuous group.

1.2.6 Order of an element

If $a \in G$, then $a^2, a^3, \ldots \in G$. If all the powers of a are distinct, then a is an element of *infinite order* (then, of course, G cannot be a group of finite order). If some of the powers of a coincide, this means that there exists some integer m such that $a^m = e$. Let n be the smallest of such positive integers m. Then a is said to be an element *of order n*.

1.2.6.1 Observations

- In a finite group G all elements have order $\leq |G|$.
- If G is a cyclic group of order n, then the order of any element is a divisor of n.
- In a finite group, the inverse of any element a is a power of a: indeed, $a^{-1} = a^{n-1}$, where n is the order of a.

1.2.7 The multiplication table of a finite group

A group is abstractly defined by describing completely (i.e., $\forall g_1, g_2 \in G$) the product law $(g_1, g_2) \mapsto g_1 g_2$, namely, by identifying the result of each possible product. For finite groups this can be encoded explicitly in a *multiplication table* whose entry in the i-th row and j-th column describes which element is the result of the product $g_i g_j$ (so we use the convention that $g_1 = e$):

e	g_2	g_3	$\cdots$
g_2	$(g_2)^2$	$(g_2 g_3)$	$\cdots$
g_3	$(g_3 g_2)$	$(g_3)^2$	$\cdots$
$\vdots$	$\vdots$	$\vdots$	$\ddots$

Notice that all the elements appearing in a line of the multiplication table are different (and therefore all elements appear in each line). Indeed, if we had $g_i g_j = g_i g_k$, we could conclude that $g_j = g_k$. The same applies to each column. These properties constrain the possible multiplication tables, especially at low orders (see Exercises 1.5, 3.23).

A finite group is abstractly defined by its multiplication table, up to relabeling of the elements (i.e., up to rearrangement of the rows and columns). A given table, i.e., a given group, may have different concrete realizations. We will shortly make more precise what we mean by this observation.

1.2.8 Homomorphisms, isomorphisms, automorphisms

We have seen that we can have different realizations of a given abstract group. We are interested in concrete realizations of the group, where the elements of the group acquire an explicit meaning as numbers, matrices, symmetry operations or other quantities with which we can perform explicit computations.

In precise terms, finding a different realization G' of a given group G means to find an *isomorphic mapping* (or *isomorphism*) between G and G'. Let us explain this terminology.

1.2.8.1 Homomorphisms

A map ϕ from a group G to a group G' is called a *homomorphism* **iff** it preserves the group structure, namely, **iff**

$$\forall g_1, g_2 \in G, \quad \phi(g_1 g_2) = \phi(g_1)\phi(g_2) \tag{1.4}$$

where the products $g_1 g_2$ and $\phi(g_1)\phi(g_2)$ are taken with the group product law of G and G', respectively. The map is not required to be invertible, i.e. one-to-one.

1.2.8.2 Isomorphisms

A homomorphism $\phi : G \to G'$ which is also *invertible* is called an *isomorphism*. Two groups G and G' such that there exists an isomorphism $\phi : G \to G'$ are said to be *isomorphic*. They correspond to different realizations of the same abstract group structure.

1.2.8.3 Automorphisms

An isomorphic mapping σ from a group G to itself is called an *automorphism* of G. The set of all automorphisms of a given group G is a group called $\mathrm{Aut}(G)$, the product law being the composition of mappings. Indeed, the composition of two automorphisms σ_1, σ_2 is still an automorphism:

$$\sigma_1(\sigma_2(h_1 h_2)) = \sigma_1(\sigma_2(h_1)\sigma_2(h_2)) = \sigma_1(\sigma_2(h_1))\sigma_1(\sigma_2(h_2)) \tag{1.5}$$

and of course the composition of mappings is associative, there is an identical automorphism and any automorphism is assumed to be invertible by definition so that the group axioms are satisfied.

For finite groups, the automorphisms of G are particular permutations of the elements of G, namely $\mathrm{Aut}(G)$ is a subgroup of $S_{|G|}$. They correspond to symmetries of the multiplication table of G, in the following sense. As we remarked, the ordering of the rows and columns of the multiplication table is irrelevant. If we apply a given map $\sigma : g \mapsto \sigma(g)$ to labels and entries of the multiplication table it may happen that the resulting table corresponds just to a rearrangement of rows and columns of the original table. In this case the permutation σ is an automorphism.

An important class of automorphic mappings is that corresponding to conjugation by a fixed element of a group:

$$\sigma_g : h \in G \mapsto \sigma_g(h) = g^{-1} h g \in G \tag{1.6}$$

Automorphisms that correspond to conjugations are called *inner automorphisms*, automorphisms which do not, *outer automorphisms*. Notice that for abelian groups the only non-trivial automorphisms are the outer ones.

1.2.9 Rank, generators, relations (a first bird's-eye view)

Let the elements of a finite group G be $g_1 = e, g_2, g_3, \ldots$. All elements can be written in the form

$$g_k = g_{i_1} g_{i_2} \cdots g_{i_s} \tag{1.7}$$

for suitable elements g_{i_j} (at worst the only possibility is $g_k = g_k$). A set of group elements (different from the identity e) which, multiplied in all possible ways, give *all* the elements of G is said to *generate* the group G. The minimal such set is a set of *generators* of G. The minimal number of generators is the *rank* of G.

If a group G is denumerable rather than finite, then we say that it is generated by a finite subset B of elements iff every element $g \in G$ can be written as

$$g = g_{i_1}^{\pm 1} g_{i_2}^{\pm 1} \ldots g_{i_s}^{\pm 1} \tag{1.8}$$

with all the g_{i_l} belonging to B. So, with respect to the case of finite groups, now not only positive but also negative powers of generators may appear. In the case of finite groups, all generators have finite order, so their negative powers can be re-expressed in terms of the positive ones; this is not true in general.

1.2.9.1 Presentation of a group: relations

A group can be described by listing the *generators* and (if present) the *relations* that such generators satisfy. Such a description of a group is called a *presentation* of the group.

Indeed, starting with the generators g_a ($a = 1, \ldots, \text{rank}\, G$) one can construct *words* of increasing length: g_a, then $g_a g_b$, etc. In this way one obtains the further elements of G. In the process, however, me must take into account the *relations* to which the generators may be subject, which may be cast in the form $R_i(\{g_a\}) = e$, the R_i being a set of specific words. If G is a finite group, then each generator must be of finite order and therefore we have at least the relations $g_a^{n_a} = e$, where n_a is the order of g_a; there can be others.

We return to this issue in Chapter 3.

1.2.9.2 Examples

- The group S_3 is defined by its presentation consisting of two generators: $s = p_{12}$ and $t = p_{13}$ subject to the relations: $s^2 = e$, $t^2 = e$ and $(st)^3 = e$ (see Exercise 1.6).

For continuous groups, the analog of generators is given, as we will see, by the existence of a set of *infinitesimal generators* closing a *Lie algebra*.

1.2.10 Subgroups

A subset $H \subset G$ is a *subgroup* of G if it is a group, with the same product law defined in G. To this effect it is sufficient that
i) $\forall h_1, h_2 \in H,\ h_1 h_2 \in H$
ii) $\forall h_1 \in H,\ h_1^{-1} \in H$

While for an infinite (or continuous) group both requirements have to be separately checked, if G is a finite group, then i) suffices since all elements are of finite order.

The relation "being subgroup of" is transitive:

$$\begin{cases} H \subset G, \\ K \subset H \end{cases} \Rightarrow \quad K \subset G \tag{1.9}$$

where $H \subset G$ means H *is a subgroup of* G. In general, a given group G admits chains of subgroups

$$G \supset H_1 \supset H_2 \supset \cdots \supset \{e\} \tag{1.10}$$

G itself and the group containing only the identity e are *trivial* subgroups of G; other subgroups are called *proper* subgroups. One of the most important problems in group theory is the determination (up to conjugation) of all proper subgroups of a given group (see Exercises 1.4, 1.6, 1.7).

Infinite groups may admit infinite sequences of subgroups.

1.3 Rings

Let us now turn to the definition of Rings.

Definition 1.3.1. A ring $\mathscr{R}$ is a set equipped with two binary composition laws respectively named the *sum* and the *product*:

$$\begin{aligned} \forall x, y \in \mathscr{R}: &\quad x + y \in \mathscr{R} \\ \forall x, y \in \mathscr{R}: &\quad x \cdot y \in \mathscr{R} \end{aligned} \tag{1.11}$$

which satisfy the following properties.
R1 With respect to the *sum* $\mathscr{R}$ being an abelian group, namely

$$x + y = y + x$$

there exists a neutral element $0 \in \mathscr{R}$ such that $0 + x = x$, $\forall x \in \mathscr{R}$ and an inverse $-x$ of each element $x \in \mathscr{R}$ such that $x + \{-x\} = 0$.

R2 The product has the *distributive* property with respect to the sum:

$$\forall x, y, z \in \mathcal{R}: \quad x \cdot (y + z) = x \cdot y + x \cdot z$$

R3 The product is *associative*:

$$\forall x, y, z \in \mathcal{R}: \quad (x \cdot y) \cdot z = x \cdot (y \cdot z)$$

R4 There exists an identity element $e \in \mathcal{R}$ for the product such that:

$$\forall x \in \mathcal{R}: \quad x \cdot e = e \cdot x = x$$

A typical example of a ring is provided by the relative integers $\mathbb{Z}$. The identity element for the product is obviously the number 1. Note that with respect to the product the existence of an inverse element is not required. This is precisely the case of integer numbers. The inverse of an integer is not integer in general; indeed, the only exception is precisely the number 1.

Another example of a ring is provided by the continuous and finite functions over an interval, for instance $[0, 1]$ (see Exercise 1.8).

1.4 Fields

Next we introduce the definition of a field.

Definition 1.4.1. A field $\mathbb{K}$ is a ring with the additional property that

K $\forall x \in \mathbb{K}$ different from the neutral element with respect to the sum $x \neq 0$ there is an inverse element, namely $\exists x^{-1} \in \mathbb{K}$, such that:

$$x \cdot x^{-1} = x^{-1} \cdot x = e$$

If the product is commutative, then we say that $\mathbb{K}$ is a commutative field.

Examples of commutative fields are provided by the set of rational numbers $\mathbb{Q}$, by the set of real numbers $\mathbb{R}$ and by the set of complex numbers $\mathbb{C}$.

There are also fields with a finite number of elements. For instance $\mathbb{Z}_q$, the set of integer numbers modulo a prime q, is a commutative field with exactly q elements. An element of $[a] \in \mathbb{Z}_q$ is the entire equivalence class of integers of the form:

$$[a] = a + nq; \quad a, n, q \in \mathbb{Z} \tag{1.12}$$

With respect to the sum $\mathbb{Z}_q$ is an abelian cyclic group as we have already seen. It suffices to note that the product operation goes to the quotient with respect to the equivalence classes. Indeed,

$$(a + qn)(b + mq) = ab + q(ma + nb + qnm) \sim [ab] \tag{1.13}$$

Furthermore, for each equivalence class $[a]$ there is always another equivalence class $[b] = [a]^{-1}$ such that $ab = 1 \bmod q$, namely the inverse element. As an example consider $\mathbb{Z}_5$. With respect to the product we have the following multiplication table of equivalence classes:

$$\begin{array}{c|cccc} & 1 & 2 & 3 & 4 \\ \hline 1 & 1 & 2 & 3 & 4 \\ 2 & 2 & 4 & 1 & 3 \\ 3 & 3 & 1 & 4 & 2 \\ 4 & 4 & 3 & 2 & 1 \end{array} \tag{1.14}$$

Further examples are provided in Exercises 1.9 and 1.10.

1.5 Vector spaces

Definition 1.5.1. A vector space is defined giving a set V, whose elements are named vectors $\mathbf{v} \in V$, $\mathbf{w} \in V, \ldots$ and a commutative field $\mathbb{K}$ (either $\mathbb{R}$ or $\mathbb{C}$), whose elements are named scalars. Two binary operations are then introduced. The first, named sum $+$ is a map:

$$V \times V \to V \tag{1.15}$$

With respect to the sum, V is a commutative group:

V1 $\forall \mathbf{v}, \mathbf{w} \in V$, we have $\mathbf{v} + \mathbf{w} = \mathbf{w} + \mathbf{v} \in V$

V2 $\exists \mathbf{0} \in V$ such that $\forall \mathbf{v} \in V$, the following is a true statement: $\mathbf{0} + \mathbf{v} = \mathbf{v} + \mathbf{0}$

V3 $\forall \mathbf{v} \in V$ there exists the opposite element $-\mathbf{v} \in V$ such that $\mathbf{v} + (-\mathbf{v}) = \mathbf{0}$

The second operation, named multiplication times scalars, associates to each pair $(\lambda, \mathbf{v})$ of a scalar $\lambda \in \mathbb{K}$ and a vector $\mathbf{v} \in V$ a new vector simply denoted $\lambda\mathbf{v} \in V$, namely it is the map:

$$\mathbb{K} \times V \to V \tag{1.16}$$

Naming 1 and 0 the neutral elements of the field, respectively for its internal multiplication and addition, the operation (1.16) has the following defining properties:

V4 $\forall \mathbf{v} \in V$: $1\mathbf{v} = \mathbf{v}$

V5 $\forall \lambda, \mu \in \mathbb{K}$ and $\forall \mathbf{v} \in V$: $\lambda(\mu\mathbf{v})\mathbf{v} = (\lambda\mu)\mathbf{v}$

V6 $\forall \lambda, \mu \in \mathbb{K}$ and $\forall \mathbf{v} \in V$: $(\lambda + \mu)\mathbf{v} = \lambda\mathbf{v} + \mu\mathbf{v}$

V7 $\forall \lambda \in \mathscr{K}$ and $\forall \mathbf{v}, \mathbf{w} \in V$: $\lambda(\mathbf{v} + \mathbf{w}) = \lambda\mathbf{v} + \lambda\mathbf{w}$

From the above axioms it follows that $0\mathbf{v} = \mathbf{0}$.

Familiar examples of vector spaces are $\mathbb{R}^n$ and $\mathbb{C}^n$.

We also recall that vector spaces can be of finite or infinite dimension. Finite-dimensional vector spaces are those V for which a set of m linearly independent vectors $\mathbf{e}_i$ ($i = 0, 1, \ldots, m$) can be found such that

B1) $\sum_{i=1}^{m} \alpha^i \mathbf{e}_i = \mathbf{0}$ with $\alpha^i \in \mathbb{F}$ necessarily implies $\alpha_i = 0$

B2) $\forall \mathbf{v} \in V\ \exists v^i \in \mathbb{F}$ such that $v = \sum_{i=1}^{m} v^i \mathbf{e}_i$

The number m is named the dimension of the vector space.

1.5.1 Dual vector spaces

Definition 1.5.2. A linear functional f on a vector space V is a linear map $f : V \to \mathbb{F}$ such that

$$f(\alpha \mathbf{v}_1 + \beta \mathbf{v}_2) = \alpha f(\mathbf{v}_1) + \beta f(\mathbf{v}_2)$$

Definition 1.5.3. Linear maps on a vector space V can be linearly composed $(\alpha f_1 + \beta f_2)(\mathbf{v}) = \alpha f_1(\mathbf{v}) + \beta f_2(\mathbf{v})$ and therefore they form a vector space which is denoted $V^\star$ and it is named the dual vector space.

For finite-dimensional vector spaces V, the dual space $V^\star$ has the same dimension:

$$\dim V = \dim V^\star \tag{1.17}$$

Indeed, it suffices to note that given a basis $\mathbf{e}_i$ ($i = 1, \ldots, n = \dim V$), of V, we can introduce a dual basis for $V^\star$ composed by the n linear functionals defined by:

$$f^j(\mathbf{e}_i) = \delta_i^j \tag{1.18}$$

1.5.2 Inner products

Definition 1.5.4. Let V be a finite-dimensional vector space over a field $\mathbb{F} = \mathbb{R}$ or $\mathbb{C}$; then an inner product on V is a map:

$$\langle , \rangle : V \times V \to \mathbb{F} \tag{1.19}$$

fulfilling the following properties:

a) Conjugate symmetry:

$$\forall x, y \in V : \quad \langle x, y \rangle = \overline{\langle y, x \rangle}$$

b) Linearity in the second argument

$$\forall x, y \in V, \forall a, b \in \mathbb{F} : \quad \langle x, ay + bz \rangle = a\langle x, y \rangle + a\langle x, z \rangle$$

c) Non-degeneracy

$$\forall y \in V : \quad \langle x, y \rangle = 0 \quad \Rightarrow \quad x = 0 \in V$$

If in addition to property c) we have also the following one
d) Positive definiteness

$$\forall x \in V : \quad \langle x, x \rangle \geq 0; \quad \langle x, x \rangle = 0 \quad \Rightarrow \quad x = 0 \in V$$

then we say that the inner product is positive definite.

Given a basis $\mathbf{e}_i$ of the vector space the inner product is completely characterized by the following matrix:

$$M_{ij} \equiv \langle \mathbf{e}_i, \mathbf{e}_j \rangle \tag{1.20}$$

which in force of the above axioms obeys the following properties:

$$M_{ji} = \overline{M_{ij}} \quad \Leftrightarrow \quad M = M^\dagger, \quad \det M \neq 0 \tag{1.21}$$

In other words, the inner product is characterized by a **non-degenerate Hermitian matrix.** The inner product is positive definite if all the eigenvalues of the matrix M are positive.

1.6 Algebras

Definition 1.6.1. An algebra $\mathscr{A}$ is a vector space over a field $\mathbb{K}$ on which it is defined a further binary operation of internal composition named product:

$$\cdot : \mathscr{A} \otimes \mathscr{A} \to \mathscr{A} \tag{1.22}$$

The product is distributive with respect to the operations characterizing a vector space:
A1) $\forall a, b, c \in \mathscr{A}$ and $\forall \alpha, \beta \in \mathbb{K}$

$$a \cdot (\alpha b + \beta c) = \alpha a \cdot b + \beta a \cdot c \tag{1.23}$$

If one imposes the extra property of associativity:

$$(a \cdot b) \cdot c = a \cdot (b \cdot c) \tag{1.24}$$

one obtains an associative algebra.

For instance the $n \times n$ matrices form an associative algebra. Indeed the linear combination with real or complex numbers of two matrices is still a matrix which shows that matrices form a vector space. Furthermore matrix multiplication provides the additional binary composition law which completes the axioms of an algebra.

1.7 Lie algebras

A Lie algebra $\mathbb{G}$ is an algebra where the product is endowed with special properties.

Definition 1.7.1. A Lie algebra $\mathbb{G}$ is an algebra over a commutative field $\mathbb{K}$ where the product operation $\mathbb{G} \times \mathbb{G} \to \mathbb{G}$, named *Lie product* or *Lie bracket* and denoted as follows:

$$\forall x, y \in \mathbb{G}, \quad x, y \mapsto [x, y] \tag{1.25}$$

has the following properties.

i) Linearity:

$$[x, \alpha x + \beta z] = \alpha[x, y] + \beta[x, z] \tag{1.26}$$

where $\alpha, \beta \in \mathbb{K}$.

ii) Antisymmetry:

$$[x, y] = -[y, x] \tag{1.27}$$

iii) Jacobi identity:

$$[x, [y, z]] + [y, [z, x]] + [z, [x, y]] = 0 \tag{1.28}$$

Notice that i) and ii) imply the linearity also in the first argument and that the Lie product is *not associative*. Indeed, we have $[x, [y, z]] \neq [[x, y], z]$, as from the Jacobi identity we get $[x, [y, z]] - [[x, y], z] = -[y, [z, x]]$, which is generically non-zero.

The dimension of $\mathbb{G}$ as a vector space is also named the *dimension* or order of the Lie algebra, and is denoted $\dim \mathbb{G}$.

Fixing a basis $\{t_i\}$ of vectors, on which every element of $\mathbb{G}$ can be expanded: $x = x^i t_i$, the Lie product structure of $\mathbb{G}$ is encoded in a set of $(\dim \mathbb{G})^3$ constants $c_{ij}{}^k$, called the *structure constants* of the algebra. The basis vectors t_i are called the *generators* of the Lie algebra.

As a consequence of the properties of the Lie product, the structure constants obey the following relations.

i) Antisymmetry:

$$c_{ij}{}^k = -c_{ji}{}^k \tag{1.29}$$

ii) Jacobi identity:

$$c_{im}{}^{s}c_{jk}{}^{m} + c_{jm}{}^{s}c_{ki}{}^{m} + c_{km}{}^{s}c_{ij}{}^{m} = 0 \tag{1.30}$$

The explicit set of structure constants for a given Lie algebra is unique only up to the effect of changes of basis.

1.7.1 Homomorphism

Two Lie algebras $\mathbb{G}$ and $\mathbb{K}$ are *homomorphic* if there exists an *homomorphism* between the two, namely, a map $\phi : \mathbb{G} \to \mathbb{K}$ that preserves the Lie product:

$$[\phi(x), \phi(y)] = \phi([x, y]) \tag{1.31}$$

1.7.2 Subalgebras and ideals

Definition 1.7.2. A vector subspace $\mathbb{H} \subset \mathbb{G}$ of a Lie algebra $\mathbb{G}$ is named a **subalgebra** iff $\forall \mathbf{X}, \mathbf{Y} \in \mathbb{H}$ we have $[\mathbf{X}, \mathbf{Y}] \in \mathbb{H}$ which is usually abbreviated as $[\mathbb{H}, \mathbb{H}] \subset \mathbb{H}$.

Definition 1.7.3. A subalgebra $\mathbb{I} \subset \mathbb{G}$ of a Lie algebra $\mathbb{G}$ is named an **ideal** iff $\forall \mathbf{X} \in \mathbb{I}$ and $\forall \mathbf{Y} \in \mathbb{G}$ we have $[\mathbf{X}, \mathbf{Y}] \in \mathbb{I}$ which is usually abbreviated as $[\mathbb{I}, \mathbb{G}] \subset \mathbb{I}$.

1.7.3 Representations

A Lie algebra can be realized by a set of square matrices (forming a vector space), the Lie product being defined as the commutator in the matrix sense. If a Lie algebra $\mathbb{G}$ is *homomorphic* to some *matrix Lie algebra* $\mathscr{D}(\mathbb{G})$, then $\mathscr{D}(\mathbb{G})$ is said to give a matrix *representation* of $\mathbb{G}$.

1.7.4 Isomorphism

Two Lie algebras between which there exists a *homomorphism* that is *invertible*, i.e. an *isomorphism*, are said to be *isomorphic*. They correspond to two different realizations of the same abstract Lie algebra. An isomorphism of $\mathbb{G}$ into some matrix Lie algebra $\mathscr{D}(\mathbb{G})$ is called a *faithful* representation.

1.7.5 Adjoint representation

We have seen that, given a Lie algebra $\mathbb{G}$, we can associate to it a set of structure constants (unique only up to changes of basis). The converse is also true. Indeed, given a set of n^3 constants $c_{ij}{}^k$, the conditions in eq. (1.29) and eq. (1.30) are necessary and sufficient for the set of $c_{ij}{}^k$ to be the structure constant of a Lie algebra. The necessity was argued before. That it is sufficient is exhibited by constructing explicitly a *matrix* Lie algebra with structure constants $c_{ij}{}^k$. Indeed, define the $n \times n$ matrices $\mathbf{T}_i$, with matrix elements

$$(\mathbf{T}_i)_j{}^k \equiv c_{ji}^k \tag{1.32}$$

The matrix commutator of two such matrices can be computed by making use of the antisymmetry and of the Jacobi identity:

$$[\mathbf{T}_i, \mathbf{T}_j]_p{}^q = (\mathbf{T}_i)_p{}^s (\mathbf{T}_j)_s{}^q - (i \leftrightarrow j) = \cdots = c_{ij}{}^k (\mathbf{T}_k)_p{}^q \tag{1.33}$$

Thus, the matrices obtained as real linear combinations of the $\mathbf{T}_i$'s form a Lie algebra of dimension n, with structure constants $c_{ij}{}^k$.

The above construction tells us also that, given a Lie algebra $\mathbb{G}$ with structure constants $c_{ij}{}^k$, it always possesses a faithful representation in terms of $\dim \mathbb{G} \times \dim \mathbb{G}$ matrices, called the *adjoint representation*, generated by the matrices $\mathbf{T}_i$ of eq. (1.32).

1.8 Moduli and representations

Next we come to the proper mathematical setup underlying the notion of Lie algebra representations which is that of *module*.

Definition 1.8.1. Let $\mathbb{L}$ be a Lie algebra and V a vector space over the field $\mathbb{F} = \mathbb{R}$ or $\mathbb{C}$, endowed with an operation:

$$\mathbb{L} \otimes V \mapsto V$$

denoted $(x, v) \mapsto x.v$. The vector space V is named an $\mathbb{L}$-module if the conditions provided by the following axioms are satisfied:

M1) $\forall x, y \in \mathbb{L}$, $\forall v \in V$ and $\forall a, b \in \mathbb{F}$

$$(ax + by).v = a(x.v) + b(y.v)$$

M2) $\forall x \in \mathbb{L}$, $\forall v, w \in V$ and $\forall a, b \in \mathbb{F}$

$$x.(av + bw) = a(x.v) + b(x.w)$$

M3) $\forall x, y \in \mathbb{L}$, $\forall v \in V$

$$[x, y].v = x.(y.v) - y.(x.v)$$

As we see from axiom (*M2*), to each element $x \in \mathbb{L}$ one associates a linear map $\mathscr{D}(x) \in \mathrm{Hom}(V, V)$:

$$\forall x \in \mathbb{L} \quad \mathscr{D}(x) : V \to V \tag{1.34}$$

Choosing a basis $\mathbf{e}_1, \ldots, \mathbf{e}_m$ of the vector space V we obtain:

$$x.\mathbf{e}_i = \mathbf{e}_j \mathscr{D}_{ji}(x) \tag{1.35}$$

where $\mathscr{D}_{ij}(x)$ is an $m \times m$ matrix. Hence to each element $x \in \mathbb{L}$ we associate a matrix $\mathscr{D}(x)$ and in force of the axiom *M3* we have:

$$\mathscr{D}([x,y]) = \mathscr{D}(x)\mathscr{D}(y) - \mathscr{D}(y)\mathscr{D}(x) = [\mathscr{D}(x), \mathscr{D}(y)] \tag{1.36}$$

which shows the equivalence of the previously introduced notion of linear representation of a Lie algebra with that of module. The notion of module is mathematically more precise since it does not make reference to any particular basis in the vector space V. Two linear representations $\mathscr{D}_1$ and $\mathscr{D}_2$ of the same Lie algebra are equivalent if they can be related one to the other by an overall similarity transformation $\mathscr{S}$:

$$\forall x \in \mathbb{L} \quad \mathscr{D}_1(x) = \mathscr{S}\mathscr{D}_2(x)\mathscr{S}^{-1} \tag{1.37}$$

The above happens when $\mathscr{D}_1$ and $\mathscr{D}_2$ are obtained from the same module in two different systems of basis vectors.

Summarizing, the notion of module encodes the notion of equivalence classes of equivalent linear representations of a Lie algebra.

1.9 Bibliographical note

The basic notions of algebra summarized in the present chapter are covered in various combinations in many elementary and less elementary textbooks. A short list of suggestions for further reading is the following one:

1. Gilmore's textbook [72].
2. Helgason's monography [84].
3. Jacobson's monography [92].
4. Cornwell's treatise [29].
5. Guido Fano's textbook [45] for a comprehensive introduction to vector spaces and the basic notions of topology.
6. Nakahara's advanced textbook [116] for a review of many different items.

2 Groups: a bestiary for beginners

Esse quam videri
Arthur Cayley

2.1 Survey of the contents of this chapter

In Section 1.2 we introduced the notion of a group, we illustrated it with a few simple examples and we listed the main essential concepts relative to the discussion of such an algebraic structure. In the present chapter we introduce many typical classes of groups which are very frequently encountered in Physics. This should *substantiate* the definitions given in the previous chapter and provide a set of important concrete examples with which we will work extensively in the sequel. For us concrete means written in terms of matrices.

2.2 Groups of matrices

In this section we display a finite list of groups G whose elements are just $n \times n$ matrices M characterized by some defining property that is preserved by matrix multiplication. All these matrix groups are instances of what we shall name *Lie groups*, namely groups that are also differentiable analytic manifolds. Actually, when we come to the classification of Lie groups via the classification of *Lie algebras* we will see that the classical matrix groups exhaust the list of non-exceptional Lie algebras. Matrix groups appear ubiquitously in all branches of Physics.

2.2.1 General linear groups

The group of *all* $n \times n$ invertible matrices with complex entries is called the (complex) *general linear group* in n dimensions and is denoted as $\mathrm{GL}(n, \mathbb{C})$. If the entries are real, we have the real general linear group $\mathrm{GL}(n, \mathbb{R})$, which is a subgroup of the former.

An element of $\mathrm{GL}(n, \mathbb{C})$ is parameterized by n^2 complex numbers, the entries of the matrix (of course the n^2 parameters are real for $\mathrm{GL}(n, \mathbb{R})$).

One can define further matrix groups by placing restrictions, typically in the form of matrix equations or of conditions on the determinant, that are preserved by the matrix product.

2.2.2 Special linear groups

The group of all $n \times n$ matrices with complex entries and *determinant equal to* 1 is named the *special linear group* and is indicated as $\mathrm{SL}(n, \mathbb{C})$. It is obviously a subgroup

https://doi.org/10.1515/9783110551204-002

of $GL(n, \mathbb{C})$: the condition of having unit determinant is preserved by the product. Similarly, one defines $SL(n, \mathbb{R})$. One can also define the group $SL(n, \mathbb{Z})$: indeed, the inverse matrices have also integer entries since the determinant that would appear in the denominator of these entries is just 1.

An element of $SL(n, \mathbb{C})$ depends on $n^2 - 1$ complex parameters, as the relation $\det M = 1$ has to be imposed on the n^2 entries of any matrix M. Such parameters are real (integers) for $SL(n, \mathbb{R})$ or $SL(n, \mathbb{Z})$.

2.2.3 Unitary groups

The group of *unitary matrices* $U(n, \mathbb{C}) \subset GL(n, \mathbb{C})$ contains all the complex matrices U such that

$$U^\dagger U = \mathbf{1} \tag{2.1}$$

This property is preserved by matrix multiplication and this guarantees that $U(n, \mathbb{C})$ is a group. Similarly one could define $U(n, \mathbb{R}) \subset GL(n, \mathbb{R})$, containing real unitary matrices: $U^\dagger U = U^T U = \mathbf{1}$, but these are nothing else than real orthogonal matrices, to be introduced shortly. So the group of complex unitary matrices is usually simply denoted as $U(n)$. Complex unitary matrices are parameterized by $2n^2 - n^2 = n^2$ *real* parameters (we have to subtract the n^2 real conditions corresponding to the entries of the equation $U^\dagger U - \mathbf{1} = 0$ from the n^2 complex parameters of the matrix U). So the condition of unitarity halves the number of parameters, with respect to a generic complex matrix.

2.2.4 Special unitary groups

The subgroup $SU(n, \mathbb{C}) \subset U(n, \mathbb{C})$ contains the unitary matrices with unit determinant. Usually it is simply denoted by $SU(n)$. It is determined by $n^2 - 1$ real parameters. We have to subtract the real condition of having determinant 1 from the parameters of a unitary matrix; recall that by itself the determinant of a unitary matrix can assume a continuous range of values $\exp(2\pi i\theta)$, $\theta \in [0, 1]$.

2.2.5 Orthogonal groups

The group of *orthogonal matrices* $O(n, \mathbb{C}) \subset GL(n, \mathbb{C})$ contains all the complex matrices O such that

$$O^T O = \mathbf{1} \tag{2.2}$$

More frequently encountered are the *real* orthogonal matrices $O(n, \mathbb{R}) \subset GL(n, \mathbb{R})$. Usually these groups are simply denoted as $O(n)$. Real orthogonal matrices are parameterized by $n(n-1)/2$ real numbers. Indeed, from the n^2 parameters of a general real matrix, we have to subtract the $n(n+1)/2$ conditions given by the entries of the matrix condition $O^T O = \mathbf{1}$, which is *symmetric*.

2.2.6 Special orthogonal groups

The group $\mathrm{SO}(n)$ contains the real orthogonal matrices with unit determinant. Analogously one could define its complex extension $\mathrm{SO}(n, \mathbb{C})$. They have the same number of parameters, $n(n-1)/2$, as the orthogonal matrices. Indeed, the determinant of an orthogonal matrix O can have only a finite set of values: $\det O = \pm 1$; imposing that $\det O = 1$ does not alter the dimensionality of the parameter space.

2.2.7 Symplectic groups

The group of *symplectic matrices* $\mathrm{Sp}(n, \mathbb{C})$ contains the $2n \times 2n$ matrices A that preserve the "symplectic[1] form" Ω, namely the matrices such that

$$A^T \Omega A = \Omega, \quad \Omega = \begin{pmatrix} \mathbf{0} & \mathbf{1} \\ -\mathbf{1} & \mathbf{0} \end{pmatrix} \tag{2.3}$$

Similarly one defines $\mathrm{Sp}(n, \mathbb{R})$. Since the restriction eq. (2.3) is an antisymmetric matrix equation, it poses $(2n)(2n-1)/2$ conditions. Thus the symplectic matrices depend on $(2n)^2 - (2n)(2n-1)/2 = n(2n+1)$ parameters (complex or real for $\mathrm{Sp}(n, \mathbb{C})$ or $\mathrm{Sp}(n, \mathbb{R})$).

The groups $\mathrm{U}(n)$, $\mathrm{O}(n)$ and $\mathrm{Sp}(n)$ form the three families of *classical matrix groups*.

2.2.8 Groups of transformations

Square $n \times n$ matrices represent homomorphisms of an n-dimensional vector space V into itself. This is what, according to the tale told in [53], took such a long time to recognize. Thus the matrix groups are in fact groups of linear transformations of vector spaces. In many physical applications the elements of G are interpreted as transformations $\tau \in \mathrm{Hom}(V, V)$ acting on some space V:

$$\tau \in G : v \in V \mapsto \tau(v) \in V \tag{2.4}$$

and the group composition is the composition of these transformations:

$$\tau_1 \tau_2 \in G : v \in V \mapsto \tau_1(\tau_2(v)) \in V \tag{2.5}$$

1 The symplectic form is the non-positive quadratic form often appearing in analytical mechanics, e.g., in the definition of the Poisson brackets: $\{F, G\} = \frac{\partial F}{\partial y^I} \Omega^{IJ} \frac{\partial F}{\partial y^J}$, where $y^I = (q^i, p^i)$ are the phase-space coordinates.

In this case, the associativity is automatically satisfied. Notice the convention that in the product $\tau_1\tau_2$ one acts first with τ_2 and then with $\tau_1 : v \overset{\tau_2}{\mapsto} \tau_2(v) \overset{\tau_1}{\mapsto} \tau_1(\tau_2(v)) \equiv \tau_1\tau_2(v)$. Of course, this is the same convention that arises in taking the matrix product as the group product law for groups of linear transformations on vector spaces.

A major conceptual step-forward is the identification of the group or of another algebraic structure with its isomorphism class. Two isomorphic structures are just the same structure, in the sense that there is a sort of abstract *platonic idea* of the structure and the actual instances of its realizations by means of matrices or transformations are just *representations*.[2]

Transformations groups are abelian when the order in which two transformations are subsequently performed does not affect the final result.

2.3 Some examples of finite groups

As we already emphasized all the groups considered in the previous section are instances of continuous (actually Lie) groups where the group elements fill a continuous space and cannot be enumerated. In this section we consider instead instances of *discrete groups*, whose elements can be enumerated. Discrete groups subdivide into two main classes: *discrete infinite groups* that contain a denumerable infinity of elements and *finite groups* whose elements constitute a finite set. The theory of finite groups is a very rich and fascinating chapter of mathematics with extremely relevant applications both in Physics and Chemistry: we will come to it in Chapter 3; here we begin by presenting some primary examples.

2.3.1 The permutation groups S_n

Consider a finite set A. The automorphisms (i.e., the *bijective mappings* $P : A \leftrightarrow A$) form a group $S(A)$, called the *symmetric group* of A. The nature of the objects in the set A does not matter, only their number $|A|$ does. So, if $|A| = n$, we can label the objects with the integer numbers $1, 2, \dots, n$ and indicate the symmetric group, also called the *permutation group* on n objects, as S_n. An element of this group, a *permutation* P, is explicitly defined by its action on the elements $1, \dots, n$ of the set:

$$P = \begin{pmatrix} 1 & 2 & \dots & n \\ P(1) & P(2) & \dots & P(n) \end{pmatrix} \tag{2.6}$$

The product law of the symmetric group is the composition of permutations (with the convention described above: PQ means effecting first the permutation Q and then the

2 For further details on the conceptual history of the notion of group, see [53].

permutation P). While S_2 is abelian, all S_n with $n > 2$ are not abelian. The symmetric group S_n has $n!$ elements (see Exercise 2.2).

We can give an explicit expression to a permutation $P \in S_n$ as an $n \times n$ matrix defined by

$$(P)_{ij} = \delta_{i,P(j)}, \quad i,j = 1, \dots, n \tag{2.7}$$

In this way the composition of permutations corresponds to the product of the defining matrix representatives (2.7):

$$(PQ)_{ij} = \sum_k \delta_{i,P(k)} \delta_{k,P(j)} = \sum_k \delta_{i,P(k)} \delta_{P(k),P(Q(j))} = \delta_{i,P(Q(j))} \tag{2.8}$$

Notice that the matrix representatives of permutations are unitary and real, that is, they are orthogonal matrices. So S_n can be seen as a subgroup of $O(n)$.

2.3.1.1 Cycle decomposition

Let us illustrate the notion of *cycle* of a permutation by means of an example. Consider the permutation $P \in S_8$ displayed below:

$$P = \begin{pmatrix} 1 & 2 & 3 & 4 & 5 & 6 & 7 & 8 \\ 2 & 3 & 1 & 5 & 4 & 7 & 6 & 8 \end{pmatrix} \tag{2.9}$$

Let us follow what would happen to the various elements $1, \dots, 8$ if we were to apply the permutation repeatedly. We would have $1 \to 2 \to 3 \to 1$, after which everything repeats again. We say that $1, 2, 3$ form a cycle of order 3 in P, and we denote this cycle compactly as (123). We also have that $4 \to 5 \to 4$, so that we have a cycle (45). Also, $6 \to 7 \to 6$ and thus there's the cycle (67). Finally 8 is invariant, i.e., it is a trivial cycle (8). The permutation P has a the *cycle decomposition*

$$P = (123)(45)(67)(8) \tag{2.10}$$

Often the cycles of length 1, such as (8) above, are omitted when writing the cycle decomposition. It is quite evident that the example we took is not limited in any way, and every permutation of any S_n group will admit a cycle decomposition (see Exercise 2.2). As we will see, the (type of) cycle decompositions of permutations is of fundamental importance in the analysis of permutation groups. Let us make some simple observations.

i) The sum of the lengths of the cycles for a $P \in S_n$ equals n: every element is in some cycle, and cycles do not have elements in common.
ii) Having no common elements, two cycles in the decomposition of a given permutation commute. For instance, in eq. (2.10), $(123)(45) = (45)(123)$.
iii) Just by their definition, the cycles can be shifted freely without being affected: $(123) = (231) = (312)$ (but $(123) \neq (213)$!)

iv) Every cycle can in turn be decomposed into a product of cycles of order 2, called also *transpositions* or *exchanges*. However, the latter now have elements in common. For instance, $(123) = (13)(12)$. In general, $(12\dots n) = (1n)(1, n-1)\dots(13)(12)$.

2.3.1.2 Odd and even permutations

As discussed above, every permutation can be decomposed into a product of transpositions. A permutation is called *odd* or *even* depending on whether in such a decomposition an odd or an even number of transposition appears.

2.3.1.3 The alternating groups

The *even* permutations form a subgroup of S_n (the odd ones clearly do not form a subgroup) called the *alternating group* on n elements, denoted as A_n. Its order is $|A_n| = |S_n|/2 = n!/2$.

2.4 Groups as the invariances of a given geometry

After two thousand years of Euclidean Geometry, elevated by Immanuel Kant to the rank of an *a priori fundament of all perceptions*, in the middle XIXth century non-Euclidean geometries finally worked out their way into mathematics and philosophy. As discussed at length [53], in a famous programme named the *Erlangen Programme* after the German town where it was announced, Felix Klein proposed a new vision where all possible geometries are classified according to the group of transformations with respect to which the relations and the propositions existing in that geometry are kept invariant. Euclidean geometries are simply the geometries invariant with respect to the Euclidean groups.

2.4.1 The Euclidean groups

The so-called Euclidean group in d dimensions is the group of *isometry* transformations in an Euclidean space $\mathbb{R}^d$. It consists of translations, of rotations around some axis (proper rotations) and reflections with respect to hyperplanes. All such transformations leave unaltered the Euclidean distance between any two points of $\mathbb{R}^d$.

2.4.1.1 The Euclidean group in two dimensions

Consider the transformations of a plane $\mathbb{R}^2$ into itself given by rigid rotations around a perpendicular axis through the origin. They clearly form a group. The elements R_θ of the group are identified by an angle θ (the angle of, e.g., clockwise rotation) defined mod 2π; that is, the elements of the group correspond to the points of a circle S^1. The composition of two rotations results in $\mathcal{R}(\theta_1)\mathcal{R}(\theta_2) = R(\theta_1 + \theta_2)$; the group is abelian. We can describe a transformation $R(\theta)$ via its effects on the Cartesian coordinates $\mathbf{x} \equiv (x, y)$ of a point: $\mathbf{x} \overset{\mathscr{R}(\theta)}{\mapsto} \mathbf{x}'$, with

$$\begin{cases} x' = \cos\theta x + \sin\theta y, \\ y' = -\sin\theta x + \cos\theta y \end{cases} \tag{2.11}$$

that is, $\mathbf{x}' = \mathscr{R}(\theta)\mathbf{x}$, with $\mathscr{R}(\theta)$ an orthogonal 2×2 matrix with unit determinant: $\mathscr{R}(\theta) \in \mathrm{SO}(2)$. In fact, this correspondence between rotations and matrices of SO(2) is an isomorphism. Thus, with a slight abuse of language, we can say that SO(2) is the (proper) rotation group in two dimensions.

Also the translations $T(\mathbf{v})$ by a two-vector vector $\mathbf{v}$ acting on the Euclidean space $\mathbb{R}^2$, $T(\mathbf{v}) : \mathbf{x} \mapsto \mathbf{x} + \mathbf{v}$, $\forall \mathbf{x} \in \mathbb{R}^2$ form a group, with the composition of two translations resulting in $T(\mathbf{v}_1)T(\mathbf{v}_2) = T(\mathbf{v}_1 + \mathbf{v}_2)$; this group is abelian.

Consider now the group of all transformations of $\mathbb{R}^2$ onto itself given by simultaneous rotations and/or translations with arbitrary parameters. Let us denote such transformations as $(\mathscr{R}(\theta), T(\mathbf{v}))$. They act on the coordinate vectors by

$$(\mathscr{R}(\theta), T(\mathbf{v})) : \mathbf{x} \mapsto \mathscr{R}(\theta)\mathbf{x} + \mathbf{v} \tag{2.12}$$

Notice that the translation parameters $\mathbf{v}$, being vectors, are acted on by the rotations. You can easily verify that the product law resulting from the composition of two such transformations is

$$(\mathscr{R}(\theta_1), T(\mathbf{v}_1)) \cdot (\mathscr{R}(\theta_2), T(\mathbf{v}_2)) = (\mathscr{R}(\theta_1 + \theta_2), T(\mathbf{v}_1 + \mathscr{R}(\theta_1)\mathbf{v}_2)) \tag{2.13}$$

With respect to this composition law the set of elements $(\mathscr{R}(\theta), T(\mathbf{v}))$ close a group which is named the *inhomogeneous rotation group* in two dimensions denoted ISO(2).

A proper rotation maps an oriented orthogonal frame into a new orthogonal frame with the same orientation. Inversions (or reflections) of the plane with respect to a line through the origin also map it to an orthogonal frame but with the opposite orientation. For instance, reflection with respect to the x-axis maps (x, y) to $(x, -y)$. The set of all transformations obtained as compositions of proper rotations and inversions is a larger group. The inversion with respect to a direction forming an angle θ with the x axis is effected by the matrix:

$$\mathscr{I}(\theta) = \begin{pmatrix} \cos 2\theta & \sin 2\theta \\ \sin 2\theta & -\cos 2\theta \end{pmatrix} = \mathscr{I}(0)\mathscr{R}(2\theta) \tag{2.14}$$

where $\mathscr{I}(0) = \mathrm{diag}(1, -1)$ is orthogonal but with determinant -1. Thus rotations plus inversions of the plane are represented by all orthogonal 2×2 matrices, i.e., by elements of O(2).

Translations, rotations and inversions form a group, $\mathrm{Eucl}_2 \supset \mathrm{ISO}(2)$, which is named the *Euclidean group* in two dimensions.

2.4.1.2 The Euclidean group in three dimensions

The translations acting on $\mathbb{R}^3$ form an abelian group (there is no difference with respect to the $\mathbb{R}^2$ case).

Then consider the rotations around any axis through the origin. Such transformations form the group of *proper orthogonal rotations* in three dimensions; the effect of performing two subsequent rotations around two different axes is a new rotation around a third axis. The proper rotations map an orthonormal frame into a new orthonormal frame with the same orientation, and are represented on the vectors by means of orthogonal 3×3 matrices with unit determinant. In fact, the group of proper rotations is isomorphic to SO(3). The group of three-dimensional rotations SO(3) is *non-abelian*.

The group ISO(3) of roto-translations in three dimensions can be defined with no formal modification with respect to the two-dimensional case.

Reflections with respect to a plane and total spatial reflection $\mathbf{x} \mapsto -\mathbf{x}$ map orthogonal frames to orthogonal frames with the opposite orientation. They are represented by orthogonal 3×3 matrices with determinant -1. Thus rotations and reflections form a group, which is isomorphic to the group O(3) of orthogonal matrices acting on the vectors.

Translations, rotations and reflections form the Euclidean group in three dimensions Eucl_3.

2.4.2 Projective geometry

The primary example of a non-Euclidean geometry is provided by Lobachevsky hyperbolic geometry analytically realized on the upper complex plane named in this context after *Henri Poincaré*. This will form the object of Section 10.8.2. Here we consider the wider case of projective geometry. Figures in the complex planes can change their size and shape by compression and stretching, yet the angles between the tangents to any two lines composing the figure have to be preserved. This is done by the group of Möbius transformations.

2.4.2.1 The Möbius group

Consider the *conformal transformations* of the compactified complex plane (or Riemann sphere) $\tilde{\mathbb{C}} \equiv \mathbb{C} \cup \{\infty\}$. Conformal mappings $z \mapsto w$ are represented by *analytic* functions $w(z)$. We require that the transformations should be invertible, i.e., one-to-one. Hence the function $w(z)$ can have at most one simple pole, otherwise the point at infinity would have several preimages, and, for an analogous motivation at most one zero; furthermore, the Jacobian $\partial w/\partial z$ should not vanish. Thus the most general transformation M fulfilling all the requirements is of the following form

$$z \overset{M}{\mapsto} w(z) = \frac{az+b}{cz+d}, \quad a, b, c, d \in \mathbb{C},\ ad - bc \neq 0 \tag{2.15}$$

where the last condition follows from the invertibility requirement $\partial w/\partial z \neq 0$. Notice that all transformations of parameters (ka, kb, kc, kd) with $k \in \mathbb{C} \setminus \{0\}$ are equivalent. This scale invariance can be used to fix $ad - bc = 1$. In mathematical and physical literature the transformations defined by eq. (2.15) are named either *fractional linear transformations* or *Möbius transformations*.

The composition of two Möbius transformations M, M' is again a Möbius transformation M'':

$$z \overset{M}{\mapsto} w = \frac{az+b}{cz+d} \overset{M'}{\mapsto} x = \frac{a'w+b'}{c'w+d'} = \frac{(a'a+b'c)z+a'b+b'd}{(c'a+d'c)z+c'b+d'd} = \frac{a''z+b''}{c''z+d''} \tag{2.16}$$

The product transformation $z \overset{M''}{\mapsto} x$ satisfies $a''d'' - b''c'' = 1$.

Thus, it is natural to associate a Möbius transformation M with a 2×2 matrix $\mathscr{M}$ with unit determinant

$$\mathscr{M} = \begin{pmatrix} a & b \\ c & d \end{pmatrix}, \quad \det \mathscr{M} = ad - bc = 1 \tag{2.17}$$

namely, with an element of $\mathrm{SL}(2, \mathbb{C})$. To the product of two transformations $M'' = M'M$ corresponds a matrix which is the matrix product of the two factors: $\mathscr{M}'' = \mathscr{M}'\mathscr{M}$. However, the mapping from the group of Möbius transformations to $\mathrm{SL}(2, \mathbb{C})$ is one-to-two, since the matrices $\pm\mathscr{M}$ correspond to the same Möbius transformation M.

2.5 Matrix groups and vector spaces

We turn to analyze classical groups as groups made of homorphisms $\mu : V \to V$ of a vector space into itself.

2.5.1 General linear groups and basis changes in vector spaces

As already recalled, the non-singular $n \times n$ matrices A with real or complex entries, i.e. the elements of $\mathrm{GL}(n, \mathbb{R})$ or $\mathrm{GL}(n, \mathbb{C})$, are *automorphisms* of real or complex vector spaces. Namely, they describe the possible change of basis in an n-dimensional vector space $\mathbb{R}^n$ or $\mathbb{C}^n$: $\mathbf{e}_i{}' = \mathbf{e}_j A^j{}_i$ (in matrix notation, $\mathbf{e}' = \mathbf{e}A$). Two subsequent changes of basis, first with A and then with B, result in a change described by the matrix product BA. The basis elements $\mathbf{e}_i$ are said to transform *contravariantly*. The components v^i of a vector $\mathbf{v} = v^i \mathbf{e}_i$ transform instead *covariantly*: from $\mathbf{v} = v^{i\prime} \mathbf{e}_i{}' = v^i \mathbf{e}_i$ it follows $v^i = A^i{}_j v^{j\prime}$ (in matrix notation, $v = Av'$, or $v' = A^T v$).

2.5.2 Tensor product spaces

Starting from two vector spaces, say V_1 (n-dimensional, with basis $\mathbf{e}_i$) and V_2 (m-dimensional, with basis $\mathbf{f}_j$), one can construct bigger vector spaces, for instance by taking the *direct sum* $V_1 \oplus V_2$ of the two spaces or their *direct product* $V_1 \otimes V_2$. The direct product has dimension mn and basis $\mathbf{e}_i \otimes \mathbf{f}_j$. A change of basis A in V_1 and B in V_2 induces a change of basis in the direct product space described[3] by $(\mathbf{e}_i \otimes \mathbf{f}_j)' = A_i{}^k B_j{}^l \mathbf{e}_k \otimes \mathbf{f}_l$. An element of a direct product vector space is called a *tensor*.

In particular, one can take the direct product of a vector space V of dimension n with itself (in general, m times): $V \otimes \cdots \otimes V$, which we may indicate for shortness as $\otimes^m V$. Its n^m basis vectors are $\mathbf{e}_{i_1} \otimes \cdots \otimes \mathbf{e}_{i_m}$. An element of $\otimes^m V$ is called a *tensor of order m*. An element A of the general linear group G of V (a change of basis in V) induces a change of basis on $\otimes^m V$:

$$\mathbf{e}_{i_1}{}' \otimes \cdots \otimes \mathbf{e}_{i_m}{}' = A_{i_1}{}^{j_1} \cdots A_{i_m}{}^{j_m} \mathbf{e}_{j_1} \otimes \cdots \otimes \mathbf{e}_{j_m} \tag{2.18}$$

Such changes of bases on $\otimes^m V$ form a group which is the direct product of G with itself, m times, $\otimes^m G$. Its elements are $n^m \otimes n^m$ matrices, and the association of an element A of G with an element of $\otimes^m G$ given in eq. (2.18) preserves the product; thus $\otimes^m G$ is a representation of G, called an m-th order tensor product representation. As we shall see in the sequel, this way of constructing representations by means of tensor products is of the utmost relevance.

2.5.3 (Anti)-symmetrized product spaces

Within the direct product space $V \otimes V$, we can single out two subspaces respectively spanned by the *symmetric* and by the *antisymmetric* combinations[4] of the basis vectors $\mathbf{e}_i \otimes \mathbf{e}_j$:

$$\begin{aligned} \mathbf{e}_i \vee \mathbf{e}_j &\equiv \mathbf{e}_i \otimes \mathbf{e}_j + \mathbf{e}_j \otimes \mathbf{e}_i = (\mathbf{1} + (12))(\mathbf{e}_i \otimes \mathbf{e}_j) \\ \mathbf{e}_i \wedge \mathbf{e}_j &\equiv \mathbf{e}_i \otimes \mathbf{e}_j - \mathbf{e}_j \otimes \mathbf{e}_i = (\mathbf{1} - (12))(\mathbf{e}_i \otimes \mathbf{e}_j) \end{aligned} \tag{2.19}$$

where in the second equality by $\mathbf{1}$ and (12) we have respectively denoted the identity and the exchange element of the symmetric group S_2 acting on the positions of the two elements in the direct product basis. The symmetric subspace has dimension $n(n+1)/2$, the antisymmetric one has $n(n-1)/2$.

Similarly, within $\otimes^m V$ the *fully symmetric* and *fully antisymmetric* subspaces can easily be defined. Their basis elements can be written as

3 The subgroup of all possible changes of basis obtained in this way is the *direct product* $G_1 \otimes G_2$ (a concept we will deal with in next section) of the general linear groups (groups of basis changes) G_1 and G_2 of V_1 and V_2, respectively.

4 The denotation $\mathbf{e}_i \vee \mathbf{e}_j$, though logical, is not used too much in the literature. Often, with a bit of abuse, a symmetric combination is simply indicated as $\mathbf{e}_i \otimes \mathbf{e}_j$; we will sometimes do so, when no confusion can arise.

$$\begin{aligned} \mathbf{e}_{i_1} \vee \cdots \vee \mathbf{e}_{i_m} &\equiv \sum_{P \in S_m} P(\mathbf{e}_{i_1} \otimes \cdots \otimes \mathbf{e}_{i_m}) \\ \mathbf{e}_{i_1} \wedge \cdots \wedge \mathbf{e}_{i_m} &\equiv \sum_{P \in S_m} (-)^{\delta_P} P(\mathbf{e}_{i_1} \otimes \cdots \otimes \mathbf{e}_{i_m}) \end{aligned} \tag{2.20}$$

where δ_P is 0 or 1 if the permutation P is even or odd, respectively. The fully symmetric subspace has dimension $n(n+1)\dots(n+m-1)/m!$, the fully antisymmetric one has dimension $n(n-1)\dots(n-m+1)/m! = \binom{n}{m}$.

2.5.4 Exterior forms

Let W be a vector space of finite dimension over a field $\mathbb{F}$ ($\mathbb{F}$ can be either $\mathbb{R}$ or $\mathbb{C}$, depending on the case). In this section we show how we can construct a sequence of vector spaces $\Lambda_k(W)$ with $k = 0, 1, 2, \dots, n = \dim W$ defined in the following way:

$$\begin{aligned} &\Lambda_0(W) = \mathbb{F} \\ &\Lambda_1(W) = W^\star \\ &\dots\dots\dots \\ &\Lambda_k(W) = \text{vector space of } k\text{-linear antisymmetric functionals over } W \end{aligned} \tag{2.21}$$

The spaces $\Lambda_k(W)$ contain the linear functionals on the k-th exterior powers of the vector space W. Such functionals are called *exterior forms* of degree k on W.

Let $\phi^{(k)} \in \Lambda_k(W)$ be a k-form. It describes a map:

$$\phi^{(k)} : \underbrace{W \otimes W \otimes \cdots \otimes W}_{k \text{ times}} \to \mathbb{F} \tag{2.22}$$

with the following properties:

$$\begin{aligned} &i) \quad \phi^{(k)}(w_1, \dots, w_i, \dots, w_j, \dots, w_k) = -\phi^{(k)}(w_1, \dots, w_j, \dots, w_i, \dots, w_k) \\ &ii) \quad \phi^{(k)}(w_1, \dots, \alpha x + \beta y, \dots, w_k) = \alpha\phi^{(k)}(w_1, \dots, x, \dots, w_k) \\ &\qquad\qquad + \beta\phi^{(k)}(w_1, \dots, y, \dots, w_k) \\ &\text{where } \alpha, \beta \in \mathbb{F} \text{ and } w_i, x, y \in W. \end{aligned} \tag{2.23}$$

The first of properties (2.23) guarantees that the map $\phi^{(k)}$ is antisymmetric in any two arguments. The second property states that $\phi^{(k)}$ is linear in each argument.

The sequence of vector spaces $\Lambda_k(W)$:

$$\Lambda(W) \equiv \bigcup_{k=0}^{n} \Lambda_k(W) \tag{2.24}$$

can be equipped with an additional operation, named exterior product that to each pair of a k_1 and a k_2 form $(\phi^{(k_1)}, \phi^{(k_2)})$ associates a new $k_1 + k_2$ form. Namely, we have:

$$\wedge : \Lambda_{k_1} \otimes \Lambda_{k_2} \to \Lambda_{k_1+k_2} \tag{2.25}$$

More precisely, we set:

$$\phi^{(k_1)} \wedge \phi^{(k_2)} \in \Lambda_{k_1+k_2}(W) \tag{2.26}$$

and we write:

$$\begin{aligned}\phi^{(k_1)} \wedge \phi^{(k_2)}(w_1, w_2, \ldots, w_{k_1+k_2}) = \sum_P (-)^{\delta_P} \frac{1}{(k_1+k_2)!} (\phi^{(k_1)}(w_{P(1)}, \ldots, w_{P(k_1)}) \\ \times \phi^{(k_2)}(w_{P(k_1+1)}, \ldots, w_{P(k_1+k_2)}))\end{aligned} \tag{2.27}$$

where P are the permutations of $k_1 + k_2$ objects, namely the elements of the symmetric group $\mathscr{S}_{k_1+k_2}$ and δ_P is the parity of the permutation P ($\delta_P = 0$ if P contains an even number of exchanges with respect to the identity permutation, while $\delta_P = 1$ if such a number is odd).

In order to make this definition clear, let us consider the explicit example where $k_1 = 2$ and $k_2 = 1$. We have:

$$\phi^{(2)} \wedge \phi^{(1)} = \phi^{(3)} \tag{2.28}$$

and we find

$$\begin{aligned}\phi^{(3)}(w_1, w_2, w_3) &= \frac{1}{3!}(\phi^{(2)}(w_1, w_2)\phi^{(1)}(w_3) - \phi^{(2)}(w_2, w_1)\phi^{(1)}(w_3) \\ &\quad - \phi^{(2)}(w_1, w_3)\phi^{(1)}(w_2) - \phi^{(2)}(w_3 1, w_2)\phi^{(1)}(w_1) \\ &\quad + \phi^{(2)}(w_2, w_3)\phi^{(1)}(w_1) + \phi^{(2)}(w_3, w_1)\phi^{(1)}(w_2)) \\ &= \frac{1}{3}(\phi^{(2)}(w_1, w_2)\phi^{(1)}(w_3) + \phi^{(2)}(w_2, w_3)\phi^{(1)}(w_1) \\ &\quad + \phi^{(2)}(w_3, w_1)\phi^{(1)}(w_2))\end{aligned}$$

The exterior product we have just defined has the following formal property:

$$\phi^{(k)} \wedge \phi^{(k')} = (-)^{kk'} \phi^{(k')} \wedge \phi^{(k)} \tag{2.29}$$

which can be immediately verified starting from the definition (2.27). Indeed, assuming for instance that $k_2 > k_1$, it is sufficient to consider the parity of the permutation:

$$\Pi = \begin{pmatrix} 1, & 2, & \ldots, & k_1, & k_1+1, & \ldots, & k_2, & k_2+1, & \ldots, & k_1+k_2 \\ k_1, & k_1+1, & \ldots, & k_1+k_1, & 2k_1+1, & \ldots, & k_1+k_2, & 1, & \ldots, & k_1 \end{pmatrix} \tag{2.30}$$

which is immediately seen to be:

$$\delta_\Pi = k_1 k_2 \bmod 2 \tag{2.31}$$

Setting $P = P'\Pi$ (which implies $\delta_P = \delta_{P'} + \delta_\Pi$) we obtain:

$$\begin{aligned}
\phi^{(k_2)} \wedge \phi^{(k_1)}(w_1, \dots, w_{k_1+k_2}) &= \frac{1}{(k_1+k_2)!} \sum_P (-)^{\delta_P} \phi^{(k_2)}(w_{P(1)}, \dots, w_{P(k_2)}) \\
&\quad \times \phi^{(k_1)}(w_{P(k_2+1)}, \dots, w_{P(k_1+k_2)}) \\
&= \frac{1}{(k_1+k_2)!} \sum_{P'} (-)^{\delta_{P'}+\delta_\Pi} \phi^{(k_2)}(w_{P'\Pi(1)}, \dots, w_{P'\Pi(k_2)}) \\
&\quad \times \phi^{(k_1)}(w_{P'\Pi(k_2+1)}, \dots, w_{P'\Pi(k_2+k_1)}) \\
&= \frac{1}{(k_1+k_2)!} (-)^{\delta_\Pi} \sum_{P'} (-)^{\delta_{P'}} \phi^{(k_2)}(w_{P'(k_1+1)}, \dots, w_{P'(k_1+k_2)}) \\
&\quad \times \phi^{(k_1)}(w_{P'(1)}, \dots, w_{P'(k_1)}) \\
&= (-)^{\delta_\Pi} \phi^{(k_1)} \wedge \phi^{(k_2)}(w_1, \dots, w_{k_1+k_2})
\end{aligned} \tag{2.32}$$

2.5.5 Special linear groups as volume-preserving transformations

In the particular case of $\otimes^n V$ (with V of dimension n), the fully antisymmetric subspace has dimension $\binom{n}{n} = 1$, and its basis element is $\mathbf{e}_1 \wedge \mathbf{e}_2 \cdots \wedge \mathbf{e}_n$. It is named[5] the *volume element*. A change of basis A on V induces a transformation of the volume element that can occur only by means of a multiplicative factor:

$$(\mathbf{e}_1 \wedge \mathbf{e}_2 \cdots \wedge \mathbf{e}_n)' = (\det A)\mathbf{e}_1 \wedge \mathbf{e}_2 \cdots \wedge \mathbf{e}_n \tag{2.33}$$

where the determinant arises from the application of eq. (2.20):

$$\det A = \sum_{P \in S_n} (-1)^{\delta_P} A_1^{j_1} A_2^{j_2} \cdots A_n^{j_n} \tag{2.34}$$

The permutations P act by exchanging the indices j_i. We have defined the *special linear group* SL(n) (real or complex) to be the subgroup of the general linear group GL(n) (real or complex) containing the matrices A such that $\det A = 1$. We see now by eq. (2.33) that the special linear group is the subset of basis changes on V that *preserve the volume* element of $\otimes^n V$.

2.5.6 Metric-preserving changes of basis

A *metric* on a vector space V is a functional from $V \otimes V$ into the field $\mathbb{F}$ ($\mathbb{R}$ or $\mathbb{C}$ for us) associated with the vector space v. That is, a metric is the assignment of a value

5 Though this name is in fact appropriate for the *dual* basis element of $\Lambda_n(V)$, the space of *n-forms* or n-linear antisymmetric functionals on V that we have discussed in the previous section.

$(\mathbf{v}_1, \mathbf{v}_2) \in \mathbb{F}$ to every pair of vectors, namely, $\forall \mathbf{v}_{1,2} \in V$. The metric can be required to be *bilinear*, in which case

$$\begin{aligned}(\mathbf{v}_1, \alpha\mathbf{v}_2 + \beta\mathbf{v}_3) &= \alpha(\mathbf{v}_1, \mathbf{v}_2) + \beta(\mathbf{v}_1, \mathbf{v}_3)\\ (\alpha\mathbf{v}_1 + \beta\mathbf{v}_2, \mathbf{v}_3) &= \alpha(\mathbf{v}_1, \mathbf{v}_3) + \beta(\mathbf{v}_2, \mathbf{v}_3)\end{aligned} \tag{2.35}$$

or *sesquilinear*, in which case

$$\begin{aligned}(\mathbf{v}_1, \alpha\mathbf{v}_2 + \beta\mathbf{v}_3) &= \alpha(\mathbf{v}_1, \mathbf{v}_2) + \beta(\mathbf{v}_1, \mathbf{v}_3)\\ (\alpha\mathbf{v}_1 + \beta\mathbf{v}_2, \mathbf{v}_3) &= \alpha^*(\mathbf{v}_1, \mathbf{v}_3) + \beta^*(\mathbf{v}_2, \mathbf{v}_3)\end{aligned} \tag{2.36}$$

Bilinearity and sesquilinearity are different only if the field $\mathbb{F}$ is $\mathbb{C}$ rather than $\mathbb{R}$.

The above notion of metric is identical with the notion of inner product introduced in Section 1.5.2. As noted there, a metric, or inner product, is specified by its action on a pair of basis vectors. Let us denote

$$(\mathbf{e}_i, \mathbf{e}_j) = g_{ij} \tag{2.37}$$

For a sesquilinear metric we have: $(\mathbf{v}, \mathbf{u}) = v^{i*} g_{ij} u^j$. We assume that the metric is *non-degenerate*, namely, that $\det g \neq 0$ (where g is the matrix of the elements g_{ij}). Under a change of basis A, a sesquilinear metric transforms as follows:

$$g'_{ij} = (A_i{}^k)^* g_{kl} A_j{}^l \tag{2.38}$$

Notice that g_{ij} transforms covariantly.

If a (sesquilinear) metric is *Hermitian*: $g_{ij} = g^*_{ji}$, then it is always possible to find a basis change that puts it into a canonical form

$$g_{ij} \to \operatorname{diag}(\underbrace{1, \dots, 1}_{p}, \underbrace{-1, \dots, -1}_{q}) \tag{2.39}$$

Indeed, we can first diagonalize g_{ij} to $\lambda_i \delta_{ij}$ changing basis with its eigenvector matrix $\mathbf{e}_i{}' = S_i{}^j \mathbf{e}_j$. Since g_{ij} is Hermitian by hypothesis, then the eigenvalues λ_i are all real. Then we can further change basis by rescaling the basis vectors to $\mathbf{f}_i = |\lambda_i|^{-1/2} \mathbf{e}_i{}'$ so as to obtain eq. (2.39). A metric whose canonical form is that presented in eq. (2.39) is said to have signature (p, q).

The procedure outlined above is the content of a theorem proved in 1852 by James Joseph Sylvester and named by him the *Law of Inertia of Quadratic Forms* [136]. Indeed, according to Sylvester's theorem, a symmetric non-degenerate $m \times m$ A matrix can always be transformed into a diagonal one with ± 1 entries by means of a substitution $A \mapsto B^T \cdot A \cdot B$. On the other hand, no such transformation can alter the signature $(p, m - p)$, which is intrinsic to the matrix A. This result in matrix algebra turned out to be very important for differential geometry and also for Relativity as we are going to discuss in Section 10.5.1.

Also a bilinear *antisymmetric* metric $g_{ij} = -g_{ij}$ can be put into a canonical form. Such a metric is non-degenerate only if the dimension n of the space is even, $n = 2m$. Indeed, $\det g = \det g^T = \det(-g) = (-)^n \det g$. If $n = 2m$, g_{ij} can be first *skew-diagonalized* by a change of basis:

$$g_{ij} \to \begin{pmatrix} \begin{matrix} 0 & \lambda_1 \\ -\lambda_1 & 0 \end{matrix} & \mathbf{0} & \cdots \\ \mathbf{0} & \begin{matrix} 0 & \lambda_2 \\ -\lambda_2 & 0 \end{matrix} & \cdots \\ \vdots & \vdots & \ddots \end{pmatrix} \tag{2.40}$$

and then brought to a canonical form by rescaling suitably the basis vector. The canonical form can be that of eq. (2.40), with all λ_i reduced to 1, or the so-called symplectic form Ω already introduced in eq. (2.3), obtained by a further reordering of the basis vectors.

Having endowed a vector space with a metric, we can consider those automorphisms that *preserve that metric*. It is not difficult to see that such changes of basis form a subgroup of the general linear group. Indeed, if A and B are two changes of bases that preserve the metric, then the product change of basis BA also preserves it, so closure is verified. Also the inverse of a metric-preserving automorphism preserves it, and the identity certainly does.

We now can identify the classical matrix groups as those subgroups of the general linear group that preserve certain types of metrics.

- The *pseudo-unitary group*[6] $\mathrm{U}(p,q;\mathbb{C})$ is the subgroup of $\mathrm{GL}(p+q,\mathbb{C})$ that preserves an *Hermitian* sesquilinear metric of signature (p,q). The prefix pseudo- is dropped when the metric is positive definite, i.e. when $q = 0$. In this case the metric in canonical form $g = \mathbf{1}$ is preserved by a basis change U **iff** $\mathrm{U}^\dagger \mathrm{U} = \mathbf{1}$.
- The *pseudo-orthogonal group*[7] $\mathrm{O}(p,q;\mathbb{R})$ is the subgroup of $\mathrm{GL}(p+q,\mathbb{R})$ that preserves a *symmetric* bilinear metric of signature (p,q). For a positive definite metric, the condition to be preserved by a basis change O is just $O^T O = \mathbf{1}$.
- The *symplectic group* $\mathrm{Sp}(m,\mathbb{R})$ is the subgroup of $\mathrm{GL}(2m,\mathbb{R})$ that preserves an *antisymmetric* bilinear metric (also called symplectic form). One can similarly define $\mathrm{Sp}(m,\mathbb{C})$.

We may further restrict the automorphisms to preserve the volume element, i.e., to have unit determinant. In this case the various groups acquire the denomination "special" and an S is prepended to their notation. For instance, $\mathrm{SL}(n,\mathbb{C}) \cap \mathrm{U}(n) = \mathrm{SU}(n)$, the special unitary group.

6 One usually writes simply $\mathrm{U}(p,q)$, as the real unitary groups coincide with orthogonal real groups (in real vector spaces there is no difference between sesquilinear and bilinear).

7 One usually writes simply $\mathrm{O}(p,q)$, as the complex orthogonal groups are not so frequently used.

2.5.6.1 Example

Let us compare the groups SO(2) and SO(1, 1). A generic matrix $R \in SO(2)$, namely, a matrix satisfying $A^T A = \mathbf{1}$ and $\det A = 1$, can be parameterized by an angle θ as

$$R = \begin{pmatrix} \cos\theta & \sin\theta \\ -\sin\theta & \cos\theta \end{pmatrix} \tag{2.41}$$

A matrix $\Lambda \in SO(1,1)$ must satisfy the following equations: $\Lambda^T \eta \Lambda = \eta$, where $\eta = \mathrm{diag}(-1, 1)$, and $\det\Lambda = 1$. Writing Λ as a generic real 2×2 matrix $\Lambda = \left(\begin{smallmatrix} a & b \\ c & d \end{smallmatrix}\right)$, these equation read:

$$\begin{pmatrix} a & c \\ b & d \end{pmatrix} \begin{pmatrix} -1 & 0 \\ 0 & 1 \end{pmatrix} \begin{pmatrix} a & b \\ c & d \end{pmatrix} = \begin{pmatrix} -a^2 + c^2 & -ab + cd \\ -ab + cd & -b^2 + d^2 \end{pmatrix} = \begin{pmatrix} -1 & 0 \\ 0 & 1 \end{pmatrix} \tag{2.42}$$

and $ad - bc = 1$. The solution to these constraints turns out to be the following:

$$\Lambda = \begin{pmatrix} \cosh\nu & \sinh\nu \\ -\sinh\nu & \cosh\nu \end{pmatrix} \tag{2.43}$$

with ν a real parameter, called the "rapidity." A possible alternative is to introduce a parameter β related to the rapidity by $\cosh\nu = 1/\sqrt{1-\beta^2}$, $\sinh\nu = \nu/\sqrt{1-\beta^2}$.

2.5.7 Isometries

The Euclidean group Eucl_d in d dimensions is the group of invertible transformations of the Euclidean space $\mathbb{R}^d$ into itself that preserve the Euclidean distance: for any transformation $\mathcal{E} \in \mathrm{Eucl}_d$, if $\mathbf{x}_1', \mathbf{x}_2'$ are the images under $\mathcal{E}$ of $\mathbf{x}_1, \mathbf{x}_2$, we have $|\mathbf{x}_2' - \mathbf{x}_1'| = |\mathbf{x}_2 - \mathbf{x}_1|$, for any couple $\mathbf{x}_1, \mathbf{x}_1$. Notice that the Euclidean distance is that arising from having endowed the vector space $\mathbb{R}^d$ with a symmetric bilinear positive definite metric $g_{ij} = \delta_{ij}$. This defines the scalar product $(\mathbf{x}, \mathbf{y}) = x^i g_{ij} y^j = \mathbf{x} \cdot \mathbf{y}$, and hence the distance $|\mathbf{x} - \mathbf{y}| = \sqrt{(\mathbf{x} - \mathbf{y}, \mathbf{x} - \mathbf{y})}$.

More generally, a notion of distance can be introduced not only in vector spaces, but in manifolds. Then one talks of Riemannian manifolds, i.e., differentiable manifolds of dimension d equipped with a positive definite quadratic form, that is a metric locally expressible as

$$ds^2 = g_{\alpha\beta}\, dx^\alpha \vee dx^\beta, \quad g_{\alpha\beta} = g_{\beta\alpha} \ (\alpha, \beta = 1, \dots, d) \tag{2.44}$$

with $g_{\alpha\beta}(x)$ a differentiable function of the coordinates (that transform as a two-tensor under coordinate changes). Then ds^2 defines the square length of the minimal arc connecting two points whose coordinates differ by dx^α.

To the notion of metric in a differentiable manifold we return extensively in Chapter 10.

For Euclidean spaces $\mathbb{R}^d$, the metric can always be chosen to be constant. Thus the possible transformations are of the form $x'^{\alpha} = \mathscr{R}^{\alpha}{}_{\beta} x^{\beta} + v^{\alpha}$, that is, in matrix notation, $\mathbf{x}' = \mathscr{R}\mathbf{x} + \mathbf{v}$. It is easily seen that the metric, which in matrix notation is written as $ds^2 = d\mathbf{x}^T g\, d\mathbf{x}$, is invariant iff

$$\mathscr{R}^T g \mathscr{R} = g \tag{2.45}$$

In a coordinate choice where $g_{\alpha\beta} = \delta_{\alpha\beta}$, the isometry condition becomes simply $\mathscr{R}^T \mathscr{R} = \mathbf{1}$, namely, $\mathscr{R} \in O(d)$. We retrieve thus the description of Euclidean isometries as products of translations and orthogonal transformations (rotations plus inversions).

2.6 Bibliographical note

The introductory topics on the notion of a group and the diverse examples collected and illustrated in this chapter are covered in a vast literature, both mathematical and physical.

As one of us (P.F.) explains at length in the forthcoming historical book [53], the very notions of a group and of a vector space were slowly developed throughout a historical process that embraces almost an entire century starting from Lagrange and Galois until the classical book of Hermann Weyl on *Classical Groups* [139]. Fundamental contributions in this process were provided by Jordan [95], Cayley [23], Sylvester [136], Klein [100, 99], Hurwitz [90], Lie [103–107], Frobenius [67], Killing [97] and finally by Cartan [16, 17] as far as finite and continuous groups are concerned. As far as vector spaces and matrices are concerned the major contributors to the development of these by now so familiar mathematical structures are Cayley [24], Sylvester [136], Grassmann [76], and conclusively Peano [124].

Modern and less modern textbooks for further reading on the topics touched upon in this chapter are:

1. For the use of group theory in quantum mechanics [137, 141, 83, 140].
2. For the multifaceted applications of Group Theory to Solid State Physics [30, 102, 135, 37].
3. For the uses of Group Theory in Particle Physics [27, 70, 31, 130].
4. For general illustrations of the uses of groups while treating the symmetries of physical systems [112, 131, 94, 108, 127, 91, 145, 134, 143, 8, 39, 86, 144].

3 Basic elements of finite group theory

Ancora indietro un poco ti rivolvi,
diss'io, là dove di' ch'usura offende
la divina bontade, e 'l groppo solvi.
Dante, Inferno XI, 94

3.1 Introduction

In Chapter 2 we already presented many different examples of groups, both continuous and discrete.

Once the conception of groups as abstract mathematical objects is completely integrated into the fabrics of Mathematics and Physics, the next two natural questions are:

a) The classification issue. Which are the possible groups G?

b) The representation issue. Which are the possible explicit realizations of each given abstract group G.

Let us comment on the first issue. If the *classification* of groups were accomplished, given a specific realization of a group arising, e.g., in some physical system, one might just identify its isomorphic class in the general classification and know a priori, via the isomorphism, all the relevant symmetry properties of the considered physical systems. For finite groups, this means the classification of all possible distinct (non-isomorphic) groups of a given finite order n. Such a goal is too ambitious and cannot be realized, yet there is a logical way to proceed.

One is able to single out certain types of groups (the so-called *simple groups*) which are the *hard core* of the possible different group structures. The complete classification of simple finite groups is one notable achievement of modern mathematics in the seventies and eighties of the XXth century which became possible thanks to massive computer calculations. We do not dwell on this very much specialized topic and we confine ourselves to mention some series of simple groups that are easily defined.

Assuming the list of simple groups as given one can study the possible *extensions* which allow the construction of new groups having the simple groups as building blocks. We do not plan an extensive discussion of this problem, except for some simple instances of extensions provided by the direct and semi-direct products.

Also for groups of infinite order there are some general results in the line of a classification, mainly regarding abelian groups. We shall touch briefly upon this issue.

For Lie groups the quest of classification follows a pattern very similar to the case of finite groups, involving the definition of *simple Lie groups* to be classified first. This we will discuss in full detail in Chapters 5 and 7.

https://doi.org/10.1515/9783110551204-003

In order to address the issue of group classification, we need to introduce several concepts and general theorems related to the inner structure of a given group; for instance, essential is the concept of *conjugacy classes* and of *invariant* subgroups.

Those concepts and theorems that allow us to discuss in much finer detail the structure of groups are the subject of Section 3.2.

The issue of representations of finite groups is addressed in detail in Section 3.3.

As we are going to see, the intrinsic structure of the abstract G determines the possible linear representations of G.

3.2 Basic notions and structural theorems for finite groups

Let us start with a theorem that strongly delimits the operational ground for finite group classification.

3.2.1 Cayley's theorem

Theorem 3.2.1. *Any group G of* finite *order $|G|$ is* isomorphic to a subgroup of the permutation group *on $|G|$ objects, $S_{|G|}$.*

This theorem was not stated in this way by Cayley but it is rightly named after him since it streams from Cayley's paper of 1854 which is extensively reviewed in [53]. An obvious consequence of Cayley's theorem is that the number of distinct groups of order G is finite, as the number of subgroups of $S_{|G|}$ certainly is. It can also be used to determine the possible group structures of low orders. However, though the permutation groups are easily defined, the structure of their subgroups is far from obvious, and the problem of classifying finite groups is far from being solved by this simple token.

Proof of Theorem 3.2.1. The proof of Cayley's theorem relies on the observation that each row of the group multiplication table defines a distinct permutation of the elements of the group. Thus to any element $g \in G$ we can associate the permutation $\pi \in S_{|G|}$ that acts as $g_k \overset{\pi_g}{\longmapsto} (gg_k)$ $(k = 1, \dots, |G|)$. Distinct group elements are associated with distinct permutations. The identity e is mapped into the identical permutation π_e. The product is preserved by the mapping: indeed, for any $c, b \in G$, we have

$$\pi_c \pi_b = \begin{pmatrix} bg_1 & \dots & bg_n \\ cbg_1 & \dots & cbg_n \end{pmatrix} \begin{pmatrix} g_1 & \dots & g_n \\ bg_1 & \dots & bg_n \end{pmatrix} = \begin{pmatrix} g_1 & \dots & g_n \\ cbg_1 & \dots & cbg_n \end{pmatrix} = \pi_{cb} \tag{3.1}$$

where, for convenience, we have described π_c by its action on the elements bg_i; this amounts just to a relabeling of the elements g_i. All in all, the set $\{\pi_b : b \in G\}$ is a subgroup of $S_{|G|}$ isomorphic to G. □

3.2.1.1 Regular permutations

The permutations π_g associated with the elements $g \in G$ in the previously described isomorphism can be read directly from the multiplication table of the group. Such permutations are called *regular* permutations, and the subgroups of S_n isomorphic to groups G of order n are subgroups of regular permutations. Let us summarize the properties of such subgroups.

i) Apart from the identical permutation, all other π_g do not leave any *symbol* (any of the objects on which the permutation acts) invariant; this corresponds to the property of the rows of the group multiplication table.[1]
ii) Any of the n permutations π_g maps a given symbol into a different symbol; this corresponds to the property of the columns of the multiplication table.
iii) All the cycles in the cycle decomposition of a regular permutation have the same length. Indeed, if a regular permutation π_g had two cycles of lengths $l_1 < l_2$, then $(\pi_g)^{l_1}$ would leave the elements of the first cycle invariant, but not those of the second, which however cannot be the case for a regular permutation.

Cayley's theorem is useful in determining the possible group structures of low order. In this respect a very convenient tool is provided by Cayley graphs described in the solution to Exercise 3.1. The following result is a corollary of Cayley's theorem.

3.2.1.2 Groups of prime order

Lemma 3.2.1. *The only finite group of order p, where p is a prime, is the cyclic group $\mathbb{Z}_p$.*

Proof of Lemma 3.2.1. By Cayley's theorem, any group is isomorphic to a subgroup of S_p made of regular permutations. Since all cycles of a regular permutation must have the same length ℓ, this latter must be a divisor of p. If p is prime, the only possible regular permutations have either p cycles of length 1 (identical permutation) or one cycle of length p. This is the case of cyclic permutations which form a group isomorphic to $\mathbb{Z}_p$. □

3.2.2 Left and right cosets

Let $H = \{e = h_1, h_2, \ldots, h_m\}$ be a subgroup of G of order $|H| = m$. Given an element a_1 not in H, $a_1 \in G \setminus H$, define its *left coset*:

$$a_1 H = \{a_1, a_1 h_2, \ldots, a_1 h_m\} \tag{3.2}$$

1 Concerning this property one might wonder if there cannot be idempotent group elements such that $\gamma \cdot \gamma = \gamma$ is a true statement. If such an element existed in the group $\gamma \in G$, the corresponding permutation π_γ would leave one symbol invariant, namely, γ itself. Yet from the axioms of a group it follows that $\gamma \cdot \gamma \cdot \gamma^{-1} = e$ which implies $\gamma = e$. Hence in any group the only idempotent element is the unique neutral element e and all the permutations induced by group elements are regular.

Since $h_i \neq h_j$, $\forall i,j = 1,\dots,m$, we have also $a_1 h_i \neq a_1 h_j$ (otherwise $a_1 = e$, but then a_1 would belong to H). Moreover, $\forall i$, $a_1 h_i \notin H$, otherwise $a_1 h_i = h_j$ for some j, and therefore $a_1 = h_j (h_i)^{-1}$ would belong to H. We can now take another element a_2 of $G \setminus H$ not contained in $a_1 H$. The m elements of its left coset $a_2 H$ are again all distinct, for the same reasoning as above. Moreover, $\forall i$, $a_2 h_i \notin H$, as above, but also $a_2 h_i \notin a_1 H$, otherwise we would have $a_2 h_i = a_1 h_j$, for some j, so that $a_2 = a_1 h_j (h_i)^{-1}$ would belong to $a_1 H$. We can iterate the reasoning until we exhaust all the elements of G.

Thus G decomposes into a *disjoint union* of left cosets with respect to any subgroup H:

$$G = H \cup a_1 H \cup a_2 H \cup \cdots \cup a_l H \tag{3.3}$$

Another way to reach the same result relies on the concept of equivalence classes. First note that we can introduce the following relation:

$$\forall g_1, g_2 \in G: \quad g_1 \sim g_2 \quad \Leftrightarrow \quad \exists h \in H \subset G / g_1 = g_2 h \tag{3.4}$$

which satisfies the axioms of an equivalence relation since it is both reflexive and transitive. Hence the entire set G decomposes into a finite set of disjoint equivalence classes and these are the cosets mentioned in eq. (3.3).

In complete analogy, given a subgroup H, we can introduce the *right cosets*

$$H a_1 = \{a_1, h_2 a_1, \dots, h_m a_1\} \tag{3.5}$$

and we can repeat the entire reasoning for the right cosets (see Exercises 3.2, 3.3).

3.2.3 Lagrange's theorem

An immediate consequence of the above reasoning is:

Theorem 3.2.2. *The order of a subgroup H of a finite group G is a divisor of the order of G:*

$$\exists l \in \mathbb{N} \text{ such that } |G| = l|H| \tag{3.6}$$

The integer $l \equiv [G:H]$ is named the index *of H in G.*

The above theorem, named after Lagrange, is very important for the classification of the possible subgroups of given groups. For instance, if $|G| = p$ is a prime, G does not admit any proper subgroup. This is the case for $G = \mathbb{Z}_p$ (the only group of order p, as we saw before).

Corollary 3.2.1. *The* order *of any element of a finite group G is a* divisor *of the order of G.*

Proof of Corollary 3.2.1. Indeed, if the order of an element $a \in G$ is h, then a generates a cyclic subgroup of order h $\{e, a, a^2, \dots, a^{h-1}\}$. This being a subgroup of G, its order h must be a divisor of $|G|$. □

3.2.4 Conjugacy classes

We have already introduced the conjugacy relation between elements of a group G in Section 1.2.4 ($g' \sim g \Leftrightarrow \exists h \in G$ such that $g' = h^{-1}gh$) and we have noted that it is an equivalence relation. Therefore we can consider the quotient of G (as a set) by means of this equivalence relation. The elements of the quotient set are named the *conjugacy classes*. Any group element g defines a conjugacy class $[g]$:

$$[g] \equiv \{g' \in G \text{ such that } g' \sim g\} = \{h^{-1}gh, \text{ for } h \in G\} \tag{3.7}$$

Basically, conjugation is the implementation of an inner automorphism of the group; we may think of it as a *change of basis* in the group (it is indeed so for matrix groups); in many instances, one is interested in those properties and those quantities that are independent from conjugation. Such properties pertain to the conjugacy classes rather than to the individual elements.

3.2.4.1 Example: conjugacy classes of the symmetric groups

The key feature that allows for an efficient description of the conjugacy classes of the symmetric group S_n is the invariance under conjugation of the *cycle decomposition* of any permutation $P \in S_n$.

Indeed, suppose that P contains a cycle $(p_1, p_2, \dots, p_k)$ of length k. Then a conjugate permutation $Q^{-1}PQ$ contains a cycle of the same length, namely,

$$(Q^{-1}(p_1), Q^{-1}(p_2), \dots, Q^{-1}(p_k))$$

Indeed, we have:

$$\begin{aligned}&(Q^{-1}(p_1), Q^{-1}(p_2), \dots, Q^{-1}(p_k)) \xrightarrow{Q} (p_1, p_2, \dots, p_k) \xrightarrow{P} (p_2, p_3, \dots, p_1)\\ &\xrightarrow{Q^{-1}} (Q^{-1}(p_2), Q^{-1}(p_3), \dots, Q^{-1}(p_1))\end{aligned} \tag{3.8}$$

Thus, *conjugacy classes* of S_n are in one-to-one correspondence with the possible *structures of cycle decompositions*. Let a permutation P be decomposed into cycles and name r_l the number of cycles of length l $(l = 1, \dots, n)$; then we have the sum rule

$$\sum_{l=1}^{n} r_l l = n \tag{3.9}$$

Hence the conjugacy class to which the permutation P belongs is determined by the set of integers $\{r_l\}$ describing how many cycles of each length l appear in the decomposition. Thus the possible conjugacy classes are in one-to-one correspondence with the set of solutions of eq. (3.9). These solutions, in turn, correspond to the set of *partitions* of the integer n into integers. A partition[2] of n is a set of integers $\{\lambda_i\}$, with

$$\sum_i \lambda_i = n, \quad \lambda_1 \geq \lambda_2 \geq \cdots \geq \lambda_n \geq 0 \tag{3.11}$$

Indeed, a set $\{r_l\}$ of integers satisfying eq. (3.9) is obtained from a partition $\{\lambda_i\}$ by setting

$$r_1 = \lambda_1 - \lambda_2, \quad r_2 = \lambda_2 - \lambda_3, \quad \ldots, \quad r_{n-1} = \lambda_{n-1} - \lambda_n, \quad r_n = \lambda_n \tag{3.12}$$

Thus, conjugacy classes of S_n are in one-to-one correspondence with partitions of n, which in turn can be graphically represented by means of *Young tableaux* with n boxes. In a Young tableaux, the boxes are distributed in rows of non-increasing length. The length of the i-th row is λ_i; the label r_l (the number of cycles of length l instead corresponds to the difference between the lengths of the l-th and the $(l+1)$-th row). As an example we display below the Young tableau $(8, 6, 6, 5, 3, 2)$ corresponding to a partition of 30.

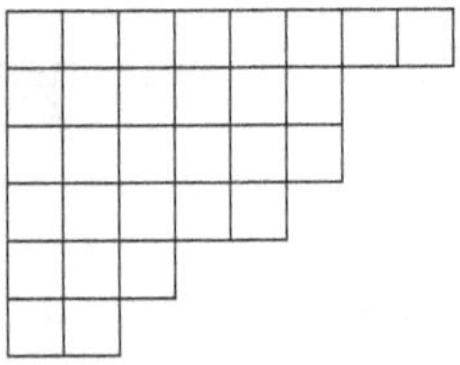

(3.13)

The corresponding cycle structure is $r_1 = 2, r_2 = 0, r_3 = 1, r_4 = 2, r_5 = 1, r_6 = 2$.

The number of elements in a given conjugacy class $\{r_l\}$, that we call the order of the class and denote as $|\{r_l\}|$, is obtained as follows. The n elements $1, \ldots, n$ must be distributed in the collection $\{r_l\}$ of cycles, ordered as follows:

$$\underbrace{(\cdot)\ldots(\cdot)}_{r_1}\ \underbrace{(\cdot\cdot)\ldots(\cdot\cdot)}_{r_2}\ \underbrace{(\cdot\cdot\cdot)\ldots(\cdot\cdot\cdot)}_{r_3}\ldots \tag{3.14}$$

There are n possible positions, so $n!$ possibilities. However, distributions differing for a permutation between cycles of the same length correspond to the same element (of course, (12)(45) is the same as (45)(12)); thus we must divide by $r_1!r_2!\ldots$. Moreover, in each cycle of length l we can make l periodic shifts (by 1, by 2, ..., by $l-1$) that leave the cycle invariant. Thus we must divide by $1^{r_1}2^{r_2}3^{r_3}\ldots$. Altogether we have obtained

$$|\{r_l\}| = \frac{n!}{r_1!2^{r_2}r_2!3^{r_3}r_3!\ldots} \tag{3.15}$$

2 Recall that $p(n)$, the number of partitions of n, is expressed through the generating function

$$P(q) \equiv \sum_{n=0}^{\infty} p(n)q^n = \prod_{k=1}^{\infty} \frac{1}{1-q^k} \tag{3.10}$$

The coefficient of q^n in the expansion of the infinite product gives the number of partitions of n.

3.2.5 Conjugate subgroups

Let H be a subgroup of a group G. Let us consider

$$H_g \equiv \{h_g \in G : h_g = g^{-1}hg, \text{ for } h \in H\} \tag{3.16}$$

which we simply write as $H_g = g^{-1}Hg$. It is easy to see that H_g is a subgroup. The subgroups H_g are called *conjugate subgroups* to H.

3.2.6 Invariant subgroups

A subgroup H of a group G is called an *invariant* (or *normal*) subgroup if it coincides with all its conjugate subgroups: $\forall g \in G,\ H_g = H$.

3.2.6.1 Left and right cosets

Given any subgroup H of G, we can define two equivalence relations in G:

$$\begin{aligned} g_1 \sim_L g_2 \quad &\Leftrightarrow \quad \exists h \in H : g_1 = hg_2 \quad \text{(left equivalence)} \\ g_1 \sim_R g_2 \quad &\Leftrightarrow \quad \exists h \in H : g_1 = g_2h \quad \text{(left equivalence)} \end{aligned} \tag{3.17}$$

Hence we can consider the set of equivalence classes with respect to either the left or the right equivalence, namely, the *left* (*right*) *cosets*. The left coset $H\backslash G$ contains the left classes already introduced that we write simply as gH, the right coset contains the right classes Hg.

If H is a normal subgroup, then the two equivalence relations of eq. (3.17) coincide:

$$g_1 \sim_L g_2 \quad \Leftrightarrow \quad g_2 \sim_R g_1 \tag{3.18}$$

It follows that the left and the right coset coincide: $H\backslash G = G/H$. This is the same as saying that the two equivalence relations are *compatible* with the group structure:

$$\begin{cases} g_1 \sim g_2 \\ g_3 \sim g_4 \end{cases} \quad \Leftrightarrow \quad g_1g_3 \sim g_2g_4 \tag{3.19}$$

where $\sim$ stands for $\sim_L$ (or $\sim_R$).

3.2.7 Factor groups

If H is a normal subgroup of G, then G/H ($= H\backslash G$) is a group, with respect to the product of classes defined as follows:

$$(g_1H)(g_2H) = g_1g_2H \tag{3.20}$$

This product is well-defined. Indeed, since $g_2H = Hg_2$ for H invariant, we have $g_1Hg_2H = g_1g_2H$. The subgroup H is the identity of G/H, and the inverse of an element gH is given $g^{-1}H$.

The converse of the above statement is also true: if $\phi : G \to G'$ is a homomorphism, then there exists a normal subgroup $H \subset G$ such that $G' = G/H$.

3.2.7.1 Example

Consider the group $n\mathbb{Z}$, namely, the set of multiples of n: $\{\dots, -2n, -n, 0, n, 2n, \dots\}$, with the addition as group law. The factor group $n\mathbb{Z}/\mathbb{Z}$ is isomorphic to the cyclic group $\mathbb{Z}_n$.

3.2.8 Center, centralizers, normalizers

3.2.8.1 The center of a group

The center $Z(G)$ of a group G is the set of all those elements of G that commute (in the group sense) with all the elements of G:

$$Z(G) = \{f \in G : g^{-1}fg = f, \ \forall g \in G\} \tag{3.21}$$

$Z(G)$ is an abelian subgroup of G.

3.2.8.2 The centralizer of a subset

The *centralizer* $C(A)$ of a subset $A \subset G$ is the subset of G containing all those elements that commute with all the elements of A:

$$C(A) = \{g \in G : \forall a \in A, \ g^{-1}ag = a\} \tag{3.22}$$

If A contains a single element a, then $C(A)$ is simply called the centralizer of a and it is denoted $C(a)$.

For any fixed element g, the product of the order of the conjugacy class of g and of its centralizer equals the order of G:

$$|[g]||C(g)| = |G| \tag{3.23}$$

Indeed, let $g' = u^{-1}gu$ be an element of $[g]$ different from g. Also the conjugation of g by $w = ut$, where $t \in C(g)$, gives g': in fact, $w^{-1}gw = u^{-1}t^{-1}gtu = u^{-1}gu = g'$. Thus, constructing $[g]$ as the set $\{u^{-1}gu : u \in G\}$ we obtain $|C(g)|$ times each distinct element.

3.2.8.3 Example: centralizers of permutations

Let a permutation $P \in S_n$ be decomposed into a set of $\{r_l\}$ cycles. It is easy to convince oneself that any permutation that i) permutes between themselves the cycles of equal length in P or ii) effects arbitrary periodic shifts within any cycle *commutes* with P. Thus, the number of permutations commuting with P is given by

$$|C(P)| = \prod_{l=1}^{n} r_l! l^{r_l} \tag{3.24}$$

We see that this expression, together with eq. (3.15) giving the order of the conjugacy class of P, is consistent with eq. (3.23).

3.2.8.4 The normalizer of a subset

The *normalizer* $N(A)$ of a subset $A \subset G$ is the subgroup of elements of G with respect to which A is invariant:

$$N(A) = \{g \in G : g^{-1}Ag = A\} \tag{3.25}$$

If A contains a single element a, then $N(A)$ is simply called the normalizer of a and it is denoted as $N(a)$.

3.2.9 The derived group

The commutator subgroup, or the *derived group* $\mathscr{D}(G)$ of a group G, is the group *generated* by the set of all group commutators in G (that is, it contains all group commutators and products thereof).

The derived group $\mathscr{D}(G)$ is normal in G, i.e., it is a *normal subgroup*. Indeed, take an element of $\mathscr{D}(G)$ which is a commutator, say $ghg^{-1}h^{-1}$. Then, any conjugate of it by an element $f \in G$, $f^{-1}ghg^{-1}h^{-1}f$ is still a commutator, that of $f^{-1}gf$ and $f^{-1}hf$. To an element of $\mathscr{D}(G)$ that is a product of commutators, the reasoning applies with little modification.

The factor group $G/\mathscr{D}(G)$ is *abelian*: it is the group obtained from G by *pretending* it is abelian. Another property is that any subgroup $H \subset G$ that contains $\mathscr{D}(G)$ is *normal*: since $\forall g \in G$ and $\forall h \in H$, we have $g^{-1}hg = [g^{-1}, h]h$.

3.2.10 Simple, semisimple, solvable groups

In general, a group G admits a chain of invariant subgroups, called its *subnormal series*:[3]

$$G = G_r \triangleright G_{r-1} \triangleright G_{r-2} \triangleright \cdots \triangleright G_1 \triangleright \{e\} \tag{3.26}$$

where every G_i is a normal subgroup.

3 Following a convention widely utilized in finite group theory we make a distinction between subgroups and normal subgroups. The denotation $G \supset H$ simply means that H is a subgroup of G, not necessarily an invariant one. On the other hand $G \triangleright N$ means that N is a normal (invariant) subgroup of G.

Definition 3.2.1. G is a *simple* group if it has *no proper normal subgroup.* For simple groups, the subnormal series is minimal:

$$G \triangleright \{e\} \tag{3.27}$$

Simple groups are the *hard core* of possible group structures. There is no factor group G/H smaller than G out of which the group G could be obtained by some *extension*, because there is no normal subgroup H other than the trivial one $\{e\}$ or G itself.

Definition 3.2.2. G is a *semisimple* group if it has *no proper normal subgroup* which is *abelian.*

Definition 3.2.3. A group G is *solvable* if it admits a subnormal series as in eq. (3.26) and all the factor groups $G/G_1, G_1/G_2, \dots, G_{k-1}/G_k, \dots$ are *abelian.*

3.2.11 Examples of simple groups

3.2.11.1 Cyclic groups of prime order

Cyclic groups $\mathbb{Z}_p$ with p a prime are simple. Indeed, they are abelian, so every subgroup would be a normal subgroup. However, by Lagrange's theorem, the order of any subgroup of $\mathbb{Z}_p$ should be a divisor of p, which leaves only the improper subgroups $\{e\}$ and $\mathbb{Z}_p$ itself. What is absolutely non-trivial is that the cyclic groups of prime order are the *only simple groups of odd order.*

3.2.11.2 The alternating groups A_n

One can show that the alternating groups A_n with $n \geq 5$ are simple. Here we omit the proof that can be found in the solution to Exercise 3.12.

3.2.12 Homomorphism theorems

Next we collect a few simple but fundamental theorems concerning homomorphisms and isomorphisms of groups that are extremely useful to understand the possible group structures.

Theorem 3.2.3 (First isomorphism theorem). *Given a homomorphism ϕ of G onto G':*

i) *the kernel of the homomorphism, $\ker\phi$, is an* invariant subgroup, *namely, $\ker\phi \triangleleft G$;*
ii) *the map ϕ gives rise to an* isomorphism *between the factor group $G/\ker\phi$ and G'.*

Proof of Theorem 3.2.3. By definition the kernel of ϕ is the subset of G that is mapped onto the identity element e' of G':

$$\ker\phi = \{g \in G : \phi(g) = \mathbf{e}'\} \tag{3.28}$$

It is immediate to see that $\ker\phi$ is a subgroup. It is also normal, because if $g \in \ker\phi$, then any of its conjugates belongs to it: $\phi(u^{-1}gu) = \phi(u^{-1})\phi(g)\phi(u) = [\phi(u)]^{-1}e'\phi(u) = e'$; this proves i). In case of finite groups, if $\ker\phi$ has order m, then ϕ is an m-to-one mapping. Indeed, if k_i $(i = 1, \dots, m)$ are the elements of $\ker\phi$, then the image $\phi(g)$ of a given element coincides with that of the elements $\phi(k_i g)$: The kernel being a normal subgroup, we can define the factor group $G/\ker\phi$. Since the kernel of the map $\phi : G/\ker\phi \to G'$ contains now only the identity of the factor group, this map is an isomorphism. This proves ii). □

There are other theorems concerning homomorphisms that we mention, without proof.

The *correspondence theorem* states that, if $\phi : G \to G'$ is an homomorphism, then:

i) The *preimage* $H = \phi^{-1}(H')$ of any subgroup H' of G' is a subgroup of G containing $\ker\phi$ (this generalizes the property of $\ker\phi = \phi^{-1}(e')$ being a subgroup). If H' is normal in G', then so is H in G.
ii) If there is any other subgroup H_1 of G, containing $\ker\phi$, that is mapped onto H' by ϕ, then $H_1 = H$.

The above statements can be rephrased (via the first isomorphism theorem) in terms of factor groups:

i) Let L be a subgroup of a factor group G/N. Then $L = H/N$ for H a subgroup of G (containing N). If L is normal in G/N, then H is normal in G.
ii) If $H/N = H_1/N$, with H and H_1 being subgroups of G containing N, then $H = H_1$.

The *factor of a factor* theorem states that if in the factor group G/N there is a normal subgroup of the form M/N, with $M \supseteq N$, then M is a normal subgroup of G, and

$$G/M \sim (G/N)/(M/N) \tag{3.29}$$

3.2.13 Direct products

As a set, the direct product $G \otimes F$ of two groups G and F is the Cartesian product of G and F:

$$G \times F = \{(g,f) : g \in G, f \in F\} \tag{3.30}$$

Elements of $G \otimes F$ are pairs, and $|G \otimes F| = |G|\,|F|$. The group operation is defined as follows. Elements of $G \otimes F$ have to be multiplied “independently” in each entry, in the first entry with the product law of G, in the second with the product law of F:

$$(g,f)(g',f') = (gg',ff') \tag{3.31}$$

Conversely, given a group G, we say that it is the direct product of certain subgroups:

$$G = H_1 \otimes H_2 \otimes \cdots \otimes H_n \tag{3.32}$$

if and only if
i) elements belonging to different subgroups H_i commute;
ii) the only element common to the various subgroups H_i is the identity;
iii) any element $g \in G$ can be expressed as product

$$g = h_1 h_2 \ldots h_n, \quad (h_1 \in H_1, \ldots, h_n \in H_n) \tag{3.33}$$

From ii), iii) it follows that the decomposition in eq. (3.33) is uniquely defined. Condition i) is equivalent to all the subgroups H_i being normal.

3.2.13.1 Example

The orthogonal group in three dimensions, O(3) is the direct product of the special orthogonal group SO(3) and of the matrix group (isomorphic to $\mathbb{Z}_2$) given by the two 3×3 matrices $\{\mathbf{1}, -\mathbf{1}\}$.

The direct product is the simplest way to build larger groups out of smaller building blocks. We can start from, say, two simple groups G and F to obtain a larger group $G \otimes F$ which is no longer simple as it admits G and F as normal subgroups. These normal subgroups are embedded into $G \otimes F$ in the simplest way; the homomorphism $\phi : G \otimes F \to G$ is provided by neglecting the F component, namely $\phi : (g,f) \mapsto g$.

3.2.14 Action of a group on a set

Before proceeding it is convenient to summarize a few notions and basic definitions about the action of a group G as a group of transformation on a set $\mathscr{S}$ that can be infinite or finite, continuous or countable.

Definition 3.2.4. Let G be a group and $\mathscr{S}$ be a set. We say that G acts as a transformation group on a set $\mathscr{S}$, if to each group element $\gamma \in G$ we can associate a map $\mu(\gamma)$ of the set onto itself:

$$\forall \gamma \in G: \quad \mu(\gamma) : \mathscr{S} \to \mathscr{S} \tag{3.34}$$

Furthermore, given two group elements $\gamma_{1,2} \in G$ we must have that:

$$\mu(\gamma_1 \cdot \gamma_2) = \mu(\gamma_1) \circ \mu(\gamma_2) \tag{3.35}$$

where $\circ$ denotes the composition of two maps, and also

$$\mu(\mathbf{e}) = \mathbf{id} \tag{3.36}$$

where $\mathbf{e}$ is the neutral element of G and $\mathbf{id}$ denotes the identity map which associates to each element $s \in \mathscr{S}$ the same element s.

Next we introduce the following notions.

Definition 3.2.5. Let G be a group that acts on the set $\mathscr{S}$ as a transformation group. For simplicity we denote by γ the map $\mu(\gamma) : \mathscr{S} \to \mathscr{S}$ associated with the group element $\gamma \in G$. Let $s \in S$ be an element in the set. We name *stability subgroup* of s and we denote $H_s \subset G$ the collection of all those transformations of G which map s onto itself:

$$H_s = \{h \in G \mid h(s) = s\} \tag{3.37}$$

That H_s defined as above is a subgroup, is immediately evident.

Conversely:

Definition 3.2.6. Let G be a group that acts on the set $\mathscr{S}$ as a transformation group and consider an element $s \in \mathscr{S}$. We name *orbit* of s with respect to G and we denote $\mathrm{Orbit}_G(s)$ the collection of all those elements of $\mathscr{S}$ which are reached from s by a suitable element of G:

$$\mathrm{Orbit}_G(s) = \{p \in \mathscr{S} \mid \exists \gamma \in G \setminus \gamma(s) = p\} \tag{3.38}$$

Furthermore,

Definition 3.2.7. Let G be a group that acts on the set $\mathscr{S}$ as a transformation group. We say that the action of G on $\mathscr{S}$ is *transitive* if, given any two elements $s_{1,2} \in \mathscr{S}$, there exists a suitable group element that maps the first onto the second.

$$\forall s_1, s_2 \in \mathscr{S} : \quad \exists \gamma_{12} \in G \setminus \gamma_{12}(s_1) = s_2 \tag{3.39}$$

When a group G has a transitive action on a set $\mathscr{S}$ we can identify it with the orbit of any of its elements:

Lemma 3.2.2. *Let G have a transitive action on the set $\mathscr{S}$. Then*

$$\forall s \in \mathscr{S} : \quad \mathrm{Orbit}_G(s) = \mathscr{S} \tag{3.40}$$

The proof of the above lemma is obvious from the very definitions of orbit and of transitivity.

Consider now the case of finite groups.

Let the finite group G act on the set $\mathscr{S}$ (finite or infinite, continuous or discrete does not matter). Consider an element $s \in \mathscr{S}$ and its orbit $\mathscr{O}_s \equiv \mathrm{Orbit}_G(s)$. By definition $\mathscr{O}_s \subset \mathscr{S}$ is a finite subset and G has a transitive action on it. The natural question is: what is the cardinality of this subset? The immediate answer is:

$$|\mathscr{O}_s| = \frac{|G|}{|H_s|} \tag{3.41}$$

where H_s is the stability subgroup of the element s. Indeed, only the elements of G that are not in H_s map s to new elements of the orbit. Yet any two elements $\gamma_1, \gamma_2 \in G$ that differ by right multiplication by any element of H_s map s to the same element of the orbit. Hence we can conclude that:

$$\mathcal{O}_s \sim G/H_s \tag{3.42}$$

and eq. (3.41) immediately follows. Furthermore, any element $s' \in \mathcal{O}_s$ has a stability subgroup $H_{s'}$ which is isomorphic to H_s being conjugate to it by means of any group element $\gamma \in G$ that maps s into s'. We conclude that the possible orbits of G are in one-to-one correspondence with the possible cosets G/H. Whether the orbit corresponding to a given subgroup H does exist or not depends on the explicit form of the G-action on $\mathcal{S}$, yet there is no orbit $\mathcal{O}$ of a certain cardinality $n = |\mathcal{O}|$ if there is no subgroup H of the required cardinality $|H| = |G|/n$.

Examples of stability subgroups are the normalizer and centralizer considered above.

3.2.15 Semi-direct products

A slightly more complicated construction, to obtain a larger group from two building blocks is the *semi-direct product*. Let G and K be two groups, and assume that G acts as a *group of transformations* on K:

$$\forall g \in G, \quad g : k \in K \mapsto g(k) \in K \tag{3.43}$$

Let us use the symbols $k_1 \circ k_2$ and $g_1 \cdot g_2$ for the group products in K and G respectively. The *semi-direct product* of G and K, denoted as $G \ltimes K$, is the Cartesian product of G and K as sets,

$$G \ltimes K = \{(g, k) : g \in G,\ k \in K\} \tag{3.44}$$

but the product in $G \ltimes K$ is defined as follows:

$$(g_1, k_1)(g_2, k_2) = (g_1 \cdot g_2, k_1 \circ g_1(k_2)) \tag{3.45}$$

That is, before being multiplied by k_1, the element k_2 is acted upon by g_1. The inverse of an element of $G \ltimes K$ is then given by:

$$(g, k)^{-1} = (g^{-1}, [g^{-1}(k)]^{-1}) \tag{3.46}$$

where the inverse g^{-1} is such with respect to the product in G, while the “external” inverse in $[g^{-1}(k)]^{-1}$ is with respect to the product in K.

The semi-direct product $G \ltimes K$ possesses a normal subgroup $\tilde{K}$, isomorphic to K, given by the elements of the form (e, k), with e the identity of G and $k \in K$. Indeed, any conjugate of such an element is again in the subgroup $\tilde{K}$:

$$(g,k)^{-1}(e,h)(g,k) = (g^{-1},[g^{-1}(k)]^{-1})(g,h\circ k) = (e,[g^{-1}(k)]^{-1}\circ g^{-1}(h\circ k)) \qquad (3.47)$$

Instead, the subgroup, isomorphic to G, containing elements of the form (g,e), where e is the identity in K and $g \in G$, is *not* a normal subgroup. The above discussion is summarized by the following:

Definition 3.2.8. Conversely, given a group G, we say that it is the semi-direct product $G = G_1 \ltimes G_2$ of two subgroups G_1 and G_2 iff:
i) G_2 is a normal subgroup of G;
ii) G_1 and G_2 have only the identity in common;
iii) every element of G can be written as a product of an element of G_1 and one of G_2.

From ii) and iii) it follows that the decomposition iii) is unique.

3.2.15.1 The Euclidean groups

The Euclidean groups Eucl_d are semi-direct products of the orthogonal group $\text{O}(d)$ and of the abelian group of d-dimensional translations. Their structure as semi-direct product has been discussed in eqs (2.12), (2.13) for the inhomogeneous rotation group ISO(2), the extension to larger Euclidean groups being immediate.

3.2.15.2 The Poincaré group

Consider a Poincaré transformation of a quadri-vector x^μ:

$$x^\mu \to \Lambda^\mu{}_\nu x^\nu + c^\mu \qquad (3.48)$$

Here Λ is a pseudo-orthogonal matrix, $\Lambda \in \text{O}(1,3)$, namely it is a matrix such that $\Lambda^T\eta\Lambda = \eta$, where $\eta = \text{diag}\,(-1,1,1,1)$ is the Minkowski metric. These matrices encode all the Lorenz transformations. Instead the 4-vector c^μ is a translation parameter. Poincaré transformations are the *isometries* of Minkowski space $\mathbb{R}^{1,3}$; they are the analog of the transformations of the Euclidean groups, the metric that they preserve, $\eta_{\mu\nu}$, being non-positive definite. Notice that the translation parameters, c^μ are 4-vectors and as such they are acted on by Lorenz transformations: $c^\mu \to \Lambda^\mu{}_\nu c^\nu$. The composition of two Poincaré transformations is the following one:

$$\begin{aligned} x^\mu &\xrightarrow{(2)} \Lambda^\mu_{(2)\nu}x^\nu + c^\nu_{(2)} \xrightarrow{(1)} \Lambda^\mu_{(1)\nu}(\Lambda^\nu_{(2)\rho}x^\rho + c^\nu_{(2)}) + c^\mu_{(1)} \\ &= (\Lambda_{(1)}\Lambda_{(2)})^\mu{}_\rho x^\rho + (\Lambda_{(1)}c_{(2)})^\mu + c^\mu_{(1)} \end{aligned} \qquad (3.49)$$

We see that the product law for the Poincaré group is

$$(\Lambda_{(1)},c_{(1)})(\Lambda_{(2)},c_{(2)}) = (\Lambda_{(1)}\Lambda_{(2)},\Lambda_{(1)}c_{(2)} + c_{(1)}) \qquad (3.50)$$

and the Poincaré group is the *semi-direct product* of the Lorenz group $\text{O}(1,3)$ with the translation group.

3.2.16 Extensions of a group

We stressed that we can regard simple groups as the building blocks of general groups. Indeed, the structure of a non-simple group G, admitting a normal subgroup H, depends only on the structure of H and on that of the factor group G/H.

Definition 3.2.9. We say that a group G is an *extension* of a group H by a group K iff there exists $\tilde{H}$, normal in G, such that $G/\tilde{H} = \tilde{K}$, with $\tilde{H}, \tilde{K}$ isomorphic to H, K, i.e., $\tilde{H} \simeq H$, $\tilde{K} \simeq K$.

The question is: how can G be built out of $\tilde{H}$ and $\tilde{K}$? We do not examine the general case, but we confine ourselves to the relevant subcase of so-called *splitting extensions*.

3.2.16.1 Splitting of a group

Let H be a normal subgroup of G, and let X be a so-called (left) *transversal* of H, namely, a set containing one and only one element from each coset of H in G. Let us assume that X is a subgroup of G. This corresponds to the following situation:

$$H \text{ normal in } G, \quad X \text{ subgroup of } G, \quad XH = G, \quad H \cap X = \{e\} \tag{3.51}$$

In such a case, G is said to *split over* H, and X is called a *complement* of H. Indeed, one can show that any element $g \in G$ can be written uniquely as a product $g = xh$, with $x \in X$ and $h \in H$, so $XH = G$, and every element of X lies in a distinct coset of H, so $H \cap X = \{e\}$. Conversely, if G splits over H, any complement X of H can be taken as a left transversal for H in G.

If G splits over H and X is a complement of H, then

$$G/H = HX/H = X/(H \cap X) = X \tag{3.52}$$

that is, G is an extension of H by X. In other words, if G splits over H, then G/H is isomorphic to any complement of H.

Definition 3.2.10. G is a *splitting extension* of H by X if there exists $\tilde{H}$ normal in G and isomorphic to H such that $\tilde{X} = G/\tilde{H}$ is isomorphic to X.

3.2.16.2 Reconstructing a group by splitting extension

All the products in a group G obtained by the splitting extension of H by X are determined from:

a) the product law in H;
b) the product law in X;
c) an "adjoint" action of X on H, that is, an homomorphism $\phi : X \to \mathrm{Aut}(H)$.

In simple words, take two elements $a, b \in G$. They will decompose as $a = A\alpha$ and $b = B\beta$, with $A, B \in X$ and $\alpha, \beta \in H$. Then we have

$$ab = A\alpha B\beta = ABB^{-1}\alpha B\beta \tag{3.53}$$

where now AB is a product in X. Moreover, $B^{-1}\alpha B$ is again in H as H is normal, so $B \in X$ defines an automorphism of H. Assigning to each $B \in X$ an automorphism of H, namely point c) above, is thus a key point to determine the products in G. Finally, the product $B^{-1}\alpha B\beta$ is then within H.

3.2.16.3 Direct and semi-direct products

It is not difficult to see that a direct product group $G = H \otimes K$ is in fact a particular splitting extension of H by K, in which the “adjoint” action of K on H is trivial (namely, $B^{-1}\alpha B = \alpha$, $\forall B \in K$ and $\forall \alpha \in H$). Similarly, one can see that a semi-direct product group $G = K \ltimes H$ is also a splitting extension of H by K.

3.2.17 Abelian groups

Finitely generated abelian groups are isomorphic to direct products (sums) of cyclic groups, finite or infinite (see Exercise 3.18).

3.3 The linear representations of finite groups

We turn next to the group-theory issue which is most relevant in physical and geometrical applications, namely, that of *linear representations*. In pure mathematics the notion of linear representations is usually replaced by the equivalent one of *module* (see Section 1.8).

Definition 3.3.1. Let G be a group and let V be a vector space of dimension n. Any homomorphism

$$D : G \to \mathrm{Hom}(V, V) \tag{3.54}$$

is named a linear representation of dimension n of the group G.

If the vectors $\{\mathbf{e}_i\}$ form a basis of the vector space V, then each group element $\gamma \in G$ is mapped into an $n \times n$ matrix $D_{ij}(\gamma)$ such that:

$$D(\gamma).\mathbf{e}_i = D_{ij}(\gamma)\mathbf{e}_j \tag{3.55}$$

and we see that the homomorphism D can also be rephrased as the following one:

$$D : G \to \mathrm{GL}(n, \mathbb{F}) \quad \mathbb{F} = \begin{cases} \mathbb{R} \\ \mathbb{C} \end{cases} \tag{3.56}$$

where the field $\mathbb{F}$ is that of the real or complex numbers, depending on whether V is a real or a complex vector space. Correspondingly, we say that D is a *real* or a *complex representation.*

Definition 3.3.1 applies in the same way to finite, infinite countable, and continuous groups. The same is true for the concept of irreducible representations introduced by the following two definitions:

Definition 3.3.2. Let $D : G \to \mathrm{Hom}(V, V)$ be a linear representation of a group G. A vector subspace $W \subset V$ is said to be **invariant** iff:

$$\forall \gamma \in G, \ \forall \mathbf{w} \in W : \quad D(\gamma).\mathbf{w} \in W \tag{3.57}$$

Definition 3.3.3. A linear representation $D : G \to \mathrm{Hom}(V, V)$ of a group G is named **irreducible** iff the only invariant subspaces of V are $\mathbf{0}$ and V itself.

In other words a representation is irreducible if it does not admit any proper invariant subspace.

Definition 3.3.4. A linear representation $D : G \to \mathrm{Hom}(V, V)$ that admits at least one proper invariant subspace $W \subset V$ is named **reducible.** A reducible representation $D : G \to \mathrm{Hom}(V, V)$ is named **fully reducible** iff the orthogonal complement $W^{\perp}$ of any invariant subspace W is also invariant.

Let D be a fully reducible representation and let $W_i \subset V$ be a sequence of invariant subspaces such that:

1)

$$V = \bigoplus_{i=1}^{r} W_i$$

2) None of the W_i contains invariant subspaces.

This situation can always be achieved by further splitting any invariant subspace W into smaller ones if it happens to admit further invariant subspaces. When we have reached 1) and 2) we see that the restriction to W_i of the representation D defines an irreducible representation D_i and we can write:

$$D = \bigoplus_{i=1}^{r} D_i; \quad D_i : G \to \mathrm{Hom}(W_i, W_i) \tag{3.58}$$

Matrix-wise we have:

$$\forall \gamma \in G; \quad D(\gamma) = \left(\begin{array}{c|c|c|c|c|c} D_1(\gamma) & 0 & 0 & \dots & 0 & 0 \\ \hline 0 & D_2(\gamma) & 0 & \dots & \dots & 0 \\ \hline \vdots & \vdots & \vdots & \vdots & \vdots & \vdots \\ \hline 0 & \dots & \dots & 0 & D_{r-1}(\gamma) & 0 \\ \hline 0 & \dots & \dots & \dots & 0 & D_r(\gamma) \end{array} \right) \tag{3.59}$$

The above is described as the decomposition of the considered reducible representation D into irreducible ones.

We see from the above discussion that irreducible representations are the building blocks for any representation and the main issue becomes the classification of irreducible representations frequently tabbed as *irreps*. Is the set of irreps an infinite set or a finite one? Here comes the main difference between continuous groups (in particular Lie groups) and finite ones. While the set of irreducible representations of Lie groups is infinite (nevertheless, as we show in later chapters, they can be tamed and classified within the constructive framework of roots and weights) the irreps of a finite group G constitute a finite set whose order is just equal to $c(G)$, this latter number being that of the conjugacy classes into which the elements of G are distributed. As we are going to see, to each irrep D_i we can associate a $c(G)$-dimensional vector $\chi[D_i]$ named its character. The theory of characters is a very simple and elegant piece of Mathematics with profound implications for Geometry, Physics, Chemistry and Crystallography. We devote this section to its development, starting from two lemmas due to Schur that provide the foundation of the whole story.

3.3.1 Schur's lemmas

Lemma 3.3.1. *Let V, W be two vector spaces of dimension n and m respectively, with $n > m$. Let G be a finite group and let $D_1 : G \to \mathrm{Hom}(V,V)$ and $D_2 : G \to \mathrm{Hom}(W,W)$ be two irreducible representations of dimension n and m respectively. Consider a linear map $\mathscr{A} : W \to V$ and impose the constraint*

$$\forall \gamma \in G \, \forall \mathbf{w} \in W \quad D_1(\gamma).\mathscr{A}.\mathbf{w} = \mathscr{A}.D_2(\gamma).\mathbf{w} \tag{3.60}$$

The only element $\mathscr{A} \in \mathrm{Hom}(W,V)$ that satisfies eq. (3.60) is $\mathscr{A} = 0$.

Proof of Lemma 3.3.1. Let $\mathrm{Im}[\mathscr{A}]$ be the image of the linear map $\mathscr{A}$, namely, the subspace of V spanned by all vectors $\mathbf{v}$ that have a preimage in W:

$$\mathbf{v} \in \mathrm{Im}[\mathscr{A}] \quad \Leftrightarrow \quad \exists \mathbf{w} \in W \backslash \mathscr{A}.\mathbf{w} = \mathbf{v} \tag{3.61}$$

The vector subspace $\mathrm{Im}[\mathscr{A}]$ is a proper subspace of V of dimension $m > 0$. Equation (3.60) states that this subspace should be invariant under the action of G in the representation D_1. This contradicts the hypothesis that D_1 is an irreducible representation. Therefore, $\mathrm{Im}[\mathscr{A}] = \mathbf{0} \in V$, namely, $\mathscr{A} = 0$. □

Lemma 3.3.2. *Let $D : G \to \mathrm{Hom}(V, V)$ be an n-dimensional irreducible representation of a finite group G. Let $C \in \mathrm{Hom}(V, V)$ be such that:*

$$\forall \gamma \in G \quad CD(\gamma) = D(\gamma)C \tag{3.62}$$

Then $C = \lambda\mathbf{1}$ where $\lambda \in \mathbb{C}$ and $\mathbf{1}$ is the identity map of the n-dimensional vector space V onto itself.

Proof of Lemma 3.3.2. Consider the eigenvalue equation:

$$C\mathbf{v} = \lambda\mathbf{v}; \quad \mathbf{v} \in V \tag{3.63}$$

The eigenspace $V_\lambda \subset V$ belonging to the eigenvalue λ is invariant under the action of G by the hypothesis of the lemma, namely, for $\forall \mathbf{v} \in V_\lambda$, $\forall \gamma \in G$, we have $D(\gamma).\mathbf{v} \in V_\lambda$. As we know, given any homomorphism $C \in \mathrm{Hom}(V, V)$ the vector space splits into the direct sum of the eigenspaces pertaining to different eigenvalues:

$$V = \oplus_{\lambda \in \mathrm{Spec}(C)} V_\lambda \tag{3.64}$$

Hence if the spectrum of C is composed by several different eigenvalues, we have as many invariant subspaces of V and this contradicts the hypothesis that D is an irreducible representation. Therefore there is only one eigenvalue $\lambda \in C$ and the map C is proportional to the identity map **1**. □

3.3.1.1 Orthogonality relations

Relying on Schur's lemmas we can now derive some very useful orthogonality relations that lead to the establishment of the main result, namely, the classification of irreducible representations for finite groups.

Let $X : V \to V$ be an arbitrary homomorphism of a finite dimensional vector space into itself, $X \in \mathrm{Hom}(V, V)$. Consider:

$$C \equiv \sum_{\gamma \in G} D(\gamma).X.D(\gamma^{-1}) \tag{3.65}$$

where D is a linear representation of a finite group G. We easily prove that:

$$\forall g \in G : \quad D(g).C = C.D(g) \tag{3.66}$$

To this effect it suffices to name $\tilde{\gamma} = g.\gamma$ and observe that the sum over all group elements γ is the same as the sum over all group elements $\tilde{\gamma}$, so that:

$$D(g).C = \sum_{\gamma \in G} D(g).D(\gamma).X.D(\gamma^{-1}) = \sum_{\gamma \in G} D(g.\gamma).X.D(\gamma^{-1}) \tag{3.67}$$

$$= \sum_{\tilde{\gamma} \in G} D(\tilde{\gamma}).X.D(\tilde{\gamma}^{-1}).D(g) = C.D(g) \tag{3.68}$$

If $D = D^\mu$ is an irreducible representation it follows from the above discussion and from Schur's lemmas that:

$$X \in \mathrm{Hom}(V, V): \quad \sum_{\gamma \in G} D^\mu(\gamma).X.D^\mu(\gamma^{-1}) = \lambda \mathbf{1} \tag{3.69}$$

where **1** denotes the identity map. Let us specialize the result (3.69) to the case where, in some given basis, the homomorphism X is represented by a matrix with all vanishing entries except one in the crossing of the $\mathfrak{m}$-th row with the $\mathfrak{l}$-th column:

$$X = \begin{pmatrix} 0 & 0 & \dots & \dots & \dots & 0 & 0 \\ 0 & \dots & \dots & \dots & \dots & \dots & 0 \\ \vdots & \dots & \dots & 0 & \dots & \dots & \vdots \\ \vdots & \dots & 0 & X_{\mathfrak{ml}} & 0 & \dots & \vdots \\ 0 & \dots & \dots & 0 & \dots & \dots & 0 \\ 0 & 0 & \dots & \dots & \dots & 0 & 0 \end{pmatrix} \tag{3.70}$$

Choosing $X_{\mathfrak{ml}} = 1$, from eq. (3.69) we obtain:

$$\sum_{\gamma \in G} D^\mu_{i\mathfrak{m}}(\gamma) D^\mu_{\mathfrak{l}j}(\gamma^{-1}) = \lambda_{\mathfrak{ml}} \delta_{ij} \tag{3.71}$$

The number $\lambda_{\mathfrak{ml}}$ can be evaluated from eq. (3.71) by setting $i = j$ and summing over i. We obtain the relation:

$$|G| \times \delta_{\mathfrak{ml}} = n_\mu \times \lambda_{\mathfrak{ml}} \tag{3.72}$$

where n_μ is the dimension of the irreducible representation under consideration and $|G|$ is the order of the group. Putting together the above results we obtain the following orthogonality relations for the matrix elements of an irreducible representation of a finite group:

$$\sum_{\gamma \in G} D^\mu_{i\mathfrak{m}}(\gamma) D^\mu_{\mathfrak{l}j}(\gamma^{-1}) = \frac{|G|}{n_\mu} \times \delta_{\mathfrak{ml}} \delta_{ij} \tag{3.73}$$

As one sees, the above result is just a consequence of Schur's second lemma. If we utilize also the first lemma, by considering two inequivalent irreducible representations μ and ν, and going through the very same steps we arrive at:

$$\sum_{\gamma \in G} D^\mu_{ij}(\gamma) D^\nu_{IJ}(\gamma^{-1}) = \frac{|G|}{n_\mu} \times \delta^{\mu\nu} \times \delta_{iJ} \delta_{jI} \tag{3.74}$$

Hence, given an irreducible representation D^μ of dimension n_μ the matrix elements give rise to n_μ^2 vectors with $|G|$-components that are orthogonal to each other and to the n_ν^2 vectors provided by another irreducible representation. Since in a G-dimensional vector space the maximal number of orthogonal vectors is $|G|$ it follows that

$$\sum_\mu n_\mu^2 \le |G| \tag{3.75}$$

This is already a very strong result since it implies that the number of irreducible representations is finite and that their dimensionality is upper bounded. Actually, refining our arguments we can show that in eq. (3.75) the symbol $\leq$ can be substituted by the symbol $=$ and this will provide an even stronger result. Provisionally let us name κ the number of inequivalent irreducible representation of the finite group G and let us name r the number of conjugacy classes of the same. We shall shortly prove that $\kappa = r$.

3.3.2 Characters

Let us now introduce the following:

Definition 3.3.5. Let $D : G \to \mathrm{Hom}(V, V)$ be a linear representation of a finite group G of dimension $n = \dim V$. Let r be the number of conjugacy classes $\mathscr{C}_i$ into which the whole group is split:

$$\begin{aligned} G = \bigcup_{i=1}^{r} \mathscr{C}_i; \quad \mathscr{C}_i \cap \mathscr{C}_j = \delta_{ij} \mathscr{C}_j \\ \forall \gamma, \tilde{\gamma} \in \mathscr{C}_i \quad \exists g \in G \quad / \quad \tilde{\gamma} = g \gamma g^{-1} \end{aligned} \tag{3.76}$$

We name **character** of the representation D the following r-dimensional vector:

$$\chi[D] = \{\mathrm{Tr}[D(\gamma_1)], \mathrm{Tr}[D(\gamma_2)], \dots, \mathrm{Tr}[D(\gamma_r)]\} \tag{3.77}$$

where:

$$\gamma_i \in \mathscr{C}_i \tag{3.78}$$

is any set of representatives of the r conjugacy classes.

The above definition of characters makes sense because of the following obvious lemma:

Lemma 3.3.3. *In any linear representation $D : G \to Hom(V, V)$ of a finite group the matrices $D(\gamma)$ and $D(\tilde{\gamma})$ representing two conjugate elements $\gamma = g\gamma g^{-1}$ have the same trace:* $\mathrm{Tr}[D(\gamma)] = \mathrm{Tr}[D(\tilde{\gamma})]$.

Proof of Lemma 3.3.3. Indeed, from the fundamental cyclic property of the trace we have:

$$\begin{aligned} \mathrm{Tr}[D(\tilde{\gamma})] &= \mathrm{Tr}[D(g\gamma g^{-1})] = \mathrm{Tr}[D(g)D(\gamma)D(g^{-1})] \\ &= \mathrm{Tr}[D(\gamma)D(g^{-1})D(g)] = \mathrm{Tr}[D(\gamma)] \end{aligned} \tag{3.79}$$

□

Keeping in mind the definition of characters let us reconsider eq. (3.74) and put $i = j$ and $I = J$. We obtain:

$$\sum_{\gamma \in G} D^{\mu}_{ii}(\gamma) D^{\nu}_{JJ}(\gamma^{-1}) = \frac{|G|}{n_{\mu}} \times \delta^{\mu\nu} \times \delta_{iJ} \tag{3.80}$$

Summing over all i and J in eq. (3.80) we get:

$$\sum_{\gamma \in G} \mathrm{Tr}[D^{\mu}(\gamma)] \mathrm{Tr}[D^{\nu}(\gamma^{-1})] = |G| \times \delta^{\mu\nu} \tag{3.81}$$

Let us now assume that on top of being irreducible the representations $D^{\mu} : G \to \mathrm{Hom}(V, V)$ are unitary according to the following standard:

Definition 3.3.6. A linear representation $D : G \to \mathrm{Hom}(V, V)$ of a finite group G of dimension $n = \dim V$ is named **unitary** iff:

$$\gamma \in G : \quad D(\gamma^{-1}) = D^{\dagger}(\gamma) \tag{3.82}$$

The above definition requires that on the vector space V it should be defined a sesquilinear scalar product, according to the definitions in Sections 1.5.2 and 2.5.6.

If the considered irreducible representations are all unitary, then eq. (3.81) becomes:

$$\sum_{\gamma \in G} \mathrm{Tr}[D^{\mu}(\gamma)] \mathrm{Tr}[D^{\nu\dagger}(\gamma)] = |G| \times \delta^{\mu\nu} \tag{3.83}$$

$$\Downarrow$$

$$\sum_{i=1}^{r} g_i \overline{\chi}^{\nu}_i \chi^{\mu}_i = |G| \delta^{\nu\mu} \tag{3.84}$$

where we have introduced the following standard notation:

$$g_i \equiv |\mathscr{C}_i| = \#\text{ of elements in the conj. class } i \tag{3.85}$$

3.3.3 Decomposition of a representation into irreducible representations

Given any representation D we can decompose it into irreducible ones setting:

$$\forall \gamma \in G : \quad D(\gamma) = \bigoplus_{\mu=1}^{\kappa} a_{\mu} D^{\mu}(\gamma) \tag{3.86}$$

where a_{ν} are necessarily all integers (the multiplicity of occurrence of the ν-th representation). Correspondingly taking the traces of the matrices on the left and on the right of the relation (3.86), it follows that:

$$\chi_i[D] = \sum_{\mu=1}^{\kappa} a_{\mu} \chi^{\mu}_i \tag{3.87}$$

namely, the character of the representation D is decomposed in a linear combination with integer coefficients of the characters of the irreducible representations that are named *simple characters*.

Utilizing the orthogonality relation (3.84) we immediately obtain a formula which expresses the multiplicity of each irreducible representation contained in a given one D in terms of the character of such a representation:

$$a_\mu = \frac{1}{|G|}\sum_{i=1}^{r} g_i \overline{\chi}_i^\mu \chi[D] \tag{3.88}$$

3.3.4 The regular representation

In order to complete our proof of the main theorem concerning group characters, we have to introduce the so-called regular representation. Let us label the elements of the finite group G as follows:

$$\underbrace{\gamma_1}_{=\mathbf{e}}, \gamma_2, \dots, \gamma_{g-1}, \gamma_g; \quad g = |G| \tag{3.89}$$

where the first is the neutral element and all the others are listed in some arbitrarily chosen order. Multiplying the entire list on the left by any element γ_α reproduces the same list in a permuted order:

$$\gamma_\alpha \cdot \{\gamma_1, \gamma_2, \dots, \gamma_{g-1}, \gamma_g\} = \{\gamma_{\pi_\alpha(1)}, \gamma_{\pi_\alpha(2)}, \dots, \gamma_{\pi_\alpha(g-1)}, \gamma_{\pi_\alpha(g)}\} \tag{3.90}$$

The permutation π_α is represented by a $g \times g$ matrix as already specified in eq. (2.7), namely:

$$D^R_{i\ell}(\gamma_\alpha) = \delta_{i,\pi_\alpha(\ell)} \tag{3.91}$$

and the set of these matrices for $\alpha = 1, \dots, g$ form the *regular representation* of the group.

The character of the regular representation is very particular. The matrices $D^R(\gamma)$ have no non-vanishing entry on the diagonal except in the case of the identity element $\gamma = e$. This follows from the discussion of Section 3.2.1.1 which states that no permutation π_α leaves any symbol unchanged unless $\alpha = 1$. From this simple observation it follows that the character of the regular representation is the following one:

$$\chi[D^R] = \{g, 0, 0, \dots, 0\} \tag{3.92}$$

Relying on (3.92) and applying eq. (3.87) to the case of the regular representation we find:

$$a_\nu[D^R] = \frac{1}{g}\sum_{i=1}^{r} g_i \overline{\chi}_i^\nu \chi_i[D^R] = \overline{\chi}_1^\nu = n_\nu \tag{3.93}$$

On the other hand, by definition of decomposition of the regular representation into irreducible ones we must have:

$$g = \sum_{\nu=1}^{\kappa} a_\nu [D^R] n_\nu \tag{3.94}$$

Combining the two above results we conclude:

$$\sum_{\nu=1}^{\kappa} n_\nu^2 = g \equiv |G| \tag{3.95}$$

This is the promised refinement of eq. (3.75). It remains only to prove that $\kappa = r$, namely, that the number of irreducible inequivalent representations is equal to the number of conjugacy classes of the group.

The proof of this last statement, which we omit for its slightly boring details, provided for instance in [81], relies on showing the complementary orthogonality relations:

$$\sum_{\nu=1}^{\kappa} \overline{\chi}_i^\nu \chi_j^\nu = \frac{g_i}{g} \delta_{ij} \tag{3.96}$$

from which it follows that $r \leq \kappa$, since in a space there cannot be orthonormal vectors in a number larger than that of its dimension. From eq. (3.84) it follows instead, for the same reason that $\kappa \leq r$. Hence $r = \kappa$ and the characters form a square matrix.

3.4 Strategy to construct the irreducible representations of a solvable group

In general, the derivation of the irreps and of the ensuing character table of a finite group G is a quite hard task. Yet a definite constructive algorithm can be devised if G is solvable and if one can establish a chain of normal subgroups ending with an abelian one, whose index is, at each step, a prime number q_i, namely, if we have the following situation:

$$\begin{aligned} & G = \mathrm{G}_{\mathrm{N_p}} \rhd \mathrm{G}_{\mathrm{N_{p-1}}} \rhd \cdots \rhd \mathrm{G}_{\mathrm{N_1}} \rhd \mathrm{G}_{\mathrm{N_0}} = \text{abelian group} \\ & \left| \frac{\mathrm{G}_{\mathrm{N_i}}}{\mathrm{G}_{\mathrm{N_{i-1}}}} \right| = \frac{\mathrm{N_i}}{\mathrm{N_{i-1}}} \equiv q_i = \text{prime integer number} \end{aligned} \tag{3.97}$$

The algorithm for the construction of the irreducible representations is based on an inductive procedure that allows to derive the irreps of the group $\mathrm{G}_{\mathrm{N_i}}$ if we know those of the group $\mathrm{G}_{\mathrm{N_{i-1}}}$ and if the index q_i is a prime number. The first step of the induction is immediately solved because any abelian finite group is necessarily a direct product of cyclic groups $\mathbb{Z}_k$, whose irreps are all one-dimensional and obtained by assigning to their generator one of the k-th roots of unity. When one consider crystallographic groups in three dimensions, the index q_i is always either 2 or 3. Hence we sketch the inductive algorithms with particular reference to the two cases of $q = 2$ and $q = 3$.

3.4.1 The inductive algorithm for irreps

To simplify notation we name $\mathscr{G} = G_{N_i}$ and $\mathscr{H} = G_{N_{i-1}}$. By hypothesis, $\mathscr{H} \lhd \mathscr{G}$ is a normal subgroup. Furthermore, $q \equiv |\frac{\mathscr{G}}{\mathscr{H}}| = \textit{prime number}$ (in particular $q = 2, \text{or } 3$). Let us name $D_\alpha[\mathscr{H}, d_\alpha]$ the irreducible representations of the subgroup. The index α (with $\alpha = 1, \dots, r_H \equiv \#$ of conj. classes of $\mathscr{H}$) enumerates them. In each case d_α denotes the dimension of the corresponding carrying vector space or, in mathematical jargon, of the corresponding module.

The first step to be taken is to distribute the $\mathscr{H}$ irreps into conjugation classes with respect to the bigger group. Conjugation classes of irreps are defined as follows. First one observes that, given an irreducible representation $D_\alpha[\mathscr{H}, d_\alpha]$, for every $g \in \mathscr{G}$ we can create another irreducible representation $D_\alpha^{(g)}[\mathscr{H}, d_\alpha]$, named the conjugate of $D_\alpha[\mathscr{H}, d_\alpha]$ with respect to g. The new representation is as follows:

$$\forall h \in \mathscr{H}: \quad D_\alpha^{(g)}[\mathscr{H}, d_\alpha](h) = D_\alpha[\mathscr{H}, d_\alpha](g^{-1} h g) \tag{3.98}$$

That the one defined above is a homomorphism of $\mathscr{H}$ onto $\mathrm{GL}(d_\alpha, \mathbb{F})$ is obvious and, as a consequence, it is also obvious that the new representation has the same dimension as the first. Secondly, if $g = \tilde{h} \in \mathscr{H}$ is an element of the subgroup, we get:

$$D_\alpha^{(\tilde{h})}[\mathscr{H}, d_\alpha](h) = A^{-1} D_\alpha[\mathscr{H}, d_\alpha](h) A \quad \text{where } A = D_\alpha[\mathscr{H}, d_\alpha](\tilde{h}) \tag{3.99}$$

so that conjugation amounts simply to a change of basis (a similarity transformation) inside the same representation. This does not alter the character vector and the new representation is equivalent to the old one. Hence the only non-trivial conjugations to be considered are those with respect to representatives of the different equivalence classes in $\frac{\mathscr{G}}{\mathscr{H}}$. Let us name γ_i $(i = 0, \dots, q-1)$ a set of representatives of such equivalence classes and define the orbit of each irrep $D_\alpha[\mathscr{H}, d_\alpha]$ as follows:

$$\mathrm{Orbit}_\alpha \equiv \{D_\alpha^{(\gamma_0)}[\mathscr{H}, d_\alpha], D_\alpha^{(\gamma_1)}[\mathscr{H}, d_\alpha], \dots, D_\alpha^{(\gamma_{q-1})}[\mathscr{H}, d_\alpha]\} \tag{3.100}$$

Since the available irreducible representations are a finite set, every $D_\alpha^{(\gamma_i)}[\mathscr{H}, d_\alpha]$ necessarily is identified with one of the existing $D_\beta[\mathscr{H}, d_\beta]$. Furthermore, since conjugation preserves the dimension, it follows that $d_\alpha = d_\beta$. It follows that $\mathscr{H}$-irreps of the same dimensions d arrange themselves into $\mathscr{G}$-orbits:

$$\mathrm{Orbit}_\alpha[d] = \{D_{\alpha_1}[\mathscr{H}, d], D_{\alpha_2}[\mathscr{H}, d], \dots, D_{\alpha_q}[\mathscr{H}, d]\} \tag{3.101}$$

and there are only two possibilities, either all $\alpha_i = \alpha$ are equal (self-conjugate representations) or they are all different (non-conjugate representations).

Once the irreps of $\mathscr{H}$ have been organized into conjugation orbits, we can proceed to promote them to irreps of the big group $\mathscr{G}$ according to the following scheme:

A) Each self-conjugate $\mathscr{H}$-irrep $D_\alpha[\mathscr{H}, d]$ is uplifted to q distinct irreducible $\mathscr{G}$-representations of the same dimension d, namely $D_{\alpha_i}[\mathscr{G}, d]$ where $i = 1, \dots, q$.

B) From each orbit β of q distinct but conjugate $\mathscr{H}$-irreps $\{D_{\alpha_1}[\mathscr{H}, d], D_{\alpha_2}[\mathscr{H}, d], \dots, D_{\alpha_q}[\mathscr{H}, d]\}$ one extracts a single $(q \times d)$-dimensional $\mathscr{G}$-representation.

3.4.2 A) Uplifting of self-conjugate representations

Let $D_\alpha[\mathscr{H},d]$ be a self-conjugate irrep. If the index q of the normal subgroup is a prime number, this means that $\frac{\mathscr{G}}{\mathscr{H}} \simeq \mathbb{Z}_q$. In this case the representatives γ_j of the q equivalence classes that form the quotient group can be chosen in the following way:

$$\gamma_1 = \mathbf{e}, \quad \gamma_2 = g, \quad \gamma_3 = g^2, \quad \dots, \quad \gamma_q = g^{q-1} \tag{3.102}$$

where $g \in \mathscr{G}, g \notin \mathscr{H}$ is a single group element satisfying $g^q = \mathbf{e}$. The key-point in uplifting the representation $D_\alpha[\mathscr{H},d]$ to the bigger group resides in the determination of a $d \times d$ matrix U that should satisfy the following constraints:

$$U^q = \mathbf{1} \tag{3.103}$$

$$\forall h \in \mathscr{H}: \quad D_\alpha[\mathscr{H},d](g^{-1}hg) = U^{-1}D_\alpha[\mathscr{H},d](h)U \tag{3.104}$$

These algebraic equations have exactly q distinct solutions $U_{[j]}$ and each of the solutions leads to one of the irreducible $\mathscr{G}$-representations induced by $D_\alpha[\mathscr{H},d]$. Any element $\gamma \in \mathscr{G}$ can be written as $\gamma = g^p h$ with $p = 0,1,\dots,q-1$ and $h \in \mathscr{H}$. Then it suffices to write:

$$D_{\alpha_j}[\mathscr{G},d](\gamma) = D_{\alpha_j}[\mathscr{G},d](g^p h) = U^p_{[j]} D_\alpha[\mathscr{H},d](h) \tag{3.105}$$

3.4.2 B) Uplifting of not self-conjugate representations

In the case of not self-conjugate representations, the induced representation of dimensions $q \times d$ is constructed relying once again on the possibility to write all group elements in the form $\gamma = g^p h$ with $p = 0,1,\dots,q-1$ and $h \in \mathscr{H}$. Furthermore, chosen one representation $D_\alpha[\mathscr{H},d]$ in the q-orbit (3.100), the other members of the orbit can be represented as $D_\alpha^{(g^j)}[\mathscr{H},d_\alpha]$ with $j = 1,\dots,q-1$. In view of this one writes:

$$\begin{aligned}
\forall h \in \mathscr{H}: \quad & D_\alpha[\mathscr{G},d](h) \\
= & \begin{pmatrix}
D_\alpha[\mathscr{H},d](h) & 0 & 0 & \cdots & 0 \\
0 & D_\alpha^{(g)}[\mathscr{H},d](h) & 0 & \cdots & 0 \\
0 & 0 & D_\alpha^{(g^2)}[\mathscr{H},d](h) & \cdots & 0 \\
\vdots & \vdots & \vdots & \vdots & \vdots \\
0 & 0 & \cdots & 0 & D_\alpha^{(g^{q-1})}[\mathscr{H},d](h)
\end{pmatrix} \\
g: \quad & D_\alpha[\mathscr{G},d](g) = \begin{pmatrix}
0 & \mathbf{1} & 0 & \cdots & 0 \\
0 & 0 & \mathbf{1} & \cdots & 0 \\
\vdots & \vdots & \vdots & \vdots & \vdots \\
0 & 0 & \cdots & 0 & \mathbf{1} \\
\mathbf{1} & 0 & \cdots & 0 & 0
\end{pmatrix} \\
\gamma = g^p h \quad & D_\alpha[\mathscr{G},d](g) = (D_\alpha[\mathscr{G},d](g))^p D_\alpha[\mathscr{G},d](h)
\end{aligned} \tag{3.106}$$

3.4.3 The octahedral group $O_{24} \sim S_4$ and its irreps

In Chapter 4, while discussing crystallographic lattices and their point-groups, we will meet the octahedral group and derive its irreducible representations using its interpretation as a transformation group in three-dimensional Euclidean space. In the present section, as an application of the methods described in the previous section, we want to derive the irreducible representations of O_{24} simply from the knowledge of its multiplication table and from the fact that it is solvable (see also Exercises 2.14 and 3.40, point d). To the octahedral group is also devoted the MATHEMATICA NoteBook described in Section B.7.

Abstractly the octahedral Group $O_{24} \sim S_4$ is isomorphic to the symmetric group of permutations of four objects. It is defined by the following generators and relations:

$$A, B: \quad A^3 = \mathbf{e}; \quad B^2 = \mathbf{e}; \quad (BA)^4 = \mathbf{e} \tag{3.107}$$

Since O_{24} is a finite, discrete subgroup of the three-dimensional rotation group, any $\gamma \in O_{24} \subset SO(3)$ of its 24 elements can be uniquely identified by its action on the coordinates x, y, z, as it is displayed below:

Class	Element	Class	Element
e	$1_1=\{x,y,z\}$	C_2	$4_1=\{-x,-z,-y\}$
C_3	$2_1=\{-y,-z,x\}$	C_2	$4_2=\{-x,z,y\}$
C_3	$2_2=\{-y,z,-x\}$	C_2	$4_3=\{-y,-x,-z\}$
C_3	$2_3=\{-z,-x,y\}$	C_2	$4_4=\{-z,-y,-x\}$
C_3	$2_4=\{-z,x,-y\}$	C_2	$4_5=\{z,-y,x\}$
C_3	$2_5=\{z,-x,-y\}$	C_2	$4_6=\{y,x,-z\}$
C_3	$2_6=\{z,x,y\}$	C_4	$5_1=\{-y,x,z\}$
C_3	$2_7=\{y,-z,-x\}$	C_4	$5_2=\{-z,y,x\}$
C_3	$2_8=\{y,z,x\}$	C_4	$5_3=\{z,y,-x\}$
C_4^2	$3_1=\{-x,-y,z\}$	C_4	$5_4=\{y,-x,z\}$
C_4^2	$3_2=\{-x,y,-z\}$	C_4	$5_5=\{x,-z,y\}$
C_4^2	$3_3=\{x,-y,-z\}$	C_4	$5_6=\{x,z,-y\}$

(3.108)

As one sees from the above list, the 24 elements are distributed into 5 conjugacy classes mentioned in the first column of the table, according to a nomenclature which is standard in the chemical literature on crystallography. The relation between the abstract and concrete presentation of the octahedral group is obtained by identifying in the list (3.108) the generators A and B mentioned in eq. (3.107). Explicitly we have:

$$A = 2_8 = \begin{pmatrix} 0 & 1 & 0 \\ 0 & 0 & 1 \\ 1 & 0 & 0 \end{pmatrix}; \quad B = 4_6 = \begin{pmatrix} 0 & 1 & 0 \\ 1 & 0 & 0 \\ 0 & 0 & -1 \end{pmatrix} \tag{3.109}$$

All other elements are reconstructed from the above two using the multiplication table of the group which is displayed below.

	1_1	2_1	2_2	2_3	2_4	2_5	2_6	2_7	2_8	3_1	3_2	3_3	4_1	4_2	4_3	4_4	4_5	4_6	5_1	5_2	5_3	5_4	5_5	5_6
1_1	1_1	2_1	2_2	2_3	2_4	2_5	2_6	2_7	2_8	3_1	3_2	3_3	4_1	4_2	4_3	4_4	4_5	4_6	5_1	5_2	5_3	5_4	5_5	5_6
2_1	2_1	2_5	2_4	3_3	3_2	1_1	3_1	2_6	2_3	2_7	2_2	2_8	5_3	4_4	5_6	4_6	5_4	4_2	4_1	4_3	5_1	5_5	4_5	5_2
2_2	2_2	2_6	2_3	1_1	3_1	3_3	3_2	2_5	2_4	2_8	2_1	2_7	4_5	5_2	5_5	5_4	4_6	4_1	4_2	5_1	4_3	5_6	5_3	4_4
2_3	2_3	3_2	1_1	2_2	2_8	2_7	2_1	3_3	3_1	2_4	2_6	2_5	4_6	5_1	5_3	5_6	4_1	4_5	5_2	4_2	5_5	4_4	4_3	5_4
2_4	2_4	3_1	3_3	2_1	2_7	2_8	2_2	1_1	3_2	2_3	2_5	2_6	5_4	4_3	4_5	5_5	4_2	5_3	4_4	4_1	5_6	5_2	5_1	4_6
2_5	2_5	1_1	3_2	2_8	2_2	2_1	2_7	3_1	3_3	2_6	2_4	2_3	5_1	4_6	5_2	4_2	5_5	4_4	5_3	5_6	4_1	4_5	5_4	4_3
2_6	2_6	3_3	3_1	2_7	2_1	2_2	2_8	3_2	1_1	2_5	2_3	2_4	4_3	5_4	4_4	4_1	5_6	5_2	4_5	5_5	4_2	5_3	4_6	5_1
2_7	2_7	2_3	2_6	3_1	1_1	3_2	3_3	2_4	2_5	2_1	2_8	2_2	5_2	4_5	4_2	5_1	4_3	5_6	5_5	5_4	4_6	4_1	4_4	5_3
2_8	2_8	2_4	2_5	3_2	3_3	3_1	1_1	2_3	2_6	2_2	2_7	2_1	4_4	5_3	4_1	4_3	5_1	5_5	5_6	4_6	5_4	4_2	5_2	4_5
3_1	3_1	2_8	2_7	2_6	2_5	2_4	2_3	2_2	2_1	1_1	3_3	3_2	5_6	5_5	4_6	5_3	5_2	4_3	5_4	4_5	4_4	5_1	4_2	4_1
3_2	3_2	2_7	2_8	2_5	2_6	2_3	2_4	2_1	2_2	3_3	1_1	3_1	5_5	5_6	5_4	4_5	4_4	5_1	4_6	5_3	5_2	4_3	4_1	4_2
3_3	3_3	2_2	2_1	2_4	2_3	2_6	2_5	2_8	2_7	3_2	3_1	1_1	4_2	4_1	5_1	5_2	5_3	5_4	4_3	4_4	4_5	4_6	5_6	5_5
4_1	4_1	5_4	4_6	4_5	5_3	5_2	4_4	5_1	4_3	5_5	5_6	4_2	1_1	3_3	2_8	2_6	2_3	2_2	2_7	2_5	2_4	2_1	3_1	3_2
4_2	4_2	4_6	5_4	5_3	4_5	4_4	5_2	4_3	5_1	5_6	5_5	4_1	3_3	1_1	2_7	2_5	2_4	2_1	2_8	2_6	2_3	2_2	3_2	3_1
4_3	4_3	5_3	5_2	5_6	4_2	5_5	4_1	4_5	4_4	4_6	5_1	5_4	2_6	2_4	1_1	2_8	2_7	3_1	3_2	2_2	2_1	3_3	2_5	2_3
4_4	4_4	4_2	5_5	5_1	5_4	4_6	4_3	5_6	4_1	5_2	4_5	5_3	2_8	2_1	2_6	1_1	3_2	2_5	2_3	3_1	3_3	2_4	2_2	2_7
4_5	4_5	5_6	4_1	4_6	4_3	5_1	5_4	4_2	5_5	5_3	4_4	5_2	2_2	2_7	2_4	3_2	1_1	2_3	2_5	3_3	3_1	2_6	2_8	2_1
4_6	4_6	4_4	4_5	4_1	5_5	4_2	5_6	5_2	5_3	4_3	5_4	5_1	2_3	2_5	3_1	2_1	2_2	1_1	3_3	2_7	2_8	3_2	2_4	2_6
5_1	5_1	4_5	4_4	5_5	4_1	5_6	4_2	5_3	5_2	5_4	4_3	4_6	2_5	2_3	3_3	2_7	2_8	3_2	3_1	2_1	2_2	1_1	2_6	2_4
5_2	5_2	4_1	5_6	4_3	4_6	5_4	5_1	5_5	4_2	4_4	5_3	4_5	2_7	2_2	2_5	3_3	3_1	2_6	2_4	3_2	1_1	2_3	2_1	2_8
5_3	5_3	5_5	4_2	5_4	5_1	4_3	4_6	4_1	5_6	4_5	5_2	4_4	2_1	2_8	2_3	3_1	3_3	2_4	2_6	1_1	3_2	2_5	2_7	2_2
5_4	5_4	5_2	5_3	4_2	5_6	4_1	5_5	4_4	4_5	5_1	4_6	4_3	2_4	2_6	3_2	2_2	2_1	3_3	1_1	2_8	2_7	3_1	2_3	2_5
5_5	5_5	4_3	5_1	4_4	5_2	5_3	4_5	4_6	5_4	4_1	4_2	5_6	3_2	3_1	2_2	2_4	2_5	2_8	2_1	2_3	2_6	2_7	3_3	1_1
5_6	5_6	5_1	4_3	5_2	4_4	4_5	5_3	5_4	4_6	4_2	4_1	5_5	3_1	3_2	2_1	2_3	2_6	2_7	2_2	2_4	2_5	2_8	1_1	3_3

(3.110)

This observation is important in relation to representation theory. Any linear representation of the group is uniquely specified by giving the matrix representation of the two generators $A = 2_8$ and $B = 4_6$.

3.4.3.1 The solvable structure of O_{24}

The group O_{24} is solvable since there exists the following chain of normal subgroups:

$$O_{24} \triangleright N_{12} \triangleright N_4 \qquad (3.111)$$

where the mentioned normal subgroups are given by the following lists of elements:

$$N_{12} \equiv \{1_1, 2_1, 2_2, \ldots, 2_8, 3_1, 3_2, 3_3\} \qquad (3.112)$$

$$N_4 \equiv \{1_1, 3_1, 3_2, 3_3\} \qquad (3.113)$$

See Exercise 3.22, point b). The group N_4 is abelian and we have:

$$N_4 \sim \mathbb{Z}_2 \times \mathbb{Z}_2 \tag{3.114}$$

since all of its elements are of order two. On the other hand, the indices of the two normal subgroups are respectively $q = 2$ and $q = 3$ and we have:

$$\frac{O_{24}}{N_{12}} \sim \mathbb{Z}_2; \quad \frac{N_{12}}{N_4} \sim \mathbb{Z}_3 \tag{3.115}$$

so that we can apply the strategy outlined in the previous section for the construction of the irreducible representations.

3.4.3.2 Irreps of N_4 (Klein group)

Since N_4 is abelian, its irreducible representations are all one-dimensional and they are immediately determined from its $\mathbb{Z}_2 \times \mathbb{Z}_2$ structure. Naming $3_1 = a$, $3_2 = b$, we obtain $3_3 = ab$ and the multiplication table of this normal subgroup is:

	1	a	b	ab
1	1	a	b	ab
a	a	1	ab	b
b	b	ab	1	a
ab	ab	b	a	1

(3.116)

The representations are obtained by assigning to a, b the values ± 1 in all possible ways (see Exercise 3.38). We have the following representation table which is also the character table:

	1	3_1	3_2	3_3
Δ_1	1	1	1	1
Δ_2	1	−1	1	−1
Δ_3	1	1	−1	−1
Δ_4	1	−1	−1	1

(3.117)

3.4.3.3 The irreps of N_{12} by induction

Next we construct the irreducible representation of the subgroup N_{12} by induction from those of its normal subgroup N_4. First we analyze the conjugacy class structure of N_{12}. Looking at the multiplication table (3.110) we determine four conjugacy classes:

$$\begin{aligned}
\mathscr{C}_1 &= \{1\} \\
\mathscr{C}_2 &= \{3_1, 3_2, 3_3\} \\
\mathscr{C}_3 &= \{2_1, 2_2, 2_7, 2_8\} \\
\mathscr{C}_4 &= \{2_3, 2_4, 2_5, 2_6\}
\end{aligned} \tag{3.118}$$

Hence we expect four irreducible representations. In order to determine them by induction we consider also the three equivalence classes in the quotient N_{12}/N_4. We find:

$$\begin{aligned}\mathfrak{C}_1 &= \mathscr{C}_1 \cup \mathscr{C}_2 \sim N_4\\ \mathfrak{C}_2 &= \mathscr{C}_3 \sim 2_8 \cdot N_4\\ \mathfrak{C}_2 &= \mathscr{C}_4 \sim 2_6 \cdot N_4 = (2_8)^2 \cdot N_4\end{aligned} \tag{3.119}$$

According to the strategy outlined in Section 3.4, we analyze the orbits of the four irreducible representations of N_4 with respect to the action of N_{12}. According to eq. (3.119) it suffices to consider conjugation with respect to 2_8 and its powers. We find two orbits:

$$\mathscr{O}_1 = \{\Delta_1, \Delta_1, \Delta_1\}; \quad \mathscr{O}_2 = \{\Delta_2, \Delta_3, \Delta_4\} \tag{3.120}$$

Hence there is one self-conjugate representation (the identity representation) of the normal subgroup N_4 which gives rise to three inequivalent one-dimensional representations ($\Pi_{1,2,3}$) of the big group N_{12} and there is an orbit of three conjugate representations of the normal subgroup which gives rise to a 3-dimensional representation (Π_4) of the big group. The summation rule is easily verified: $12 = 1 + 1 + 1 + 3^2$.

Let us construct these representations according to the algorithm outlined in Section 3.4. We begin with the one-dimensional representations.

3.4.3.4 Representation Π_1 = identity

In this representation all 12 elements of the group are represented by 1.

3.4.3.5 Representation Π_2

In this representation the elements of the group in $\mathfrak{C}_1$ are represented by 1, those in $\mathfrak{C}_2$ by $\exp[\frac{2\pi}{3}\mathrm{i}]$, those in $\mathfrak{C}_3$ by $\exp[\frac{4\pi}{3}i]$.

3.4.3.6 Representation Π_3

In this representation the elements of the group in $\mathfrak{C}_1$ are represented by 1, those in $\mathfrak{C}_2$ by $\exp[\frac{4\pi}{3}\mathrm{i}]$, those in $\mathfrak{C}_3$ by $\exp[\frac{2\pi}{3}i]$.

3.4.3.7 The 3-dimensional representation Π_4

In order to construct this representation we first of all write the block-diagonal matrices representing the elements of the normal subgroup, namely:

$$
\begin{aligned}
1_1 &= \begin{pmatrix} 1 & 0 & 0 \\ 0 & 1 & 0 \\ 0 & 0 & 1 \end{pmatrix} \\
3_1 &= \begin{pmatrix} -1 & 0 & 0 \\ 0 & 1 & 0 \\ 0 & 0 & -1 \end{pmatrix} \\
3_2 &= \begin{pmatrix} 1 & 0 & 0 \\ 0 & -1 & 0 \\ 0 & 0 & -1 \end{pmatrix} \\
3_3 &= \begin{pmatrix} -1 & 0 & 0 \\ 0 & -1 & 0 \\ 0 & 0 & 1 \end{pmatrix}
\end{aligned}
\tag{3.121}
$$

Furthermore, we introduce the representation of the element 2_8 according to the recipe provided in eq. (3.106):

$$
2_8 = \begin{pmatrix} 0 & 1 & 0 \\ 0 & 0 & 1 \\ 1 & 0 & 0 \end{pmatrix}
\tag{3.122}
$$

All the other elements of the order-12 group N_{12} are obtained by multiplication of the above five matrices according to the multiplication table, schematically summarized in the structure of the equivalence classes (3.119).

3.4.3.8 The irreps of the octahedral group

As already anticipated, the irreducible representations of the octahedral group will be directly constructed in Section 4.3.6. What we did above should have sufficiently illustrated the method of construction by induction which is available when the group is solvable. We briefly outline the next step, namely the construction of the irreps of O_{24} from those of N_{12} if we were to do it by induction instead that directly as we will do.

The group O_{24} has five conjugacy classes, so we expect five irreducible representations. Since there is only one 3-dimensional representation of N_{12}, it is necessarily self-conjugate. As the index of N_{12} in O_{24} is 2, it follows that the self-conjugate three-dimensional representation gives rise to two 3-dimensional irreps of O_{24}; let's name them D_4 and D_5, respectively. The identity representation is also self-conjugate for any group. Hence by the same argument as before there must be two one-dimensional representations D_1 and D_2 of which the first is necessarily the identity representation. It remains only one irrep at our disposal D_3 ($\dim D_3 = x$) and this is necessarily two-dimensional ($x = 2$) since, by numerology, we must have:

$$
1 + 1 + x^2 + 3^2 + 3^2 = 24
\tag{3.123}
$$

Indeed, we can verify that (Π_2, Π_3) form an orbit of conjugate representation and they are combined in the unique two-dimensional representation D_3 of the octahedral group.

The previous discussion clarifies how the method of induction allows to determine all the irreducible representations in the case of solvable groups.

3.5 Bibliographical note

The elements of finite group theory contained in this chapter are covered in different combinations in several textbooks and monographs. Our treatment of representation theory follows closely, yet modernized both in notation and in presentation the classical exposition of Hamermesh [81] and, in some instances, the mathematical textbook by James and Liebeck [93]. Furthermore, the induction method for the construction of solvable group representations presented here was systematized by one of us in collaboration with Alexander Sorin in a recent paper on Arnold–Beltrami Flows [56].

Suggested further references are the same already mentioned in the Bibliographical note 2.6.

4 Finite subgroups of SO(3) and crystallographic groups

χαλεπὰ, τὰ καλά
Nothing beautiful without struggle
Plato

4.1 Introduction

The geometrical structures advocated by supergravity, superstrings and other modern unified theories are strongly related to the theory of symmetric spaces and of Lie algebras, the exceptional ones being of utmost relevance in such a context.

At various stages of these constructions also the finite groups play an important role and, among them, those that are crystallographic in certain dimensions. This is not too much surprising since there exists a profound relation among the classification of simple, simply-laced, complex Lie algebras and the classification of finite subgroups of the three-dimensional rotation group, the so-called ADE classification.[1]

This ADE correspondence, known for a long time, finds a deeper and fertile interpretation in the McKay correspondence that is crucial for the Kronheimer construction of gravitational instantons as HyperKähler quotients. The McKay correspondence admits a generalization to finite subgroups $\Gamma \subset SU(n)$, in particular for the case $n = 3$, which has a significant role to play in the context of the AdS_4/CFT_3 correspondence.

We introduce here the discussion of the joint ADE classification of binary extensions $\Gamma_b \subset SU(2)$ of finite subgroups $\Gamma \subset SO(3)$ and of simple, simply-laced, complex Lie algebras.

After obtaining the classification of these finite subgroups of SU(2) we use them as main examples where we illustrate the theorems and the constructions of finite group theory discussed in the previous chapter. We introduce also the notion of crystallographic lattices and of their symmetry groups: this constitutes an important part of Physical Chemistry and Crystallography. The crystallographic groups are strongly related to the ADE finite groups mentioned above.

---

1 The ADE classification, according to the name frequently utilized in the physical literature, is based on a diophantine inequality that we spell out in the sequel of the present chapter. It encompasses in just one scheme the classification of several different types of mathematical objects:
1. the finite rotation groups;
2. the simple simply-laced complex Lie algebras;
3. the locally Euclidean gravitational instantons;
4. the singularities $\mathbb{C}^2/\Gamma$;
5. the modular invariant partition functions of $2D$-conformal field theories.

https://doi.org/10.1515/9783110551204-004

In this chapter we present also the full-fledged theory of the simple group $L_{168} \equiv \mathrm{PSL}(2,\mathbb{Z}_7)$ which fits into the discussion of crystallographic groups and provides an excellent non-trivial illustrative example.

4.2 The ADE classification of the finite subgroups of SU(2)

Let us start by considering the homomorphism:

$$\omega: \quad \mathrm{SU}(2) \to \mathrm{SO}(3) \tag{4.1}$$

between the group SU(2) of unitary 2×2 matrices, each of which can be written as follows

$$\mathrm{SU}(2) \ni \mathcal{U} = \begin{pmatrix} \alpha & i\beta \\ i\overline{\beta} & \overline{\alpha} \end{pmatrix} \tag{4.2}$$

in terms of two complex numbers α, β satisfying the constraint:

$$|\alpha|^2 + |\beta|^2 = 1 \tag{4.3}$$

and the group SO(3) of 3×3 of orthogonal matrices with unit determinant:

$$\mathcal{O} \in \mathrm{SO}(3) \quad \Leftrightarrow \quad \mathcal{O}^T \mathcal{O} = \mathbf{1} \quad \text{and} \quad \det \mathcal{O} = 1 \tag{4.4}$$

The homomorphism ω can be explicitly constructed utilizing the so-called triplet σ^x of Hermitian Pauli matrices:

$$\sigma^1 = \begin{pmatrix} 0 & 1 \\ 1 & 0 \end{pmatrix}; \quad \sigma^2 = \begin{pmatrix} 0 & -i \\ i & 0 \end{pmatrix}; \quad \sigma^3 = \begin{pmatrix} 1 & 0 \\ 0 & -1 \end{pmatrix} \tag{4.5}$$

Using the above we can define:

$$\mathcal{H} = \sum_{x=1}^{3} h_x \sigma^x \tag{4.6}$$

where h_x is a three-vector with real components. The matrix $\mathcal{H} = \mathcal{H}^\dagger$ is Hermitian by construction and we have:

$$\frac{1}{2}\mathrm{Tr}[\mathcal{H}^2] = \sum_{x=1}^{3} h_x^2 \tag{4.7}$$

Consider next the following matrix transformed by means of an SU(2) element:

$$\begin{aligned} \tilde{\mathcal{H}} &= \mathcal{U}^\dagger \mathcal{H} \mathcal{U} = \tilde{h}_x \sigma^x \\ \tilde{h}_x &= \mathcal{O}_x{}^y h_y \end{aligned} \tag{4.8}$$

The first line of eq. (4.8) can be written since the Pauli matrices form a complete basis for the space of 2×2 Hermitian traceless matrices. The second line can be written since the matrix $\tilde{\mathscr{H}}$ depends linearly on the matrix $\mathscr{H}$. Next we observe that because of its definition the matrix $\tilde{\mathscr{H}}$ has the following property:

$$\frac{1}{2}\mathrm{Tr}[\tilde{\mathscr{H}}^2] = \sum_{x=1}^{3} \tilde{h}_x^2 = \sum_{x=1}^{3} h_x^2 \tag{4.9}$$

This implies that the matrix $\mathscr{O}_x{}^y$ is orthogonal and, by definition, it is the image of $\mathscr{U}$ through the homomorphism ω. We can write an explicit formula for the matrix elements $\mathscr{O}_x{}^y$ in terms of $\mathscr{U}$:

$$\forall \mathscr{U} \in \mathrm{SU}(2): \quad \omega[\mathscr{U}] = \mathscr{O} \in \mathrm{SO}(3) \quad / \quad \mathscr{O}_x{}^y = \frac{1}{2}\mathrm{Tr}[\mathscr{U}^\dagger \sigma_x \mathscr{U} \sigma^y] \tag{4.10}$$

which follows from the trace-orthogonality of the Pauli matrices $\frac{1}{2}\mathrm{Tr}[\sigma^y \sigma_x] = \delta^y_x$.

We named the map defined above a homomorphism rather than an isomorphism since it has a non-trivial kernel of order two. Indeed, the following two SU(2) matrices constitute the kernel of ω since they are both mapped into the identity element of SO(3).

$$\begin{aligned} \ker\omega &= \left\{ \mathbf{e} = \begin{pmatrix} 1 & 0 \\ 0 & 1 \end{pmatrix}, \; \mathscr{Z} = \begin{pmatrix} -1 & 0 \\ 0 & -1 \end{pmatrix} \right\} \\ \mathbf{1} &= \omega[\mathbf{e}] = \omega[\mathscr{Z}] \end{aligned} \tag{4.11}$$

We will now obtain the classification of all finite subgroups of SU(2), that we collectively name G^b_{2n}, denoting by $2n$ their necessarily even order. Through the isomorphism ω each of them maps into a finite subgroup $\mathrm{G}_n \subset \mathrm{SO}(3)$, whose order is just n because of the two-dimensional kernel mentioned above:

$$\omega[\mathrm{G}^b_{2n}] = \mathrm{G}_n \tag{4.12}$$

The groups G^b_{2n} are named the binary extensions of the finite rotation groups G_n.

4.2.1 The argument leading to the Diophantine equation

We begin by considering one-parameter subgroups of SO(3). These are singled out by a rotation axis, namely by a point on the two-sphere S^2. Explicitly let us consider a solution (ℓ, m, n) to the sphere equation:

$$\ell^2 + m^2 + n^2 = 1 \tag{4.13}$$

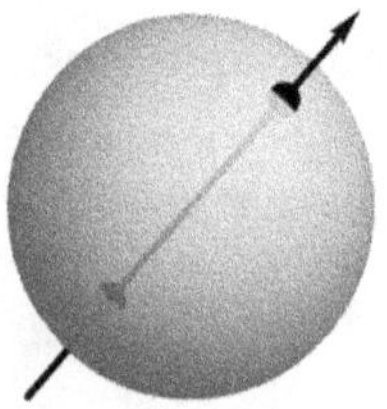

Figure 4.1: Every element of the rotation group $\mathcal{O}_{(\ell,m,n)} \in \mathrm{SO}(3)$ corresponds to a rotation around some axis $\mathbf{a} = \{\ell, m, n\}$. On the surface of the two-sphere $\mathbb{S}^2$ this rotation has two fixed points, a North Pole and a South Pole, that do not rotate to any other point. The rotation $\mathcal{O}_{(\ell,m,n)}$ is the image, under the homomorphism ω of either one of the 2×2-matrices $\mathcal{U}^{\pm}_{\ell,m,n}$ that, acting on the space $\mathbb{C}^2$, admit two eigenvectors $\mathbf{z}_1$ and $\mathbf{z}_2$. The one-dimensional complex spaces $p_{1,2} \equiv \lambda_{1,2}\mathbf{z}_{1,2}$ are named the two poles of the unitary rotation.

The triplet of real numbers (ℓ, m, n) parameterize the direction of a possible rotation angle (see Figure 4.1). The generator of infinitesimal rotations around such an axis is given by the following matrix

$$A_{\ell,m,n} = \begin{pmatrix} 0 & -n & m \\ n & 0 & -\ell \\ -m & \ell & 0 \end{pmatrix} = -A^T_{\ell,m,n} \tag{4.14}$$

which being antisymmetric belongs to the SO(3) Lie algebra. The matrix A has the property that $A^3 = -A$, and explicitly we have:

$$A^2_{\ell,m,n} = \begin{pmatrix} -1+\ell^2 & \ell m & \ell n \\ \ell m & -1+m^2 & mn \\ \ell n & mn & -1+n^2 \end{pmatrix} \tag{4.15}$$

Hence a finite element of the group SO(3) corresponding to a rotation of an angle θ around this axis is given by:

$$\mathcal{O}_{(\ell,m,n)} = \exp[\theta A_{\ell,m,n}] = \mathbf{1} + \sin\theta A_{\ell,m,n} + (1-\cos\theta)A^2_{\ell,m,n} \tag{4.16}$$

Setting

$$\lambda = \ell \sin\frac{\theta}{2}; \quad \mu = m\sin\frac{\theta}{2}; \quad \nu = n\sin\frac{\theta}{2}; \quad \rho = \cos\frac{\theta}{2} \tag{4.17}$$

the corresponding SU(2) finite group elements, realizing the double covering, are:

$$\mathcal{U}^{\pm}_{\ell,m,n} = \pm\begin{pmatrix} \rho + \mathrm{i}\nu & \mu - \mathrm{i}\lambda \\ -\mu - \mathrm{i}\lambda & \rho - \mathrm{i}\nu \end{pmatrix} \tag{4.18}$$

namely, we have:

$$\omega[\mathcal{U}^{\pm}_{\ell,m,n}] = \mathcal{O}_{(\ell,m,n)} \tag{4.19}$$

We can now consider the argument that leads to the ADE classification of the finite subgroups of SU(2). Let us consider the action of the SU(2) matrices on $\mathbb{C}^2$. A generic $\mathcal{U} \in \mathrm{SU}(2)$ acts on a $\mathbb{C}^2$-vector $\mathbf{z} = \binom{z_1}{z_2}$ by usual matrix multiplication $\mathcal{U}\mathbf{z}$. Each element $\mathcal{U} \in \mathrm{SU}(2)$ has two eigenvectors $\mathbf{z}_1$ and $\mathbf{z}_2$, such that

$$\begin{aligned} \mathcal{U}\mathbf{z}_1 &= \exp[\mathrm{i}\theta]\mathbf{z}_1 \\ \mathcal{U}\mathbf{z}_2 &= \exp[-\mathrm{i}\theta]\mathbf{z}_2 \end{aligned} \tag{4.20}$$

where θ is some (half-)rotation angle. Namely, for each $\mathcal{U} \in \mathrm{SU}(2)$ we can find an orthogonal basis where $\mathcal{U}$ is diagonal and given by:

$$\mathcal{U} = \begin{pmatrix} \exp[\mathrm{i}\theta] & 0 \\ 0 & \exp[-\mathrm{i}\theta] \end{pmatrix} \tag{4.21}$$

for some angle θ. Then let us consider the rays $\{\lambda \mathbf{z}_1\}$ and $\{\mu \mathbf{z}_2\}$ where $\lambda, \mu \in \mathbb{C}$ are arbitrary complex numbers. Since $\mathbf{z}_1 \cdot \mathbf{z}_2 = \mathbf{z}_1^\dagger \mathbf{z}_2 = 0$ it follows that each element of SU(2) singles out two rays, hereafter named *poles*, that are determined one from the other by the orthogonality relation. This concept of pole is the basic item in the argument leading to the classification of finite rotation groups.

Let $H \subset \mathrm{SO}(3)$ be a finite, discrete subgroup of the rotation group and let $\hat{H} \subset \mathrm{SU}(2)$ be its preimage in SU(2) with respect to the homomorphism ω. Then the order of H is some positive integer number:

$$|H| = n \in \mathbb{N} \tag{4.22}$$

The total number of poles associated with H is:

$$\#\text{ of poles} = 2n - 2 \tag{4.23}$$

since $n - 1$ is the number of elements in H that are different from the identity. Let us then adopt the notation

$$p_i \equiv \{\lambda \mathbf{z}_i\} \tag{4.24}$$

for the pole or ray singled out by the eigenvector $\mathbf{z}_i$. We say that two poles are equivalent if there exists an element of the group H that maps one onto the other:

$$p_i \sim p_j \quad \text{iff} \quad \exists \gamma \in H / \gamma p_i = p_j \tag{4.25}$$

Let us distribute the poles p_i into orbits under the action of the group H:

$$\mathcal{Q}_\alpha = \{p_1^\alpha, \dots, p_{m_\alpha}^\alpha\}; \quad \alpha = 1, \dots, r \tag{4.26}$$

and name m_α the cardinality of the orbit class $\mathcal{Q}_\alpha$, namely, the number of poles it contains. Hence we have assumed that there are r orbits and that each orbit $\mathcal{Q}_\alpha$ contains m_α elements.

Each pole $p \in \mathcal{Q}_\alpha$ has a stability subgroup $K_p \subset H$:

$$\forall h \in K_p : \quad hp = p \tag{4.27}$$

that is finite, abelian and cyclic of order k_α. Indeed, it must be finite since it is a subgroup of a finite group, it must be abelian since in the basis $\mathbf{z}_1, \mathbf{z}_2$ the SU(2) matrices that preserve the poles $\lambda \mathbf{z}_1$ and $\mu \mathbf{z}_2$ are of the form (4.21) and therefore it is cyclic of some order. The H group can be decomposed into cosets according to the subgroup K_p:

$$H = K_p \cup v_1 K_p \cup \cdots \cup v_{m_\alpha} K_p \quad m_\alpha \in \mathbb{N} \tag{4.28}$$

Consider now an element $x_i \in v_i K_p$ belonging to one of the cosets and define the group conjugate to K_p through x_i:

$$K_{(xp)_i} = x_i K_p x_i^{-1} \tag{4.29}$$

Each element $h \in K_{(xp)_i}$ admits a pole p_x:

$$hp_x = p_x \tag{4.30}$$

that is given by:

$$p_x = x_i p \tag{4.31}$$

since

$$hp_x = xh_p xx^{-1} p = xh_p p = xp = p_x \tag{4.32}$$

Hence the set of poles $\{p, v_1 p, v_2 p, \dots, v_{m_\alpha} p\}$ are equivalent forming an orbit. Each of them has a stability group K_{p_i} conjugate to K_p which implies that all K_{p_i} are finite of the same order:

$$\forall v_i p \quad |K_{p_i}| = k_\alpha \tag{4.33}$$

By this token we have proven that in each orbit $\mathcal{Q}_\alpha$ the stability subgroups of each element are isomorphic, and cyclic of the same order k_α which is a property of the orbit. Hence we must have:

$$\forall \mathcal{Q}_\alpha; \quad k_\alpha m_\alpha = n \tag{4.34}$$

The total number of poles we have in the orbit $\mathcal{Q}_\alpha$ (counting coincidences) is:

$$\#\text{ of poles in the orbit } \mathcal{Q}_\alpha = m_\alpha (k_\alpha - 1) \tag{4.35}$$

since the number of elements in K_p differently from the identity is $k_\alpha - 1$. Hence we find

$$2n - 2 = \sum_{\alpha=1}^{r} m_\alpha(k_\alpha - 1) \tag{4.36}$$

Dividing by n we obtain:

$$2\left(1 - \frac{1}{n}\right) = \sum_{\alpha=1}^{r}\left(1 - \frac{1}{k_\alpha}\right) \tag{4.37}$$

We consider next the possible solutions to the Diophantine equation (4.37) and to this effect we rewrite it as follows:

$$r + \frac{2}{n} - 2 = \sum_{\alpha=1}^{r} \frac{1}{k_\alpha} \tag{4.38}$$

We observe that $k_\alpha \geq 2$. Indeed, each pole admits at least two group elements that keep it fixed, the identity and the non-trivial group element that defines it by diagonalization. Hence we have the bound:

$$r + \frac{2}{n} - 2 \leq \frac{r}{2} \tag{4.39}$$

which implies:

$$r \leq 4 - \frac{4}{n} \quad \Rightarrow \quad r = 1, 2, 3 \tag{4.40}$$

On the other hand, we also have $k_\alpha \leq n$ so that:

$$r + \frac{2}{n} - 2 \geq \frac{r}{n} \quad \Rightarrow \quad r\left(1 - \frac{1}{n}\right) \geq 2\left(1 - \frac{1}{n}\right) \quad \Rightarrow \quad r \geq 2 \tag{4.41}$$

Therefore there are only two possible cases:

$$r = 2 \quad \text{or} \quad r = 3 \tag{4.42}$$

Let us now consider the solutions of the Diophantine equation (4.39) and identify the finite rotation groups and their binary extensions.

Taking into account the conclusion (4.42) we have two cases.

4.2.2 Case $r = 2$: the infinite series of cyclic groups $\mathbb{A}_n$

Choosing $r = 2$, the Diophantine equation (4.38) reduces to:

$$\frac{2}{n} = \frac{1}{k_1} + \frac{1}{k_2} \tag{4.43}$$

Since we have $k_{1,2} \leq n$, the only solution of (4.43) is $k_1 = k_2 = n$, with n arbitrary. Since the order of the cyclic stability subgroup of the two poles coincides with the order of the full group H, it follows that H itself is a cyclic subgroup of SU(2) of order n. We name it $\Gamma_b[n,n,1]$. The two orbits are given by the two eigenvectors of the unique cyclic group generator:

$$\mathscr{A} \in \mathrm{SU}(2): \quad \mathscr{Z} \equiv \mathscr{A}^n \tag{4.44}$$

The finite subgroup of SU(2), isomorphic to the abstract group $\mathbb{Z}_{2n}$, is composed by the following $2n$ elements:

$$\mathbb{Z}_{2n} \sim \Gamma_b[n,n,1] = \{\mathbf{1}, \mathscr{A}, \mathscr{A}^2, \dots, \mathscr{A}^{n-1}, \mathscr{Z}, \mathscr{Z}\mathscr{A}, \mathscr{Z}\mathscr{A}^2, \dots, \mathscr{Z}\mathscr{A}^{n-1}\} \tag{4.45}$$

Under the homomorphism ω, the SU(2)-element $\mathscr{Z}$ maps into the identity and both $\mathscr{A}$ and $\mathscr{Z}\mathscr{A}$ map into the same 3×3 orthogonal matrix $\mathrm{A} \in \mathrm{SO}(3)$ with the property $\mathrm{A}^n = \mathbf{1}$. Hence we have:

$$\omega[\Gamma_b[n,n,1]] = \Gamma[n,n,1] \sim \mathbb{Z}_n \tag{4.46}$$

In conclusion we can define the cyclic subgroups of SO(3) and their binary extensions in SU(2) by means of the following presentation in terms of generators and relations:

$$\mathbb{A}_n \quad \Leftrightarrow \quad \begin{cases} \Gamma_b[n,n,1] = (\mathscr{A}, \mathscr{Z} \mid \mathscr{A}^n = \mathscr{Z};\ \mathscr{Z}^2 = \mathbf{1}) \\ \Gamma[n,n,1] = (\mathrm{A} \mid \mathrm{A}^n = \mathbf{1}) \end{cases} \tag{4.47}$$

where $\mathscr{Z}$, being by definition a central extension, commutes with the other generators and henceforth with all the group elements. The nomenclature $\mathbb{A}_n$ introduced in the above equation is just for future comparison. As we will see, in the ADE-classification of simply laced Lie algebras the case of cyclic groups corresponds to that of $\mathbb{A}_n$ algebras.

4.2.3 Case $r = 3$ and its solutions

In the $r = 3$ case the Diophantine equation becomes:

$$\frac{1}{k_1} + \frac{1}{k_2} + \frac{1}{k_3} = 1 + \frac{2}{n} \tag{4.48}$$

In order to analyze its solutions in a unified way and inspired by the above case it is convenient to introduce the following notation:

$$\mathscr{R} = 1 + \sum_{\alpha}^{r} k_\alpha \tag{4.49}$$

and consider the abstract groups that turn out to be of finite order, associated with each triple of integers $\{k_1, k_2, k_3\}$ satisfying (4.48) and defined by the following presentation:

$$\begin{aligned} \Gamma_b[k_1,k_2,k_3] &= (\mathscr{A},\mathscr{B},\mathscr{Z} \mid (\mathscr{A}\mathscr{B})^{k_1} = \mathscr{A}^{k_2} = \mathscr{B}^{k_3} = \mathscr{Z};\mathscr{Z}^2 = \mathbf{1}) \\ \Gamma[k_1,k_2,k_3] &= (\mathrm{A},\mathrm{B} \mid (\mathrm{AB})^{k_1} = \mathrm{A}^{k_2} = \mathrm{B}^{k_3} = \mathbf{1}) \end{aligned} \tag{4.50}$$

We will see that the finite subgroups of SU(2) are indeed isomorphic to the above defined abstract groups $\Gamma_b[k_1, k_2, k_3]$ and that their image under the homomorphism ω is isomorphic to $\Gamma[k_1, k_2, k_3]$.

4.2.3.1 The solution $(k, 2, 2)$ and the dihedral groups Dih_k

One infinite class of solutions of the Diophantine equation (4.48) is given by

$$\{k_1, k_2, k_3\} = \{k, 2, 2\}; \quad 2 < k \in \mathbb{Z} \tag{4.51}$$

The corresponding subgroups of SU(2) and SO(3) are:

$$\mathrm{Dih}_k \quad \Leftrightarrow \quad \begin{cases} \Gamma_b[k,2,2] = (\mathscr{A},\mathscr{B},\mathscr{Z} \mid (\mathscr{A}\mathscr{B})^k = \mathscr{A}^2 = \mathscr{B}^2 = \mathscr{Z};\ \mathscr{Z}^2 = \mathbf{1}) \\ \Gamma[k,2,2] = (\mathrm{A},\mathrm{B} \mid (\mathrm{AB})^k = \mathrm{A}^2 = \mathrm{B}^2 = \mathbf{1}) \end{cases} \tag{4.52}$$

whose structure we illustrate next.

$\Gamma_b[k, 2, 2] \simeq \mathrm{Dih}_k^b$ is the binary dihedral subgroup. Its order is

$$|\mathrm{Dih}_k^b| = 4k \tag{4.53}$$

and it contains a cyclic subgroup of order k that we name K. Its index in Dih_k^b is two. The elements of Dih_k^b that are not in K are of period equal to two since $k_2 = k_3 = 2$. Altogether the elements of the dihedral group are the matrices given below:

$$\begin{aligned} F_l &= \begin{pmatrix} e^{il\pi/k} & 0 \\ 0 & e^{-il\pi/k} \end{pmatrix}; \quad (l = 0, 1, 2, \ldots, 2k-1) \\ G_l &= \begin{pmatrix} 0 & ie^{-il\pi/k} \\ ie^{il\pi/k} & 0 \end{pmatrix}; \quad (l = 0, 1, 2, \ldots, 2k-1) \end{aligned}$$

In terms of them the generators are identified as follows:

$$F_0 = 1; \quad F_1 G_0 = \mathscr{A}; \quad F_k = \mathscr{Z}; \quad G_0 = \mathscr{B} \tag{4.54}$$

There are exactly $\mathscr{R} = k + 3$ conjugacy classes:

1. K_e contains only the identity F_0
2. K_Z contains the central extension $\mathscr{Z}$
3. $K_{G\,\text{even}}$ contains the elements $G_{2\nu}$ $(\nu = 1, \ldots, k-1)$
4. $K_{G\,\text{odd}}$ contains the elements $G_{2\nu+1}$ $(\nu = 1, \ldots, k-1)$
5. the $k - 1$ classes K_{F_μ}: each of these classes contains the pair of elements F_μ and $F_{2k-\mu}$ for $(\mu = 1, \ldots, k-1)$.

Correspondingly the group Dih_k^b admits $k+3$ irreducible representations, four of which are 1-dimensional while $k-1$ are 2-dimensional. We name them as follows:

$$\begin{cases} D_e;\ D_Z\ ;\ D_{G\,\mathrm{even}};\ D_{G\,\mathrm{odd}}; & 1-\text{dimensional} \\ D_{F_1};\ldots;D_{F_{k-1}}; & 2-\text{dimensional} \end{cases} \tag{4.55}$$

The combinations of the $\mathbf{C}^2$ vector components (z_1,z_2) that transform in the four 1-dimensional representations are easily listed:

$$\begin{aligned} D_e &\longrightarrow |z_1|^2+|z_2|^2 \\ D_Z &\longrightarrow z_1z_2 \\ D_{G\,\mathrm{even}} &\longrightarrow z_1^k+z_2^k \\ D_{G\,\mathrm{odd}} &\longrightarrow z_1^k-z_2^k \end{aligned} \tag{4.56}$$

The matrices of the $k-1$ two-dimensional representations are obtained in the following way. In the DF_s representation, $s=1,\ldots,k-1$, the generator $\mathscr{A}$, namely the group element F_1, is represented by the matrix F_s. The generator $\mathscr{B}$ is instead represented by $(i)^{s-1}G_0$ and the generator $\mathscr{Z}$ is given by F_{sk}, so that:

$$\begin{aligned} DF_s(F_j) &= F_{sj} \\ DF_s(G_j) &= (i)^{s-1}G_{sj} \end{aligned} \tag{4.57}$$

The character table is immediately obtained and it is displayed in Table 4.1. This concludes the discussion of the binary dihedral groups.

Table 4.1: Character table of the group Dih_k^b.

.	KE	KZ	KG_e	KG_o	KF_1	…	KF_{k-1}
DE	1	1	1	1	1	…	1
DZ	1	1	−1	−1	1	…	1
DG_e	1	$(-1)^k$	i^k	$-i^k$	$(-1)^1$	…	$(-1)^{k-1}$
DG_o	1	$(-1)^k$	$-i^k$	i^k	$(-1)^1$	…	$(-1)^{k-1}$
DF_1	2	$(-2)^1$	0	0	$2\cos\frac{\pi}{k}$	…	$2\cos\frac{(k-1)\pi}{k}$
⋮	⋮	⋮	⋮	⋮	⋮	⋱	⋮
DF_{k-1}	2	$(-2)^{k-1}$	0	0	$2\cos\frac{(k-1)\pi}{k}$	…	$2\cos\frac{(k-1)^2\pi}{k}$

4.2.3.2 The three isolated solutions corresponding to the tetrahedral, octahedral and icosahedral groups

There remain three isolated solutions of the Diophantine equation (4.48), namely:

$$\{k_1,k_2,k_3\} = \{3,3,2\} \tag{4.58}$$

$$\{k_1,k_2,k_3\} = \{4,3,2\} \tag{4.59}$$

$$\{k_1,k_2,k_3\} = \{5,3,2\} \tag{4.60}$$

They respectively correspond to the tetrahedral T_{12}, octahedral O_{24}, and icosahedral I_{60} groups and to their binary extensions, namely:

$$\Gamma[3,3,2] \simeq T_{12} \tag{4.61}$$

$$\Gamma[4,3,2] \simeq O_{24} \tag{4.62}$$

$$\Gamma[5,3,2] \simeq I_{60} \tag{4.63}$$

As their name reveals these three groups have 12, 24, and 60 elements, respectively. The corresponding binary extensions have 24, 48, and 120 elements, respectively. With a procedure completely analogous to the one utilized in the case of the dihedral groups we might reconstruct all these elements and organize them into conjugacy classes. We do not do this explicitly; in the next section, while discussing crystallographic groups, we will rather study in full detail the example of the octahedral group O_{24} and we will do that starting from the three-dimensional realization in SO(3).

4.2.4 Summary of the ADE classification of finite rotation groups

Here we prepare the stage for the illustration of the deep and surprising relation, several times already anticipated, between the platonic classification of finite rotation groups and that of semisimple Lie algebras. To this effect let us consider Figure 4.2 and diagrams of the sort displayed there. Such diagrams are called Dynkin diagrams and will obtain a well-defined interpretation while studying root spaces and the classification of simple Lie algebras (see Chapter 7). For the time being let us note that Dynkin diagrams such as that in Figure 4.2 are characterized by three integer numbers $\{k_1, k_2, k_3\}$, denoting the lengths of three chains of dots, linked one to the other and departing from a central node which belongs to each of the three chains. In the case one of the numbers k_α is equal to one (say k_3), the corresponding chain disappears and we are left with a simple chain of length $k_1 + k_2 - 1$. In Chapter 7 we will see that the admissible Dynkin diagrams with one node are those and only those where the numbers $\{k_1, k_2, k_3\}$ satisfy the Diophantine equation (4.48). Hence each solution of that equation has a double interpretation: it singles out a finite rotation group and labels a simple Lie algebra. The anticipated correspondence is the following one:

$$\Gamma[\ell,\ell,1] \simeq \mathbb{Z}_\ell \quad \Leftrightarrow \quad \mathfrak{a}_\ell \tag{4.64}$$

$$\Gamma[\ell,2,2] \simeq \mathrm{Dih}_\ell \quad \Leftrightarrow \quad \mathfrak{d}_\ell \tag{4.65}$$

$$\Gamma[3,3,2] \simeq T_{12} \quad \Leftrightarrow \quad \mathfrak{e}_6 \tag{4.66}$$

$$\Gamma[4,3,2] \simeq O_{24} \quad \Leftrightarrow \quad \mathfrak{e}_7 \tag{4.67}$$

$$\Gamma[5,3,2] \simeq I_{60} \quad \Leftrightarrow \quad \mathfrak{e}_8 \tag{4.68}$$

where $\mathfrak{a}_\ell$ is the Lie algebra associated with the Lie group $\mathrm{SL}(\ell+1,\mathbb{C})$, $\mathbb{D}_\ell$ is the Lie algebra associated with the Lie group $\mathrm{SO}(2\ell,\mathbb{C})$, and $\mathbb{E}_{6,7,8}$ are the Lie algebras of three

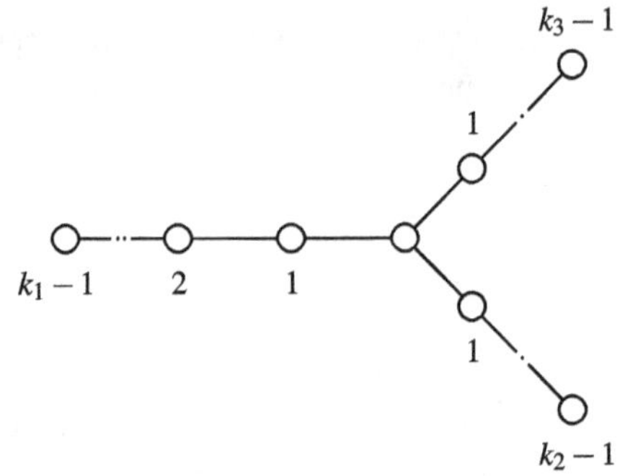

	Simple Lie Algebras	Finite subgroups of $\Gamma_b \subset \mathrm{SU}(2)$
r	number of simple chains in the Dynkin diagram	# of different group–element orders in $\Gamma \equiv \omega[\Gamma_b]$
k_α	$k_\alpha - 1$ = lengths of the simple chains in the Dynkin diagram	group-element orders in $\Gamma \equiv \left(\mathrm{A},\mathrm{B} \mid (\mathrm{AB})^{k_1} = \mathrm{A}^{k_2} = \mathrm{B}^{k_3} = \mathbf{1}\right)$
$\mathscr{R} - 1 \equiv \Sigma_{\alpha=1}^{r}(k_\alpha - 1)$	$\mathscr{R}$ = rank of Lie algebra	$\mathscr{R} + 1$ = # of conj. classes in Γ_b

Figure 4.2: Interpretation of the solutions of the same Diophantine equation in the case of finite subgroups of $\Gamma_b \subset \mathrm{SU}(2)$ and of simply laced Lie algebras.

exceptional Lie groups of dimensions 78, 133 and 248, respectively. As we will see in Chapter 6, a very important concept in Lie algebra theory is that of *rank* that is the maximal number of mutually commuting and diagonalizable elements of the algebra. As we see from Figure 4.2, the rank has a counterpart in the binary extension of the corresponding finite rotation group: it is the number of non-trivial conjugacy classes of the group, except the class of the identity element. The property of Lie algebras that in Dynkin diagrams there are no nodes with more than three converging lines corresponds on the finite rotation group side to the property that in such a group there are at most three different types of group-element orders.

A further challenging reinterpretation of the ADE-classification regards the construction of the so-called ALE-manifolds, that are 4-dimensional spaces with a self-dual curvature and asymptotic flatness.[2]

2 For this topic we refer the reader to [54].

An equivalent derivation of the three exceptional finite rotation subgroups T_{12}, O_{24}, I_{60} is presented in the solution to Exercise 3.31 which starts from consideration of the five regular platonic polyhedra and looks for the rotational symmetries. It is suggested that the reader should compare the ADE classification in eqs (4.64)–(4.68) with Table A.1.

4.3 Lattices and crystallographic groups

In this section we consider the finite rotation groups from the point of view of crystallography, namely as groups of automorphisms of certain lattices. To this effect we need first to introduce the very notion of lattice and then introduce the notion of crystallographic group.

4.3.1 Lattices

We begin by fixing our notations for space and momentum lattices that define an n-torus T^n endowed with a flat metric structure, namely, with a symmetric positive definite inner product (see definition (1.19)).

Let us consider the standard $\mathbb{R}^n$ manifold[3] and introduce a basis of n linearly independent n-vectors that are not necessarily orthogonal to each other and of equal length:

$$\mathbf{w}_\mu \in \mathbb{R}^n \quad \mu = 1, \dots, n \tag{4.69}$$

Any vector in $\mathbb{R}$ can be decomposed along such a basis and we have:

$$\mathbf{r} = r^\mu \mathbf{w}_\mu \tag{4.70}$$

The flat (constant) metric on $\mathbb{R}^n$ is defined by:

$$g_{\mu\nu} = \langle \mathbf{w}_\mu, \mathbf{w}_\nu \rangle \tag{4.71}$$

where $\langle\, ,\, \rangle$ denotes the standard Euclidean scalar product. The space lattice Λ consistent with the metric (4.71) is the free abelian group (with respect to sum) generated by the n basis vectors (4.69), namely:

$$\mathbf{q} \in \Lambda \subset \mathbb{R}^n \quad \Leftrightarrow \quad \mathbf{q} = q^\mu \mathbf{w}_\mu \quad \text{where } q^\mu \in \mathbb{Z} \tag{4.72}$$

The dual lattice $\Lambda^\star$ is defined by the property:

$$\mathbf{p} \in \Lambda^\star \subset \mathbb{R}^n \quad \Leftrightarrow \quad \langle \mathbf{p}, \mathbf{q} \rangle \in \mathbb{Z} \quad \forall \mathbf{q} \in \Lambda \tag{4.73}$$

3 For the mathematically precise notion of manifold we refer the reader to Chapter 5.

A basis for the dual lattice is provided by a set of n *dual vectors* $\mathbf{e}^\mu$ defined by the relations:[4]

$$\langle \mathbf{w}_\mu, \mathbf{e}^\nu \rangle = \delta_\mu^\nu \tag{4.74}$$

so that

$$\forall \mathbf{p} \in \Lambda^\star \quad \mathbf{p} = p_\mu \mathbf{e}^\mu \quad \text{where } p_\mu \in \mathbb{Z} \tag{4.75}$$

4.3.2 The *n*-torus T^n

The n-torus is topologically defined as the product of n-circles, namely:

$$\mathrm{T}^n \equiv \underbrace{\mathbb{S}^1 \times \cdots \times \mathbb{S}^1}_{n\text{-times}} \equiv \underbrace{\frac{\mathbb{R}}{\mathbb{Z}} \times \cdots \times \frac{\mathbb{R}}{\mathbb{Z}}}_{n\text{-times}} \tag{4.76}$$

Alternatively we can define the n-torus by modding $\mathbb{R}^n$ with respect to an n-dimensional lattice. In this case the n-torus comes out automatically equipped with a flat constant metric:

$$\mathrm{T}^n_g = \frac{\mathbb{R}^n}{\Lambda} \tag{4.77}$$

According to (4.77) the flat Riemannian space[5] T^n_g is defined as the set of equivalence classes with respect to the following equivalence relation:

$$\mathbf{r}' \sim \mathbf{r} \quad \text{iff} \quad \mathbf{r}' - \mathbf{r} \in \Lambda \tag{4.78}$$

The metric (4.71) defined on $\mathbb{R}^n$ is inherited by the quotient space and therefore it endows the topological torus (4.76) with a flat Riemannian structure. Seen from another point of view the space of flat metrics on T^3 is just the coset manifold $\mathrm{SL}(3,\mathbb{R})/\mathrm{O}(3)$ encoding all possible symmetric matrices, alternatively all possible space lattices, each lattice being spanned by an arbitrary triplet of basis vectors (4.69).

4.3.3 Crystallographic groups and the Bravais lattices for $n = 3$

Every lattice Λ yields a metric g and every metric g singles out an isomorphic copy $\mathrm{SO}_g(3)$ of the continuous rotation group $\mathrm{SO}(n)$, which leaves it invariant:

$$M \in \mathrm{SO}_g(n) \quad \Leftrightarrow \quad M^T g M = g \tag{4.79}$$

---

4 In the sequel for the scalar product of two vectors we utilize also the equivalent shorter notation $\mathbf{a} \cdot \mathbf{b} = \langle \mathbf{a} \cdot \mathbf{b} \rangle$.

5 For the precise mathematical definition of Riemannian space we refer the reader to Chapter 10.

By definition $\mathrm{SO}_g(n)$ is the conjugate of the standard $\mathrm{SO}(n)$ in $\mathrm{GL}(n,\mathbb{R})$:

$$\mathrm{SO}_g(n) = \mathcal{S}\,\mathrm{SO}(n)\mathcal{S}^{-1} \tag{4.80}$$

with respect to the matrix $\mathcal{S} \in \mathrm{GL}(n,\mathbb{R})$ which reduces the metric g to the Kronecker delta:

$$\mathcal{S}^T g \mathcal{S} = \mathbf{1} \tag{4.81}$$

Notwithstanding this, a generic lattice Λ is not invariant with respect to any proper subgroup of the rotation group $\mathrm{G} \subset \mathrm{SO}_g(n) \equiv \mathrm{SO}(n)$. Indeed, by invariance of the lattice one understands the following condition:

$$\forall \gamma \in \mathrm{G} \text{ and } \forall \mathbf{q} \in \Lambda : \quad \gamma \cdot \mathbf{q} \in \Lambda \tag{4.82}$$

For $n = 3$ lattices that have a non-trivial symmetry group $\mathrm{G} \subset \mathrm{SO}(3)$ are those relevant to Solid State Physics and Crystallography. There are 14 of them grouped in 7 classes that were already classified in the XIX century by Bravais. The symmetry group G of each of these Bravais lattices Λ_B is necessarily one of the well-known finite subgroups of the three-dimensional rotation group O(3). In the language universally adopted by Chemistry and Crystallography, for each Bravais lattice Λ_B the corresponding invariance group $\mathrm{G_B}$ is named the *Point Group*.

According to a standard nomenclature the seven classes of Bravais lattices are respectively named *Triclinic, Monoclinic, Orthorombic, Tetragonal, Rhombohedral, Hexagonal, and Cubic*. Such classes are specified by giving the lengths of the basis vectors $\mathbf{w}_\mu$ and the three angles between them, in other words, by specifying the six components of the metric (4.71).

In general we have the following

Definition 4.3.1. An abstract group Γ is named crystallographic in n dimensions if there exists an n-dimensional lattice Λ_n with basis vectors $\mathbf{w}_\mu$ such that:

1. there is a isomorphism:

$$\omega : \Gamma \to H \subset \mathrm{SO}_g(n) \tag{4.83}$$

   where $\mathrm{SO}_g(n)$ is the conjugate of the n-dimensional group rotation group respecting a metric g (see eq. (4.81));
2. the metric g is that defined by the basis vectors of the lattice Λ_n (see eq. (4.71));
3. all elements of H are $n \times n$ matrices with integer valued entries.

This is equivalent to the statement that Γ has an orthogonal action in $\mathbb{R}^n$ and preserves the lattice Λ_n.

When a group Γ is crystallographic with respect to a given n-dimensional lattice Λ_n we say that it is the *Point Group* of Λ_n.

4.3.4 The proper Point Groups

Restricting one's attention to $n = 3$, it was shown in the classical crystallographic literature that the proper Point Groups that appear in the seven lattice classes are either the cyclic groups $\mathbb{Z}_h$ with $h = 2, 3, 4$ or the dihedral groups Dih_k with $k = 3, 4, 6$ or the tetrahedral group T_{12} or the octahedral group O_{24}. Indeed, the $n = 3$ crystallographic Point Groups are, by definition, finite subgroups of the rotation group, hence they must fall in the ADE-classification. Yet not every finite rotation group is crystallographic. For instance there is no lattice that is invariant under the icosahedral group and in general in a $n = 3$ Point Group there are no elements with orders different from 2, 3, 4, 6.

In this section, for the sake of illustration by means of a well structured example, we restrict our attention to the largest possible Point Group, namely that of the cubic lattice which has O_{24} symmetry.

4.3.5 The cubic lattice and the octahedral Point Group

Let us now consider, within the general frame presented above the cubic lattice.

The cubic lattice is displayed in Figure 4.3.

The basis vectors of the cubic lattice Λ_{cubic} are:

$$\mathbf{w}_1 = \{1, 0, 0\}; \quad \mathbf{w}_2 = \{0, 1, 0\}; \quad \mathbf{w}_3 = \{0, 0, 1\} \tag{4.84}$$

which implies that the metric is just the Kronecker delta:

$$g_{\mu\nu} = \delta_{\mu\nu} \tag{4.85}$$

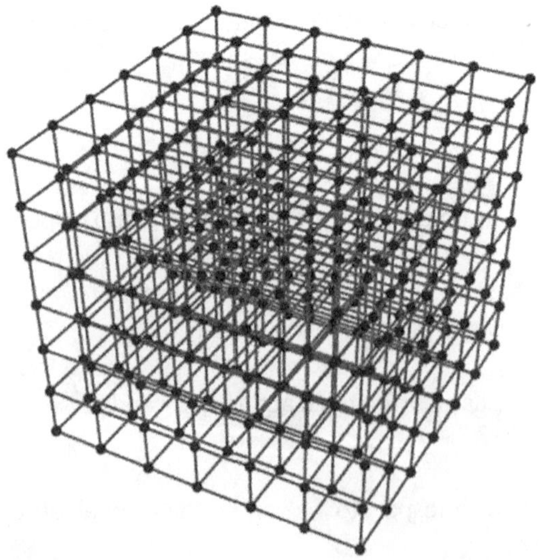

Figure 4.3: A view of the self-dual cubic lattice.

and the basis vectors $\mathbf{e}^\mu$ of the dual lattice $\Lambda^\star_{\text{cubic}}$ coincide with those of the lattice Λ. Hence the cubic lattice is self-dual:

$$\mathbf{w}_\mu = \mathbf{e}^\mu \quad \Rightarrow \quad \Lambda_{\text{cubic}} = \Lambda^\star_{\text{cubic}} \tag{4.86}$$

The subgroup of the proper rotation group which maps the cubic lattice onto itself is the octahedral group O whose order is 24. The structure of O_{24} was described in Section 3.4.3. There, showing that it is a solvable group, we used the induction algorithm to deduce the main properties of its irreducible representation. In the next subsection we construct them directly.

4.3.6 Irreducible representations of the octahedral group

There are five conjugacy classes in O_{24} and therefore according to theory there are five irreducible representations of the same group, that we name D_i, $i = 1, \dots, 5$. We know from Section 3.4.3.8 that they have dimensions:

$$\dim D_1 = 1; \quad \dim D_2 = 1; \quad \dim D_3 = 2; \quad \dim D_4 = 3; \quad \dim D_5 = 4 \tag{4.87}$$

Let us briefly describe them.

4.3.6.1 D_1: the identity representation

The identity representation which exists for all groups is that one where to each element of O we associate the number 1:

$$\forall \gamma \in O_{24} : \quad D_1(\gamma) = 1 \tag{4.88}$$

Obviously the character of such a representation is:

$$\chi_1 = \{1, 1, 1, 1, 1\} \tag{4.89}$$

4.3.6.2 D_2: the quadratic Vandermonde representation

The representation D_2 is also one-dimensional. It is constructed as follows. Consider the following polynomial of order 6 in the coordinates of a point in $\mathbb{R}^3$ or T^3:

$$\mathfrak{B}(x, y, z) = (x^2 - y^2)(x^2 - z^2)(y^2 - z^2) \tag{4.90}$$

As one can explicitly check under the transformations of the octahedral group listed in eq. (3.108), the polynomial $\mathfrak{B}(x, y, z)$ is always mapped onto itself modulo an overall sign. Keeping track of such a sign provides the form of the second one-dimensional representation whose character is explicitly calculated to be the following one:

$$\chi_1 = \{1, 1, 1, -1, -1\} \tag{4.91}$$

4.3.6.3 D_3: the two-dimensional representation

The representation D_3 is two-dimensional and it corresponds to a homomorphism:

$$D_3: \quad O_{24} \rightarrow GL(2, \mathbb{Z}) \tag{4.92}$$

which associates to each element of the octahedral group a 2×2 integer-valued matrix. The homomorphism is completely specified by giving the two matrices representing the two generators:

$$D_3(A) = \begin{pmatrix} 0 & 1 \\ -1 & -1 \end{pmatrix}; \quad D_3(B) = \begin{pmatrix} 0 & 1 \\ 1 & 0 \end{pmatrix} \tag{4.93}$$

The character vector of D_3 is easily calculated from the above information and we have:

$$\chi_3 = \{2, -1, 2, 0, 0\} \tag{4.94}$$

4.3.6.4 D_4: the three-dimensional defining representation

The three dimensional representation D_4 is simply the defining representation, where the generators A and B are given by the matrices in eq. (3.109).

$$D_4(A) = A; \quad D_4(B) = B \tag{4.95}$$

From this information the characters are immediately calculated and we get:

$$\chi_3 = \{3, 0, -1, -1, 1\} \tag{4.96}$$

4.3.6.5 D_5: the three-dimensional unoriented representation

The three-dimensional representation D_5 is simply that where the generators A and B are given by the following matrices:

$$D_5(A) = \begin{pmatrix} 0 & 1 & 0 \\ 0 & 0 & 1 \\ 1 & 0 & 0 \end{pmatrix}; \quad D_5(B) = \begin{pmatrix} 0 & 1 & 0 \\ 1 & 0 & 0 \\ 0 & 0 & 1 \end{pmatrix} \tag{4.97}$$

From this information the characters are immediately calculated and we get:

$$\chi_5 = \{3, 0, -1, 1, -1\} \tag{4.98}$$

The table of characters is summarized in Table 4.2.

4.4 A simple crystallographic point-group in seven dimensions

In the previous section we analyzed the possible crystallographic Point Groups in our familiar three-dimensional Euclidean space.

Table 4.2: Character table of the proper octahedral group.

Irrep		Class				
		$\{e, 1\}$	$\{C_3, 8\}$	$\{C_4^2, 3\}$	$\{C_2, 6\}$	$\{C_4, 6\}$
D_1, χ_1	=	1	1	1	1	1
D_2, χ_2	=	1	1	1	−1	−1
D_3, χ_3	=	2	−1	2	0	0
D_4, χ_4	=	3	0	−1	−1	1
D_5, χ_5	=	3	0	−1	1	−1

Summarizing our discussion we point out some group-theoretical features that follow from the ADE classification, combined with the further compatibility constraints which emerge when you impose the crystallographic condition that a lattice should be left invariant by the action of the Point Group:

a) The Point Group $\mathfrak{P}$ must be a finite rotation group in $d = 3$ hence it must belong to the list:

$$\mathfrak{P} \in \{\mathbb{Z}_k, \mathrm{Dih}_k, \mathrm{T}_{12}, \mathrm{O}_{24}, \mathrm{I}_{60}\} \tag{4.99}$$

b) The order of any element $\gamma \in \mathfrak{P}$ belonging to the Point Group must be in the range $2, 3, 4, 6$.

The intersection of these two conditions leads to the conclusion that:

$$\mathfrak{P} \in \{\mathbb{Z}_{2,3,4,6}, \mathrm{Dih}_{3,4,6}, \mathrm{T}_{12}, \mathrm{O}_{24}\} \tag{4.100}$$

A detailed discussion of the Point Groups listed above in eq. (4.100) is provided in the solution of Exercise 3.41 (see eq. (A.2) and the following lines).

The classification of Bravais lattices, which is responsible for so many chemical-physical properties of matter, is essentially encoded in eq. (4.100). In this list of candidate Point Groups there is no simple one which is non-abelian. They are all either solvable or abelian and this implies that their irreducible representations can be constructed by means of the induction strategy explained in Section 3.4. A simple group which occurs in the ADE classification is the icosahedral group I_{60} which is isomorphic to the simple alternating group A_5 (the even permutations of five objects). It is barred out by the crystallographic condition because it contains elements of order 5.

In many respects this is the analog of what happens with algebraic equations. The algebraic equations of order $2, 3, 4$ are always solvable by radicals since their Galois group is solvable. In degree $d \geq 5$ the generic equation is not solvable because the Galois group is generically not solvable.

A natural question arises at this point. Is the condition b) on the possible orders of the Point Group elements intrinsic to the crystallographic constraint in any dimension or is it a specific feature of $d = 3$?

The correct answer to the above question is the second option and in this section we show a counterexample of a crystallographic group in seven dimensions that has group elements of order 7. Not only that. Ours is an example of a simple non-abelian crystallographic Point Group!

It is quite remarkable that, up to the knowledge of these authors, the analog of the ADE classification of finite rotation groups in $d > 5$ is so far non-existing. Even less is known about higher dimensional crystallographic groups.

It is philosophically quite challenging to imagine what Chemistry, Geology and even Molecular Biology and Genetics might be in a world where the Point Group is a simple non-abelian group!

4.4.1 The simple group L_{168}

The finite group:

$$L_{168} \equiv PSL(2, \mathbb{Z}_7) \tag{4.101}$$

is the second smallest simple group after the alternating group A_5 which has 60 elements and coincides with the symmetry group of the regular icosahedron or dodecahedron. As anticipated by its given name, L_{168} has 168 elements: they can be identified with all the possible 2×2 matrices with determinant one whose entries belong to the finite field $\mathbb{Z}_7$, counting them up to an overall sign. In projective geometry, L_{168} is classified as a *Hurwitz group* since it is the automorphism group of a Hurwitz Riemann surface, namely, a surface of genus g with the maximal number $84(g-1)$ of conformal automorphisms.[6] The Hurwitz surface pertaining to the Hurwitz group L_{168} is the Klein[7] quartic [99], namely, the locus $\mathcal{K}_4$ in $\mathbb{P}_2(\mathbb{C})$ cut out by the following quartic polynomial constraint on the homogeneous coordinates $\{x, y, z\}$:

$$x^3 y + y^3 z + z^3 x = 0 \tag{4.102}$$

Indeed, $\mathcal{K}_4$ is a genus $g = 3$ compact Riemann surface and it can be realized as the quotient of the hyperbolic Poincaré plane $\mathbb{H}_2$ by a certain group Γ that acts freely on $\mathbb{H}_2$ by isometries.

The L_{168} group, which is also isomorphic to $GL(3, \mathbb{Z}_2)$ [14], has received a lot of attention in Mathematics and it has important applications in algebra, geometry, and

6 Hurwitz's automorphisms theorem proved in 1893 [90] states that the order $|\mathcal{G}|$ of the group $\mathcal{G}$ of orientation-preserving conformal automorphisms, of a compact Riemann surface of genus $g > 1$ admits the following upper bound: $|\mathcal{G}| \leq 84(g-1)$.

7 We already mentioned Felix Klein in Chapter 2. Much more about him and about his role in the development of Geometry and Group Theory in the XIX century is presented in [53].

number theory: for instance, besides being associated with the Klein quartic, L_{168} is the automorphism group of the Fano plane [44] (see Exercise 3.37 and its solution).

The reason why we consider L_{168} in this section is associated with another property of this finite simple group which was proved fifteen years ago in [98], namely:

$$L_{168} \subset G_{2(-14)} \tag{4.103}$$

This means that L_{168} is a finite subgroup of the compact form of the exceptional Lie group G_2 and the 7-dimensional fundamental representation of the latter is irreducible upon restriction to L_{168}.

The reader of this chapter has not yet learned what exceptional Lie groups are, yet he will, when he comes to Chapters 7 and 8, so we ask him to be patient. When he will reach that level of knowledge he will also be aware of the fact that associated with every simple Lie group there is, by means of its Lie algebra, a lattice in dimensions equal to its rank, **the root lattice.** The key reason to consider L_{168} in this section is that it happens to be crystallographic in $d = 7$, the preserved lattice being the root lattice of either the simple Lie algebra $\mathfrak{a}_7$ or, even more inspiringly, of the exceptional Lie algebra $\mathfrak{e}_7$. Actually when the reader will have digested the notion he will also appreciate that L_{168} is a subgroup of the $\mathfrak{e}_7$ Weyl group. Because of the role of $\mathfrak{e}_7$ in supergravity related special geometries, it emerges to prominence in contemporary literature on the super World. Here we are interested in its properties in order to illustrate the case of **simple crystallographic non-abelian group.**

4.4.2 Structure of the simple group $L_{168} = PSL(2, \mathbb{Z}_7)$

For the reasons outlined above we consider the simple group (4.101) and its crystallographic action in $d = 7$. The Hurwitz simple group L_{168} is abstractly presented as follows:[8]

$$L_{168} = (R, S, T \parallel R^2 = S^3 = T^7 = RST = (TSR)^4 = \mathbf{e}) \tag{4.104}$$

and, as its name implicitly advocates, it has order 168:

$$|L_{168}| = 168 \tag{4.105}$$

The elements of this simple group are organized in six conjugacy classes according to the scheme displayed below:

Conjugacy class	$\mathscr{C}_1$	$\mathscr{C}_2$	$\mathscr{C}_3$	$\mathscr{C}_4$	$\mathscr{C}_5$	$\mathscr{C}_6$
representative of the class	$\mathbf{e}$	R	S	TSR	T	SR
order of the elements in the class	1	2	3	4	7	7
number of elements in the class	1	21	56	42	24	24

(4.106)

8 In the rest of this section we follow closely the results obtained by one of us in a recent paper [52].

As one sees from the above table (4.106), the group contains elements of order 2, 3, 4 and 7 and there are two inequivalent conjugacy classes of elements of the highest order. According to the general theory of finite groups, there are six different irreducible representations of dimensions 1, 6, 7, 8, 3, 3, respectively. The character table of the group L_{168} can be found in the mathematical literature, for instance in the book [93]. It reads as follows:

Representation	$\mathscr{C}_1$	$\mathscr{C}_2$	$\mathscr{C}_3$	$\mathscr{C}_4$	$\mathscr{C}_5$	$\mathscr{C}_6$
$D_1[L_{168}]$	1	1	1	1	1	1
$D_6[L_{168}]$	6	2	0	0	−1	−1
$D_7[L_{168}]$	7	−1	1	−1	0	0
$D_8[L_{168}]$	8	0	−1	0	1	1
$DA_3[L_{168}]$	3	−1	0	1	$\frac{1}{2}(-1+i\sqrt{7})$	$\frac{1}{2}(-1-i\sqrt{7})$
$DB_3[L_{168}]$	3	−1	0	1	$\frac{1}{2}(-1-i\sqrt{7})$	$\frac{1}{2}(-1+i\sqrt{7})$

(4.107)

Soon we will retrieve it by constructing explicitly all the irreducible representations.

4.4.3 The 7-dimensional irreducible representation

For many purposes the most interesting representation is the 7-dimensional one. Indeed, its properties are the very reason to consider the group L_{168} in the present context. The following three statements are true:

1. The 7-dimensional irreducible representation is crystallographic since all elements $\gamma \in L_{168}$ are represented by integer valued matrices $D_7(\gamma)$ in a basis of vectors that span a lattice, namely, the root lattice Λ_{root} of the $\mathfrak{a}_7$ simple Lie algebra.
2. The 7-dimensional irreducible representation provides an immersion $L_{168} \hookrightarrow SO(7)$ since its elements preserve the symmetric Cartan matrix of $\mathfrak{a}_7$:

$$\begin{aligned} \forall \gamma \in L_{168} : \quad & D_7^T(\gamma)\mathscr{C}D_7(\gamma) = \mathscr{C} \\ & \mathscr{C}_{i,j} = \alpha_i \cdot \alpha_j \quad (i,j = 1,\dots,7) \end{aligned} \tag{4.108}$$

defined in terms of the simple roots α_i whose standard construction in terms of the unit vectors ϵ_i of $\mathbb{R}^8$ is recalled below:[9]

$$\begin{aligned} &\alpha_1 = \epsilon_1 - \epsilon_2; \quad \alpha_2 = \epsilon_2 - \epsilon_3; \quad \alpha_3 = \epsilon_3 - \epsilon_4 \\ &\alpha_4 = \epsilon_4 - \epsilon_5; \quad \alpha_5 = \epsilon_5 - \epsilon_6; \quad \alpha_6 = \epsilon_6 - \epsilon_7 \\ &\alpha_7 = \epsilon_7 - \epsilon_8 \end{aligned} \tag{4.109}$$

9 We refer the reader to Chapter 7 and in particular to Section 7.5.1 for the explicit form of the Cartan matrices associated with $\mathfrak{a}_\ell$ algebras.

3. Actually the 7-dimensional representation defines an embedding $L_{168} \hookrightarrow G_2 \subset SO(7)$ since there exists a three-index antisymmetric tensor ϕ_{ijk} satisfying the relations of octonionic structure constants[10] that is preserved by all the matrices $D_7(\gamma)$:

$$\forall \gamma \in L_{168}: \quad D_7(\gamma)_{ii'} D_7(\gamma)_{jj'} D_7(\gamma)_{kk'} \phi_{i'j'k'} = \phi_{ijk} \tag{4.110}$$

Let us prove the above statements. It suffices to write the explicit form of the generators R, S and T in the crystallographic basis of the considered root lattice:

$$\mathbf{v} \in \Lambda_{\text{root}} \quad \Leftrightarrow \quad \mathbf{v} = n_i \alpha_i \quad n_i \in \mathbb{Z} \tag{4.111}$$

Explicitly if we set:

$$\begin{aligned}
\mathscr{R} &= \begin{pmatrix} 0 & 0 & 0 & 0 & 0 & 0 & -1 \\ 0 & 0 & 0 & 0 & 0 & -1 & 0 \\ 0 & 0 & -1 & 1 & 0 & -1 & 0 \\ 0 & -1 & 0 & 1 & 0 & -1 & 0 \\ 0 & -1 & 0 & 1 & -1 & 0 & 0 \\ 0 & -1 & 0 & 0 & 0 & 0 & 0 \\ -1 & 0 & 0 & 0 & 0 & 0 & 0 \end{pmatrix} \\
\mathscr{S} &= \begin{pmatrix} 0 & 0 & 0 & 0 & 0 & 0 & -1 \\ 1 & 0 & 0 & 0 & 0 & 0 & -1 \\ 1 & 0 & 0 & -1 & 1 & 0 & -1 \\ 1 & 0 & -1 & 0 & 1 & 0 & -1 \\ 1 & 0 & -1 & 0 & 1 & -1 & 0 \\ 1 & 0 & -1 & 0 & 0 & 0 & 0 \\ 1 & -1 & 0 & 0 & 0 & 0 & 0 \end{pmatrix} \\
\mathscr{T} &= \begin{pmatrix} 0 & 0 & 0 & 0 & 0 & -1 & 1 \\ 1 & 0 & 0 & 0 & 0 & -1 & 1 \\ 0 & 1 & 0 & 0 & 0 & -1 & 1 \\ 0 & 0 & 1 & 0 & 0 & -1 & 1 \\ 0 & 0 & 0 & 1 & 0 & -1 & 1 \\ 0 & 0 & 0 & 0 & 1 & -1 & 1 \\ 0 & 0 & 0 & 0 & 0 & 0 & 1 \end{pmatrix}
\end{aligned} \tag{4.112}$$

we find that the defining relations of L_{168} are satisfied:

$$\mathscr{R}^2 = \mathscr{S}^3 = \mathscr{T}^7 = \mathscr{R}\mathscr{S}\mathscr{T} = (\mathscr{T}\mathscr{S}\mathscr{R})^4 = \mathbf{1}_{7\times 7} \tag{4.113}$$

10 For the history of quaternions and octonions we refer the reader to [53].

and furthermore we have:

$$\mathscr{R}^{\mathrm{T}}\mathscr{C}\mathscr{R} = \mathscr{S}^{\mathrm{T}}\mathscr{C}\mathscr{S} = \mathscr{T}^{\mathrm{T}}\mathscr{C}\mathscr{T} = \mathscr{C} \tag{4.114}$$

where the explicit form of the $\mathfrak{a}_7$ Cartan matrix is recalled below:

$$\mathscr{C} = \begin{pmatrix} 2 & -1 & 0 & 0 & 0 & 0 & 0 \\ -1 & 2 & -1 & 0 & 0 & 0 & 0 \\ 0 & -1 & 2 & -1 & 0 & 0 & 0 \\ 0 & 0 & -1 & 2 & -1 & 0 & 0 \\ 0 & 0 & 0 & -1 & 2 & -1 & 0 \\ 0 & 0 & 0 & 0 & -1 & 2 & -1 \\ 0 & 0 & 0 & 0 & 0 & -1 & 2 \end{pmatrix} \tag{4.115}$$

This proves statements 1 and 2.

In order to prove statement 3 we proceed as follows. In $\mathbb{R}^7$ we consider the antisymmetric three-index tensor ϕ_{ABC} that, in the standard orthonormal basis, has the following components:

$$\begin{aligned} \phi_{1,2,6} &= \frac{1}{6} \\ \phi_{1,3,4} &= -\frac{1}{6} \\ \phi_{1,5,7} &= -\frac{1}{6} \\ \phi_{2,3,7} &= \frac{1}{6} \quad ; \quad \text{all other components vanish} \\ \phi_{2,4,5} &= \frac{1}{6} \\ \phi_{3,5,6} &= -\frac{1}{6} \\ \phi_{4,6,7} &= -\frac{1}{6} \end{aligned} \tag{4.116}$$

This tensor satisfies the algebraic relations of octonionic structure constants, namely:[11]

$$\phi_{ABM}\phi_{CDM} = \frac{1}{18}\delta^{AB}_{CD} + \frac{2}{3}\Phi_{ABCD} \tag{4.117}$$

$$\phi_{ABC} = -\frac{1}{6}\epsilon_{ABCPQRS}\Phi_{ABCD} \tag{4.118}$$

and the subgroup of SO(7) which leaves ϕ_{ABC} invariant is, by definition, the compact section $G_{(2,-14)}$ of the complex G_2 Lie group (see for instance [57]). A particular matrix

11 In this equation the indices of the G_2-invariant tensor are denoted with capital letter of the Latin alphabet, as was the case in the quoted literature on weak G_2-structures. In the following we will use lower case Latin letters, the upper Latin letters being reserved for $d = 8$.

that transforms the standard orthonormal basis of $\mathbb{R}^7$ into the basis of simple roots α_i is the following one:

$$\mathfrak{M} = \begin{pmatrix} \sqrt{2} & -\frac{1}{\sqrt{2}} & 0 & 0 & 0 & 0 & 0 \\ 0 & -\frac{1}{\sqrt{2}} & \sqrt{2} & -\frac{1}{\sqrt{2}} & 0 & 0 & 0 \\ 0 & 0 & 0 & -\frac{1}{\sqrt{2}} & \sqrt{2} & -\frac{1}{\sqrt{2}} & 0 \\ 0 & 0 & 0 & 0 & 0 & -\frac{1}{\sqrt{2}} & \sqrt{2} \\ 0 & -\frac{1}{\sqrt{2}} & 0 & \frac{1}{\sqrt{2}} & 0 & -\frac{1}{\sqrt{2}} & 0 \\ 0 & 0 & 0 & -\frac{1}{\sqrt{2}} & 0 & 0 & 0 \\ 0 & \frac{1}{\sqrt{2}} & 0 & 0 & 0 & -\frac{1}{\sqrt{2}} & 0 \end{pmatrix} \tag{4.119}$$

since:

$$\mathfrak{M}^T \mathfrak{M} = \mathscr{C} \tag{4.120}$$

Defining the transformed tensor:

$$\varphi_{ijk} \equiv (\mathfrak{M}^{-1})_i^{\ I} (\mathfrak{M}^{-1})_j^{\ J} (\mathfrak{M}^{-1})_k^{\ K} \phi_{IJK} \tag{4.121}$$

we can explicitly verify that:

$$\begin{aligned} \varphi_{ijk} &= (\mathscr{R})_i^{\ p} (\mathscr{R})_j^{\ q} (\mathscr{R})_k^{\ r} \varphi_{pqr} \\ \varphi_{ijk} &= (\mathscr{S})_i^{\ p} (\mathscr{S})_j^{\ q} (\mathscr{S})_k^{\ r} \varphi_{pqr} \\ \varphi_{ijk} &= (\mathscr{T})_i^{\ p} (\mathscr{T})_j^{\ q} (\mathscr{T})_k^{\ r} \varphi_{pqr} \end{aligned} \tag{4.122}$$

Hence, being preserved by the three generators $\mathscr{R}$, $\mathscr{S}$ and $\mathscr{T}$, the antisymmetric tensor φ_{ijk} is preserved by the entire discrete group L_{168} which, henceforth, is a subgroup of $G_{(2,-14)} \subset SO(7)$, as it was shown by intrinsic group theoretical arguments in [98]. The other representations of the group L_{168} were explicitly constructed about ten years ago by Pierre Ramond and his younger collaborators in [109]. They are completely specified by giving the matrix form of the three generators R, S, T satisfying the defining relations (4.104).

4.4.4 The 3-dimensional complex representations

The two 3-dimensional irreducible representations are complex and they are conjugate to each other. It suffices to give the form of the generators for one of them. The generators of the conjugate representation are the complex conjugates of the same matrices.

Setting:

$$\rho \equiv e^{\frac{2i\pi}{7}} \tag{4.123}$$

we have the following form for the representation **3**:

$$
\begin{aligned}
D[R]_3 &= \begin{pmatrix} \frac{i(\rho^2-\rho^5)}{\sqrt{7}} & \frac{i(\rho-\rho^6)}{\sqrt{7}} & \frac{i(\rho^4-\rho^3)}{\sqrt{7}} \\ \frac{i(\rho-\rho^6)}{\sqrt{7}} & \frac{i(\rho^4-\rho^3)}{\sqrt{7}} & \frac{i(\rho^2-\rho^5)}{\sqrt{7}} \\ \frac{i(\rho^4-\rho^3)}{\sqrt{7}} & \frac{i(\rho^2-\rho^5)}{\sqrt{7}} & \frac{i(\rho-\rho^6)}{\sqrt{7}} \end{pmatrix} \\
D[S]_3 &= \begin{pmatrix} \frac{i(\rho^3-\rho^6)}{\sqrt{7}} & \frac{i(\rho^3-\rho)}{\sqrt{7}} & \frac{i(\rho-1)}{\sqrt{7}} \\ \frac{i(\rho^2-1)}{\sqrt{7}} & \frac{i(\rho^6-\rho^5)}{\sqrt{7}} & \frac{i(\rho^6-\rho^2)}{\sqrt{7}} \\ \frac{i(\rho^5-\rho^4)}{\sqrt{7}} & \frac{i(\rho^4-1)}{\sqrt{7}} & \frac{i(\rho^5-\rho^3)}{\sqrt{7}} \end{pmatrix} \\
D[T]_3 &= \begin{pmatrix} -ie^{\frac{3i\pi}{14}} & 0 & 0 \\ 0 & -ie^{-\frac{i\pi}{14}} & 0 \\ 0 & 0 & -e^{-\frac{i\pi}{7}} \end{pmatrix}
\end{aligned}
\tag{4.124}
$$

4.4.5 The 6-dimensional representation

Introducing the following shorthand notation:

$$
\begin{aligned}
c_n &= \cos\left[\frac{2\pi}{7}n\right] \\
s_n &= \sin\left[\frac{2\pi}{7}n\right]
\end{aligned}
\tag{4.125}
$$

The generators of the group L_{168} in the 6-dimensional irreducible representation can be explicitly written as is displayed below:

$$
\begin{aligned}
D[R]_6 &= \left(\begin{array}{c|c|c|c|c|c}
\frac{c_3-1}{\sqrt{2}} & \frac{c_2-1}{\sqrt{2}} & \frac{c_1-1}{\sqrt{2}} & c_3-c_1 & c_1-c_2 & c_2-c_3 \\ \hline
\frac{c_2-1}{\sqrt{2}} & \frac{c_1-1}{\sqrt{2}} & \frac{c_3-1}{\sqrt{2}} & c_2-c_3 & c_3-c_1 & c_1-c_2 \\ \hline
\frac{c_1-1}{\sqrt{2}} & \frac{c_3-1}{\sqrt{2}} & \frac{c_2-1}{\sqrt{2}} & c_1-c_2 & c_2-c_3 & c_3-c_1 \\ \hline
c_3-c_1 & c_2-c_3 & c_1-c_2 & \frac{c_1-1}{\sqrt{2}} & \frac{c_2-1}{\sqrt{2}} & \frac{c_3-1}{\sqrt{2}} \\ \hline
c_1-c_2 & c_3-c_1 & c_2-c_3 & \frac{c_2-1}{\sqrt{2}} & \frac{c_3-1}{\sqrt{2}} & \frac{c_1-1}{\sqrt{2}} \\ \hline
c_2-c_3 & c_1-c_2 & c_3-c_1 & \frac{c_3-1}{\sqrt{2}} & \frac{c_1-1}{\sqrt{2}} & \frac{c_2-1}{\sqrt{2}}
\end{array}\right) \\
D[S]_6 &= \left(\begin{array}{c|c|c|c|c|c}
\frac{(c_3-1)\rho^2}{\sqrt{2}} & \frac{(c_2-1)\rho^4}{\sqrt{2}} & \frac{(c_1-1)\rho}{\sqrt{2}} & (c_3-c_1)\rho^3 & (c_1-c_2)\rho^5 & (c_2-c_3)\rho^6 \\ \hline
\frac{(c_2-1)\rho^2}{\sqrt{2}} & \frac{(c_1-1)\rho^4}{\sqrt{2}} & \frac{(c_3-1)\rho}{\sqrt{2}} & (c_2-c_3)\rho^3 & (c_3-c_1)\rho^5 & (c_1-c_2)\rho^6 \\ \hline
\frac{(c_1-1)\rho^2}{\sqrt{2}} & \frac{(c_3-1)\rho^4}{\sqrt{2}} & \frac{(c_2-1)\rho}{\sqrt{2}} & (c_1-c_2)\rho^3 & (c_2-c_3)\rho^5 & (c_3-c_1)\rho^6 \\ \hline
(c_3-c_1)\rho^2 & (c_2-c_3)\rho^4 & (c_1-c_2)\rho & \frac{(c_1-1)\rho^3}{\sqrt{2}} & \frac{(c_2-1)\rho^5}{\sqrt{2}} & \frac{(c_3-1)\rho^6}{\sqrt{2}} \\ \hline
(c_1-c_2)\rho^2 & (c_3-c_1)\rho^4 & (c_2-c_3)\rho & \frac{(c_2-1)\rho^3}{\sqrt{2}} & \frac{(c_3-1)\rho^5}{\sqrt{2}} & \frac{(c_1-1)\rho^6}{\sqrt{2}} \\ \hline
(c_2-c_3)\rho^2 & (c_1-c_2)\rho^4 & (c_3-c_1)\rho & \frac{(c_3-1)\rho^3}{\sqrt{2}} & \frac{(c_1-1)\rho^5}{\sqrt{2}} & \frac{(c_2-1)\rho^6}{\sqrt{2}}
\end{array}\right) \\
D[T]_6 &= (D[R]_6 \cdot D[S]_6)^{-1}
\end{aligned}
\tag{4.126}
$$

4.4.6 The 8-dimensional representation

Utilizing the same notation as before we can write the matrix form of the generators also in the irreducible 8-dimensional representation.

$$D[R]_8$$

$$= \begin{pmatrix} 2-2c_1 & 0 & 2c_1+2c_2-4c_3 & 2-2c_2 & 0 & 2-2c_3 & 0 & 2\sqrt{3}c_1-2\sqrt{3}c_2 \\ 0 & -2c_1+4c_2-2 & 0 & 0 & 2c_2-4c_3+2 & 0 & 4c_1-2c_3-2 & 0 \\ 2c_1+2c_2-4c_3 & 0 & -c_1+2c_2-c_3 & -4c_1+2c_2+2c_3 & 0 & 2c_1-4c_2+2c_3 & 0 & \sqrt{3}c_1-\sqrt{3}c_3 \\ 2-2c_2 & 0 & -4c_1+2c_2+2c_3 & 2-2c_3 & 0 & 2-2c_1 & 0 & 2\sqrt{3}c_2-2\sqrt{3}c_3 \\ 0 & 2c_2-4c_3+2 & 0 & 0 & 4c_1-2c_3-2 & 0 & 2c_1-4c_2+2 & 0 \\ 2-2c_3 & 0 & 2c_1-4c_2+2c_3 & 2-2c_1 & 0 & 2-2c_2 & 0 & 2\sqrt{3}c_3-2\sqrt{3}c_1 \\ 0 & 4c_1-2c_3-2 & 0 & 0 & 2c_1-4c_2+2 & 0 & -2c_2+4c_3-2 & 0 \\ 2\sqrt{3}c_1-2\sqrt{3}c_2 & 0 & \sqrt{3}c_1-\sqrt{3}c_3 & 2\sqrt{3}c_2-2\sqrt{3}c_3 & 0 & 2\sqrt{3}c_3-2\sqrt{3}c_1 & 0 & c_1-2c_2+c_3 \end{pmatrix}$$

$$D[S]_8 = \begin{pmatrix} c_1 & s_1 & 0 & 0 & 0 & 0 & 0 & 0 \\ -s_1 & c_1 & 0 & 0 & 0 & 0 & 0 & 0 \\ 0 & 0 & 1 & 0 & 0 & 0 & 0 & 0 \\ 0 & 0 & 0 & c_3 & s_3 & 0 & 0 & 0 \\ 0 & 0 & 0 & -s_3 & c_3 & 0 & 0 & 0 \\ 0 & 0 & 0 & 0 & 0 & c_2 & s_2 & 0 \\ 0 & 0 & 0 & 0 & 0 & -s_2 & c_2 & 0 \\ 0 & 0 & 0 & 0 & 0 & 0 & 0 & 1 \end{pmatrix} \tag{4.127}$$

$$D[T]_8 = (D[R]_8 \cdot D[S]_8)^{-1}$$

4.4.7 The proper subgroups of L_{168}

From the complexity of the other irreps, in relation to the simplicity of the 7-dimensional one, it is already clear that this latter should be considered the natural defining representation. The crystallographic nature of the group in $d = 7$ has already been stressed. Anticipating a notion that will become clear to the reader after studying Chapter 8, we introduce the $\mathfrak{a}_7$ weight lattice which, by definition, is just the dual of the root lattice. Explicitly,

$$\Lambda_w \ni \mathbf{w} = n_i \lambda^i : \quad n^i \in \mathbb{Z} \tag{4.128}$$

is spanned by the simple weights that are implicitly defined by the relations:

$$\lambda^i \cdot \alpha_j = \delta^i_j \quad \Rightarrow \quad \lambda^i = (\mathscr{C}^{-1})^{ij} \alpha_j \tag{4.129}$$

Since the group L_{168} is crystallographic on the root lattice, by necessity it is crystallographic also on the weight lattice. Given the generators of the group L_{168} in the basis of simple roots we obtain the same in the basis of simple weights through the following transformation:

$$\mathscr{R}_W = \mathscr{C}\mathscr{R}\mathscr{C}^{-1}; \quad \mathscr{S}_W = \mathscr{C}\mathscr{S}\mathscr{C}^{-1}; \quad \mathscr{T}_W = \mathscr{C}\mathscr{T}\mathscr{C}^{-1} \tag{4.130}$$

Explicitly we find:

$$\mathscr{R}_W = \begin{pmatrix} 0 & 0 & 0 & 0 & 0 & 0 & -1 \\ 0 & 0 & 0 & -1 & -1 & -1 & 0 \\ 0 & 0 & -1 & 0 & 0 & 0 & 0 \\ 0 & 0 & 1 & 1 & 1 & 0 & 0 \\ 0 & 0 & 0 & 0 & -1 & 0 & 0 \\ 0 & -1 & -1 & -1 & 0 & 0 & 0 \\ -1 & 0 & 0 & 0 & 0 & 0 & 0 \end{pmatrix} \tag{4.131}$$

$$\mathscr{S}_W = \begin{pmatrix} -1 & -1 & -1 & -1 & -1 & -1 & -1 \\ 1 & 1 & 1 & 1 & 0 & 0 & 0 \\ 0 & 0 & 0 & -1 & 0 & 0 & 0 \\ 0 & 0 & 0 & 1 & 1 & 1 & 0 \\ 0 & 0 & 0 & 0 & 0 & -1 & 0 \\ 0 & 0 & -1 & -1 & -1 & 0 & 0 \\ 0 & -1 & 0 & 0 & 0 & 0 & 0 \end{pmatrix} \tag{4.132}$$

$$\mathscr{T}_W = \begin{pmatrix} -1 & -1 & -1 & -1 & -1 & -1 & 0 \\ 1 & 0 & 0 & 0 & 0 & 0 & 0 \\ 0 & 1 & 0 & 0 & 0 & 0 & 0 \\ 0 & 0 & 1 & 0 & 0 & 0 & 0 \\ 0 & 0 & 0 & 1 & 0 & 0 & 0 \\ 0 & 0 & 0 & 0 & 1 & 0 & 0 \\ 0 & 0 & 0 & 0 & 0 & 1 & 1 \end{pmatrix} \tag{4.133}$$

Given the weight basis, which is useful in several constructions, let us conclude our survey of the remarkable simple group L_{168} by a brief discussion of its subgroups, none of which, obviously, is normal.

L_{168} contains maximal subgroups only of index 8 and 7, namely, of order 21 and 24. The order 21 subgroup G_{21} is the unique non-abelian group of that order and abstractly it has the structure of the semidirect product $\mathbb{Z}_3 \ltimes \mathbb{Z}_7$. Up to conjugation there is only one subgroup G_{21} as we have explicitly verified with the computer (see the MATHEMATICA NoteBook in Section B.6). On the other hand, up to conjugation, there are two different groups of order 24 that are both isomorphic to the octahedral group O_{24}.

4.4.7.1 The maximal subgroup G_{21}

The group G_{21} has two generators $\mathscr{X}$ and $\mathscr{Y}$ that satisfy the following relations:

$$\mathscr{X}^3 = \mathscr{Y}^7 = \mathbf{1}; \quad \mathscr{X}\mathscr{Y} = \mathscr{Y}^2\mathscr{X} \tag{4.134}$$

The organization of the 21 group elements into conjugacy classes is displayed below:

Conjugacy class	C_1	C_2	C_3	C_4	C_5
representative of the class	e	$\mathcal{Y}$	$\mathcal{X}^2\mathcal{Y}\mathcal{X}\mathcal{Y}^2$	$\mathcal{Y}\mathcal{X}^2$	$\mathcal{X}$
order of the elements in the class	1	7	7	3	3
number of elements in the class	1	3	3	7	7

(4.135)

As we see there are five conjugacy classes which implies that there should be five irreducible representations the square of whose dimensions should sum up to the group order 21. The solution of this problem is:

$$21 = 1^2 + 1^2 + 1^2 + 3^2 + 3^2 \tag{4.136}$$

and the corresponding character table is mentioned below:

	e	$\mathcal{Y}$	$\mathcal{X}^2\mathcal{Y}\mathcal{X}\mathcal{Y}^2$	$\mathcal{Y}\mathcal{X}^2$	$\mathcal{X}$
$D_1[G_{21}]$	1	1	1	1	1
$DX_1[G_{21}]$	1	1	1	$-(-1)^{1/3}$	$(-1)^{2/3}$
$DY_1[G_{21}]$	1	1	1	$(-1)^{2/3}$	$-(-1)^{1/3}$
$DA_3[G_{21}]$	3	$\frac{1}{2}i(i+\sqrt{7})$	$-\frac{1}{2}i(-i+\sqrt{7})$	0	0
$DB_3[G_{21}]$	3	$-\frac{1}{2}i(-i+\sqrt{7})$	$\frac{1}{2}i(i+\sqrt{7})$	0	0

(4.137)

In the weight-basis the two generators of the G_{21} subgroup of L_{168} can be chosen to be the following matrices and this fixes our representative of the unique conjugacy class:

$$\begin{aligned}
\mathcal{X} &= \begin{pmatrix} 1 & 1 & 1 & 1 & 1 & 1 & 1 \\ 0 & 0 & 0 & 0 & 0 & 0 & -1 \\ 0 & -1 & -1 & -1 & -1 & -1 & 0 \\ 0 & 1 & 1 & 1 & 0 & 0 & 0 \\ 0 & 0 & -1 & -1 & 0 & 0 & 0 \\ 0 & 0 & 1 & 1 & 1 & 0 & 0 \\ 0 & 0 & 0 & -1 & -1 & 0 & 0 \end{pmatrix} \\
\mathcal{Y} &= \begin{pmatrix} 0 & 1 & 1 & 0 & 0 & 0 & 0 \\ 0 & 0 & 0 & 1 & 1 & 1 & 1 \\ 0 & 0 & -1 & -1 & -1 & -1 & -1 \\ 0 & 0 & 1 & 1 & 0 & 0 & 0 \\ -1 & -1 & -1 & -1 & 0 & 0 & 0 \\ 1 & 1 & 1 & 1 & 1 & 0 & 0 \\ 0 & 0 & 0 & 0 & 0 & 1 & 0 \end{pmatrix}
\end{aligned} \tag{4.138}$$

4.4.7.2 The maximal subgroups O_{24A} and O_{24B}

As we know from Section 3.4.3, the octahedral group O_{24} has two generators S and T that satisfy the following relations:

$$S^2 = T^3 = (ST)^4 = \mathbf{1} \tag{4.139}$$

The 24 elements are organized in five conjugacy classes according to the scheme displayed below:

Conjugacy class	C_1	C_2	C_3	C_4	C_5
representative of the class	e	T	$STST$	S	ST
order of the elements in the class	1	3	2	2	4
number of elements in the class	1	8	3	6	6

(4.140)

The irreducible representations of O_{24} were explicitly constructed in Section 4.3.6. We repeat here the corresponding character table mentioning also a standard representative of each conjugacy class:

	e	T	$STST$	S	ST
$D_1[O_{24}]$	1	1	1	1	1
$D_2[O_{24}]$	1	1	1	−1	−1
$D_3[O_{24}]$	2	−1	2	0	0
$D_4[O_{24}]$	3	0	−1	−1	1
$D_5[O_{24}]$	3	0	−1	1	−1

(4.141)

By computer calculations we have verified that there are just two disjoint conjugacy classes of O_{24} maximal subgroups in L_{168} that we have named A and B, respectively (see MATHEMATICA NoteBook described in Section B.6). We have chosen two standard representatives, one for each conjugacy class, that we have named O_{24A} and O_{24B} respectively. To fix these subgroups it suffices to mention the explicit form of the their generators in the weight basis.

For the group O_{24A}, we chose:

$$T_A = \begin{pmatrix} 1 & 1 & 1 & 1 & 1 & 1 & 1 \\ 0 & 0 & 0 & 0 & 0 & 0 & -1 \\ 0 & -1 & -1 & -1 & -1 & -1 & 0 \\ 0 & 1 & 1 & 1 & 0 & 0 & 0 \\ 0 & 0 & -1 & -1 & 0 & 0 & 0 \\ 0 & 0 & 1 & 1 & 1 & 0 & 0 \\ 0 & 0 & 0 & -1 & -1 & 0 & 0 \end{pmatrix}$$

$$S_A = \begin{pmatrix} 0 & 0 & 0 & 1 & 1 & 1 & 0 \\ 0 & 0 & 0 & 0 & -1 & -1 & 0 \\ -1 & -1 & -1 & -1 & 0 & 0 & 0 \\ 1 & 1 & 0 & 0 & 0 & 0 & 0 \\ 0 & 0 & 1 & 1 & 1 & 1 & 1 \\ 0 & -1 & -1 & -1 & -1 & -1 & -1 \\ 0 & 1 & 1 & 1 & 1 & 0 & 0 \end{pmatrix} \tag{4.142}$$

For the group O_{24B}, we chose:

$$T_B = \begin{pmatrix} 1 & 1 & 1 & 1 & 0 & 0 & 0 \\ 0 & -1 & -1 & -1 & 0 & 0 & 0 \\ 0 & 1 & 1 & 1 & 1 & 0 & 0 \\ 0 & 0 & -1 & -1 & -1 & 0 & 0 \\ 0 & 0 & 1 & 1 & 1 & 1 & 0 \\ 0 & 0 & 0 & -1 & -1 & -1 & 0 \\ 0 & 0 & 0 & 1 & 1 & 1 & 1 \end{pmatrix}$$
$$S_B = \begin{pmatrix} 0 & 0 & 1 & 1 & 1 & 0 & 0 \\ -1 & -1 & -1 & -1 & -1 & 0 & 0 \\ 1 & 1 & 1 & 1 & 1 & 1 & 1 \\ 0 & 0 & 0 & 0 & 0 & 0 & -1 \\ 0 & -1 & -1 & -1 & -1 & -1 & 0 \\ 0 & 1 & 1 & 1 & 0 & 0 & 0 \\ 0 & 0 & 0 & -1 & 0 & 0 & 0 \end{pmatrix} \tag{4.143}$$

4.4.7.3 The tetrahedral subgroup $T_{12} \subset O_{24}$

Every octahedral group O_{24} has, up to O_{24}-conjugation, a unique tetrahedral subgroup T_{12} whose order is 12. The abstract description of the tetrahedral group is provided by the following presentation in terms of two generators:

$$T_{12} = (s, t \mid s^2 = t^3 = (st)^3 = 1) \tag{4.144}$$

The 12 elements are organized into four conjugacy classes as displayed below:

Classes	C_1	C_2	C_3	C_4
standard representative	1	s	t	t^2s
order of the elements in the conjugacy class	1	2	3	3
number of elements in the conjugacy class	1	3	4	4

(4.145)

We do not display the character table which is discussed in the solution of Exercise 3.40, point c). The two tetrahedral subgroups $T_{12A} \subset O_{24A}$ and $T_{12B} \subset O_{24B}$ are not conjugate under the big group L_{168}. Hence we have two conjugacy classes of tetrahedral subgroups of L_{168}.

4.4.7.4 The dihedral subgroup $\mathrm{Dih}_3 \subset O_{24}$

Every octahedral group O_{24} has a dihedral subgroup Dih_3 whose order is 6. The abstract description of the dihedral group Dih_3 is provided by the following presentation in terms of two generators:

$$\mathrm{Dih}_3 = (A, B \mid A^3 = B^2 = (BA)^2 = 1) \tag{4.146}$$

The six elements are organized into three conjugacy classes as displayed below:

Conjugacy classes	C_1	C_2	C_3
standard representative of the class	1	A	B
order of the elements in the class	1	3	2
number of elements in the class	1	2	3

(4.147)

The character table of this group is discussed in Exercise 3.39. Differently from the case of the tetrahedral subgroups the two dihedral subgroups $\mathrm{Dih}_{3A} \subset \mathrm{O}_{24A}$ and $\mathrm{Dih}_{3B} \subset \mathrm{O}_{24B}$ turn out to be conjugate under the big group L_{168}. Actually there is just one L_{168}-conjugacy class of dihedral subgroups Dih_3.

4.4.7.5 Enumeration of the possible subgroups and orbits

In $d = 3$ the orbits of the octahedral group acting on the cubic lattice are the vertices of regular geometrical figures. Since L_{168} has a crystallographic action on the mentioned 7-dimensional weight lattice, its orbits $\mathcal{O}$ in Λ_w correspond to the analogous regular geometrical figures in $d = 7$. Every orbit is in correspondence with a coset G/H where G is the big group and H is one of its possible subgroups. Indeed, H is the stability subgroup of an element of the orbit.

Since the maximal subgroups of L_{168} are of index 7 or 8 we can have subgroups $\mathrm{H} \subset \mathrm{L}_{168}$ that are either G_{21} or O_{24} or subgroups thereof. Furthermore, as we know, the order $|\mathrm{H}|$ of any subgroup $\mathrm{H} \subset \mathrm{G}$ must be a divisor of $|\mathrm{G}|$. Hence we conclude that

$$|\mathrm{H}| \in \{1, 2, 3, 4, 6, 7, 8, 12, 21, 24\} \tag{4.148}$$

Correspondingly, we might have L_{168}-orbits $\mathcal{O}$ in the weight lattice Λ_w, whose length is one of the following nine numbers:

$$\ell_{\mathcal{O}} \in \{168, 84, 56, 42, 28, 24, 21, 14, 8, 7\} \tag{4.149}$$

Combining the information about the possible group orders (4.148) with the information that the maximal subgroups are of index 8 or 7, we arrive at the following list of possible subgroups H (up to conjugation) of the group L_{168}:

Order 24) Either $\mathrm{H} = \mathrm{O}_{24A}$ or $\mathrm{H} = \mathrm{O}_{24B}$.

Order 21) The only possibility is $\mathrm{H} = \mathrm{G}_{21}$.

Order 12) The only possibilities are $\mathrm{H} = \mathrm{T}_{12A}$ or $\mathrm{H} = \mathrm{T}_{12B}$ where T_{12} is the tetrahedral subgroup of the octahedral group O_{24}.

Order 8) Either $\mathrm{H} = \mathbb{Z}_2 \times \mathbb{Z}_2 \times \mathbb{Z}_2$ or $\mathrm{H} = \mathbb{Z}_2 \times \mathbb{Z}_4$.

Order 7) The only possibility is $\mathbb{Z}_7$.

Order 6) Either $\mathrm{H} = \mathbb{Z}_2 \times \mathbb{Z}_3$ or $\mathrm{H} = \mathrm{Dih}_3$, where Dih_3 denotes the dihedral subgroup of index 3 of the octahedral group O_{24}.

Order 4) Either $\mathrm{H} = \mathbb{Z}_2 \times \mathbb{Z}_2$ or $\mathrm{H} = \mathbb{Z}_4$.

Order 3) The only possibility is $H = \mathbb{Z}_3$.
Order 2) The only possibility is $H = \mathbb{Z}_2$.

We suggest that the reader compares the above result with the solution of Exercise 3.29 where all the groups of order less than 12 have been listed.

4.4.7.6 Synopsis of the L_{168} orbits in the weight lattice Λ_w

In [52], one of us has presented his results, obtained by means of computer calculations, on the orbits of the considered simple group acting on the $\mathfrak{a}_7$ weight lattice. They are briefly summarized below

1. Orbits of length 8 (one parameter **n**; stability subgroup $H^s = G_{21}$)
2. Orbits of length 14 (two types A & B) (one parameter **n**; stability subgroup $H^s = T_{12A,B}$)
3. Orbits of length 28 (one parameter **n**; stability subgroup $H^s = \mathrm{Dih}_3$)
4. Orbits of length 42 (one parameter **n**; stability subgroup $H^s = \mathbb{Z}_4$)
5. Orbits of length 56 (three parameters **n, m, p**; stability subgroup $H^s = \mathbb{Z}_3$)
6. Orbits of length 84 (three parameters **n, m, p**; stability subgroup $H^s = \mathbb{Z}_2$)
7. Generic orbits of length 168 (seven parameters; stability subgroup $H^s = \mathbf{1}$)

As we already said, the above list is in some sense the 7-dimensional analog of Platonic solids. It is such only in some sense, since it is a complete classification for the group L_{168} yet we are not aware of a classification of the other crystallographic subgroups of SO(7), if any.

Notwithstanding this ignorance, the piece of knowledge we have summarized above is already impressively complicated and demonstrates how even flat geometry becomes more sophisticated in higher dimensions.

The next natural question is why just $d = 7$ should attract our geometrical attention. There are several reasons for the number 7. They all are probably interrelated:

1. The possible division algebras are $\mathbb{R}$, $\mathbb{C}$, $\mathbb{H}$, $\mathbb{O}$, the real numbers, the complex numbers, the quaternions, and the octonions. The corresponding number of imaginary units are 0, 1, 3, 7. The automorphisms groups of these division algebras are 1, U(1), SU(2), $G_{2(-14)}$.
2. The spheres that are globally parallelizable are $\mathbb{S}^1$, $\mathbb{S}^3$, $\mathbb{S}^7$.
3. The manifolds of restricted holonomy are the complex ones, the Kähler ones, the quaternionic ones, that exist in all dimensions $d = 2n$, respectively $d = 4n$, and then, just in $d = 7$ we have the G_2 manifolds and in $d = 8$ we have the Spin(7) manifolds.
4. Seven are the dimensions that one has to compactify in order to step down from the 11-dimensional M-theory to our $d = 4$ space-time and many solutions of the theory naturally perform the splitting $11 = 4 + 7$.

4.5 Bibliographical note

The ADE classification is a widely discussed topic in contemporary algebraic and differential geometry and is the subject of a vast specialized literature. An in-depth review of its application to ALE manifolds and to advanced issues in supersymmetric field theories is presented in the forthcoming book [54] by one of us. The here presented derivation of the ADE classification of SO(3) finite subgroups is based on thirty-year-old lecture notes of one of us (P.F.) who can no longer remember what was his inspiration source at the time.

Lattices and Crystallographic groups are discussed in many textbooks not only in Mathematics, but also in Chemistry and Crystallography. Two excellent mathematical books are the monographs by Humphrey [88, 89]. The explicit constructions concerning the octahedral group are taken from the already quoted paper by one of us and Alexander Sorin [56]. The bulk of material on the group PSL(2, 7) comes instead from the paper of P.F. [52], several times quoted above.

5 Manifolds and Lie groups

The analysts try in vain to conceal the fact that they do not deduce: they combine, they compose ... when they do arrive at the truth they stumble over it after groping their way along.
Evariste Galois

5.1 Introduction

In Chapter 1 we focused on algebraic structures and we reviewed the basic algebraic concepts that apply both to discrete and to continuous groups. In the present chapter we turn to basic concepts of differential geometry preparing the stage for the study of Lie groups. These latter, which constitute one of the main goals of this course, arise from the consistent merging of two structures:

1. an algebraic structure, since the elements of a Lie group G can be composed via an internal binary operation, generically called product, that obeys the axioms of a group;
2. a differential geometric structure since G is an analytic differentiable manifold and the group operations are infinitely differentiable in such a topology.

General Relativity is founded on the concept of *differentiable manifolds*. The mathematical model of *space-time* that we adopt is given by a pair $(\mathcal{M}, g)$ where $\mathcal{M}$ is a differentiable manifold of dimension $D = 4$ and g is a *metric*, that is, a rule to calculate the length of curves connecting points of $\mathcal{M}$. In physical terms the points of $\mathcal{M}$ take the name of *events* while every physical process is a continuous succession of events. In particular the motion of a *point-like particle* is represented by a *world-line*, namely a curve in $\mathcal{M}$ while the motion of an *extended object* of dimension p is given by a $d = p+1$ dimensional *world-volume* obtained as a continuous succession of p-dimensional hypersurfaces $\Sigma_p \subset \mathcal{M}$.

Therefore, the discussion of such *physical concepts* is necessarily based on a collection of *geometrical concepts* that constitute the backbone of differential geometry. The latter is at the basis not only of General Relativity but of all Gauge Theories by means of which XXth century Physics obtained a consistent and experimentally verified description of all Fundamental Interactions.

The central notions are those which fix the geometric environment:

- Differentiable Manifolds
- Fiber-Bundles

and those which endow such environment with structures accounting for the measure of lengths and for the rules of parallel transport, namely:

- Metrics
- Connections

https://doi.org/10.1515/9783110551204-005

Once the geometric environments are properly mathematically defined, the metrics and connections one can introduce over them turn out to be the structures which encode the Fundamental Forces of Nature.

On the other hand, *differential geometry* and *Lie group theory*
- are **intimately and inextricably related**, and
- have a **much wider range of applications** in all branches of physics and other sciences

since that of a manifold is the appropriate mathematical concept of a continuous space whose points can have the most disparate interpretations and that of a group is the appropriate mathematical framework to deal with symmetry operations acting on that space.

In the following sections we introduce Differentiable Manifolds and then Lie Groups.

5.2 Differentiable manifolds

First and most fundamental in the list of geometrical concepts we need to introduce is that of a *manifold* which corresponds, as we already explained, to our intuitive idea of a *continuous space*. In mathematical terms this is, to begin with, a *topological space*, namely a set of elements where one can define the notion of *neighborhood* and *limit*. This is the correct mathematical description of our intuitive ideas of vicinity and close-by points. Secondly, the characterizing feature that distinguishes a manifold from a simple topological space is the possibility of labeling its points with a set of coordinates. Coordinates are a set of real numbers $x_1(p), \dots, x_D(p) \in \mathbb{R}$ associated with each point $p \in \mathcal{M}$ that tell us *where* we are. Actually, in General Relativity each point is an event so that coordinates specify not only its *where* but also its *when*. In other applications the coordinates of a point can be the most disparate parameters specifying the state of some complex system of the most general kind (dynamical, biological, economical or whatever).

In classical physics the laws of motion are formulated as a set of differential equations of the second order where the unknown functions are the three Cartesian coordinates x, y, z of a particle and the variable t is time. Solving the dynamical problem amounts to determining the continuous functions $x(t), y(t), z(t)$, that yield a parametric description of a curve in $\mathbb{R}^3$ or better define a curve in $\mathbb{R}^4$, having included the time t in the list of coordinates of each event. Coordinates, however, are not uniquely defined. Each observer has its own way of labeling space points and the laws of motion take a different form if expressed in the coordinate frame of different observers. There is however a privileged class of observers in whose frames the laws of motion have always the same form: these are the inertial frames, that are in rectilinear relative motion with constant velocity. The existence of a privileged class of inertial frames is common to classical Newtonian physics and to special relativity: the only difference is the form of coordinate transformations connecting them, Galileo transformations in the first case and Lorentz transformations in the second. This goes hand in hand with the fact that the space-time manifold is the *flat affine*[1] *manifold* $\mathbb{R}^4$ in both cases. By definition, all points of $\mathbb{R}^N$ can be covered by one coordinate frame $\{x^i\}$ and all frames with such a property are related to each other by general linear transformations, that is, by the elements of the general linear group $\mathrm{GL}(N,\mathbb{R})$:

$$x^{i'} = A^{i}{}_{j}x^{j}; \quad A^{i}{}_{j} \in \mathrm{GL}(N,\mathbb{R}) \tag{5.1}$$

The restriction to the Galileo or Lorentz subgroups of $\mathrm{GL}(4,\mathbb{R})$ is a consequence of the different *scalar product* on $\mathbb{R}^4$ vectors one wants to preserve in the two cases, but the relevant common feature is the fact that the space-time manifold has a vector-space structure. The privileged coordinate frames are those that use the corresponding vectors as labels of each point.

A different situation arises when the space-time manifold is not flat, like, for instance, the surface of a hypersphere $\mathbb{S}^N$. As cartographers know very well, there is no way of representing all points of a curved surface in a single coordinate frame, namely, in a single *chart*. However, we can succeed in representing all points of a curved surface by means of an *atlas*, namely, by a collection of charts, each of which maps one open region of the surface and such that the union of all these regions covers the entire surface. Knowing the transition rule from one chart to the next one, in the regions where they overlap, we obtain a complete coordinate description of the curved surface by means of our atlas.

The intuitive idea of an *atlas* of *open charts*, suitably reformulated in mathematical terms, provides the very definition of a differentiable manifold, the geometrical concept that generalizes our notion of space-time, from $\mathbb{R}^N$ to more complicated non-flat situations.

There are many possible *atlases* that describe the same manifold $\mathcal{M}$, related to each other by more or less complicated transformations. For a generic $\mathcal{M}$ no privileged choice of the atlas is available differently from the case of $\mathbb{R}^N$: here the inertial frames are singled out by the additional *vector space* structure of the manifold, which allows to label each point with the corresponding vector. Therefore if the laws of physics have to be universal and have to accommodate non-flat space-times, then they must be formulated in such a way that they have the same form in whatsoever atlas. This is the principle of *general covariance* at the basis of General Relativity: all observers see the same laws of physics.

Similarly, in a wider perspective, the choice of a particular set of parameters to describe the state of a complex system should not be privileged with respect to any

---

1 A manifold (defined in this section) is named *affine* when it is also a vector space.

other choice. The laws that govern the dynamics of a system should be intrinsic and should not depend on the set of variables chosen to describe it.

5.2.1 Homeomorphisms and the definition of manifolds

A fundamental ingredient in formulating the notion of differential manifolds is that of homeomorphism:[2]

Definition 5.2.1. Let X and Y be two topological spaces and let h be a map:

$$h : X \to Y \tag{5.2}$$

If h is one-to-one and if both h and its inverse h^{-1} are continuous, then we say that h is a **homeomorphism.**

As a consequence of the theorems proved in all textbooks about elementary topology and calculus, homeomorphisms preserve all topological properties. Indeed, let h be a homeomorphism mapping X onto Y and let $A \subset X$ be an open subset. Its image through h, namely $h(A) \subset Y$, is also an open subset in the topology of Y. Similarly the image $h(C) \subset Y$ of a closed subset $C \subset X$ is a closed subset. Furthermore, for all $A \subset X$ we have:

$$h(\overline{A}) = \overline{h(A)} \tag{5.3}$$

namely, the closure of the image of a set A coincides with the image of the closure.

Definition 5.2.2. Let X and Y be two topological spaces. If there exists a homeomorphism $h : X \to Y$ then we say that X and Y are homeomorphic.

It is easy to see that given a topological space X, the set of all homeomorphisms $h : X \to X$ constitutes a group, usually denoted $\mathrm{Hom}(X)$. Indeed, if $h \in \mathrm{Hom}(X)$ is a homeomorphism, then also $h^{-1} \in \mathrm{Hom}(X)$ is a homeomorphism. Furthermore, if $h \in \mathrm{Hom}(X)$ and $h' \in \mathrm{Hom}(X)$, then also $h \circ h' \in \mathrm{Hom}(X)$. Finally the identity map:

$$\mathbf{1} : X \to X \tag{5.4}$$

is certainly one-to-one and continuous and it coincides with its own inverse. Hence $\mathbf{1} \in \mathrm{Hom}(X)$. As we discuss later on, for any manifold X the group $\mathrm{Hom}(X)$ is an example of an infinite and continuous group.

---

2 We assume that the reader possesses the basic notions of general topology concerning the notions of bases of neighborhoods, open and close subsets, boundary and limit (see for instance [116, 117, 115, 45, 129]).

Let now $\mathcal{M}$ be a topological Hausdorff space. An *open chart* of $\mathcal{M}$ is a pair (U,φ) where $U \subset \mathcal{M}$ is an open subset of $\mathcal{M}$ and φ is a homeomorphism of U on an open subset $\mathbb{R}^m$ (m being a positive integer). The concept of open chart allows to introduce the notion of coordinates for all points $p \in U$. Indeed, the coordinates of p are the m real numbers that identify the point $\varphi(p) \in \varphi(U) \subset \mathbb{R}^m$.

Using the notion of open chart we can finally introduce the notion of differentiable structure.

Definition 5.2.3. Let $\mathcal{M}$ be a topological Hausdorff space. A differentiable structure of dimension m on $\mathcal{M}$ is an *atlas* $\mathcal{A} = \bigcup_{i\in A}(U_i,\varphi_i)$ of open charts (U_i,φ_i) where $\forall i \in A$, $U_i \subset \mathcal{M}$ is an open subset and

$$\varphi_i : U_i \to \varphi_i(U_i) \subset \mathbb{R}^m \tag{5.5}$$

is a homeomorphism of U_i in $\mathbb{R}^m$, namely a continuous, invertible map onto an open subset of $\mathbb{R}^m$ such that the inverse map

$$\varphi_i^{-1} : \varphi_i(U_i) \to U_i \subset \mathcal{M} \tag{5.6}$$

is also continuous (see Figure 5.1). The atlas must fulfill the following axioms:

$\mathbf{M_1}$ It covers $\mathcal{M}$, namely,

$$\bigcup_i U_i = \mathcal{M} \tag{5.7}$$

so that each point of $\mathcal{M}$ is contained at least in one chart and generically in more than one: $\forall p \in \mathcal{M} \mapsto \exists (U_i,\varphi_i)/p \in U_i$.

$\mathbf{M_2}$ Chosen any two charts (U_i,φ_i), (U_j,φ_j) such that $U_i \cap U_j \neq \emptyset$, on the intersection

$$U_{ij} \overset{\text{def}}{=} U_i \cap U_j \tag{5.8}$$

there exist two homeomorphisms:

$$\begin{aligned} \varphi_i|_{U_{ij}} &: U_{ij} \to \varphi_i(U_{ij}) \subset \mathbb{R}^m \\ \varphi_j|_{U_{ij}} &: U_{ij} \to \varphi_j(U_{ij}) \subset \mathbb{R}^m \end{aligned} \tag{5.9}$$

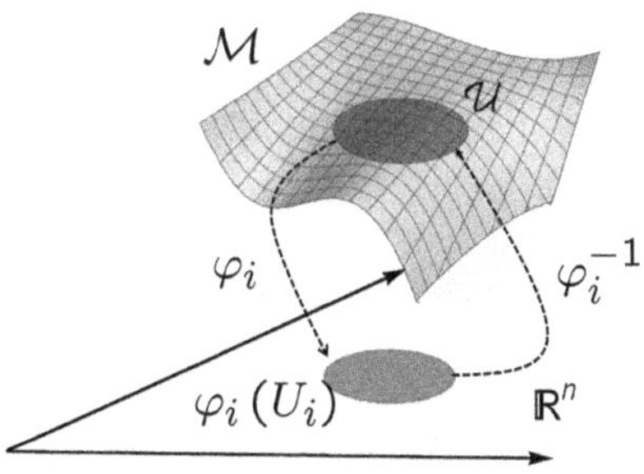

Figure 5.1: An open chart is a homeomorphism of an open subset U_i of the manifold $\mathcal{M}$ onto an open subset of $\mathbb{R}^m$.

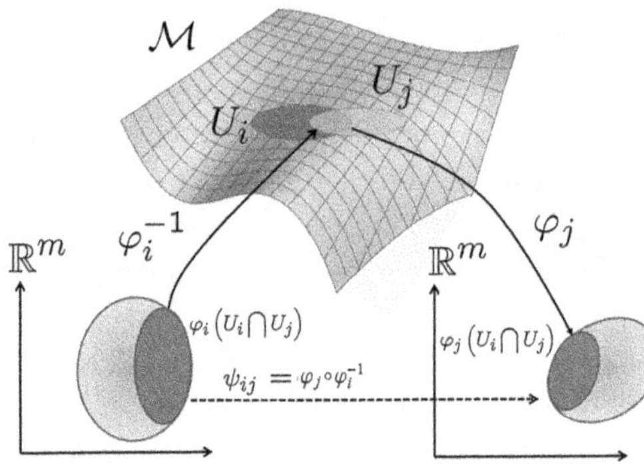

Figure 5.2: A transition function between two open charts is a differentiable map from an open subset of $\mathbb{R}^m$ to another open subset of the same.

and the composite map:

$$\psi_{ij} \stackrel{\text{def}}{=} \varphi_j \circ \varphi_i^{-1}$$
$$\psi_{ij} : \varphi_i(U_{ij}) \subset \mathbb{R}^m \to \varphi_j(U_{ij}) \subset \mathbb{R}^m \tag{5.10}$$

named the *transition function* which is actually an m-tuplet of m real functions of m real variables is requested to be *differentiable* (see Figure 5.2).

M_3 The collection $(U_i, \varphi_i)_{i \in A}$ is the maximal family of open charts for which both M_1 and M_2 hold true.

Next we can finally introduce the definition of differentiable manifold.

Definition 5.2.4. A differentiable manifold of dimension m is a topological space $\mathcal{M}$ that admits at least one differentiable structure $(U_i, \varphi_i)_{i \in A}$ of dimension m.

The definition of a differentiable manifold is constructive in the sense that it provides a way to construct it explicitly. What one has to do is to give an atlas of open charts (U_i, φ_i) and the corresponding transition functions ψ_{ij} which should satisfy the necessary consistency conditions:

$$\forall i,j \quad \psi_{ij} = \psi_{ji}^{-1} \tag{5.11}$$
$$\forall i,j,k \quad \psi_{ij} \circ \psi_{jk} \circ \psi_{ki} = \mathbf{1} \tag{5.12}$$

In other words, a general recipe to construct a manifold is to specify the open charts and how they are *glued* together. The properties assigned to a manifold are the properties fulfilled by its transition functions. In particular we have:

Definition 5.2.5. A differentiable manifold $\mathcal{M}$ is said to be *smooth* if the transition functions (5.10) are *infinitely differentiable*

$$\mathcal{M} \text{ is smooth} \quad \Leftrightarrow \quad \psi_{ij} \in \mathbb{C}^\infty(\mathbb{R}^m) \tag{5.13}$$

Similarly one has the definition of a complex manifold.

Definition 5.2.6. A real manifold of even dimension $m = 2\nu$ is *complex* of dimension ν if the 2ν real coordinates in each open chart U_i can be arranged into ν complex numbers so that eq. (5.5) can be replaced by

$$\varphi_i : U_i \to \varphi_i(U_i) \subset \mathbb{C}^\nu \tag{5.14}$$

and the transition functions ψ_{ij} are *holomorphic maps*:

$$\psi_{ij} : \varphi_i(U_{ij}) \subset \mathbb{C}^\nu \to \varphi_j(U_{ij}) \subset \mathbb{C}^\nu \tag{5.15}$$

Although the constructive definition of a differentiable manifold is always in terms of an atlas, in many occurrences we can have other intrinsic global definitions of what $\mathcal{M}$ is and the construction of an atlas of coordinate patches is an a posteriori operation. Typically this happens when the manifold admits a description as an algebraic locus. The prototype example is provided by the $\mathbb{S}^N$ sphere which can be defined as the locus in $\mathbb{R}^{N+1}$ of points with distance r from the origin:

$$\{X_i\} \in \mathbb{S}^N \quad \Leftrightarrow \quad \sum_{i=1}^{N+1} X_i^2 = r^2 \tag{5.16}$$

In particular for $N = 2$ we have the familiar $\mathbb{S}^2$ which is diffeomorphic to the compactified complex plane $\mathbb{C} \cup \{\infty\}$. Indeed, we can easily verify that $\mathbb{S}^2$ is a one-dimensional complex manifold considering the atlas of holomorphic open charts suggested by the geometrical construction named *the stereographic projection*. To this effect consider the picture in Figure 5.3 where we have drawn the two-sphere $\mathbb{S}^2$ of radius $r = 1$ centered in the origin of $\mathbb{R}^3$. Given a generic point $P \in \mathbb{S}^2$ we can construct its image on the equatorial plane $\mathbb{R}^2 \sim \mathbb{C}$ drawing the straight line in $\mathbb{R}^3$ that goes through P and through the *North Pole* of the sphere N. Such a line will intersect the equatorial plane in the point P_N whose value z_N, regarded as a complex number, we can identify with the complex coordinate of P in the open chart under consideration:

$$\varphi_N(P) = z_N \in \mathbb{C} \tag{5.17}$$

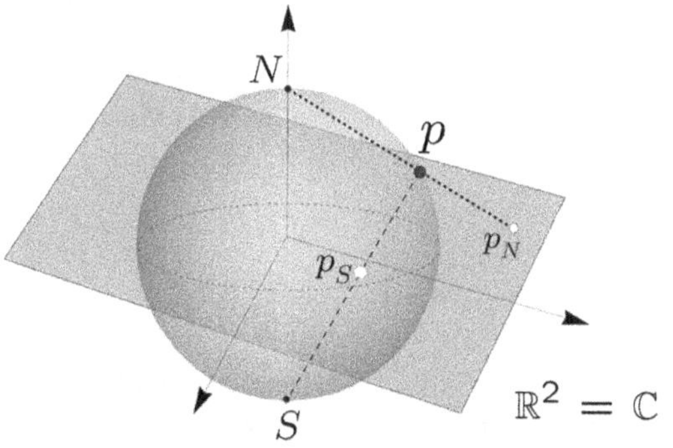

Figure 5.3: Stereographic projection of the two-sphere.

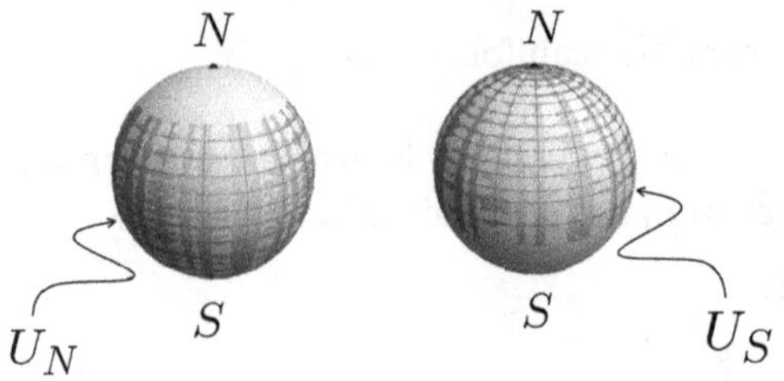

Figure 5.4: The open charts of the North and South Pole.

Alternatively we can draw the straight line through P and the *South Pole* S. This intersects the equatorial plane in another point P_S whose value as a complex number, named z_S, is just the reciprocal of z_N: $z_S = 1/z_N$. We can take z_S as the complex coordinate of the same point P. In other words, we have another open chart:

$$\varphi_S(P) = z_S \in \mathbb{C} \tag{5.18}$$

What is the domain of these two charts, namely, what are the open subsets U_N and U_S? This is rather easily established considering that the North Pole projection yields a finite result $z_N < \infty$ for all points P except the North Pole itself. Hence $U_N \subset \mathbb{S}^2$ is the open set obtained by subtracting one point (the North Pole) from the sphere. Similarly the South Pole projection yields a finite result for all points P except the South Pole itself and U_S is $\mathbb{S}^2$ minus the South Pole. More definitely, we can choose for U_N and U_S any two open neighborhoods of the South and North Pole respectively with non-vanishing intersection (see Figure 5.4). In this case the intersection $U_N \cap U_S$ is a band wrapped around the equator of the sphere and its image in the complex equatorial plane is a circular corona that excludes both a circular neighborhood of the origin and a circular neighborhood of infinity. On such an intersection we have the transition function:

$$\psi_{NS} : z_N = \frac{1}{z_S} \tag{5.19}$$

which is clearly holomorphic and satisfies the consistency conditions in eqs (5.11), (5.12). Hence we see that $\mathbb{S}^2$ is a complex 1-manifold that can be constructed with an atlas composed of two open charts related by the transition function (5.19). Obviously a complex 1-manifold is *a fortiori* a *smooth real 2-manifold*. Manifolds with infinitely differentiable transition functions are named smooth not without a reason. Indeed, they correspond to our intuitive notion of smooth hypersurfaces without conical points or edges. The presence of such defects manifests itself through the lack of differentiability in some regions.

5.2.2 Functions on manifolds

Being the mathematical model of possible space-times, manifolds are the geometrical support of physics. They are the arenas where physical processes take place and where physical quantities take values. Mathematically, this implies that calculus, originally introduced on $\mathbb{R}^N$, must be extended to manifolds. The physical entities defined over manifolds with which we have to deal are mathematically characterized as *scalar functions, vector fields, tensor fields, differential forms, sections of more general fiber-bundles*. We introduce such basic geometrical notions slowly, beginning with the simplest concept of a *scalar function*.

Definition 5.2.7. A real scalar function on a differentiable manifold $\mathcal{M}$ is a map:

$$f : \mathcal{M} \to \mathbb{R} \tag{5.20}$$

that assigns a real number $f(p)$ to every point $p \in \mathcal{M}$ of the manifold.

The properties of a scalar function, for instance its differentiability, are the properties characterizing its local description in the various open charts of an atlas. For each open chart (U_i, φ_i) let us define:

$$f_i \stackrel{\text{def}}{=} f \circ \varphi_i^{-1} \tag{5.21}$$

By construction

$$f_i : \mathbb{R}^m \supset \varphi_i(U_i) \to \mathbb{R} \tag{5.22}$$

is a map of an open subset of $\mathbb{R}^m$ onto the real line $\mathbb{R}$, namely a real function of m real variables (see Figure 5.5). The collection of the real functions $f_i(x_1^{(i)}, \dots, x_m^{(i)})$ constitute the local description of the scalar function f. The function is said to be *continuous, differentiable, infinitely differentiable* if the real functions f_i have such properties. From the definition (5.21) of the local description and from the definition (5.10) of the transition functions it follows that we must have:

$$\forall U_i, U_j : \quad f_j|_{U_i \cap U_j} = f_i|_{U_i \cap U_j} \circ \psi_{ij} \tag{5.23}$$

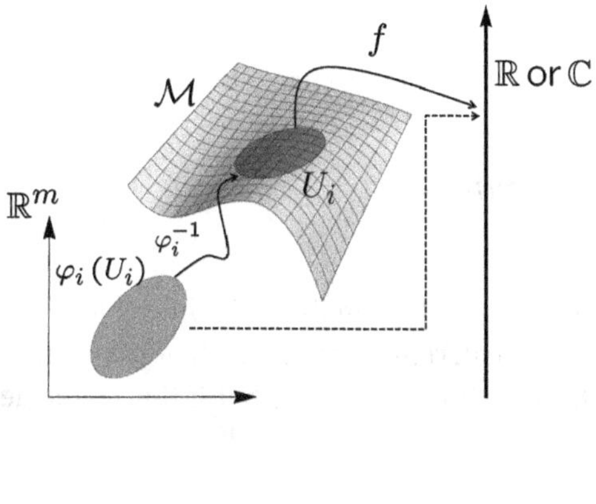

Figure 5.5: Local description of a scalar function on a manifold.

Let $x^\mu_{(i)}$ be the coordinates in the patch U_i and $x^\mu_{(j)}$ be the coordinates in the patch U_j. For points p that belong to the intersection $U_i \cap U_j$ we have:

$$x^\mu_{(j)}(p) = \psi^\mu_{ji}(x^1_{(i)}(p), \dots, x^m_{(i)}(p)) \tag{5.24}$$

and the gluing rule (5.23) takes the form:

$$f(p) = f_j(x_{(j)}) = f_j(\psi_{ji}(x_{(i)})) = f_i(x_{(i)}) \tag{5.25}$$

The practical way of assigning a function on a manifold is therefore that of writing its local description in the open charts of an atlas, taking care that the various f_i glue together correctly, namely, through eq. (5.23). Although the number of continuous and differentiable functions one can write on any open region of $\mathbb{R}^m$ is infinite, the smooth functions globally defined on a non-trivial manifold can be very few. Indeed, it is only occasionally that we can consistently glue together various local functions $f_i \in \mathbb{C}^\infty(U_i)$ into a global f. When this happens we say that $f \in \mathbb{C}^\infty(\mathcal{M})$.

All what we said about real functions can be trivially repeated for complex functions. It suffices to replace $\mathbb{R}$ by $\mathbb{C}$ in eq. (5.20).

5.2.3 Germs of smooth functions

The local geometry of a manifold is studied by considering operations not on the space of smooth functions $\mathbb{C}^\infty(\mathcal{M})$ which, as just explained, can be very small, but on the space of germs of functions defined at each point $p \in \mathcal{M}$ that is always an infinite dimensional space.

Definition 5.2.8. Given a point $p \in \mathcal{M}$, the space of germs of smooth functions at p, denoted $\mathbb{C}^\infty_p$, is defined as follows. Consider all the open neighborhoods of p, namely, all the open subsets $U_p \subset \mathcal{M}$ such that $p \in U_p$. Consider the space of smooth functions $\mathbb{C}^\infty(U_p)$ on each U_p. Two functions $f \in \mathbb{C}^\infty(U_p)$ and $g \in \mathbb{C}^\infty(U'_p)$ are said to be equivalent if they coincide on the intersection $U_p \cap U'_p$ (see Figure 5.6):

$$f \sim g \quad \Leftrightarrow \quad f|_{U_p \cap U'_p} = g|_{U_p \cap U'_p} \tag{5.26}$$

The union of all the spaces $\mathbb{C}^\infty(U_p)$ modulo by the equivalence relation (5.26) is the space of germs of smooth functions at p:

$$\mathbb{C}^\infty_p \equiv \frac{\bigcup_{U_p} \mathbb{C}^\infty(U_p)}{\sim} \tag{5.27}$$

What underlies the above definition of germs is the familiar principle of analytic continuation. Of the same function we can have different definitions that have different domains of validity: apparently we have different functions but if they coincide

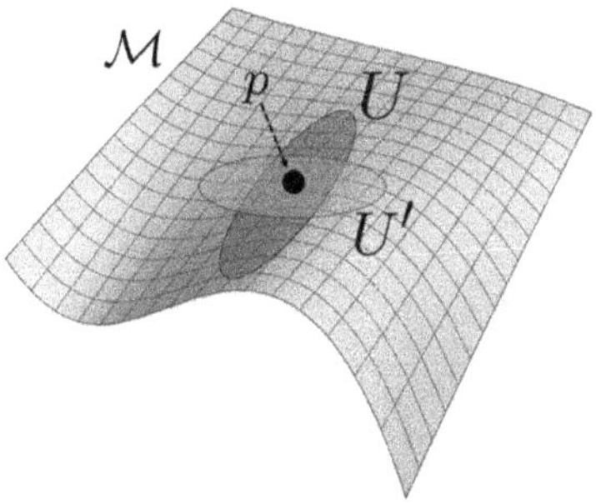

Figure 5.6: A germ of smooth function is the equivalence class of all locally defined functions that coincide in some neighborhood of a point p.

on some open region then we consider them just as different representations of a single function. Given any germ in some open neighborhood U_p we try to extend it to a larger domain by suitably changing its representation. In general there is a limit to such extension and only very special germs extend to globally defined functions on the whole manifold $\mathcal{M}$. For instance, the power series $\sum_{k\in\mathbb{N}} z^k$ defines a holomorphic function within its radius of convergence $|z| < 1$. As is known, within the convergence radius the sum of this series coincides with $1/(1-z)$ which is a holomorphic function defined on a much larger neighborhood of $z = 0$. According to our definition the two functions are equivalent and correspond to two different representatives of the same *germ*. The germ, however, does not extend to a holomorphic function on the whole Riemann sphere $\mathbb{C} \cup \infty$ since it has a singularity in $z = 1$. Indeed, as stated by Liouville theorem, the space of global holomorphic functions on the Riemann sphere contains only the constant function.

5.3 Tangent and cotangent spaces

In elementary geometry the notion of a *tangent line* is associated with the notion of a curve. Hence to introduce tangent vectors we have to begin with the notion of *curves in a manifold.*

Definition 5.3.1. A curve $\mathcal{C}$ in a manifold $\mathcal{M}$ is a continuous and differentiable map of an interval of the real line (say $[0,1] \subset \mathbb{R}$) onto $\mathcal{M}$:

$$\mathcal{C} : [0,1] \to \mathcal{M} \tag{5.28}$$

In other words, a curve is a one-dimensional submanifold $\mathcal{C} \subset \mathcal{M}$ (see Figure 5.7).

There are curves with a *boundary*, namely $\mathcal{C}(0) \cup \mathcal{C}(1)$, and open curves that do not contain their boundary. This happens if in eq. (5.28) we replace the closed interval $[0,1]$ with the open interval $]0,1[$. *Closed curves* or *loops* correspond to the case where the initial and final points coincide, that is, when $p_i \equiv \mathcal{C}(0) = \mathcal{C}(1) \equiv p_f$. Differently said:

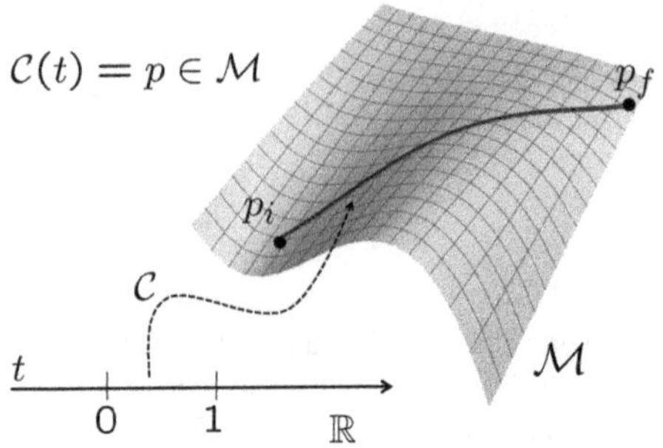

Figure 5.7: A curve in a manifold is a continuous map of an interval of the real line onto the manifold.

Definition 5.3.2. A closed curve is a continuous differentiable map of a circle onto the manifold:

$$\mathscr{C} : \mathbb{S}^1 \to \mathscr{M} \tag{5.29}$$

Indeed, identifying the initial and final point means to consider the points of the curve as being in one-to-one correspondence with the equivalence classes

$$\mathbb{R}/\mathbb{Z} \equiv \mathbb{S}^1 \tag{5.30}$$

which constitutes the mathematical definition of the circle. Explicitly eq. (5.30) means that two real numbers r and r' are declared to be equivalent if their difference $r' - r = n$ is an integer number $n \in \mathbb{Z}$. As representatives of these equivalence classes we have the real numbers contained in the interval $[0,1]$ with the proviso that $0 \sim 1$.

We can also consider *semi-open curves* corresponding to maps of the semi-open interval $[0,1[$ onto $\mathscr{M}$. In particular, in order to define tangent vectors we are interested in open branches of curves defined in the neighborhood of a point.

5.3.1 Tangent vectors at a point $p \in \mathscr{M}$

For each point $p \in \mathscr{M}$ let us fix an open neighborhood $U_p \subset \mathscr{M}$ and let us consider the semi-open curves of the following type:

$$\begin{cases} \mathscr{C}_p : [0,1[\to U_p \\ \mathscr{C}_p(0) = p \end{cases} \tag{5.31}$$

In other words, for each point p let us consider all possible curves $\mathscr{C}_p(t)$ that go through p (see Figure 5.8).

Intuitively the tangent in p to a curve that starts from p is the vector that specifies the curve's *initial* direction. The basic idea is that in an m-dimensional manifold there are as many directions in which the curve can depart as there are vectors in $\mathbb{R}^m$: furthermore, for sufficiently small neighborhoods of p we cannot tell the difference

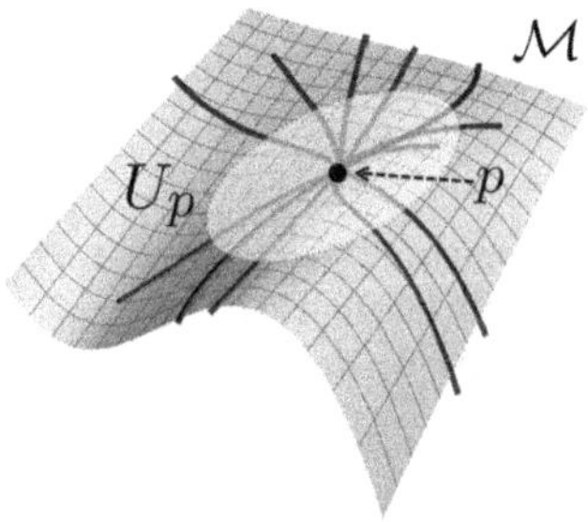

Figure 5.8: In a neighborhood U_p of each point $p \in \mathcal{M}$ we consider the curves that go through p.

between the manifold $\mathcal{M}$ and the flat vector space $\mathbb{R}^m$. Hence to each point $p \in \mathcal{M}$ of a manifold we can attach an m-dimensional real vector space

$$\forall p \in \mathcal{M}: \quad p \mapsto T_p\mathcal{M} \quad \dim T_p\mathcal{M} = m \tag{5.32}$$

which parameterizes the possible directions in which a curve starting at p can depart. This vector space is named the tangent space to $\mathcal{M}$ at the point p and is, by definition, isomorphic to $\mathbb{R}^m$, namely $T_p\mathcal{M} \sim \mathbb{R}^m$. For instance, to each point of an $\mathbb{S}^2$ sphere we attach a tangent plane $\mathbb{R}^2$ (see Figure 5.9).

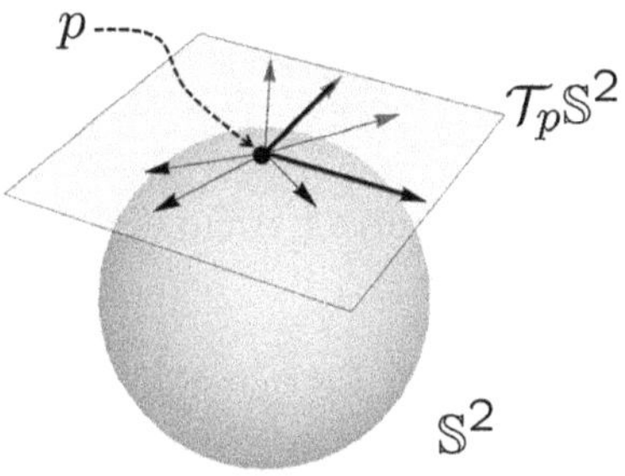

Figure 5.9: The tangent space in a generic point of an $\mathbb{S}^2$ sphere.

Let us now make this intuitive notion mathematically precise. Consider a point $p \in \mathcal{M}$ and a germ of a smooth function $f_p \in C_p^\infty(\mathcal{M})$. In any open chart $(U_\alpha, \varphi_\alpha)$ that contains the point p, the germ f_p is represented by an infinitely differentiable function of m variables:

$$f_p(x^1_{(\alpha)}, \dots, x^m_{(\alpha)}) \tag{5.33}$$

Let us now choose an open curve $\mathscr{C}_p(t)$ that lies in U_α and starts at p:

$$\mathscr{C}_p(t): \begin{cases} \mathscr{C}_p : [0,1[\to U_\alpha \\ \mathscr{C}_p(0) = p \end{cases} \tag{5.34}$$

and consider the composed map:

$$f_p \circ \mathscr{C}_p : [0,1[\subset \mathbb{R} \to \mathbb{R} \tag{5.35}$$

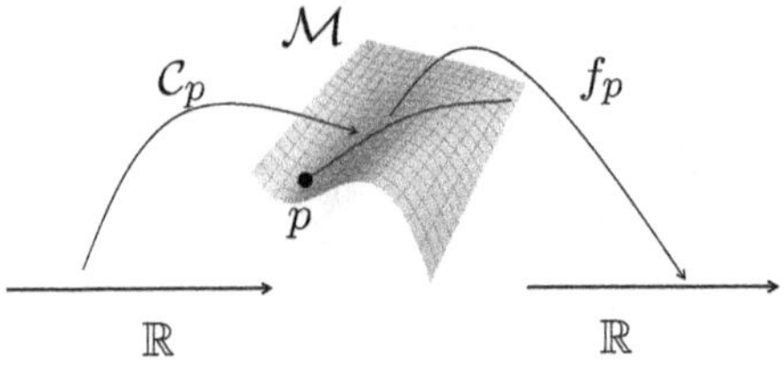

Figure 5.10: The composed map $f_p \circ \mathscr{C}_p$ where f_p is a germ of a smooth function in p and $\mathscr{C}_p$ is a curve departing from $p \in \mathscr{M}$.

which is a real function

$$f_p(\mathscr{C}_p(t)) \equiv g_p(t) \tag{5.36}$$

of one real variable (see Figure 5.10).

We can calculate its derivative with respect to t in $t = 0$ which, in the open chart $(U_\alpha, \varphi_\alpha)$, reads as follows:

$$\frac{d}{dt} g_p(t)\bigg|_{t=0} = \frac{\partial f_p}{\partial x^\mu} \cdot \frac{dx^\mu}{dt}\bigg|_{t=0} \tag{5.37}$$

We see from the above formula that the increment of any germ $f_p \in \mathbb{C}_p^\infty(\mathscr{M})$ along a curve $\mathscr{C}_p(t)$ is defined by means of the following m real coefficients:

$$c^\mu \equiv \frac{dx^\mu}{dt}\bigg|_{t=0} \in \mathbb{R} \tag{5.38}$$

which can be calculated whenever the parametric form of the curve is given: $x^\mu = x^\mu(t)$. Explicitly we have:

$$\frac{df_p}{dt} = c^\mu \frac{\partial f_p}{\partial x^\mu} \tag{5.39}$$

Equation (5.39) can be interpreted as the action of a differential operator on the space of germs of smooth functions, namely:

$$\mathbf{t}_p \equiv c^\mu \frac{\partial}{\partial x^\mu} \quad \Rightarrow \quad \mathbf{t}_p : \mathbb{C}_p^\infty(\mathscr{M}) \mapsto \mathbb{C}_p^\infty(\mathscr{M}) \tag{5.40}$$

Indeed, for any germ f and for any curve

$$\mathbf{t}_p f = \frac{dx^\mu}{dt}\bigg|_{t=0} \frac{\partial f}{\partial x^\mu} \in \mathbb{C}_p^\infty(\mathscr{M}) \tag{5.41}$$

is a new germ of a smooth function in the point p. This discussion justifies the mathematical definition of the tangent space:

Definition 5.3.3. The tangent space $T_p\mathscr{M}$ to the manifold $\mathscr{M}$ in the point p is the vector space of *first order differential operators* on the germs of smooth functions $\mathbb{C}_p^\infty(\mathscr{M})$.

Next let us observe that the space of germs $\mathbb{C}_p^\infty(\mathcal{M})$ is an *algebra* with respect to linear combinations with real coefficients $(\alpha f + \beta g)(p) = \alpha f(p) + \beta g(p)$ and pointwise multiplication $f \cdot g(p) \equiv f(p)g(p)$:

$$\begin{aligned} \forall \alpha, \beta \in \mathbb{R}\ \forall f, g \in \mathbb{C}_p^\infty(\mathcal{M}) & \qquad \alpha f + \beta g \in \mathbb{C}_p^\infty(\mathcal{M}) \\ \forall f, g \in \mathbb{C}_p^\infty(\mathcal{M}) & \qquad f \cdot g \in \mathbb{C}_p^\infty(\mathcal{M}) \\ & \qquad (\alpha f + \beta g) \cdot h = \alpha f \cdot h + \beta g \cdot h \end{aligned} \tag{5.42}$$

and a tangent vector $\mathbf{t}_p$ is a *derivation* of this algebra.

Definition 5.3.4. A *derivation* $\mathcal{D}$ of an algebra $\mathcal{A}$ is a map:

$$\mathcal{D} : \mathcal{A} \to \mathcal{A} \tag{5.43}$$

that

1. is linear

$$\forall \alpha, \beta \in \mathbb{R}\ \forall f, g \in \mathcal{A} : \quad \mathcal{D}(\alpha f + \beta g) = \alpha \mathcal{D} f + \beta \mathcal{D} g \tag{5.44}$$

2. obeys Leibnitz rule

$$\forall f, g \in \mathcal{A} : \quad \mathcal{D}(f \cdot g) = \mathcal{D} f \cdot g + f \cdot \mathcal{D} g \tag{5.45}$$

That tangent vectors fit into Definition 5.3.4 is clear from their explicit realization as differential operators (eqs (5.40), (5.41)). It is also clear that the set of *derivations* $D[\mathcal{A}]$ of an algebra constitutes a real vector space. Indeed, a linear combination of derivations is still a derivation, having set:

$$\forall \alpha, \beta \in \mathbb{R},\ \forall \mathcal{D}_1, \mathcal{D}_2 \in D[\mathcal{A}],\ \forall f \in \mathcal{A} : \quad (\alpha \mathcal{D}_1 + \beta \mathcal{D}_2) f = \alpha \mathcal{D}_1 f + \beta \mathcal{D}_2 f \tag{5.46}$$

Hence an equivalent and more abstract definition of the tangent space is the following:

Definition 5.3.5. The tangent space to a manifold $\mathcal{M}$ at the point p is the vector space of derivations of the algebra of germs of smooth functions in p:

$$T_p\mathcal{M} \equiv D[\mathbb{C}_p^\infty(\mathcal{M})] \tag{5.47}$$

Indeed, for any tangent vector (5.40) and for any pair of germs $f, g \in \mathbb{C}_p^\infty(\mathcal{M})$ we have:

$$\begin{aligned} \mathbf{t}_p(\alpha f + \beta g) &= \alpha \mathbf{t}_p(f) + \beta \mathbf{t}_p(g) \\ \mathbf{t}_p(f \cdot g) &= \mathbf{t}_p(f) \cdot g + f \cdot \mathbf{t}_p(g) \end{aligned} \tag{5.48}$$

In each coordinate patch a tangent vector is, as we have seen, a first-order differential operator singled out by its *components*, namely, by the coefficients c^μ. In the language of tensor calculus the tangent vector *is identified* with the m-tuplet of real numbers c^μ. The relevant point, however, is that such m-tuplet representing the *same tangent vector* is different in different coordinate patches. Consider two coordinate patches (U, φ) and (V, ψ) with non-vanishing intersection. Name x^μ the coordinate of a point $p \in U \cap V$ in the patch (U, φ) and y^α the coordinate of the same point in the patch (V, ψ). The transition function and its inverse are expressed by setting:

$$x^\mu = x^\mu(y); \quad y^\nu = y^\nu(x) \tag{5.49}$$

Then the same first-order differential operator can be alternatively written as:

$$\mathbf{t}_p = c^\mu \frac{\partial}{\partial x^\mu} \quad \text{or} \quad \mathbf{t}_p = c^\mu \left(\frac{\partial y^\nu}{\partial x^\mu} \right) \frac{\partial}{\partial y^\nu} = c^\nu \frac{\partial}{\partial y^\nu} \tag{5.50}$$

having defined:

$$c^\nu \equiv c^\mu \left(\frac{\partial y^\nu}{\partial x^\mu} \right) \tag{5.51}$$

Equation (5.51) expresses the transformation rule for the components of a tangent vector from one coordinate patch to another one (see Figure 5.11).

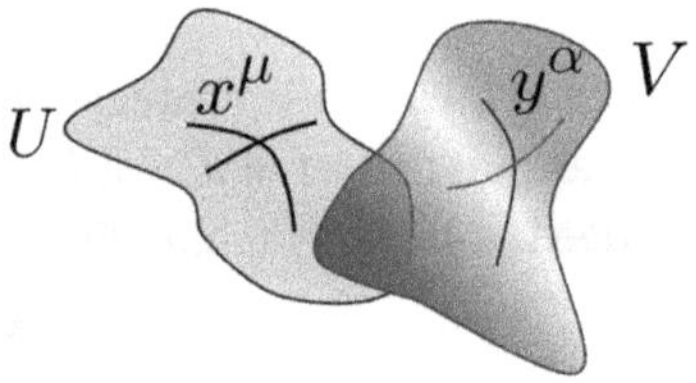

Figure 5.11: Two coordinate patches.

Such a transformation is *linear* and the matrix that realizes it is the *inverse of the Jacobian matrix* $(\partial y / \partial x) = (\partial x / \partial y)^{-1}$. For this reason we say that the components of a tangent vector constitute a *contravariant world vector*. By definition a *covariant world vector* transforms instead with the *Jacobian matrix*. We will see that covariant world vectors are the components of a differential form.

5.3.2 Differential forms at a point $p \in \mathcal{M}$

Let us now consider the total differential of a function (better of a germ of a smooth function) when we evaluate it along a curve. $\forall f \in \mathbb{C}_p^\infty(\mathcal{M})$ and for each curve $c(t)$ starting at p we have:

$$\frac{d}{dt} f(c(t)) \Big|_{t=0} = c^\mu \frac{\partial}{\partial x^\mu} f \equiv \mathbf{t}_p f \tag{5.52}$$

where we have named $\mathbf{t}_p = \frac{dc^\mu}{dt}|_{t=0}\frac{\partial}{\partial x^\mu}$ the tangent vector to the curve in its initial point p. So, fixing a tangent vector means that for any germ f we know its total differential along the curve that admits such a vector as tangent in p. Let us now reverse our viewpoint. Rather than keeping the tangent vector fixed and letting the germ f vary, let us keep the germ f fixed and let us consider all possible curves that depart from the point p. We would like to evaluate the total derivative of the germ $\frac{df}{dt}$ along each curve. The solution of such a problem is easily obtained: given the tangent vector $\mathbf{t}_p$ to the curve in p we have $df/dt = \mathbf{t}_p f$. The moral of this tale is the following: the concept of *total differential of a germ* is the *dual* of the concept of *tangent vector*. Indeed, we recall from linear algebra that the dual of a vector space is the space of linear functionals on that vector space and our discussion shows that the total differential of a germ is precisely a linear functional on the tangent space $T_p\mathcal{M}$.

Definition 5.3.6. The total differential df_p of a smooth germ $f \in \mathbb{C}_p^\infty(\mathcal{M})$ is a *linear functional* on $T_p\mathcal{M}$ such that

$$\begin{aligned} &\forall \mathbf{t}_p \in T_p\mathcal{M} && df_p(\mathbf{t}_p) = \mathbf{t}_p f \\ &\forall \mathbf{t}_p, \mathbf{k}_p \in T_p\mathcal{M},\ \forall \alpha,\beta \in \mathbb{R} && df_p(\alpha\mathbf{t}_p + \beta\mathbf{k}_p) = \alpha\, df_p(\mathbf{t}_p) + \beta\, df_p(\mathbf{k}_p) \end{aligned} \tag{5.53}$$

The linear functionals on a finite dimensional vector space $\mathcal{V}$ constitute a vector space $\mathcal{V}^\star$ (the dual) with the same dimension. This justifies the following

Definition 5.3.7. We name *cotangent space* to the manifold $\mathcal{M}$ in the point p the vector space $T_p^*\mathcal{M}$ of linear functionals (or 1-forms in p) on the tangent space $T_p\mathcal{M}$:

$$T_p^*\mathcal{M} \equiv \mathrm{Hom}(T_p\mathcal{M}, \mathbb{R}) = (T_p\mathcal{M})^\star \tag{5.54}$$

So we name differential 1-forms in p the elements of the cotangent space and $\forall \omega_p \in T_p^*\mathcal{M}$ we have:

$$\begin{aligned} &1) && \forall \mathbf{t}_p \in T_p\mathcal{M}: && \omega_p(\mathbf{t}_p) \in \mathbb{R} \\ &2) && \forall \alpha,\beta \in \mathbb{R}, \forall \mathbf{t}_p, \mathbf{k}_p \in T_p\mathcal{M}: && \omega_p(\alpha\mathbf{t}_p + \beta\mathbf{k}_p) = \alpha\omega_p(\mathbf{t}_p) + \beta\omega_p(\mathbf{k}_p) \end{aligned} \tag{5.55}$$

The reason why the above linear functionals are named differential 1-forms is that in every coordinate patch $\{x^\mu\}$ they can be expressed as linear combinations of the coordinate differentials:

$$\omega_p = \omega_\mu\, dx^\mu \tag{5.56}$$

and their action on the tangent vectors is expressed as follows:

$$\mathbf{t}_p = c^\mu \frac{\partial}{\partial x^\mu} \quad \Rightarrow \quad \omega_p(\mathbf{t}_p) = \omega_\mu c^\mu \in \mathbb{R} \tag{5.57}$$

Indeed, in the particular case where the 1-form is exact (namely, it is the differential of a germ) $\omega_p = df_p$ we can write $\omega_p = \partial f / \partial x^\mu \, dx^\mu$ and we have $df_p(\mathbf{t}_p) \equiv \mathbf{t}_p f = c^\mu \partial f / \partial x^\mu$. Hence when we extend our definition to differential forms that are not exact we continue to state the same statement, namely, that the value of the 1-form on a tangent vector is given by eq. (5.57).

Summarizing, in each coordinate patch, a differential 1-form in a point $p \in \mathcal{M}$ has the representation (5.56) and its coefficients ω_μ constitute a *contravariant vector*. Indeed, in complete analogy to eq. (5.50), we have

$$\omega_p = \omega_\mu \, dx^\mu \quad \text{or} \quad \omega_p = \omega_\mu \left(\frac{\partial x^\mu}{\partial y^\nu} \right) dy^\nu = \omega_\nu \, dy^\nu \tag{5.58}$$

having defined:

$$\omega_\nu \equiv \omega_\mu \left(\frac{\partial x^\mu}{\partial y^\nu} \right) \tag{5.59}$$

Finally the duality relation between 1-forms and tangent vectors can be summarized writing the rule:

$$dx^\mu \left(\frac{\partial}{\partial x^\nu} \right) = \delta^\mu_\nu \tag{5.60}$$

5.4 About the concept of fiber-bundle

The next step we have to take is *gluing together* all the tangent $T_p\mathcal{M}$ and cotangent spaces $T_p^*\mathcal{M}$ we have discussed in the previous sections. The result of such a gluing procedure is not a vector space, rather it is a vector bundle. Vector bundles are specific instances of the more general notion of *fiber-bundles*.

The concept of *fiber-bundle* is absolutely central in contemporary physics and provides the appropriate mathematical framework to formulate modern field theory since all the fields one can consider are either *sections* of *associated bundles* or *connections* on *principal bundles*. There are two kinds of fiber-bundles:
1. principal bundles
2. associated bundles

The notion of a principal fiber-bundle is the appropriate mathematical concept underlying the formulation of *gauge theories* that provide the general framework to describe the dynamics of all non-gravitational interactions. The concept of a connection on such principal bundles codifies the physical notion of the bosonic particles mediating the interaction, namely the gauge bosons, like the photon, the gluon or the graviton. Indeed, gravity itself is a gauge theory although of a very special type. On the other hand, the notion of associated fiber-bundles is the appropriate mathematical framework to describe *matter fields* that interact through the exchange of the *gauge bosons*.

Also from a more general viewpoint and in relation with all sort of applications the notion of fiber-bundles is absolutely fundamental. As we already emphasized, the points of a manifold can be identified with the possible states of a complex system specified by an m-tuplet of parameters $x_1, \dots, x_m$. Real or complex functions of such parameters are the natural objects one expects to deal with in any scientific theory that explains the phenomena observed in such a system. Yet, as we already anticipated, calculus on manifolds that are not trivial as the flat $\mathbb{R}^m$ cannot be confined to functions, which is a too restrictive notion. The appropriate generalization of functions is provided by the *sections* of fiber-bundles. Locally, namely in each coordinate patch, functions and sections are just the same thing. Globally, however, there are essential differences. A section is obtained by gluing together many local functions by means of non-trivial transition functions that reflect the geometric structure of the fiber-bundle.

5.5 The notion of Lie group

To introduce the mathematical definition of a fiber-bundle we need to introduce the definition of a Lie group which will be a central topic in the sequel.

Definition 5.5.1. A Lie group G is:

- A group from the algebraic point of view, namely, a set with an internal composition law, the product

$$\forall g_1 g_2 \in G \quad g_1 \cdot g_2 \in G \tag{5.61}$$

  which is associative, admits a unique neutral element e and yields an inverse for each group element.
- A smooth manifold of finite dimension $\dim G = n < \infty$ whose transition functions are not only infinitely differentiable but also real analytic, namely, they admit an expansion in power series.
- In the topology defined by the manifold structure, the two algebraic operations of taking the inverse of an element and performing the product of two elements are real analytic (admit a power series expansion).

The last point in Definition 5.5.1 deserves a more extended explanation. To each group element the product operation associates two maps of the group onto itself:

$$\begin{aligned} \forall g \in G: \quad & L_g : G \to G : g' \to L_g(g') \equiv g' \cdot g \\ \forall g \in G: \quad & R_g : G \to G : g' \to R_g(g') \equiv g \cdot g' \end{aligned} \tag{5.62}$$

respectively named the *left translation* and the *right translation*. Both maps are required to be real analytic for each choice of $g \in G$. Similarly the group structure induces a map:

$$(\cdot)^{-1} : G \to G : g \to g^{-1} \tag{5.63}$$

which is also required to be real analytic.

5.6 Developing the notion of fiber-bundle

Coming now to fiber-bundles let us begin by recalling that a pedagogical and pictorial example of such spaces is provided by the celebrated case of a Möbius strip (see Figure 5.12).

The basic idea is that if we consider a piece of the bundle, this cannot be distinguished from a trivial direct product of two spaces, an open subset of the base manifold and the fiber. In Figure 5.12 the base manifold is a circle and the fiber is a segment $I \equiv [-1, 1]$. Locally the space is the direct product of an open interval of $U =]a, b[\subset \mathbb{R}$ with the standard fiber I, as it is evident from Figure 5.13. However, the relevant point is that, *globally*, the bundle *is not a direct product of spaces*.

Hence the notion of fiber-bundle corresponds to that of a differentiable manifold P with dimension $\dim P = m + n$ that locally *looks like* the direct product $U \times F$ of an open manifold U of dimension $\dim U = m$ with another manifold F (the standard fiber) of dimension $\dim F = n$. Essential in the definition is the existence of a map:

$$\pi : P \to \mathcal{M} \tag{5.64}$$

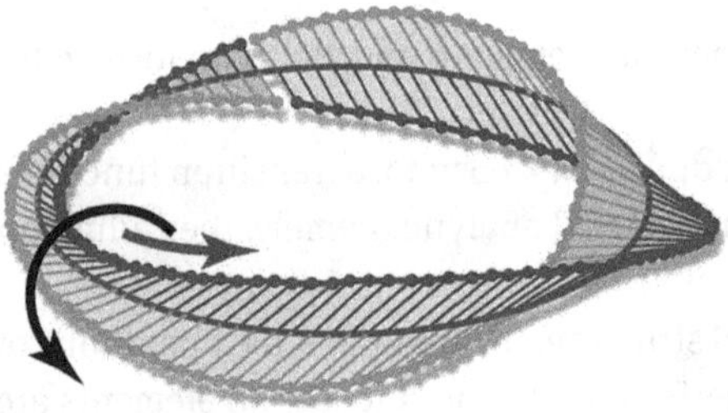

Figure 5.12: Möbius strip provides a pedagogical example of a fiber-bundle.

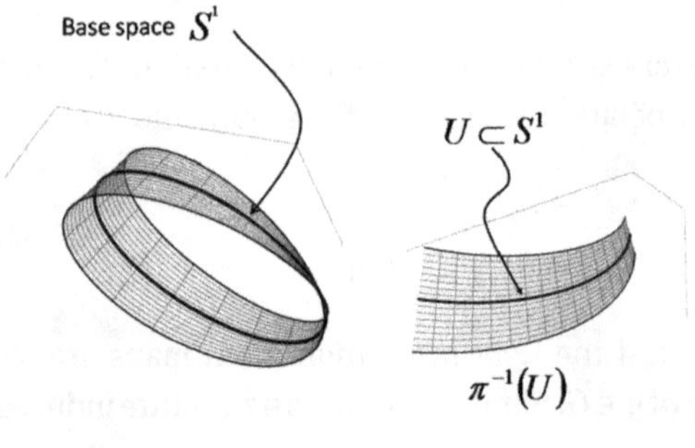

Figure 5.13: Local triviality of an open piece of the Möbius strip.

named *the projection* from the *total manifold* P of dimension $m+n$ to a manifold $\mathcal{M}$ of dimension m, named the *base manifold*. Such a map is required to be continuous. Due to the difference in dimensions the projection cannot be invertible. Indeed, to every point $\forall p \in \mathcal{M}$ of the base manifold the projection associates a submanifold $\pi^{-1}(p) \subset P$ of dimension $\dim \pi^{-1}(p) = n$ composed by those points of $x \in P$ whose projection on $\mathcal{M}$ is the chosen point p: $\pi(x) = p$. The submanifold $\pi^{-1}(p)$ is named the *fiber over p* and the basic idea is that each fiber is homeomorphic to the *standard fiber F*. More precisely, for each open subset $U_\alpha \subset \mathcal{M}$ of the base manifold we must have that the submanifold

$$\pi^{-1}(U_\alpha)$$

is homeomorphic to the direct product

$$U_\alpha \times F$$

This is the precise meaning of the statement that, locally, the bundle looks like a direct product (see Figure 5.14). Explicitly, what we require is the following: there should be a family of pairs (U_α, ϕ_α) where U_α are open charts covering the base manifold $\bigcup_\alpha U_\alpha = \mathcal{M}$ and ϕ_α are maps:

$$\phi_\alpha : \pi^{-1}(U_\alpha) \subset P \to U_\alpha \otimes F \tag{5.65}$$

that are required to be one-to-one, bicontinuous (= continuous, together with its inverse) and to satisfy the property that:

$$\pi \circ \phi_\alpha^{-1}(p, f) = p \tag{5.66}$$

Namely, the projection of the image in P of a base manifold point p *times* some fiber point f is p itself.

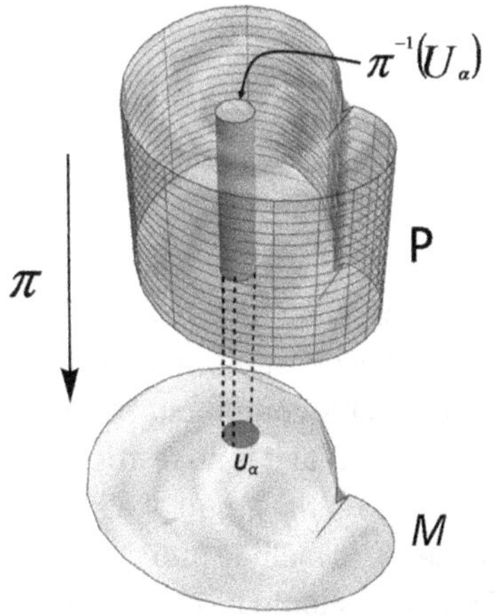

Figure 5.14: A fiber-bundle is locally trivial.

Each pair (U_α, ϕ_α) is named a *local trivialization*. As for the case of manifolds, the interesting question is what happens in the intersection of two different local trivializations. Indeed, if $U_\alpha \cap U_\beta \neq \varnothing$, then we also have $\pi^{-1}(U_\alpha) \cap \pi^{-1}(U_\beta) \neq \varnothing$. Hence each point $x \in \pi^{-1}(U_\alpha \cap U_\beta)$ is mapped by ϕ_α and ϕ_β in two different pairs $(p, f_\alpha) \in U_\alpha \otimes F$ and $(p, f_\beta) \in U_\alpha \otimes F$ with the property, however, that the first entry p is the same in both pairs. This follows from property (5.66). It implies that there must exist a map:

$$t_{\alpha\beta} \equiv \phi_\beta^{-1} \circ \phi_\alpha : (U_\alpha \cap U_\beta) \otimes F \to (U_\alpha \cap U_\beta) \otimes F \tag{5.67}$$

named *transition function*, which acts exclusively on the fiber points in the sense that:

$$\forall p \in U_\alpha \cap U_\beta,\ \forall f \in F \quad t_{\alpha\beta}(p, f) = (p, t_{\alpha\beta}(p).f) \tag{5.68}$$

where for each choice of the point $p \in U_\alpha \cap U_\beta$,

$$t_{\alpha\beta}(p) : F \mapsto F \tag{5.69}$$

is a continuous and invertible map of the standard fiber F onto itself (see Figure 5.15).

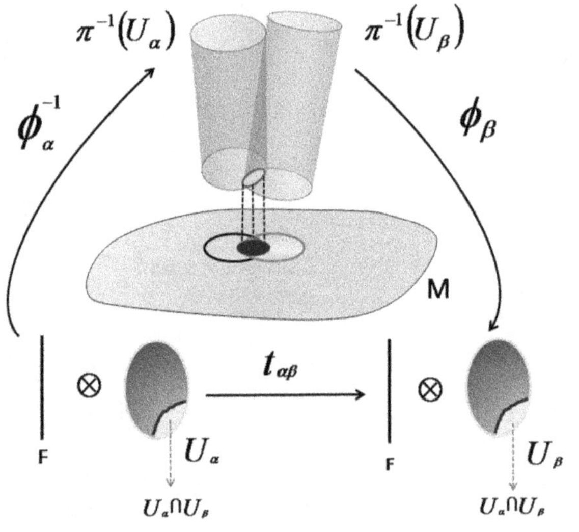

Figure 5.15: Transition function between two local trivializations of a fiber-bundle.

The last bit of information contained in the notion of fiber-bundle is related to the *structural group*. This has to do with answering the following question: Where are the transition functions chosen from? Indeed, the set of all possible continuous invertible maps of the standard fiber F onto itself constitute a group, so that it is no restriction to say that the transition functions $t_{\alpha\beta}(p)$ are group elements. Yet the group of all homeomorphisms $\mathrm{Hom}(F, F)$ is very very large and it makes sense to include into the definition of fiber-bundle the request that the transition functions should be chosen within a smaller hunting ground, namely, inside some finite dimensional Lie group G that has a well-defined action on the standard fiber F.

The above discussion can be summarized into the following technical definition of fiber-bundles.

Definition 5.6.1. A fiber-bundle $(P, \pi, \mathcal{M}, F, G)$ is a geometrical structure that consists of the following list of elements:

1. A differentiable manifold P named the **total space.**
2. A differentiable manifold $\mathcal{M}$ named the **base space.**
3. A differentiable manifold F named the **standard fiber.**
4. A Lie group G, named the **structure group,** which acts as a transformation group on the standard fiber:

$$\forall g \in G; \quad g : F \longrightarrow F \quad \{\text{i.e. } \forall f \in F\; g.f \in F\} \tag{5.70}$$

5. A surjection map $\pi \; : \; P \longrightarrow \mathcal{M}$, named the **projection.** If $n = \dim \mathcal{M}$, $m = \dim F$, then we have $\dim P = n + m$ and $\forall p \in \mathcal{M}$, $F_p = \pi^{-1}(p)$ is an m-dimensional manifold diffeomorphic to the standard fiber F. The manifold F_p is named the **fiber at the point** p.
6. A covering of the base space $\bigcup_{(\alpha \in A)} U_\alpha = \mathcal{M}$, realized by a collection $\{U_\alpha\}$ of open subsets $(\forall \alpha \in A\; U_\alpha \subset \mathcal{M})$, equipped with a homeomorphism:

$$\phi_\alpha^{-1} : U_\alpha \times F \longrightarrow \pi^{-1}(U_\alpha) \tag{5.71}$$

such that

$$\forall p \in U_\alpha, \forall f \in F : \quad \pi \cdot \phi_\alpha^{-1}(p,f) = p \tag{5.72}$$

The map ϕ_α^{-1} is named a **local trivialization** of the bundle, since its inverse ϕ_α maps the open subset $\pi^{-1}(U_\alpha) \subset P$ of the total space onto the direct product $U_\alpha \times F$.
7. If we write $\phi_\alpha^{-1}(p,f) = \phi_{\alpha,p}^{-1}(f)$, the map $\phi_{\alpha,p}^{-1} : F \longrightarrow F_p$ is the homeomorphism required by point 6 of the present definition. For all points $p \in U_\alpha \cap U_\beta$ in the intersection of two different local trivialization domains, the composite map $t_{\alpha\beta}(p) = \phi_{\alpha,p} \cdot \phi_{\beta,p}^{-1} F \longrightarrow F$ is an element of the structure group $t_{\alpha\beta} \in G$, named the **transition function.** Furthermore, the transition function realizes a smooth map $t_{\alpha\beta} : U_\alpha \cap U_\beta \longrightarrow G$. We have

$$\phi_\beta^{-1}(p,f) = \phi_\alpha^{-1}(p, t_{\alpha\beta}(p).f) \tag{5.73}$$

Just as manifolds can be constructed by gluing together open charts, fiber-bundles can be obtained by gluing together local trivializations. Explicitly one proceeds as follows.

1. First choose a base manifold $\mathcal{M}$, a typical fiber F and a structural Lie group G whose action on F must be well-defined.
2. Then choose an atlas of open neighborhoods $U_\alpha \subset \mathcal{M}$ covering the base manifold $\mathcal{M}$.
3. Next to each non-vanishing intersection $U_\alpha \cap U_\beta \neq \emptyset$ assign a transition function, namely, a smooth map:

$$\psi_{\alpha\beta} : U_\alpha \cap U_\beta \mapsto G \tag{5.74}$$

   from the open subset $U_\alpha \cap U_\beta \subset \mathcal{M}$ of the base manifold to the structural Lie group. For consistency the transition functions must satisfy the two conditions:

$$\begin{aligned} &\forall U_\alpha, U_\beta \,/\, U_\alpha \cap U_\beta \neq \emptyset : && \psi_{\beta\alpha} = \psi_{\alpha\beta}^{-1} \\ &\forall U_\alpha, U_\beta, U_\gamma \,/\, U_\alpha \cap U_\beta \cap U_\gamma \neq \emptyset : && \psi_{\alpha\beta} \cdot \psi_{\beta\gamma} \cdot \psi_{\gamma\alpha} = \mathbf{1}_G \end{aligned} \tag{5.75}$$

Whenever a set of local trivializations with consistent transition functions satisfying eq. (5.75) has been given, a fiber-bundle is defined. A different and much more difficult question to answer is to decide whether two sets of local trivializations define the same fiber-bundle or not. We do not address such a problem whose proper treatment is beyond the scope of this textbook. We just point out that the classification of inequivalent fiber-bundles one can construct on a given base manifold $\mathcal{M}$ is a problem of global geometry which can also be addressed with the techniques of algebraic topology and algebraic geometry.

Typically inequivalent bundles are characterized by topological invariants that receive the name of *characteristic classes*.

In physical language the transition functions (5.74) from one local trivialization to another one are the *gauge transformations*, namely, group transformations depending on the position in space-time (*i.e.*, the point on the base manifold).

Definition 5.6.2. A principal bundle $P(\mathcal{M}, G)$ is a fiber-bundle where the standard fiber coincides with the structural Lie group $F = G$ and the action of G on the fiber is the left (or right) multiplication (see eq. (5.62)):

$$\forall g \in G \quad \Rightarrow \quad L_g : G \mapsto G \tag{5.76}$$

The name principal is given to the fiber-bundle in Definition 5.6.2 since it is a "*father*" bundle which, once given, generates an infinity of *associated vector bundles*, one for each linear representation of the Lie group G.

Let us anticipate the notion of linear representations of a Lie group that will be extensively discussed in Chapter 8.

Definition 5.6.3. Let V be a vector space of finite dimension $\dim V = m$ and let $\mathrm{Hom}(V,V)$ be the group of all linear homomorphisms of the vector space into itself:

$$\begin{aligned} & f \in \mathrm{Hom}(V,V) \quad / \quad f : V \to V \\ & \forall \alpha, \beta \in \mathbb{R}\; \forall v_1, v_2 \in V : \quad f(\alpha \mathbf{v}_1 + \beta \mathbf{v}_2) = \alpha f(\mathbf{v}_1) + \beta f(\mathbf{v}_2) \end{aligned} \tag{5.77}$$

A linear representation of the Lie group G of dimension n is a *group homomorphism*:

$$\begin{cases} \forall g \in G & g \mapsto D(g) \in \mathrm{Hom}(V,V) \\ \forall g_1 g_2 \in G & D(g_1 \cdot g_2) = D(g_1) \cdot D(g_2) \\ & D(e) = \mathbf{1} \\ \forall g \in G & D(g^{-1}) = [D(g)]^{-1} \end{cases} \tag{5.78}$$

Whenever we choose a basis $\mathbf{e}_1, \mathbf{e}_2, \dots, \mathbf{e}_n$ of the vector space V, every element $f \in \mathrm{Hom}(V,V)$ is represented by a matrix $f^j_{\ i}$ defined by:

$$f(\mathbf{e}_i) = \mathbf{e}_j f^j_{\ i} \tag{5.79}$$

Therefore a linear representation of a Lie group associates to each abstract group element g an $n \times n$ matrix $D(g)_i^{\ j}$. As it should be known to the reader, linear representations are said to be *irreducible* if the vector space V admits *no* non-trivial vector subspace $W \subset V$ that is *invariant* with respect to the action of the group: $\forall g \in G / D(g)W \subset W$ (see Section 8.1). For simple Lie groups reducible representations can always be decomposed into a direct sum of irreducible representations, namely, $V = V_1 \oplus V_2 \oplus \cdots \oplus V_r$ (with V_i irreducible) and irreducible representations are completely defined by the structure of the group. These notions that we have recalled from group theory motivate the definition:

Definition 5.6.4. An *associated vector bundle* is a fiber-bundle where the standard fiber $F = V$ is a vector space and the action of the structural group on the standard fiber is a linear representation of G on V.

The reason why the bundles in Definition 5.6.4 are named associated is almost obvious. Given a principal bundle and a linear representation of G we can immediately construct a corresponding vector bundle. It suffices to use as transition functions the linear representation of the transition functions of the principal bundle:

$$\psi^{(V)}_{\alpha\beta} \equiv D(\psi^{(G)}_{\alpha\beta}) \in \mathrm{Hom}(V,V) \tag{5.80}$$

For any vector bundle the dimension of the standard fiber is named the *rank* of the bundle.

Whenever the base-manifold of a fiber-bundle is complex and the transition functions are holomorphic maps, we say that the bundle is *holomorphic*.

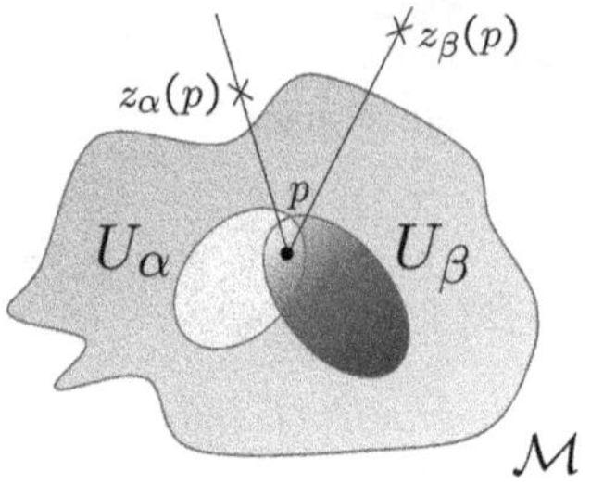

Figure 5.16: The intersection of two local trivializations of a line bundle.

A very important and simple class of holomorphic bundles are the *line bundles*. By definition these are principal bundles on a complex base manifold $\mathcal{M}$ with structural group $\mathbb{C}^\star \equiv \mathbb{C}\backslash 0$, namely, the multiplicative group of non-zero complex numbers.

Let $z_\alpha(p) \in \mathbb{C}^\star$ be an element of the standard fiber above the point $p \in U_\alpha \cap U_\beta \subset \mathcal{M}$ in the local trivialization α and let $z_\beta(p) \in \mathbb{C}^\star$ be the corresponding fiber point in the local trivialization β (see Figure 5.16). The transition function between the two trivialization is expressed by:

$$z_\alpha(p) = \underbrace{f_{\alpha\beta}(p)}_{\in \mathbb{C}^\star} \cdot z_\beta(p) \quad \Rightarrow \quad f_{\alpha\beta}(p) = \frac{z_\alpha(p)}{z_\beta(p)} \neq 0 \tag{5.81}$$

5.7 Tangent and cotangent bundles

Let $\mathcal{M}$ be a differentiable manifold of dimension $\dim \mathcal{M} = m$: in Section 5.3 we have seen how to construct the tangent spaces $T_p\mathcal{M}$ associated with each point $p \in \mathcal{M}$ of the manifold. We have also seen that each $T_p\mathcal{M}$ is a real vector space isomorphic to $\mathbb{R}^m$. Considering the definition of fiber-bundles discussed in the previous section we now realize that what we actually did in Section 5.3 was to construct a vector-bundle, the *tangent bundle* $T\mathcal{M}$ (see Figure 5.17).

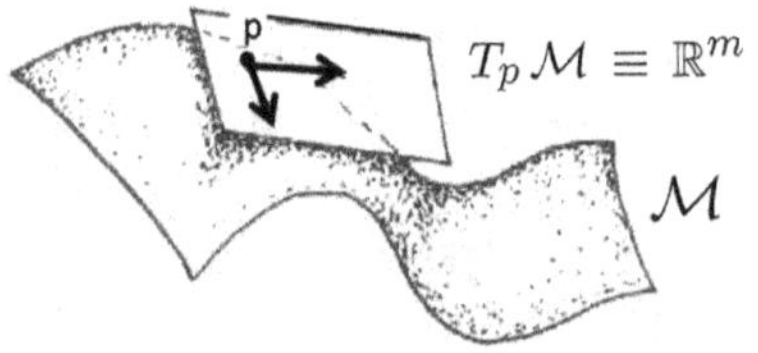

Figure 5.17: The tangent bundle is obtained by gluing together all the tangent spaces.

In the tangent bundle $T\mathcal{M}$ the *base manifold* is the differentiable manifold $\mathcal{M}$, the *standard fiber* is $F = \mathbb{R}^m$ and the structural group is $\mathrm{GL}(m, \mathbb{R})$, namely, the group of real $m \times m$ matrices. The main point is that the transition functions are not newly introduced to construct the bundle, rather they are completely determined from the transition functions relating open charts of the base manifold. In other words, whenever we define a manifold $\mathcal{M}$, associated with it there is a unique vector bundle $T\mathcal{M} \to \mathcal{M}$ which encodes many intrinsic properties of $\mathcal{M}$. Let us see how.

Consider two intersecting local charts (U_α, ϕ_α) and (U_β, ϕ_β) of our manifold. A tangent vector, in a point $p \in \mathcal{M}$, was written as:

$$\mathbf{t}_p = c^\mu(p) \left.\frac{\partial}{\partial x^\mu}\right|_p \tag{5.82}$$

Now we can consider choosing smoothly a tangent vector for each point $p \in \mathcal{M}$, namely, introducing a map:

$$p \in \mathcal{M} \mapsto \mathbf{t}_p \in T_p\mathcal{M} \tag{5.83}$$

Mathematically what we have obtained is a *section of the tangent bundle*, namely, a smooth choice of a point in the fiber for each point of the base. Explicitly this just means that the components $c^\mu(p)$ of the tangent vector are smooth functions of the base point coordinates x^μ. Since we use coordinates, we need an extra label denoting in which local patch the vector components are given:

$$\begin{cases} \mathbf{t} = c^\mu_{(\alpha)}(x) \left.\dfrac{\partial}{\partial x^\mu}\right|_p & \Rightarrow \quad \text{in chart } \alpha \\ \mathbf{t} = c^\nu_{(\beta)}(y) \left.\dfrac{\partial}{\partial y^\nu}\right|_p & \Rightarrow \quad \text{in chart } \beta \end{cases} \tag{5.84}$$

having denoted x^μ and y^ν the local coordinates in patches α and β, respectively. Since the tangent vector is the same, irrespectively of the coordinates used to describe it, we have:

$$c^\nu_{(\beta)}(y) \frac{\partial}{\partial y^\nu} = c^\mu_{(\alpha)}(x) \frac{\partial y^\nu}{\partial x^\mu} \frac{\partial}{\partial y^\nu} \tag{5.85}$$

namely:

$$c^\nu_{(\beta)}(p) = c^\mu_{(\alpha)}(p) \left(\frac{\partial y^\nu}{\partial x^\mu}\right)(p) \tag{5.86}$$

In formula (5.86) we see the explicit form of the transition function between two local trivializations of the tangent bundle: it is simply the *inverse Jacobian matrix* associated with the transition functions between two local charts of the base manifold $\mathcal{M}$. On the intersection $U_\alpha \cap U_\beta$ we have:

$$\forall p \in U_\alpha \cap U_\beta : \quad p \to \psi_{\beta\alpha}(p) = \left(\frac{\partial y}{\partial x}\right)(p) \in \mathrm{GL}(m, \mathbb{R}) \tag{5.87}$$

as it is pictorially described in Figure 5.18.

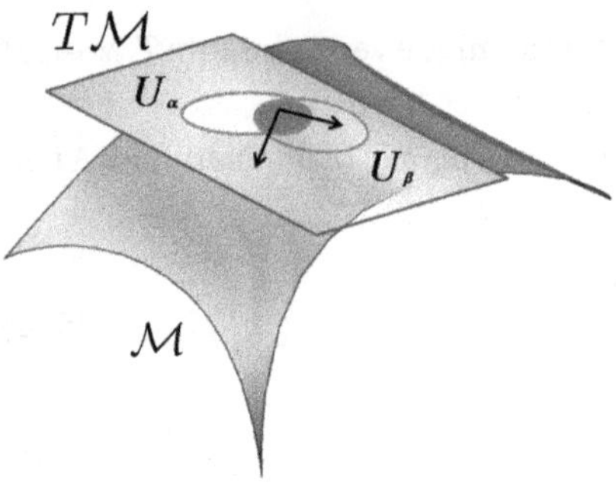

Figure 5.18: Two local charts of the base manifold $\mathscr{M}$ yield two local trivializations of the tangent bundle $T\mathscr{M}$. The transition function maps a vector onto another one.

5.7.1 Sections of a bundle

It is now the appropriate time to associate a precise definition to the notion of bundle section that we have implicitly advocated in eq. (5.83).

Definition 5.7.1. Consider a generic fiber-bundle $E \overset{\pi}{\longrightarrow} \mathscr{M}$ with generic fiber F. We name **section of the bundle** a rule s that to each point $p \in \mathscr{M}$ of the base manifold associates a point $s(p) \in F_p$ in the fiber above p, namely a map

$$s : \mathscr{M} \mapsto E \tag{5.88}$$

such that:

$$\forall p \in \mathscr{M} : s(p) \in \pi^{-1}(p) \tag{5.89}$$

The above definition is illustrated in Figure 5.19 which also clarifies the intuitive idea standing behind the chosen name for such a concept.

It is clear that sections of the bundle can be chosen to be *continuous, differentiable, smooth* or, in the case of complex manifolds, even *holomorphic*, depending on the properties of the map s in each local trivialization of the bundle. Indeed, given a

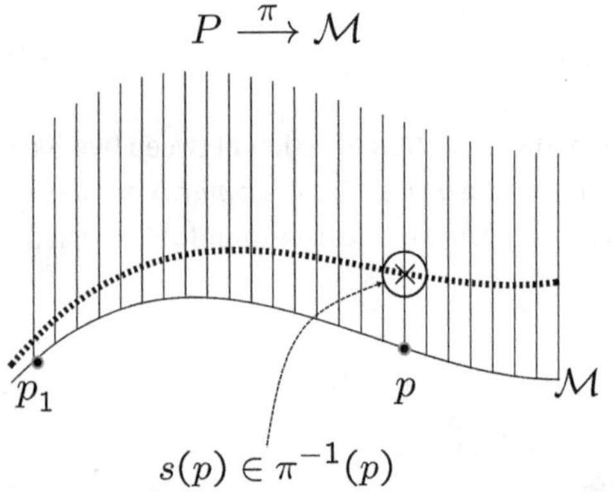

Figure 5.19: A section of a fiber-bundle.

local trivialization and given open charts for both the base manifold $\mathcal{M}$ and for the fiber F, the local description of the section reduces to a map:

$$\mathbb{R}^m \supset U \mapsto F_U \subset \mathbb{R}^n \tag{5.90}$$

where m and n are the dimensions of the base manifold and of the fiber, respectively.

We are specifically interested in smooth sections, namely, in section that are infinitely differentiable. Given a bundle $E \xrightarrow{\pi} \mathcal{M}$, the set of all such sections is denoted by:

$$\Gamma(E, \mathcal{M}) \tag{5.91}$$

Of particular relevance are the smooth sections of vector bundles. In this case to each point of the base manifold p we associate a vector $\mathbf{v}(p)$ in the vector space above the point p. In particular we can consider sections of the tangent bundle $T\mathcal{M}$ associated with a smooth manifold M. Such sections correspond to the notion of *vector fields.*

Definition 5.7.2. Given a smooth manifold $\mathcal{M}$, we name **vector field** on $\mathcal{M}$ a smooth section $\mathbf{t} \in \Gamma(T\mathcal{M}, \mathcal{M})$ of the tangent bundle. The local expression of such vector field in any open chart (U, ϕ) is

$$\mathbf{t} = t^\mu(x) \frac{\vec{\partial}}{\partial x^\mu} \quad \forall x \in U \subset \mathcal{M} \tag{5.92}$$

5.7.1.1 Example: holomorphic vector fields on $\mathbb{S}^2$

As we have seen above, the 2-sphere $\mathbb{S}^2$ is a complex one-dimensional manifold covered by an atlas composed by two charts, that of the North Pole and that of the South Pole (see Figure 5.3) and the transition function between the local complex coordinate in the two patches is the following one:

$$z_N = \frac{1}{z_S} \tag{5.93}$$

Correspondingly, in the two patches, the local description of a holomorphic vector field $\mathbf{t}$ is given by:

$$\begin{aligned} \mathbf{t} &= v_N(z_N) \frac{d}{dz_N} \\ \mathbf{t} &= v_S(z_S) \frac{d}{dz_S} \end{aligned} \tag{5.94}$$

where the two functions $v_N(z_N)$ and $v_S(z_S)$ are supposed to be holomorphic functions of their argument, namely, to admit a Taylor power series expansion:

$$\begin{aligned} v_N(z_N) &= \sum_{k=0}^{\infty} c_k z_N^k \\ v_S(z_S) &= \sum_{k=0}^{\infty} d_k z_S^k \end{aligned} \tag{5.95}$$

However, from the transition function (5.93) we obtain the relations:

$$\frac{d}{dz_N} = -z_S^2 \frac{d}{dz_S}; \quad \frac{d}{dz_S} = -z_N^2 \frac{d}{dz_N} \tag{5.96}$$

and hence:

$$\mathbf{t} = -\sum_{k=0}^{\infty} c_k z_S^{2-k} \frac{d}{dz_S} = \sum_{k=0}^{\infty} d_k z_S^k \frac{d}{dz_S} = -\sum_{k=0}^{\infty} d_k z_N^{2-k} \frac{d}{dz_N} = \sum_{k=0}^{\infty} c_k z_N^k \frac{d}{dz_N} \tag{5.97}$$

The only way for eq. (5.97) to be self-consistent is to have:

$$\forall k > 2 \quad c_k = d_k = 0,; \quad c_0 = -d_2, \quad c_1 = -d_1, \quad c_2 = -d_0 \tag{5.98}$$

This shows that the space of holomorphic sections of the tangent bundle $T\mathbb{S}^2$ is a **finite dimensional** vector space of dimension **three** spanned by the three differential operators:

$$\begin{aligned} \mathbf{L}_0 &= -z\frac{d}{dz} \\ \mathbf{L}_1 &= -\frac{d}{dz} \\ \mathbf{L}_{-1} &= -z^2\frac{d}{dz} \end{aligned} \tag{5.99}$$

What we have so far discussed can be summarized by stating the transformation rule of vector field components when we change coordinate patch form x^μ to $x^{\mu'}$:

$$t^{\mu'}(x') = t^\nu(x)\frac{\partial x^{\mu'}}{\partial x^\nu} \tag{5.100}$$

Indeed, a convenient way of defining a fiber-bundle is provided by specifying the way its sections transform from one local trivialization to another one which amounts to giving all the transition functions. This method can be used to discuss the construction of the cotangent bundle.

5.7.2 The Lie algebra of vector fields

In Section 5.3 we saw that the tangent space $T_p\mathcal{M}$ at point $p \in \mathcal{M}$ of a manifold can be identified with the vector space of derivations of the algebra of germs (see Definition 5.3.5). After gluing together all tangent spaces into the tangent bundle $T\mathcal{M}$, such an identification of tangent vectors with the derivations of an algebra can be extended from the local to the global level. The crucial observation is that the set of smooth functions on a manifold $\mathbb{C}^\infty(\mathcal{M})$ constitutes an algebra with respect to pointwise multiplication just as the set of germs at point p. The vector fields, namely, the sections of the tangent bundle, are derivations of this algebra. Indeed, each vector field $\mathbf{X} \in \Gamma(T\mathcal{M}, \mathcal{M})$ is a linear map of the algebra $\mathbb{C}^\infty(\mathcal{M})$ onto itself:

$$\mathbf{X} : \mathbb{C}^\infty(\mathcal{M}) \to \mathbb{C}^\infty(\mathcal{M}) \tag{5.101}$$

that satisfies the analogous properties of those mentioned in eqs (5.48) for tangent vectors, namely:

$$\begin{aligned} \mathbf{X}(\alpha f + \beta g) &= \alpha \mathbf{X}(f) + \beta \mathbf{X}(g) \\ \mathbf{X}(f \cdot g) &= \mathbf{X}(f) \cdot g + f \cdot \mathbf{X}(g) \\ &\left[\forall \alpha, \beta \in \mathbb{R}(\text{or } \mathbb{C});\ \forall f, g \in \mathbb{C}^\infty(\mathcal{M})\right] \end{aligned} \tag{5.102}$$

On the other hand, the set of vector fields, renamed for this reason

$$\mathbb{D}\text{iff}(\mathcal{M}) \equiv \Gamma(T\mathcal{M}, \mathcal{M}) \tag{5.103}$$

forms a Lie algebra with respect to the following Lie bracket operation:

$$[\mathbf{X}, \mathbf{Y}]f = \mathbf{X}(\mathbf{Y}(f)) - \mathbf{Y}(\mathbf{X}(f)) \tag{5.104}$$

Indeed, the set of vector fields is a vector space with respect to the scalar numbers ($\mathbb{R}$ or $\mathbb{C}$, depending on the type of manifold, real or complex), namely, we can take linear combinations of the following form:

$$\forall \lambda, \mu \in \mathbb{R} \text{ or } \mathbb{C}\ \ \forall \mathbf{X}, \mathbf{Y} \in \mathbb{D}\text{iff}(\mathcal{M}) : \quad \lambda \mathbf{X} + \mu \mathbf{Y} \in \mathbb{D}\text{iff}(\mathcal{M}) \tag{5.105}$$

having defined:

$$[\lambda \mathbf{X} + \mu \mathbf{Y}](f) = \lambda[\mathbf{X}(f)] + \mu[\mathbf{Y}(f)], \quad \forall f \in \mathbb{C}^\infty(\mathcal{M}) \tag{5.106}$$

Furthermore, the operation (5.104) is the commutator of two maps and as such it is antisymmetric and satisfies the Jacobi identity.

The Lie algebra of vector fields is named $\mathbb{D}\text{iff}(\mathcal{M})$ since each of its elements can be interpreted as the generator of an infinitesimal diffeomorphism of the manifold onto itself. As we are going to see $\mathbb{D}\text{iff}(\mathcal{M})$ is a Lie algebra of **infinite dimension**, but it can contain finite dimensional subalgebras generated by particular vector fields. The typical example will be the case of the Lie algebra of a Lie group: this is the finite dimensional subalgebra $\mathbb{G} \subset \mathbb{D}\text{iff}(G)$ spanned by those vector fields defined on the Lie group manifold that have an additional property of invariance with respect to either left or right translations (see Chapter 10).

5.7.3 The cotangent bundle and differential forms

Let us recall that a differential 1-form in the point $p \in \mathcal{M}$ of a manifold $\mathcal{M}$, namely, an element $\omega_p \in T_p^*\mathcal{M}$ of the cotangent space over such a point was defined as a real-valued linear functional over the tangent space at p, namely

$$\omega_p \in \mathrm{Hom}(T_p\mathcal{M}, \mathbb{R}) \tag{5.107}$$

which implies:

$$\forall \mathbf{t}_p \in T_p\mathcal{M} \quad \omega_p : \mathbf{t}_p \mapsto \omega_p(\mathbf{t}_p) \in \mathbb{R} \tag{5.108}$$

The expression of ω_p in a coordinate patch around p is:

$$\omega_p = \omega_\mu(p)\, dx^\mu \tag{5.109}$$

where $dx^\mu(p)$ are the differentials of the coordinates and $\omega_\mu(p)$ are real numbers. We can glue together all the cotangent spaces and construct the cotangent bundle by stating that a *generic smooth section* of such a bundle is of the form (5.109) where $\omega_\mu(p)$ are now smooth functions of the base manifold point p. Clearly if we change coordinate system, an argument completely similar to that employed in the case of the tangent bundle tells us that the coefficients $\omega_\mu(x)$ transform as follows:

$$\omega'_\mu(x') = \omega_\nu(x)\frac{\partial x^\nu}{\partial x^{\mu'}} \tag{5.110}$$

and equation (5.110) can be taken as a definition of the **cotangent bundle** $T^*\mathcal{M}$, whose sections transform with the Jacobian matrix rather than with the inverse Jacobian matrix as the sections of the tangent bundle do (see eq. (5.100)). So we can write:

Definition 5.7.3. A differential 1-form ω on a manifold $\mathcal{M}$ is a section of the cotangent bundle, namely $\omega \in \Gamma(T^*\mathcal{M}, \mathcal{M})$.

This means that a differential 1-form is a map:

$$\omega : \Gamma(T\mathcal{M}, \mathcal{M}) \mapsto \mathbb{C}^\infty(\mathcal{M}) \tag{5.111}$$

from the space of vector fields (*i.e.*, the sections of the tangent bundle) to smooth functions. Locally we can write:

$$\begin{aligned} \Gamma(T\mathcal{M}, \mathcal{M}) \ni \omega &= \omega_\mu(x)\, dx^\mu \\ \Gamma(T^*\mathcal{M}, \mathcal{M}) \ni \mathbf{t} &= t^\mu(x)\frac{\partial}{\partial x^\mu} \end{aligned} \tag{5.112}$$

and we obtain

$$\omega(\mathbf{t}) = \omega_\mu(x)t^\nu(x)\,dx^\mu\left(\frac{\partial}{\partial x^\nu}\right) = \omega_\mu(x)t^\mu(x) \tag{5.113}$$

using

$$dx^\mu\left(\frac{\partial}{\partial x^\nu}\right) = \delta^\mu_\nu \tag{5.114}$$

which is the statement that coordinate differentials and partial derivatives are dual bases for 1-forms and tangent vectors, respectively.

Since $T\mathscr{M}$ is a vector bundle it is meaningful to consider the addition of its sections, namely, the addition of vector fields and also their pointwise multiplication by smooth functions. Taking this into account we see that the map (5.111) used to define sections of the cotangent bundle, namely 1-forms, is actually an F-linear map. This means the following. Considering any F-linear combination of two vector fields, namely:

$$f_1\mathbf{t}_1 + f_2\mathbf{t}_2, f_1, f_2 \in \mathbb{C}^\infty(\mathscr{M}) \quad \mathbf{t}_1, \mathbf{t}_2 \in \Gamma(T\mathscr{M}, \mathscr{M}) \tag{5.115}$$

for any 1-form $\omega \in \Gamma(T^*\mathscr{M}, \mathscr{M})$ we have:

$$\omega(f_1\mathbf{t}_1 + f_2\mathbf{t}_2) = f_1(p)\omega(\mathbf{t}_1)(p) + f_2(p)\omega(\mathbf{t}_2)(p) \tag{5.116}$$

where $p \in \mathscr{M}$ is a any point of the manifold $\mathscr{M}$.

It is now clear that the definition of differential 1-form generalizes the concept of *total differential* of the germ of a smooth function. Indeed, in an open neighborhood $U \subset \mathscr{M}$ of a point p we have:

$$\forall f \in \mathbb{C}^\infty_p(\mathscr{M}) \quad df = \partial_\mu f\, dx^\mu \tag{5.117}$$

and the value of df at p on any tangent vector $\mathbf{t}_p \in T_p\mathscr{M}$ is defined to be:

$$df_p(\mathbf{t}_p) \equiv \mathbf{t}_p(f) = t^\mu \partial_\mu f \tag{5.118}$$

which is the directional derivative of the local function f along $\mathbf{t}_p$ in the point p. If rather than the germ of a function we take a global function $f \in \mathbb{C}^\infty(\mathscr{M})$, we realize that the concept of 1-form generalizes the concept of total differential of such a function. Indeed, the total differential df fits into the definition of a 1-form, since for any vector field $\mathbf{t} \in \Gamma(T\mathscr{M}, \mathscr{M})$ we have:

$$df(\mathbf{t}) = t^\mu(x)\partial_\mu f(x) \equiv \mathbf{t}f \in \mathbb{C}^\infty(\mathscr{M}) \tag{5.119}$$

A first obvious question is the following. Is any 1-form $\omega = \omega_\mu(x)\,dx^\mu$ the differential of some function? The answer is clearly no and in any coordinate patch there is a simple test to see whether this is the case or not. Indeed, if $\omega^{(1)}_\mu = \partial_\mu f$ for some germ

$f \in \mathbb{C}_p^\infty(\mathcal{M})$ then we must have:

$$\frac{1}{2}(\partial_\mu \omega_\nu^{(1)} - \partial_\nu \omega_\mu^{(1)}) = \frac{1}{2}[\partial_\mu, \partial_\nu]f = 0 \tag{5.120}$$

On the left-hand side of eq. (5.120) are the components of what we will name a differential 2-form

$$\omega^{(2)} = \omega_{\mu\nu}^{(2)}\, dx^\mu \wedge dx^\nu \tag{5.121}$$

and in particular the 2-form of eq. (5.120) will be identified with the exterior differential of the 1-form $\omega^{(1)}$, namely $\omega^{(2)} = d\omega^{(1)}$. In simple words, the exterior differential operator d is the generalization on any manifold and to differential forms of any degree of the concept of *curl*, familiar from ordinary tensor calculus in $\mathbb{R}^3$. Forms whose exterior differential vanishes will be named *closed forms*. All these concepts need appropriate explanations that will be provided shortly. Yet, already at this intuitive level, we can formulate the next basic question. We saw that, in order to be the total differential of a function, a 1-form must be necessarily closed. Is such a condition also sufficient? In other words, are all closed forms the differential of something? Locally the correct answer is yes, but globally it may be no. Indeed, in any open neighborhood a closed form can be represented as the differential of another differential form, but the forms that do the job in the various open patches may not glue together nicely into a globally defined one. This problem and its solution constitute an important chapter of geometry, named cohomology. Actually cohomology is a central issue in algebraic topology, the art of characterizing the topological properties of manifolds through appropriate algebraic structures.

5.7.4 Differential k-forms

Next we introduce differential forms of degree k and the exterior differential d. In a later section, after the discussion of homology we show how this relates to the important construction of cohomology. For the time being our approach is simpler and down to earth.

We have seen that the 1-forms at a point $p \in \mathcal{M}$ of a manifold are linear functionals on the tangent space $T_p\mathcal{M}$. First of all we recall the construction of exterior k-forms on any vector space W defined to be the k-th linear antisymmetric functionals on such a space (see Section 2.5.4).

5.7.4.1 Exterior differential forms

It follows that on $T_p\mathcal{M}$ we can construct not only the 1-forms but also all the higher degree k-forms. They span the vector space $\Lambda_k(T_p\mathcal{M})$. By gluing together all such vector spaces, as we did in the case of 1-forms, we obtain the vector-bundles of k-forms. More explicitly, we can set:

Definition 5.7.4. A differential k-form $\omega^{(k)}$ is a smooth assignment:

$$\omega^{(k)} : p \mapsto \omega_p^{(k)} \in \Lambda_k(T_p\mathcal{M}) \tag{5.122}$$

of an exterior k-form on the tangent space at p for each point $p \in \mathcal{M}$ of a manifold.

Let now (U, φ) be a local chart and let $\{dx_p^1, \dots, dx_p^m\}$ be the usual natural basis of the cotangent space $T_p^*\mathcal{M}$. Then in the same local chart the differential form $\omega^{(k)}$ is written as:

$$\omega^{(k)} = \omega_{i_1,\dots,i_k}(x_1, \dots, x_m)\, dx^{i_1} \wedge \dots \wedge dx^{i_k} \tag{5.123}$$

where $\omega_{i_1,\dots,i_k}(x_1, \dots, x_m) \in \mathbb{C}^\infty(U)$ are smooth functions on the open neighborhood U, completely antisymmetric in the indices $i_1, \dots, i_k$.

At this point it is obvious that the operation of exterior product, defined on exterior forms, can be extended to *exterior differential forms*. In particular, if $\omega^{(k)}$ and $\omega^{(k')}$ are a k-form and a k'-form, respectively, then $\omega^{(k)} \wedge \omega^{(k')}$ is a $(k+k')$-form. As a consequence of eq. (2.29) we have:

$$\omega^{(k)} \wedge \omega^{(k')} = (-)^{kk'}\, \omega^{(k')} \wedge \omega^{(k)} \tag{5.124}$$

and in local coordinates we find:

$$\omega^{(k)} \wedge \omega^{(k')} = \omega^{(k)}_{[i_1\dots i_k}\, \omega^{(k')}_{i_{k+1}\dots i_{k+k'}]}\, dx^1 \wedge \dots \wedge dx^{k+k'} \tag{5.125}$$

where $[\dots]$ denotes the complete antisymmetrization on the indices.

Let $\mathcal{A}_0(\mathcal{M}) = \mathbb{C}^\infty(\mathcal{M})$ and let $\mathcal{A}_k(\mathcal{M})$ be the $\mathbb{C}^\infty(\mathcal{M})$-module of differential k-forms. To justify the naming module, observe that we can construct the product of a smooth function $f \in \mathbb{C}^\infty(\mathcal{M})$ with a differential form $\omega^{(k)}$ setting:

$$[f\omega^{(k)}](\mathbf{Z}_1, \dots, \mathbf{Z}_k) = f \cdot \omega^{(k)}(\mathbf{Z}_1, \dots, \mathbf{Z}_k) \tag{5.126}$$

for each k-tuplet of vector fields $\mathbf{Z}_1, \dots, \mathbf{Z}_k \in \Gamma(T\mathcal{M}, \mathcal{M})$.

Furthermore, let

$$\mathcal{A}(\mathcal{M}) = \bigoplus_{k=0}^{m} \mathcal{A}_k(\mathcal{M}) \quad \text{where } m = \dim \mathcal{M} \tag{5.127}$$

Then $\mathcal{A}$ is an algebra over $\mathbb{C}^\infty(\mathcal{M})$ with respect to the exterior wedge product $\bigwedge$.

To introduce the exterior differential d we proceed as follows. Let $f \in \mathbb{C}^\infty(\mathcal{M})$ be a smooth function: for each vector field $\mathbf{Z} \in \mathbb{D}\mathrm{iff}(\mathcal{M})$, we have $\mathbf{Z}(f) \in \mathbb{C}^\infty(\mathcal{M})$ and therefore there is a unique differential 1-form, noted df, such that $df(\mathbf{Z}) = \mathbf{Z}(f)$. This differential form is named the total differential of the function f. In a local chart U with local coordinates $x^1, \dots, x^m$ we have:

$$df = \frac{\partial f}{\partial x^j}\, dx^j \tag{5.128}$$

More generally, we can see that there exists an endomorphism d, $(\omega \mapsto d\omega)$ of $\mathscr{A}(\mathscr{M})$ onto itself with the following properties:

$$\begin{array}{llcl} i) & \forall \omega \in \mathscr{A}_k(\mathscr{M}) & & d\omega \in \mathscr{A}_{k+1}(\mathscr{M}) \\ ii) & \forall \omega \in \mathscr{A}(\mathscr{M}) & & dd\omega = 0 \\ iii) & \forall \omega^{(k)} \in \mathscr{A}_k(\mathscr{M}) & & \forall \omega^{(k')} \in \mathscr{A}_{k'}(\mathscr{M}) \\ & d(\omega^{(k)} \wedge \omega^{(k')}) & = & d\omega^{(k)} \wedge \omega^{(k')} + (-1)^k \omega^{(k)} \wedge d\omega^{(k')} \\ iv) & \text{if } f \in \mathscr{A}_0(\mathscr{M}) & & df = \text{total differential} \end{array} \tag{5.129}$$

In each local coordinate patch the above intrinsic definition of the exterior differential leads to the following explicit representation:

$$d\omega^{(k)} = \partial_{[i_1}\omega_{i_2\dots i_{k+1}]}\, dx^{i_1} \wedge \dots \wedge dx^{i_{k+1}} \tag{5.130}$$

As already stressed the exterior differential is the generalization of the concept of curl, well known in elementary vector calculus.

5.8 Lie groups and Lie algebras

After the previous long introduction about manifolds and fiber-bundles we turn to the core of the present chapter which is the relation, firstly envisaged by Lie, between Lie Groups and Lie Algebras.

5.8.1 The Lie algebra of a Lie group

The definition of a Lie group was given in Definition 5.5.1. As already sketched there, we recall that, as a consequence of its definition, on a Lie group G one can define two transitive actions of the same group on itself, the left and the right multiplication, respectively. Indeed, to each element $\gamma \in G$ we can associate two continuous, infinitely differentiable and invertible maps of G in G, named the left and the right translation, which we introduced in eq. (5.62). Secondly, we introduce the concepts of pull-back and push-forward of any diffeomorphism mapping a differential manifold $\mathscr{M}$ onto an open submanifold of another differentiable manifold $\mathscr{N}$. Let ϕ be any such map

$$\phi : \mathscr{M} \to \mathscr{N} \tag{5.131}$$

The push-forward of ϕ, denoted ϕ_*, is a map from the space of sections of the tangent bundle $T\mathscr{M}$ to the space of sections of the tangent bundle $T\mathscr{N}$:

$$\phi_* : \Gamma(T\mathscr{M}, \mathscr{M}) \to \Gamma(T\mathscr{N}, \mathscr{N}) \tag{5.132}$$

Explicitly, if $\mathbf{X} \in \Gamma(T\mathscr{M}, \mathscr{M})$ is a vector field over $\mathscr{M}$, we can use it to define a new vector field $\phi_*\mathbf{X} \in \Gamma(T\mathscr{N}, \mathscr{N})$ over $\mathscr{N}$ using the following procedure. For any $f \in C^\infty(\mathscr{N})$,

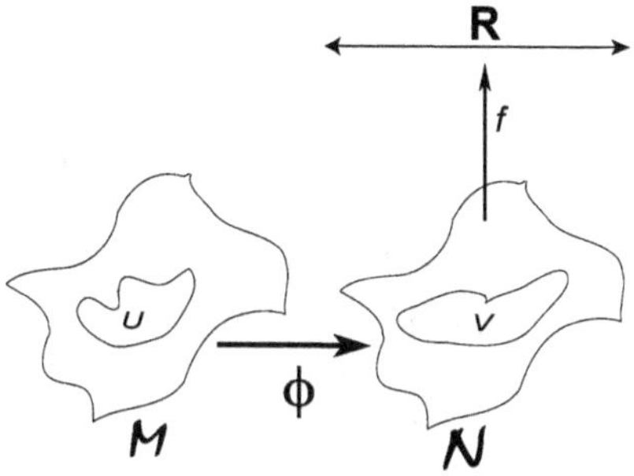

Figure 5.20: Graphical description of the concept of push-forward.

namely, for any smooth function on $\mathcal{N}$, the action of $\phi_* \mathbf{X}$ on such a function is given by:

$$\phi_* \mathbf{X}(f) \equiv [\mathbf{X}(f \circ \phi)] \circ \phi^{-1} \tag{5.133}$$

To clarify this concept let us describe the push-forward in a pair of open charts for both manifolds. Consider Figure 5.20: on the open neighborhood $U \subset \mathcal{M}$ we have coordinates x^μ while on the open neighborhood $V \subset \mathcal{N}$ we have coordinates y^ν. In this pair of local charts $\{U, V\}$ the diffeomorphism (5.131) is described by giving the coordinates y^ν as smooth functions of the coordinates x^μ:

$$y^\nu = y^\nu(x) \equiv \phi^\nu(x) \tag{5.134}$$

On the neighborhood V the smooth function $f \in C^\infty(\mathcal{N})$ is described by a real function $f(y)$ of the n-coordinates y^ν. It follows that $f \circ \phi$ is a smooth function $\tilde{f}(x)$ on the open neighborhood $U \subset \mathcal{M}$ simply given by:

$$\tilde{f}(x) = f(y(x)) \tag{5.135}$$

Any tangent vector $\mathbf{X} \in \Gamma(T\mathcal{M}, \mathcal{M})$ is locally described on the open chart U by a first-order differential operator of the form:

$$\mathbf{X} = X^\mu(x) \frac{\partial}{\partial x^\mu} \tag{5.136}$$

which therefore can act on $\tilde{f}(x)$:

$$\mathbf{X}\tilde{f}(x) = X^\mu(x) \frac{\partial y^\nu}{\partial x^\mu} \frac{\partial}{\partial y^\nu} f(y(x)) \tag{5.137}$$

Considering now the coordinates x^μ on U as functions of the y^ν on V, through the inverse of the diffeomorphism ϕ:

$$x^\mu = x^\mu(y) = (\phi^{-1}(y))^\mu \tag{5.138}$$

We realize that eq. (5.137) defines a new linear first-order differential operator acting on any function $f : V \to \mathbb{R}$. This differential operator is the push-forward of $\mathbf{X}$ through the diffeomorphism ϕ:

$$\phi_* \mathbf{X} = X^\nu(y)\frac{\partial}{\partial y^\nu} = X^\mu(x(y))\frac{\partial y^\nu}{\partial x^\mu}\frac{\partial}{\partial y^\nu} \tag{5.139}$$

Similarly the pull-back of ϕ, denoted ϕ^*, is a map from the space of sections of the cotangent bundle $T^*\mathcal{N}$ to the space of sections of the cotangent bundle $T^*\mathcal{M}$:

$$\phi^* : \Gamma(T^*\mathcal{N}, \mathcal{N}) \to \Gamma(T^*\mathcal{M}, \mathcal{M}) \tag{5.140}$$

Explicitly, if $\omega \in \Gamma(T^*\mathcal{N}, \mathcal{N})$ is a differential 1-form over $\mathcal{N}$, we can use it to define a differential 1-form $\phi^*\omega$ over $\mathcal{M}$ as follows. We recall that a 1-form is defined if we assign its value on any vector-field, hence we set:

$$\forall \mathbf{X} \in \Gamma(T\mathcal{M}, \mathcal{M}); \quad \phi^*\omega(\mathbf{X}) \equiv \omega(\phi_*\mathbf{X}) \tag{5.141}$$

Considering eq. (5.139), the local description of the pull-back on a pair of open charts $\{U, V\}$ is easily derived from the definition (5.141). If we name $\omega_\mu(y)$ the local components of the 1-form ω on the coordinate patch V:

$$\omega = \omega_\mu(y)\, dy^\mu \tag{5.142}$$

the components of the pull-back are immediately deduced:

$$\phi^*\omega = (\phi^*\omega)_\mu(x)\, dx^\mu \equiv \omega_\nu(y(x))\frac{\partial y^\nu}{\partial x^\mu}\, dx^\mu \tag{5.143}$$

5.8.1.1 Left/right invariant vector fields

Let us now consider the case where the manifold $\mathcal{M}$ coincides with manifold $\mathcal{N}$ and both are equal to a Lie group manifold G. The left and the right translations defined in (5.62) are diffeomorphisms and for each of them we can consider both the push-forward and the pull-back. This construction allows to introduce the notion of left (respectively right) invariant vector fields and 1-forms over the Lie group manifolds G.

Definition 5.8.1. A vector field $\mathbf{X} \in \Gamma(TG, G)$ defined over a Lie group-manifold G is named left-invariant (respectively right-invariant) if the following condition holds true:

$$\forall y \in G: \quad L_{y*}\mathbf{X} = \mathbf{X} \quad (\text{respectively } R_{y*}\mathbf{X} = \mathbf{X}) \tag{5.144}$$

Similarly:

Definition 5.8.2. A 1-form $\sigma \in \Gamma(T^*G, G)$ defined over a Lie group-manifold G is named left-invariant (respectively right-invariant) if the following condition holds true:

$$\forall y \in G: \quad L_y^*\sigma = \sigma \quad (\text{respectively } R_y^*\sigma = \sigma) \tag{5.145}$$

Let us recall that the space of sections of the tangent bundle has the structure of an infinite dimensional Lie algebra for any manifold $\mathcal{M}$. Indeed, given any two sections, we can compute their commutator as differential operators and this defines the necessary Lie bracket:

$$\forall \mathbf{X}, \mathbf{Y} \in \Gamma(T\mathcal{M}, \mathcal{M}); \quad [\mathbf{X}, \mathbf{Y}] = \mathbf{Z} \in \Gamma(T\mathcal{M}, \mathcal{M}) \tag{5.146}$$

Viewed as a Lie algebra the space of sections of the tangent bundle is usually denoted $\mathrm{Diff}_0(\mathcal{M})$ since every vector field can be regarded as the generator of a diffeomorphism infinitesimally close to the identity.

In the case of group manifolds we have the following simple but very fundamental theorem:

Theorem 5.8.1. *The two sets of left-invariant and of right-invariant vector fields over a Lie group manifold G close two finite dimensional Lie subalgebras of* $\mathrm{Diff}_0(G)$*, respectively named* $\mathbb{G}_{L/R}$*, which are isomorphic to each other and define the abstract Lie algebra* $\mathbb{G}$ *of the Lie group* G*. Furthermore,* $\mathbb{G}_{L/R}$ *commute with each other.*

This theorem is essentially, reformulated in modern terms, the content of Lie's work of 1874. The proof is obtained through a series of steps and through the proof of some intermediate lemmas. Let us begin with the first.

Lemma 5.8.1. *For any diffeomorphism* ϕ *the push-forward map* ϕ_* *has the following property:*

$$\forall \mathbf{X}, \mathbf{Y} \in \mathrm{Diff}_0(\mathcal{M}): \quad \phi_*[\mathbf{X}, \mathbf{Y}] = [\phi_*\mathbf{X}, \phi_*\mathbf{Y}] \tag{5.147}$$

No proof is required for this lemma since it follows straightforwardly from the definition (5.133) of the push-forward. However, the consequences of the lemma are far-reaching. Indeed, it implies that the commutator of two left-invariant (respectively right-invariant) vector fields is still left-invariant (respectively right-invariant). So we have:

Lemma 5.8.2. *The two sets* $\mathbb{G}_{L/R}$ *of left-invariant (respectively right-invariant) vector fields constitute two Lie subalgebras* $\mathbb{G}_{L/R} \subset \mathrm{Diff}_0(G)$.

$$\forall \mathbf{X}, \mathbf{Y} \in \mathbb{G}_{L/R}: \quad [\mathbf{X}, \mathbf{Y}] \in \mathbb{G}_{L/R} \tag{5.148}$$

This being established, we can now show that the left-invariant vector fields can be put into one-to-one correspondence with the elements of tangent space to the group-manifold at the identity element, namely, with T_eG. From this correspondence it will follow that the Lie subalgebra $\mathbb{G}_L$ has dimension equal to the dimension n of the Lie group. The same correspondence can be established also for the right-invariant vector fields and the same conclusion about the dimension of the Lie algebra $\mathbb{G}_R$ can be deduced. The argument goes as follows.

Let $g : [0,1] \mapsto G$ be a path in G with initial point at the identity element $g(0) = e \in G$. In local coordinates on the group manifold the path g is described as follows:

$$t \in [0,1]; \quad G \ni g(t) = \mathfrak{g}(\alpha^1(t) \dots \alpha^n(t)) \tag{5.149}$$

where α^i denote the group parameters and $\mathfrak{g}(\alpha)$ denotes the group element identified by the parameters α. Hence we obtain:

$$\frac{d}{dt}g(t) = \frac{d\alpha^i}{dt}\frac{\partial}{\partial\alpha^i}\mathfrak{g}(\alpha) \tag{5.150}$$

The set of derivatives:

$$c^i = \left.\frac{d\alpha^i}{dt}\right|_{t=0} \tag{5.151}$$

constitutes the component of a tangent vector at the identity that we name $\mathbf{X}_e$:

$$T_eG \ni \mathbf{X}_e \equiv c^i \frac{\vec{\partial}}{\partial\alpha^i} \tag{5.152}$$

Conversely, for each $\mathbf{X}_e \in T_eG$ we can construct a path $g(t)$ that admits $\mathbf{X}_e$ as tangent vector at the identity element. Given a tangent vector $\mathbf{t}$ at a point $p \in \mathcal{M}$ of a manifold, constructing a curve on $\mathcal{M}$ which goes through p and admits $\mathbf{t}$ as tangent vector in that point is a problem which admits solutions on any differentiable manifold, a fortiori on a Lie group manifold G. Let therefore $g(t)$ be such a path. To $\mathbf{X}_e$ we can associate a left-invariant vector field $\mathbf{X}_L$ defining the action of the latter on any smooth function f as it follows:

$$\forall f \in \mathbb{C}^\infty(G) \text{ and } \forall \rho \in G: \quad \mathbf{X}_L f(\rho) \equiv f(\rho g(t))|_{t=0} \tag{5.153}$$

Applying the definition (5.133) the reader can immediately verify that $\mathbf{X}_L$ defined above is left-invariant. In a completely analogous way, to the same tangent vector $\mathbf{X}_e$ we can associate a right-invariant vector field $\mathbf{X}_R$, setting:

$$\forall f \in \mathbb{C}^\infty(G) \text{ and } \forall \rho \in G: \quad \mathbf{X}_R f(\rho) \equiv f(g(t)\rho)|_{t=0} \tag{5.154}$$

In this way we have established that to each tangent vector at the identity we can associate both a left-invariant and a right-invariant vector field. On the other hand, since each vector field $\mathbf{X}$ is a section of the tangent bundle, namely a map:

$$\forall p \in \mathcal{M}: \quad p \mapsto \mathbf{X}_p \in T_p\mathcal{M} \tag{5.155}$$

it follows that each left-invariant or right-invariant vector fields singles out a tangent vector at the identity. The relevant point is that this double correspondence is an isomorphism of vector spaces. Indeed we have:

Lemma 5.8.3. *Let G be a Lie group, let $\mathbb{G}_{L/R}$ be the Lie algebra of left-invariant (respectively right-invariant) vector fields. The correspondence:*

$$\forall \mathbf{X} \in G_{L/R} \quad \pi : \mathbf{X} \mapsto \mathbf{X}_e \in T_e G \tag{5.156}$$

is an isomorphism of vector spaces.

Proof of Lemma 5.8.3. Let us first of all observe that the correspondence (5.156) is a linear map. Indeed, for $a, b \in \mathbb{R}$ and $X, Y \in \mathbb{G}_{L/R}$ we have:

$$\pi(a\mathbf{X} + b\mathbf{Y}) = a\mathbf{X}_e + b\mathbf{Y}_e \tag{5.157}$$

In order to show that π is an isomorphism, we need to prove that π is both injective and surjective, in other words that:

$$\ker \pi = 0; \quad \operatorname{Im} \pi = T_e G \tag{5.158}$$

The second of the two conditions (5.158) was already proved. Indeed, we have shown above that to each tangent vector $\mathbf{X}_e$ at the identity we can associate a left-invariant (right-invariant) vector field which reduces to that vector at $g = e$. It remains to be shown that $\ker \pi = 0$. This follows from a simple observation. If $\mathbf{X}$ is left-invariant, then its value at the group element $\gamma \in G$ is determined by its value at the identity by means of a left translation, namely:

$$\mathbf{X}_\gamma = L_{\gamma *} \mathbf{X}_e \tag{5.159}$$

Therefore if $\mathbf{X}_e = 0$ it follows that $\mathbf{X}_\gamma = 0$ for all group elements $\gamma \in G$ and hence $\mathbf{X} = 0$ as a vector field. Hence there are no non-trivial vectors in the kernel of the map π and this concludes the proof of the lemma. □

We have therefore:

Corollary 5.8.1. *The dimensions of the two Lie algebras $\mathbb{G}_L$ and $\mathbb{G}_R$ are equal among themselves and equal to those of the abstract Lie algebra $\mathbb{G}$*

$$dim\, \mathbb{G}_L = dim\, \mathbb{G}_R = dim\, \mathbb{G} = dim\, T_e G \tag{5.160}$$

In this way we have shown the isomorphism $\mathbb{G}_L \sim \mathbb{G}_R \sim \mathbb{G}$ as vector spaces. It remains to be shown the same isomorphism as Lie algebras.

Proof of Theorem 5.8.1. We can now complete the proof of Theorem 5.8.1. To this effect we argue in the following way. Utilizing the just established vector space isomorphism let us choose a basis for $T_e G$ which we denote as $\mathbf{t}_A$ ($A = 1, \ldots, \dim G$). In this way we have:

$$\forall \mathbf{t} \in T_e G : \quad \mathbf{t} = x^A \mathbf{t}_A, \quad x^A \in \mathbb{R} \tag{5.161}$$

Relying on the constructions (5.153) and (5.154) to each element of the basis $\{\mathbf{t}_A\}$ we can associate the corresponding left (right)-invariant vector field:

$$\mathbf{T}_A^{(L)}(f)(\rho) = \frac{d}{dt} f(\rho g_A(t))|_{t=0} \tag{5.162}$$

$$\mathbf{T}_A^{(R)}(f)(\rho) = \frac{d}{dt} f(g_A(t)\rho)|_{t=0} \tag{5.163}$$

where $g_A(t)$ denotes a path in G passing through the identity and there admitting the vector $\mathbf{t}_A$ as tangent. The established vector space isomorphism guarantees that $\mathbf{T}_A^{(L)}$ and $\mathbf{T}_A^{(R)}$ constitute a basis respectively for $\mathbb{G}_L$ and $\mathbb{G}_R$. Hence we can write:

$$[\mathbf{T}_A^{(L)}, \mathbf{T}_B^{(L)}] = C_{AB}^{(L)C} \mathbf{T}_C^{(L)} \tag{5.164}$$

$$[\mathbf{T}_A^{(R)}, \mathbf{T}_B^{(R)}] = C_{AB}^{(R)C} \mathbf{T}_C^{(R)} \tag{5.165}$$

where $C_{AB}^{(L)C}$ and $C_{AB}^{(L)C}$ are constants that, a priori, might be completely different. Equations follow from the fact that we have established that both $\mathbb{G}_L$ and $\mathbb{G}_R$ are dimension-n Lie algebras and the generators $\mathbf{T}_A^{(L/R)}$ constitute a basis.

Let us now calculate explicitly the commutators of the basis elements using their definitions. On any function $f : G \mapsto \mathbb{R}$, we find:

$$\begin{aligned}[\mathbf{T}_A^{(L)}, \mathbf{T}_B^{(L)}](f)(\rho) &= \frac{d}{dt}\frac{d}{d\tau}[f(\rho g_B(t) g_A(\tau)) - f(\rho g_A(t) g_B(\tau))]|_{t=\tau=0} \\ &= C_{AB}^{(L)C} \frac{d}{dt} f(\rho g_C(t))|_{t=0}\end{aligned} \tag{5.166}$$

$$\begin{aligned}[\mathbf{T}_A^{(R)}, \mathbf{T}_B^{(R)}](f)(\rho) &= \frac{d}{dt}\frac{d}{d\tau}[f(g_A(t) g_B(\tau)\rho) - f(g_B(t) g_A(\tau)\rho)]|_{t=\tau=0} \\ &= C_{AB}^{(R)C} \frac{d}{dt} f(\rho g_C(t))|_{t=0}\end{aligned} \tag{5.167}$$

The above equations (5.167) hold true at any point $\rho \in G$. Evaluating them at the identity $\rho = e$ and taking their sum we obtain:

$$(C_{AB}^{(L)C} + C_{AB}^{(R)C}) \frac{d}{dt} f(\rho g_C(t))|_{t=0} = 0 \tag{5.168}$$

which implies:

$$C_{AB}^{(L)C} = -C_{AB}^{(R)C} \tag{5.169}$$

This relation suffices to establish the Lie algebra isomorphism of $\mathbb{G}_L$ with $\mathbb{G}_R$. Indeed, they are isomorphic as vector spaces and their structure constants become identical under the very simple change of basis:

$$\mathbf{T}_A^{(L)} \leftrightarrow -\mathbf{T}_A^{(R)} \tag{5.170}$$

This concludes the proof of Theorem 5.8.1. Indeed, the last isomorphism advocated in that proposition amounts simply to a definition. □

Definition 5.8.3. The Lie algebra $\mathbb{G}_L$ of the left-invariant vector fields on the Lie group manifold G, isomorphic to that of the right-invariant ones $\mathbb{G}_R$, is named the Lie algebra $\mathbb{G}$ of the considered Lie group.

By explicit evaluation as in eq. (5.167) we can also show that any left-invariant vector field commutes with any right-invariant one and vice versa. Indeed, we find:

$$\begin{aligned}[\mathbf{T}_A^{(L)}, \mathbf{T}_B^{(R)}](f)(\rho) &= \frac{d}{dt}\frac{d}{d\tau}[f(g_B(t)\rho g_A(\tau)) - f(g_B(t)\rho g_A(\tau))]\big|_{t=\tau=0} \\ &= 0\end{aligned} \tag{5.171}$$

The interpretation of these relations is indeed very simple. The left-invariant vector fields happen to be the infinitesimal generators of the right-translations while the right-invariant ones generate the left-translations. Hence the vanishing of the above commutators just amounts to say that the left-invariant are indeed invariant under left translations while the right-invariant are insensitive to right translations.

5.8.2 Maurer–Cartan forms on Lie group manifolds

Let us now consider left-invariant (respectively right-invariant) 1-forms on the group manifold. They were defined in eq. (5.145). Starting from the construction of the left (right)-invariant vector fields it is very easy to construct an independent set of n differential forms with the invariance property (5.145) that are in one-to-one correspondence with the generators of the Lie algebra $\mathbb{G}$. Let us consider the explicit form of the $\mathbf{T}_A^{(L/R)}$ as first order differential operators:

$$\mathbf{T}_A^{(L/R)} = \overset{(l/r)}{\Sigma_A^i}(\alpha)\,\frac{\vec{\partial}}{\partial\alpha^i} \tag{5.172}$$

According to the already introduced convention α^μ are the group parameters and the square matrix:

$$\overset{(l/r)}{\Sigma_A^i}(\alpha) = \left.\underbrace{\begin{pmatrix} * & * & * & * \\ * & * & * & * \\ * & * & * & * \\ * & * & * & * \end{pmatrix}}_{A=1,\dots,n}\right\} i = 1,\dots,n \tag{5.173}$$

whose entries are functions of the parameters can be calculated starting from the constructive algorithm encoded in eqs (5.163). In terms of the inverse of the above matrix denoted $\overset{(l/r)}{\Sigma_\alpha^A}(\alpha)$ and such that:

$$\overset{(l/r)}{\Sigma_i^A}(\alpha)\;\overset{(l/r)}{\Sigma_B^i}(\alpha) = \delta_B^A \tag{5.174}$$

we can define the following set of n differential 1-forms:

$$\sigma^A_{(L/R)} = \overset{(l/r)}{\Sigma^A_i}(\alpha)\ d\alpha^i \tag{5.175}$$

which by construction satisfy the relations:

$$\sigma^A_{(L/R)}(\mathbf{T}^{(L/R)}_B) = \delta^A_B \tag{5.176}$$

From eq. (5.176) follows that all the forms $\sigma^A_{L/R}$ are left (respectively right)-invariant. Indeed, we have:

$$\forall \gamma \in G: \quad L^*_\gamma \sigma^A_{(L)}(\mathbf{T}^{(L)}_B) \equiv \sigma^A_{(L)}(L_{\gamma *}\mathbf{T}^{(L)}_B) = \sigma^A_{(L)}(\mathbf{T}^{(L)}_B) = \delta^A_B \tag{5.177}$$

which implies $L^*_\gamma \sigma^A_{(L)} = \sigma^A$ since both forms have the same values on a basis of sections of the tangent bundle as it is provided by the set of left-invariant vector fields $\mathbf{T}^{(L)}_B$. A completely identical proof holds obviously true for the right-invariant 1-forms $\sigma^A_{(R)}$ defined by the same construction.

We come therefore to the conclusion that on each Lie group manifold G we can construct both a left-invariant $\sigma_{(L)}$ and a right-invariant $\sigma_{(R)}$ Lie algebra valued 1-form defined as follows:

$$\mathbb{G} \ni \sigma_{(L)} = \sum_{A=1}^{n} \sigma^A_{(L)} \mathbf{T}_A \tag{5.178}$$

$$\mathbb{G} \ni \sigma_{(R)} = \sum_{A=1}^{n} \sigma^A_{(R)} \mathbf{T}_A \tag{5.179}$$

where, just as before, $\{\mathbf{T}_A\}$ denotes a basis of generators of the abstract Lie algebra $\mathbb{G}$.

One may wonder how the Lie algebra forms $\sigma_{(L/R)}$ could be constructed directly without going through the previous construction of the left (right)-invariant vector fields. The answer is very simple. Let $\mathfrak{g}(\alpha) \in G$ be a running element of the Lie group parameterized by the parameters α which constitute a coordinate patch for the corresponding group manifold and consider the 1-forms:

$$\theta_L = \mathfrak{g}^{-1}\, d\mathfrak{g} = \mathfrak{g}^{-1}\partial_i \mathfrak{g}\, d\alpha^i; \quad \theta_R = \mathfrak{g}\, d\mathfrak{g}^{-1} = \mathfrak{g}\partial_i \mathfrak{g}^{-1}\, d\alpha^i \tag{5.180}$$

It is immediate to check that such 1-forms are left (respectively right)-invariant. For instance, we have:

$$L^*_\gamma \theta_L = (\gamma\mathfrak{g})^{-1}\, d(\gamma\mathfrak{g}) = \mathfrak{g}^{-1}\gamma^{-1}\gamma\, d\mathfrak{g} = \mathfrak{g}^{-1}\, d\mathfrak{g} = \theta_L \tag{5.181}$$

What might not be immediately evident to the reader is why the left (right)-invariant 1-forms $\theta_{L/R}$ introduced in eq. (5.180) should be Lie algebra valued. The answer is actually very simple. The relation between Lie algebras and Lie groups is provided by the exponential map: every element of a Lie group which lies in the branch connected to the identity can be represented as the exponential of a suitable Lie algebra element:

$$\forall \gamma \in \mathrm{G}_0 \subset G, \quad \exists X \in \mathbb{G} \setminus \gamma = \exp \mathbf{X} \tag{5.182}$$

The Lie algebra element actually singles out an entire one parameter subgroup:

$$\forall t \in \mathbb{R}, \quad G_0 \ni \gamma(t) = \exp[t\mathbf{X}] \tag{5.183}$$

With an obvious calculation we obtain:

$$\gamma^{-1}(t)\, d\gamma(t) = \mathbf{X}\, dt \in \mathbb{G} \tag{5.184}$$

namely the left-invariant 1-form associated with this parameter subgroup lies in the Lie algebra. This result extends to any group element, so that indeed the previously constructed Lie algebra valued left (right)-invariant one forms $\sigma_{L/R}$ and the theta forms defined in eq. (5.180) are just the very same objects:

$$\sigma_{L/R} = \theta_{L/R} \tag{5.185}$$

All the above statements become much clearer when we consider classical groups whose elements are just matrices subject to some defining algebraic condition. Consider for instance the rotation group in N dimensions, $\mathrm{SO}(N)$. All elements of this group are orthogonal $N \times N$ matrices:

$$\mathcal{O} \in \mathrm{SO}(N) \quad \Leftrightarrow \quad \mathcal{O}^T \mathcal{O} = \mathbf{1}_{N\times N} \tag{5.186}$$

The elements of the orthogonal Lie algebra $\mathfrak{so}(N)$ are instead antisymmetric matrices:

$$\mathrm{A} \in \mathfrak{so}(N) \quad \Leftrightarrow \quad \mathrm{A}^T + \mathrm{A} = \mathbf{0}_{N\times N} \tag{5.187}$$

Calculating the transpose of the matrix $\Theta = \mathcal{O}^T d\mathcal{O}$ we immediately obtain:

$$\Theta^T = (\mathcal{O}^T d\mathcal{O})^T = d\mathcal{O}^T \mathcal{O} = -\mathcal{O}^T d\mathcal{O} = -\Theta \quad \Rightarrow \quad \Theta \in \mathfrak{so}(N) \tag{5.188}$$

which proves that the left-invariant 1-form is indeed Lie algebra valued.

5.8.3 Maurer Cartan equations

It is now of the utmost interest to consider the following identity which follows immediately from the definitions (5.180) and (5.185) of the left (right)-invariant 1-forms:

$$\mathfrak{F}[\sigma] \equiv d\sigma_{L/R} + \sigma_{L/R} \wedge \sigma_{L/R} = 0 \tag{5.189}$$

To prove the above statement it is sufficient to observe that $\mathfrak{g}^{-1}\, d\mathfrak{g} = -d\mathfrak{g}^{-1}\mathfrak{g}$.

The reason why we introduced a special name $\mathfrak{F}[\sigma]$ for the Lie algebra valued 2-form $d\sigma + \sigma \wedge \sigma$, which turns out to be zero in the case of left (right)-invariant 1-forms σ, is that precisely this combination will play a fundamental role in the theory of connections, representing their *curvature*. Equations (5.189) are named the Maurer Cartan equations of the considered Lie algebra $\mathbb{G}$ and they translate into the statement that left (right)-invariant 1-forms have *vanishing curvature*.

Since the forms σ are Lie algebra-valued, they can be expanded along a basis of generators $\mathbf{T}_A$ for $\mathbb{G}$, as we already did in eq. (5.179), and the same can be done for the curvature 2-form $\mathfrak{F}[\sigma]$, namely we get:

$$\begin{aligned} \mathfrak{F}[\sigma] &= \mathfrak{F}^A[\sigma]\mathbf{T}_A \\ \mathfrak{F}^A[\sigma] &= d\sigma^A + \frac{1}{2}C_{BC}{}^A \sigma^B \wedge \sigma^C \end{aligned} \tag{5.190}$$

where $C_{BC}{}^A$ are the structure constants of the Lie algebra:

$$[\mathbf{T}_B, \mathbf{T}_C] = C_{BC}{}^A \mathbf{T}_A \tag{5.191}$$

and the Maurer Cartan equations (5.189) amount to the statement:

$$\mathfrak{F}^A[\sigma_{L/R}] = 0 \tag{5.192}$$

5.9 Matrix Lie groups

All Lie groups can be viewed as groups whose elements are matrices. This occurs for most Lie groups in their very definition and it occurs for all Lie groups in their linear representations. For this reason we start recalling some properties of matrices that will help us very much to translate the above explained general notions into concrete cases.

5.9.1 Some properties of matrices

One can define a *matrix exponential* by means of a formal power series expansion: if A is a square matrix,

$$\exp(A) = \sum_{k=0}^{\infty} \frac{1}{k!} A^k \tag{5.193}$$

For instance, consider the following 2×2 case:

$$m(\theta) = \theta\epsilon = \begin{pmatrix} 0 & \theta \\ -\theta & 0 \end{pmatrix} \quad \Rightarrow \quad \exp[m(\theta)] = \begin{pmatrix} \cos\theta & \sin\theta \\ -\sin\theta & \cos\theta \end{pmatrix} \tag{5.194}$$

where θ is a real parameter. The above expression easily follows from eq. (5.193) and from the fact that the Levi–Civita antisymmetric symbol ϵ_{ij}, seen as a matrix, obeys $\epsilon^2 = -\mathbf{1}$.

For matrix exponentials most properties of the usual exponential hold true, yet not all of them. For instance, if T, S are square matrices, we have

$$\exp(T)\exp(S) = \exp(T+S) \quad \Leftrightarrow \quad [T,S] = 0 \tag{5.195}$$

i.e. only if the two matrices commute. Otherwise, eq. (5.195) generalizes to the so-called *Baker–Campbell–Hausdorff* formula:

$$\exp(T)\exp(S) = \exp\Big(T + S + \frac{1}{2}[T,S] + \frac{1}{12}([T,[T,S]] + [[T,S],S]) + \cdots\Big) \tag{5.196}$$

A property which is immediately verified is the following one:

$$\exp(U^{-1}TU) = U^{-1}\exp(T)U \tag{5.197}$$

A very useful property of the determinants is the following:

$$\det(\exp(m)) = \exp(\operatorname{Tr} m) \tag{5.198}$$

or equivalently, by taking the logarithm (which for matrices is defined via a formal power series)

$$\det M = \exp(\operatorname{Tr}(\ln M)) \tag{5.199}$$

These relations can be easily proven for a diagonalizable matrix; indeed, the determinant and the trace are invariant under any change of basis. Assume that M has been diagonalized. If λ_i are its eigenvalues we get:

$$\det M = \prod_i \lambda_i = \exp\Big(\sum_i \ln \lambda_i\Big) = \exp(\operatorname{Tr}(\ln M)) \tag{5.200}$$

The result can then be extended to generic matrices as it can be argued that every matrix can be approximated to any chosen accuracy by diagonalizable matrices.

Let us note that on the space of $n \times n$ matrices (with, say, complex entries) one can define a distance by

$$d(M,N) = \left(\sum_{i,j=1}^{n} |M_{ij} - N_{ij}|^2\right)^{1/2} \tag{5.201}$$

where M, M are two matrices. This is nothing else than considering the space of generic $n \times n$ matrices as the $\mathbb{C}^{n^2}$ space parameterized by the n^2 complex entries, and to endow it with the usual topology.

5.9.2 Linear Lie groups

Consider a group of matrices such that:

a) The elements of G within a neighborhood $\mathcal{S}$ of the identity matrix can be put into one-to-one correspondence with a neighborhood $\mathcal{B}$ of the origin in $\mathbb{R}^d$ (the correspondence is chosen so that the origin corresponds to the identity matrix). In other words, any matrix $M \in \mathcal{S}$ is parameterized by d *real* coordinates α^μ ($\mu = 1, \dots, d$): $M = M(\alpha)$.
b) All the matrix elements $M_{ij}(\alpha)$ of a matrix $M(\alpha) \in \mathcal{S}$ are furthermore required to be *analytic* functions of the coordinates α, namely, they ought to admit an expression as a power series of the differences $\alpha^\mu - \alpha_0^\mu$, where $\alpha_0 \in \mathcal{B}$ are the coordinates of any other matrix in the set $\mathcal{S}$. Thus all the derivatives $\partial M_{ij}/\partial\alpha^\mu, \partial^2 M_{ij}/\partial\alpha^\mu\partial\alpha^\nu, \dots$ exist at all points within $\mathcal{B}$, and in particular at $\alpha = 0$.

It is not difficult to see that from the above conditions it follows that

i) The d matrices $\mathbf{T}_\mu$ defined by
$$\mathbf{T}_\mu = \left.\frac{\partial M}{\partial\alpha^\mu}\right|_{\alpha=0} \tag{5.202}$$
form the basis of a *real* vector space of dimension d. The matrices $\mathbf{T}_\mu$ provide the explicit form of the *infinitesimal generators* of the Lie algebra described above as vector fields on the group manifold.
ii) If the product of $M(\alpha)$ and $M(\beta)$ both in $\mathcal{S}$ also belongs to $\mathcal{S}$, we have $M(\alpha)M(\beta) = M(\gamma)$ for some γ depending on α, β. The function $\gamma(\alpha, \beta)$ is analytic in both α and β. Also, the inverse matrix $[M(\alpha)]^{-1} = M(\eta)$, with $\eta(\alpha)$ analytic.

The point ii) tells us that a group of matrices such that a), b) holds satisfies the definition given above of a Lie group.

5.10 Bibliographical note

Main bibliographical sources for this chapter are following books:

1. the first volume of [51]
2. the monograph [84]
3. the textbook [138]
4. the textbook [116]
5. the textbook [29]
6. the textbook [86]
7. the textbook [144]
8. the textbook [71]

that are also suggested for further reading.

6 Structure of Lie algebras

...and when I had at last plucked it, the stalk was all frayed and the flower itself no longer seemed so fresh and beautiful. Moreover, owing to a coarseness and stiffness, it did not seem in place among the delicate blossoms of my nosegay. I threw it away feeling sorry to have vainly destroyed a flower that looked beautiful in its proper place. – But what energy and tenacity! With what determination it defended itself, and how dearly it sold its life! thought I, remembering the effort it had cost me to pluck the flower. –
L. N. Tolstoy, Hadji Murat

6.1 Introduction

The present chapter addresses the questions *what is the general structure of a Lie group* and *can we classify Lie groups?* The first step in this direction consists of reducing the question from Lie groups to their Lie algebras, secondly of describing the general form of Lie algebras in terms of a semisimple algebra (the Levi algebra) and of a solvable radical. The third step consists of classifying simple Lie algebras and will be our concern in Chapter 7. In order to establish the first two steps on a firm basis we ought to introduce a few more fundamental mathematical concepts and constructions of a very general character, in particular homology and cohomology. To these preliminaries we devote the next sections.

6.2 Homotopy, homology and cohomology

Differential 1-forms can be integrated along differentiable paths on manifolds. The higher differential p-forms can be integrated on p-dimensional submanifolds. An appropriate discussion of such integrals and of their properties requires the fundamental concepts of algebraic topology, namely, *homotopy* and *homology*. Also the global properties of Lie groups and their many-to-one relation with Lie algebras can be understood only in terms of homotopy. For this reason we devote the present section to an introductory discussion of homotopy, homology and of its dual, cohomology.

The kind of problems we are going to consider can be intuitively grasped if we consider Figure 6.1, displaying a closed two-dimensional surface with two handles (actually an oriented, closed Riemann surface of genus $g = 2$) on which we have drawn several different closed 1-dimensional paths $\gamma_1, \dots, \gamma_6$.

Consider first the path γ_5. It is an intuitive fact that γ_5 can be continuously deformed to just a point on the surface. Paths with such a property are named *homotopically trivial* or *homotopic to zero*. It is also an intuitive fact that neither γ_2, nor γ_3, nor γ_1, nor γ_4 are homotopically trivial. Paths of such a type are *homotopically non-trivial*. Furthermore, we say that two paths are homotopic if one can be continuously deformed into the other. This is for instance the case of γ_6 which is clearly homotopic to γ_3.

https://doi.org/10.1515/9783110551204-006

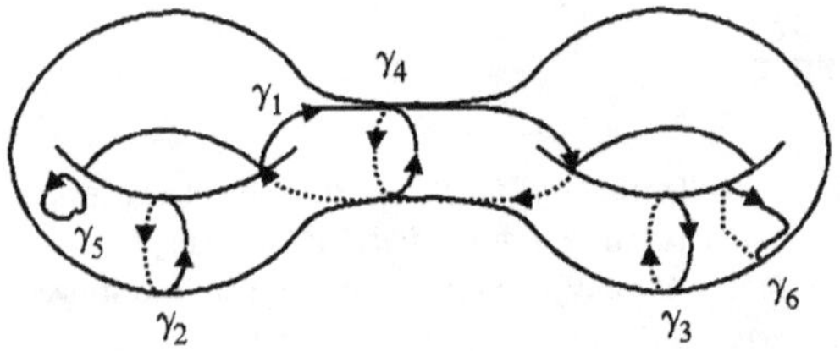

Figure 6.1: A closed surface with two handles marked by several different closed 1-dimensional paths.

Let us now consider the difference between path γ_4 and path γ_1 from another viewpoint. Imagine the result of cutting the surface along the path γ_4. After the cut the surface splits into two separate parts, R_1 and R_2, as shown in Figure 6.2. Such a splitting does not occur if we cut the original surface along the path γ_1. The reason for this different behavior resides in the following. The path γ_4 is the boundary of a region on the surface (the region R_1 or, equivalently, its complement R_2) while γ_1 is not the boundary of any region. A similar statement is true for the paths γ_2 or γ_3. We say that γ_4 is *homologically trivial* while $\gamma_1, \gamma_2, \gamma_3$ are *homologically non-trivial.*

Next let us observe that if we simultaneously cut the original surface along $\gamma_1, \gamma_2, \gamma_3$, the surface splits once again into two separate parts as shown in Figure 6.3.

This is due to the fact that the sum of the three paths is the boundary of a region: either R_1 or R_2 of Figure 6.3. In this case we say that $\gamma_2 + \gamma_3$ is *homologous* to $-\gamma_1$, since the difference $\gamma_2 + \gamma_3 - (-\gamma_3)$ is a boundary.

In order to give a rigorous formulation to these intuitive concepts, which can be extended also to higher dimensional submanifolds of any manifold, we proceed as follows.

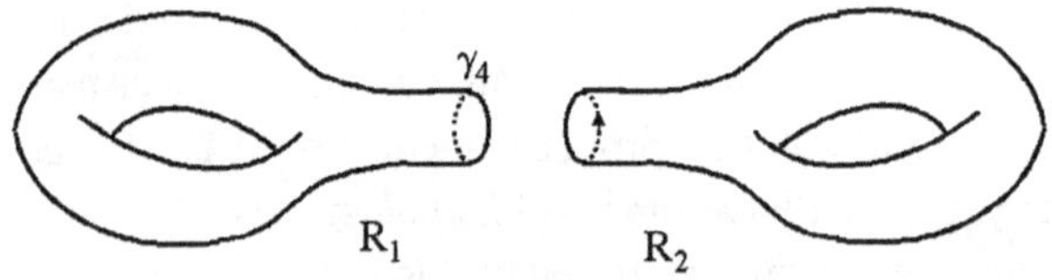

Figure 6.2: When we cut a surface along a path that is a boundary, namely it is homologically trivial, the surface splits into two separate parts.

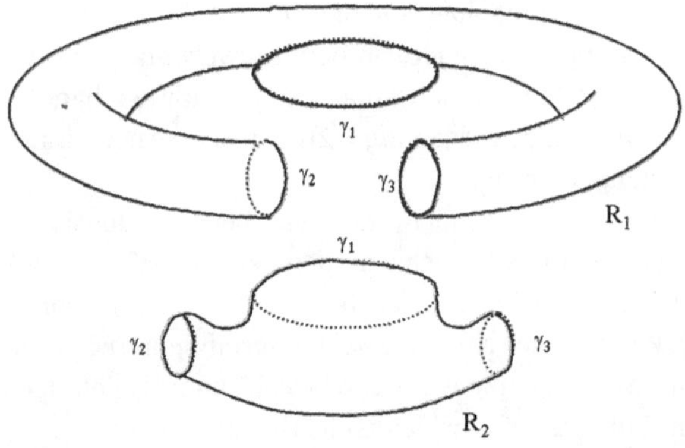

Figure 6.3: The sum of the three paths $\gamma_1, \gamma_2, \gamma_3$ is homologically trivial, namely, $\gamma_2 + \gamma_3$ is homologous to $-\gamma_1$.

6.2.1 Homotopy

Let us come back to the Definition 5.3.1 of a curve (or path) in a manifold and slightly generalize it.

Definition 6.2.1. Let $[a, b]$ be a closed interval of the real line $\mathbb{R}$ parameterized by the parameter t and subdivide it into a finite number of closed, partial intervals:

$$[a, t_1], [t_1, t_2], \dots, [t_{n-1}, t_n], [t_n, b] \tag{6.1}$$

We name **piecewise differentiable path** a continuous map:

$$\gamma : [a, b] \to \mathcal{M} \tag{6.2}$$

of the interval $[a, b]$ into a differentiable manifold $\mathcal{M}$ such that there exists a splitting of $[a, b]$ into a finite set of closed subintervals as in eq. (6.1) with the property that on each of these intervals the map γ is not only continuous but also infinitely differentiable.

Since we have parametric invariance we can always rescale the interval $[a, b]$ and reduce it to be

$$[0, 1] \equiv I \tag{6.3}$$

Let

$$\begin{aligned} \sigma &: I \to \mathcal{M} \\ \tau &: I \to \mathcal{M} \end{aligned} \tag{6.4}$$

be two piecewise differentiable paths with coinciding extrema, namely such that (see Figure 6.4):

$$\begin{aligned} \sigma(0) = \tau(0) = x_0 \in \mathcal{M} \\ \sigma(1) = \tau(1) = x_1 \in \mathcal{M} \end{aligned} \tag{6.5}$$

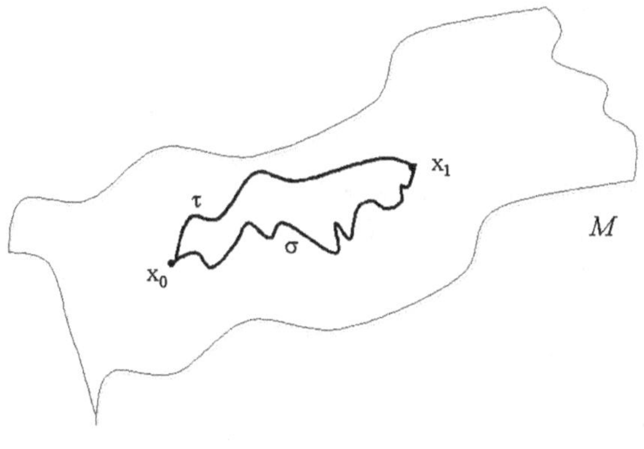

Figure 6.4: Two paths with coinciding extrema.

Definition 6.2.2. We say that σ is homotopic to τ and we write $\sigma \simeq \tau$ if there exists a continuous map:

$$F : I \times I \to \mathcal{M} \tag{6.6}$$

such that:

$$\begin{aligned} F(s,0) &= \sigma(s) \quad \forall s \in I \\ F(s,1) &= \tau(s) \quad \forall s \in I \\ F(0,t) &= x_0 \quad \forall t \in I \\ F(1,t) &= x_1 \quad \forall t \in I \end{aligned} \tag{6.7}$$

In particular if σ is a closed path, namely a loop at x_0, namely if $x_0 = x_1$ and if τ homotopic to σ is the *constant loop*, that is,

$$\forall s \in I : \quad \tau(s) = x_0 \tag{6.8}$$

then we say that σ is *homotopically trivial* and that it can be contracted to a point.

It is quite obvious that the homotopy relation $\sigma \simeq \tau$ is an equivalence relation. Hence we shall consider the homotopy classes $[\sigma]$ of paths from x_0 to x_1

Next we can define a binary product operation on the space of paths in the following way. If σ is a path from x_0 to x_1 and τ is a path from x_1 to x_2, we can define a path from x_0 to x_2 traveling first along σ and then along τ. More precisely, we set:

$$\sigma\tau(t) = \begin{cases} \sigma(2t) & 0 \leq t \leq \frac{1}{2} \\ \tau(2t-1) & \frac{1}{2} \leq t \leq 1 \end{cases} \tag{6.9}$$

What we can immediately verify from this definition is that if $\sigma \simeq \sigma'$ and $\tau \simeq \tau'$, then $\sigma\tau \simeq \sigma'\tau'$. The proof is immediate and it is left to the reader. Hence without any ambiguity we can multiply the equivalence class of σ with the equivalence class of τ always assuming that the final point of σ coincides with the initial point of τ. Relying on these definitions we have a theorem which is very easy to prove but has an outstanding relevance:

Theorem 6.2.1. *Let $\pi_1(\mathcal{M}, x_0)$ be the set of homotopy classes of loops in the manifold $\mathcal{M}$ with base in the point $x_0 \in \mathcal{M}$. If the product law of paths is defined as we just explained above, then with respect to this operation $\pi_1(\mathcal{M}, x_0)$ is a group whose identity element is provided by the homotopy class of the constant loop at x_0 and the inverse of the homotopy class $[\sigma]$ is the homotopy class of the loop σ^{-1} defined by:*

$$\sigma^{-1}(t) = \sigma(1-t) \quad 0 \leq t \leq 1 \tag{6.10}$$

(In other words, σ^{-1} is the same path followed backwards.)

Proof of Theorem 6.2.1. Clearly the composition of a loop σ with the constant loop (from now on denoted as x_0) yields σ. Hence x_0 is effectively the identity element of the group. We still have to show that $\sigma\sigma^{-1} \simeq x_0$. The explicit realization of the required homotopy is provided by the following function:

$$F(s,t) = \begin{cases} \sigma(2s) & 0 \le 2s \le t \\ \sigma(t) & t \le 2s \le 2-t \\ \sigma^{-1}(2s-1) & 2-t \le 2s \le 2 \end{cases} \tag{6.11}$$

Let us observe that having defined F as above, we have:

$$\begin{aligned} F(s,0) &= \sigma(0) = x_0 \quad \forall s \in I \\ F(s,1) &= \begin{cases} \sigma(2s) & 0 \le s \le \frac{1}{2} \\ \sigma^{-1}(2s-1) & \frac{1}{2} \le s \le 1 \end{cases} \end{aligned} \tag{6.12}$$

and furthermore:

$$\begin{aligned} F(0,t) &= \sigma(0) = x_0 \quad \forall t \in I \\ F(1,t) &= \sigma^{-1}(1) = x_0 \quad \forall t \in I \end{aligned} \tag{6.13}$$

Therefore it is sufficient to check that $F(s,t)$ is continuous. Dividing the square $[0,1] \times [0,1]$ into three triangles as in Figure 6.5 we see that $F(s,t)$ is continuous in each of the triangles and that it is consistently glued on the sides of the triangles. Hence F as defined in eq. (6.11) is continuous. This concludes the proof of the theorem. □

Theorem 6.2.2. *Let α be a path from x_0 to x_1. Then*

$$[\sigma] \xrightarrow{\alpha} [\alpha^{-1}\sigma\alpha] \tag{6.14}$$

is an isomorphism of $\pi_1(\mathscr{M}, x_0)$ into $\pi_1(\mathscr{M}, x_1)$.

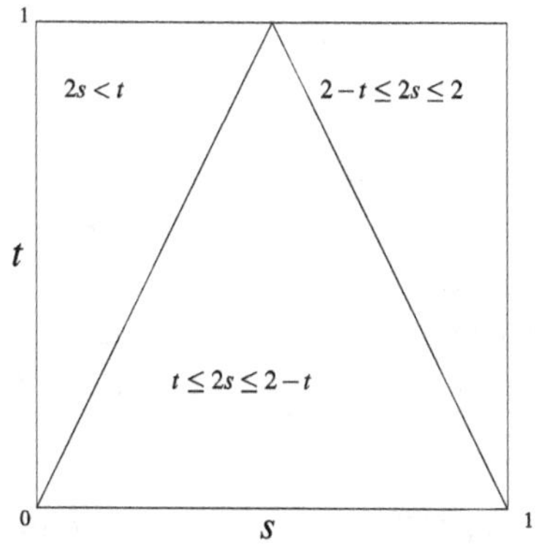

Figure 6.5: Description of the homotopy map.

Proof of Theorem 6.2.2. Indeed, since

$$[\sigma\tau] \longrightarrow [\alpha^{-1}\sigma\alpha][\alpha^{-1}\tau\alpha] = [\alpha^{-1}\sigma\tau\alpha] \tag{6.15}$$

we see that $\stackrel{\alpha}{\longrightarrow}$ is a homomorphism. Since also the inverse $\stackrel{\alpha^{-1}}{\longrightarrow}$ does exist, then the homomorphism is actually an isomorphism. □

From this theorem it follows that in an arc-wise connected manifold, namely, in a manifold where every point is connected to any other by at least one piecewise differentiable path, the group $\pi_1(\mathcal{M}, x_0)$ is independent from the choice of the base point x_0 and we can call it simply $\pi_1(\mathcal{M})$. The group $\pi_1(\mathcal{M})$ is named the first homotopy group of the manifold or simply the **fundamental group** of $\mathcal{M}$.

Definition 6.2.3. A differentiable manifold $\mathcal{M}$ which is arc-wise connected is named **simply connected** if its fundamental group $\pi_1(\mathcal{M})$ is the trivial group composed only of the identity element.

$$\pi_1(\mathcal{M}) = \mathrm{id} \quad \Leftrightarrow \quad \mathcal{M} = \text{simply connected} \tag{6.16}$$

6.2.2 Homology

The notion of homotopy led us to introduce an internal composition group for paths, the fundamental group $\pi_1(\mathcal{M})$, whose structure is a topological invariant of the manifold $\mathcal{M}$, since it does not change under continuous deformations of the latter. For this group we have used a multiplicative notation since nothing guarantees a priori that it should be abelian. Generically the fundamental homotopy group of a manifold is non-abelian. There are higher homotopy groups $\pi_n(\mathcal{M})$ whose elements are the homotopy classes of spheres $\mathbb{S}^n$ drawn on the manifold.

In this section we turn our attention to another series of groups that also codify topological properties of the manifold and are on the contrary all abelian. These are the homology groups:

$$H_k(\mathcal{M}); \quad k = 0, 1, 2, \ldots, \dim(\mathcal{M}) \tag{6.17}$$

We can grasp the notion of *homology* if we persuade ourselves that it makes sense to consider linear combinations of submanifolds or regions of dimension p of a manifold $\mathcal{M}$, with coefficients in a ring $\mathcal{R}$ that can be either $\mathbb{Z}$ or $\mathbb{R}$ or, sometimes, $\mathbb{Z}_n$. The reason is that the submanifolds of dimension p are just fit to integrate p-differential forms over them. This fact allows to give a meaning to an expression of the following form:

$$\mathcal{C}^{(p)} = m_1 S_1^{(p)} + m_2 S_2^{(p)} + \cdots + m_k S_k^{(p)} \tag{6.18}$$

where $S_i^{(p)} \subset \mathscr{M}$ are suitable p-dimensional submanifolds of the manifold $\mathscr{M}$, later on called *simplexes*, and $m_i \in \mathscr{R}$ are elements of the chosen ring of coefficients. What we systematically do is the following. For each differential p-form $\omega^{(p)} \in \Lambda_p(\mathscr{M})$ we set:

$$\int_{\mathscr{C}^{(p)}} \omega^{(p)} = \int_{m_1 S_1^{(p)} + m_2 S_2^{(p)} + \cdots + m_k S_k^{(p)} \mathscr{C}^{(p)}} \omega^{(p)} = \sum_{i=1}^{k} m_i \int_{S_i^{(p)}} \omega^{(p)} \tag{6.19}$$

and in this we define the integral of $\omega^{(p)}$ on the region $\mathscr{C}^{(p)}$. Next let us give the precise definition of the p-simplexes of which we want to take linear combinations.

Definition 6.2.4. Let us consider the Euclidean space $\mathbb{R}^{p+1}$. The standard p-simplex Δ^p is the set of all points $\{t_0, t_1, \ldots, t_p\} \in \mathbb{R}^{p+1}$ such that the following conditions are satisfied:

$$t_i \geq 0; \quad t_0 + t_1 + \cdots + t_p = 1 \tag{6.20}$$

It is just easy to see that the standard 0-simplex is a point, namely $t_0 = 1$, the standard 1-simplex is a segment of line, the standard 2-simplex is a triangle, the standard 3-simplex is a tetrahedron, and so on (see Figure 6.6).

Let us now consider the standard $(p-1)$-simplex $\Delta^{(p-1)}$ and let us observe that there are $(p+1)$ canonical maps ϕ_i that map $\Delta^{(p-1)}$ to Δ^p:

$$\phi_i : \Delta^{(p-1)} \mapsto \Delta^p \tag{6.21}$$

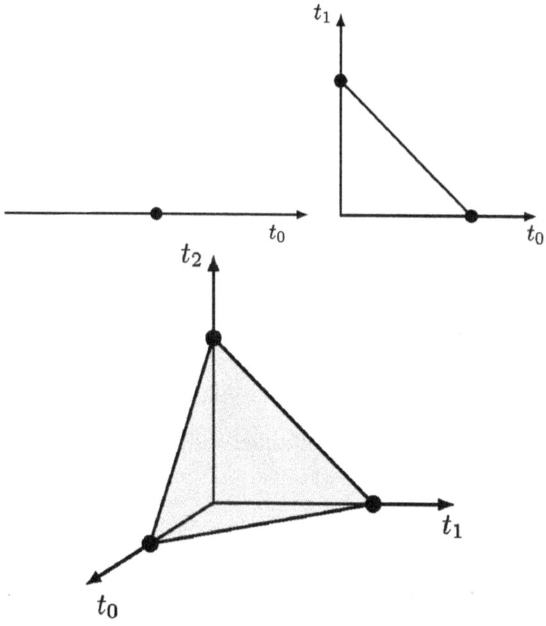

Figure 6.6: The standard p-simplexes for $p = 0, 1, 2$.

These maps are defined as follows:

$$\phi_i^{(p)}(t_0, \dots, t_{i-1}, t_{i+1}, \dots, t_p) = (t_0, \dots, t_{i-1}, 0, t_{i+1}, \dots, t_p) \tag{6.22}$$

Definition 6.2.5. The $p + 1$ standard simplexes Δ^{p-1} immersed in the standard p-simplex Δ^p by means of the $p + 1$ maps of eq. (6.22) are named the **faces** of Δ^p and the index i enumerates them. Hence the map $\phi_i^{(p)}$ yields, as a result, the ***i*-th face** of the standard p-simplex.

For instance the two faces of the standard 1-simplex are the two points ($t_0 = 0$, $t_1 = 1$) and ($t_0 = 1$, $t_1 = 0$) as shown in Figure 6.7.

Similarly, the three segments ($t_0 = 0$, $t_1 = t$, $t_2 = 1 - t$), ($t_0 = t$, $t_1 = 0$, $t_2 = 1 - t$) and ($t_0 = t$, $t_1 = 1 - t$, $t_2 = 0$) are the three faces of the standard 2-simplex (see Figure 6.8).

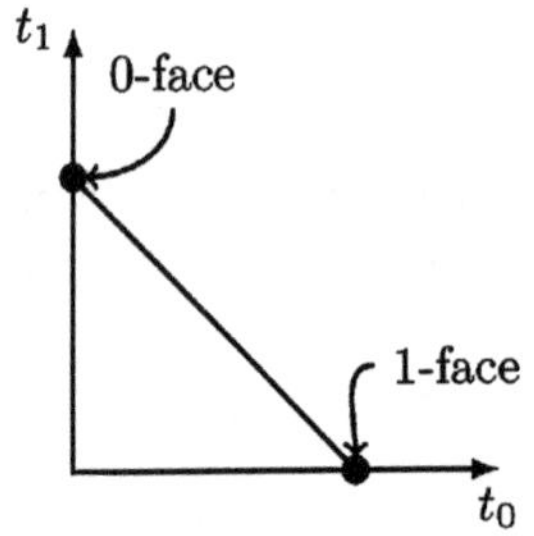

Figure 6.7: The faces of the standard 1-simplex.

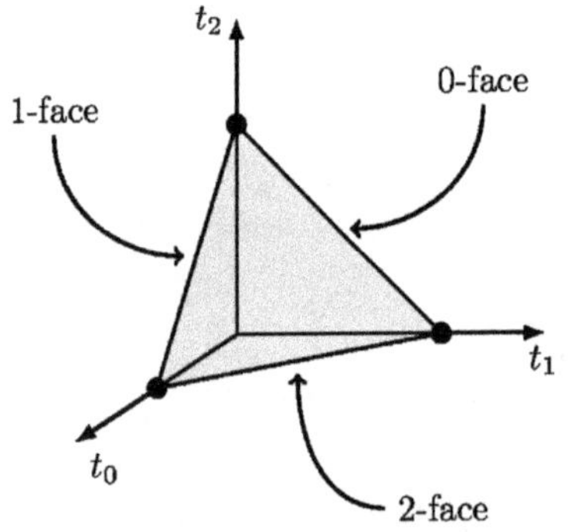

Figure 6.8: The faces of the standard 2-simplex.

Definition 6.2.6. Let $\mathscr{M}$ be a differentiable manifold of dimension m. A continuous map:

$$\sigma^{(p)} : \Delta^{(p)} \mapsto \mathscr{M} \tag{6.23}$$

of the standard p-simplex into the manifold is named a **singular *p*-simplex** or simply a **simplex** of $\mathscr{M}$.

Clearly a 1-simplex is a continuous path in $\mathcal{M}$, a 2-simplex is a portion of a surface immersed in $\mathcal{M}$, and so on. The i-th face of the simplex $\sigma^{(p)}$ is given by the $(p-1)$-simplex obtained by composing $\sigma^{(p)}$ with ϕ_i:

$$\sigma^{(p)} \circ \phi_i : \Delta^{(p-1)} \mapsto \mathcal{M} \tag{6.24}$$

Let $\mathcal{R}$ be a commutative ring.

Definition 6.2.7. Let $\mathcal{M}$ be a manifold of dimension m. For each $0 \leq n \leq m$ the group of n-chains with coefficients in $\mathcal{R}$, named $C(\mathcal{M}, \mathcal{R})$, is defined as the *free $\mathcal{R}$-module* having a generator for each n-simplex in $\mathcal{M}$.

In simple words, Definition 6.2.7 states that $C_p(\mathcal{M}, \mathcal{R})$ is the set of all possible linear combinations of p-simplexes with coefficients in $\mathcal{R}$:

$$\mathcal{C}^{(p)} = m_1 S_1^{(p)} + m_2 S_2^{(p)} + \cdots + m_k S_k^{(p)} \tag{6.25}$$

where $m_i \in \mathcal{R}$. The elements of $C_p(\mathcal{M}, \mathcal{R})$ are named **p-chains.**

The concept of p-chains gives a rigorous meaning to the intuitive idea that any p-dimensional region of a manifold can be constructed by gluing together a certain number of simplexes. For instance, a path γ can be constructed gluing together a finite number of segments (better, their homeomorphic images). In the case $p = 2$, the construction of a two-dimensional region by means of 2-simplexes corresponds to a triangulation of a surface.

As an example consider the case where the manifold we deal with is just the complex plane $\mathcal{M} = \mathbb{C}$ and let us focus on the 2-simplexes drawn in Figure 6.9.

The chain:

$$\mathcal{C}^{(2)} = S_1^{(2)} + S_2^{(2)} \tag{6.26}$$

denotes the region of the complex plane depicted in Figure 6.10, with the proviso that when we compute the integral of any 2-form on $\mathcal{C}^{(2)}$, the contribution from the simplex

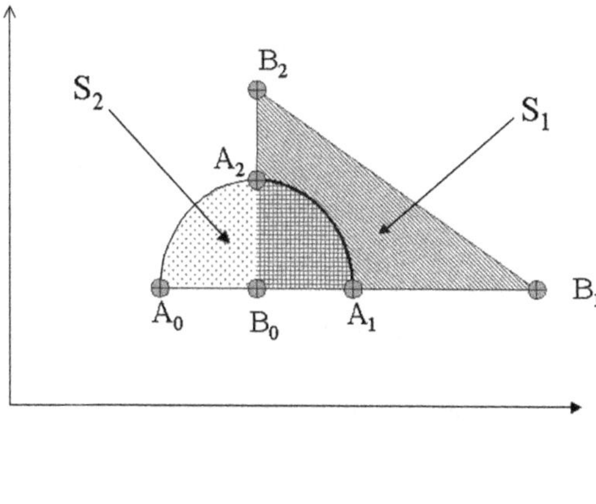

Figure 6.9: $S_1^{(2)}$ and $S_2^{(2)}$ are two distinct 2-simplexes, namely, two triangles with vertices respectively given by (A_0, A_1, A_2) and B_0, B_1, B_2. The 2-simplex $S_3^{(2)}$ with vertices B_0, A_1, A_2 is the intersection of the other two $S_3^{(2)} = S_1^{(2)} \cap S_2^{(2)}$.

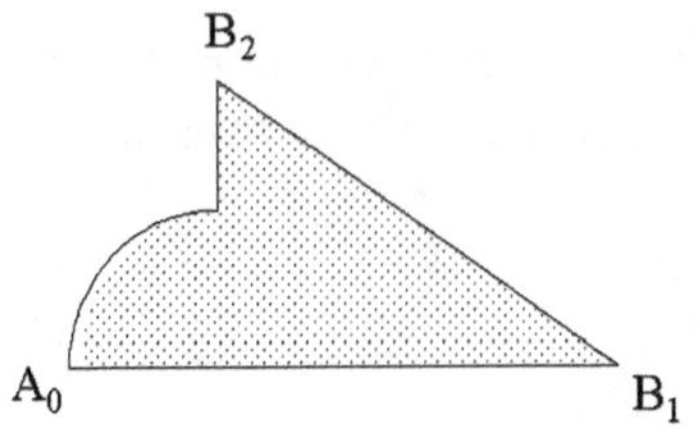

Figure 6.10: Geometrically the chain $S_1^{(2)} + S_2^{(2)}$ is the union of the two simplexes $S_1^{(2)} \cup S_2^{(2)}$.

$S_3^{(2)} = S_1^{(2)} \cap S_2^{(2)}$ (the shadowed area in Figure 6.10) has to be counted twice since it belongs both to $S_1^{(2)}$ and to $S_2^{(2)}$.

Relying on these notions we can introduce the boundary operator.

Definition 6.2.8. The boundary operator ∂ is the map:

$$\partial : C_n(\mathcal{M}, \mathcal{R}) \to C_{n-1}(\mathcal{M}, \mathcal{R}) \tag{6.27}$$

defined by the following properties:

1. **$\mathcal{R}$-linearity**

$$\begin{aligned} &\forall \mathcal{C}_1^{(p)}, \mathcal{C}_2^{(p)} \in C_p(\mathcal{M}, \mathcal{R}),\ \forall m_1, m_2 \in \mathcal{R} \\ &\partial(m_1 \mathcal{C}_1^{(p)} + m_2 \mathcal{C}_2^{(p)}) = m_1 \partial \mathcal{C}_1^{(p)} + m_2 \partial \mathcal{C}_2^{(p)} \end{aligned} \tag{6.28}$$

2. **Action on the simplexes**

$$\begin{aligned} \partial \sigma &\equiv \sigma \circ \phi_0 - \sigma \circ \phi_1 + \sigma \circ \phi_2 - \cdots \\ &= \sum_{i=1}^{p} (-)^i \sigma \circ \phi_i \end{aligned} \tag{6.29}$$

The image of a chain $\mathcal{C}$ through ∂, namely $\partial \mathcal{C}$, is called the **boundary** of the chain.

As an exercise we can compute the boundary of the 2-chain $\mathcal{C}^{(2)} = \mathcal{S}_1^{(2)} + \mathcal{S}_2^{(2)}$ of Figure 6.9, with the understanding that the relevant ring is, in this case, $\mathbb{Z}$. We have:

$$\begin{aligned} \partial C^{(2)} &= \partial S_1^{(2)} + \partial S_2^{(2)} \\ &= \overrightarrow{A_1 A_2} - \overrightarrow{A_0 A_2} + \overrightarrow{A_0 A_1} + \overrightarrow{B_1 B_2} - \overrightarrow{B_0 B_2} + \overrightarrow{B_1 B_2} \end{aligned} \tag{6.30}$$

where $\mathbf{A_1 A_2}, \ldots$ denote the oriented segments from A_1 to A_2 and so on. As one sees, the change in sign is interpreted as the change of orientation (which is the correct interpretation if one thinks of the chain and of its boundary as the support of an integral). With this convention the 1-chain:

$$\overrightarrow{A_1 A_2} - \overrightarrow{A_0 A_2} + \overrightarrow{A_0 A_1} = \overrightarrow{A_1 A_2} + \overrightarrow{A_2 A_0} + \overrightarrow{A_0 A_1} \tag{6.31}$$

is just the oriented boundary of the $S_1^{(2)}$-simplex as shown in Figure 6.11.

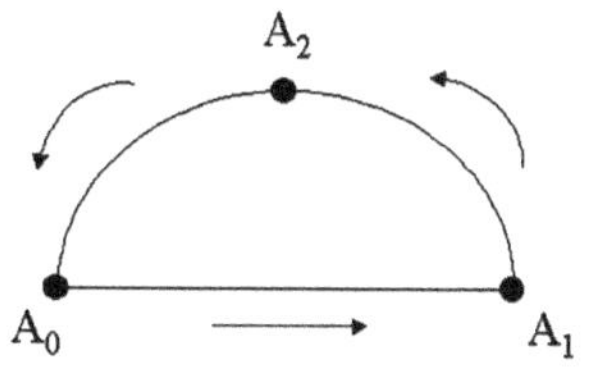

Figure 6.11: The oriented boundary of the $S^{(2)}$ simplex.

Theorem 6.2.3. *The boundary operator ∂ is nilpotent, namely, it is true that:*

$$\partial^2 \equiv \partial \circ \partial = 0 \tag{6.32}$$

Proof of Theorem 6.2.3. It is sufficient to observe that, as a consequence of their own definition, the maps ϕ_i defined in eq. (6.22) have the following property:

$$\phi_i^{(p)} \circ \phi_j^{(p-1)} = \phi_j^{(p)} \circ \phi_{i-1}^{(p-1)} \tag{6.33}$$

Then, for the p-simplex σ we have:

$$\begin{aligned}
\partial\partial\sigma &= \sum_{i=0}^{p} (-)^i \partial[\sigma \circ \phi_i] \\
&= \sum_{i=0}^{p}\sum_{j=0}^{p-1} (-)^i (-)^j \sigma \circ (\phi_i^{(p)} \circ \phi_j^{(p-1)}) \\
&= \sum_{j<i=1}^{p} (-)^{i+j} \sigma \circ (\phi_j^{(p)} \circ \phi_{i-1}^{(p-1)}) + \sum_{0=i\leq j}^{p-1} \sigma(\phi_i^{(p)} \circ \phi_j^{(p-1)})
\end{aligned} \tag{6.34}$$

We can verify that everything in the last line of eq. (6.34) cancels identically and this proves the theorem. □

As an illustration we can calculate $\partial\partial S_1^{(2)}$ for the 2-simplex $S_1^{(2)}$ described in Figure 6.9. We obtain:

$$\partial\partial S_1^{(2)} = A_2 - A_1 - A_2 + A_0 + A_1 - A_0 = 0 \tag{6.35}$$

The nilpotency of the boundary operator ∂ that acts on the *chains* is the counterpart of the nilpotency of the exterior derivative d that acts on differential forms as explained in Section 5.7.4. Consider Figure 6.12. As one sees, the sequence of the vector spaces C_m of m-chains can be put into correspondence with the sequence of vector spaces Λ_m of differential m-forms.

The operator:

$$\partial : C_k \to C_{k-1} \tag{6.36}$$

makes you travel on the sequence from left to right, while the exterior derivative operator:

$$d : \Lambda_k \to \Lambda_{k-1} \tag{6.37}$$

makes you travel along the same sequence in the opposite direction, from right to left. Both ∂ and d are nilpotent maps.

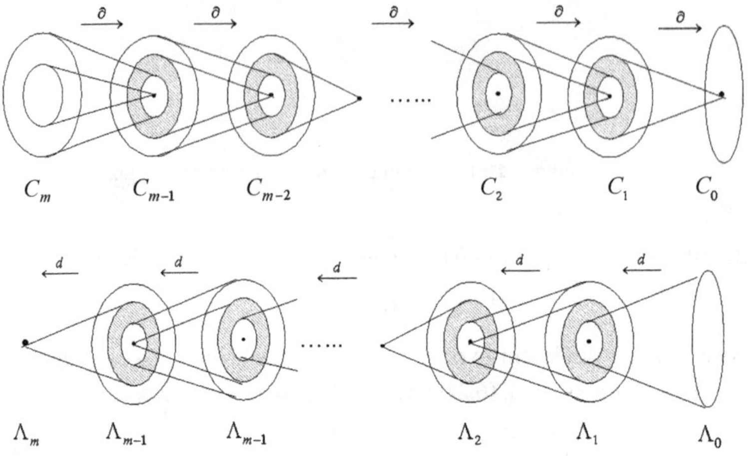

Figure 6.12: Homology versus cohomology groups.

6.2.3 Homology and cohomology groups: general construction

Let $\pi : X \to Y$ be a linear map between vector spaces. We define **kernel** of π and we denote $\ker \pi$ the subspace of X whose elements have the property of being mapped onto $0 \in Y$ by π:

$$\ker \pi = \{x \in X \,/\, \pi(x) = 0 \in Y\} \tag{6.38}$$

We call image of π and we denote $\operatorname{Im} \pi$ the subspace of Y whose elements have the property that they are the image through π of some element of X:

$$\operatorname{Im} \pi = \{y \in Y \,/\, \exists x \in X \,/\, \pi(x) = y\} \tag{6.39}$$

A nilpotent operator that acts on a sequence of vector spaces X_i defines a sequence of linear maps π_i:

$$X_1 \xrightarrow{\pi_1} X_2 \xrightarrow{\pi_2} X_3 \longrightarrow \dots X_i \xrightarrow{\pi_i} X_{i+1} \tag{6.40}$$

that have the following property:

$$\operatorname{Im} \pi_i \subset \ker \pi_{i+1} \tag{6.41}$$

The inclusion of $\operatorname{Im} \pi_i$ in $\ker \pi_{i+1}$ is what has been pictorially described in Figure 6.12 and applies both to the boundary and to the exterior derivative operator. This situation suggests the following terminology:

Definition 6.2.9. In every space $C_k(\mathcal{M}, \mathbb{R})$ we name **cycles** the elements of $\ker \partial$, namely the chains $\mathcal{C}$, whose boundary vanishes $\partial\mathcal{C} = 0$. Similarly, in every space $\Lambda_k(\mathcal{M})$ we name closed forms or **cocycles** the elements of $\ker d$, namely, the differential forms ω such that $d\omega = 0$.

At the same time:

Definition 6.2.10. In every space $C_k(\mathscr{M}, R)$ we name **boundaries** all k-chains that are the boundary of a $(k+1)$-chain:

$$\mathscr{C}^{(k)} = \text{boundary} \quad \Leftrightarrow \quad \exists \mathscr{C}^{(k+1)} \,/\, \mathscr{C}^{(k)} = \partial \mathscr{C}^{(k+1)} \tag{6.42}$$

Similarly, in every space $\Lambda_k(\mathscr{M})$ we name exact forms or **coboundaries** all differential forms $\omega^{(k)}$ such that they can be written as the exterior derivative of a $(k-1)$-form: $\omega^{(k)} = d\omega^{(k-1)}$.

Clearly, eq. (6.41) can be translated by saying that every boundary is a cycle and every coboundary is a cocycle. The reverse statement, however, is not true in general. There are cycles that are not boundaries and there are cocycles that are not coboundaries.

The concept of homology (or cohomology) previously discussed in an intuitive way can be formalized in the following way.

Definition 6.2.11. Consider the k-cycles. We say that two cycles $C_1^{(k)}$ and $C_2^{(k)}$ are **homologous** and we write $C_1^{(k)} \sim C_2^{(k)}$ if their difference is a boundary:

$$C_1^{(k)} \sim C_2^{(k)} \quad \Rightarrow \quad \exists C_3^{(k+1)} \,/\, C_1^{(k)} - C_2^{(k)} = \partial C_3^{(k+1)} \tag{6.43}$$

Clearly, homology is an equivalence relation since:

$$\left.\begin{aligned} C_1^{(k)} - C_2^{(k)} &= \partial C_a^{(k+1)} \\ C_2^{(k)} - C_3^{(k)} &= \partial C_b^{(k+1)} \end{aligned}\right\} \quad \Rightarrow \quad C_1^{(k)} - C_3^{(k)} = \partial\left[C_a^{(k+1)} + C_b^{(k+1)}\right] \tag{6.44}$$

Definition 6.2.12. We name **k-th homology group** and we denote $H_k(\mathscr{M}, \mathbb{R})$ the group of equivalence classes of the k-th cycles with respect to the k-boundaries.

Similarly we define **k-th cohomology group** and we denote $H^k(\mathscr{M}, \mathbb{R})$ the group of equivalence classes of the k-cocycles with respect to the k-th coboundaries. Indeed, we say that two closed forms ω and ω' are cohomologous if their difference is an exact form: $\omega \sim \omega' \Rightarrow \exists \phi \,/\, \omega - \omega' = d\phi$.

More generally, when we have a sequence of vector spaces X_i as in eq. (6.40) and a sequence of linear maps π_i satisfying eq. (6.41), we define the **cohomology groups** relative to the operator π as:

$$H^i_{(\pi)} \equiv \frac{\ker \pi_i}{\operatorname{Im} \pi_{i-1}} \tag{6.45}$$

The relation existing between homology and cohomology is fully contained in the following formula which generalizes to an arbitrary smooth manifold and to differential forms of any degree the familiar Gauss lemma or Stokes lemma:

$$\int_{\partial\mathscr{C}^{(k+1)}} \omega^{(k)} = \int_{\mathscr{C}^{(k+1)}} d\omega^{(k)} \tag{6.46}$$

Equation (6.46), whose general proof we omit, implies that in the case $\mathscr{C}^{(k)}$ is a cycle we have:

$$\int_{\mathscr{C}^{(k)}} [\omega^{(k)} + d\phi^{(k-1)}] = \int_{\mathscr{C}^{(k)}} \omega^{(k)} \tag{6.47}$$

namely, the integral of a closed differential form along a cycle depends only on the cohomology class and not on the choice of the representative. Similarly, if $\omega^{(k)}$ is a closed form:

$$\int_{\mathscr{C}^{(k)}+\partial\mathscr{C}^{(k+1)}} \omega^{(k)} = \int_{\mathscr{C}^{(k)}} \omega^{(k)} \tag{6.48}$$

namely, the integral of a cocycle along a cycle depends on the homology class of the class and not on the choice of the representative inside the class.

6.2.4 Relation between homotopy and homology

The relation between homotopy and homology groups of a manifold is provided by a fundamental theorem of algebraic geometry that we state without proof:

Theorem 6.2.4. *Let $\mathscr{M}$ be a smooth manifold. Then there exists a homomorphism:*

$$\chi : \pi_1(\mathscr{M}) \to H_1(\mathscr{M}, \mathbb{Z}) \tag{6.49}$$

that sends the homotopy class of each loop γ into the 1-simplex γ. If $\mathscr{M}$ is arc-wise connected, then the map χ is surjective and the kernel of χ is the subgroup of commutators in $\pi_1(\mathscr{M})$.

We recall that the subgroup of commutators of a discrete group G is the group G' generated by all elements of the form $x^{-1}y^{-1}xy$ for some $x, y \in G$.

From this theorem we have two consequences:

Corollary 6.2.1. *If $\pi_1(\mathscr{M})$ is abelian, then $H_1(\mathscr{M}) \simeq \pi_1(\mathscr{M})$, namely, the homotopy and cohomology groups coincide.*

Corollary 6.2.2. *If a manifold $\mathscr{M}$ is simply connected ($\pi_1(\mathscr{M}) = 1$) then also the first homology group is trivial $H_1(\mathscr{M}) = 0$.*

The second of the above corollaries implies that in a simply connected manifold every closed loop is homologous to zero, namely, it is the boundary of some region.

6.3 Linear algebra preliminaries

Let us consider a vector space V constructed over the field of complex numbers $\mathbb{C}$, whose dimension we denote by $\dim V = n$. We name $\mathrm{Hom}(V,V)$ the ring of all linear endomorphisms of V. In other words, an element $A \in \mathrm{Hom}(V,V)$ is a linear map:

$$\begin{aligned} &A : V \to V \\ &\forall \alpha,\beta \in \mathbb{C},\ \forall \mathbf{v},\mathbf{w} \in V : \quad A(\alpha\mathbf{v} + \beta\mathbf{w}) = \alpha A(\mathbf{v}) + \beta A(\mathbf{w}) \end{aligned} \tag{6.50}$$

As we know, if $(\mathbf{e}_1,\dots,\mathbf{e}_n)$ is a basis of V, in such a basis the endomorphism A is represented by the matrix A_{ij} determined by the condition:

$$A(\mathbf{e}_j) = \mathbf{e}_i A^i{}_j \tag{6.51}$$

Indeed, if $\{v^i\}$ are the components of the vector $\mathbf{v}$ in the basis $\{\mathbf{e}_i\}$, we have:

$$A(v^j\mathbf{e}_j) = \mathbf{e}_i A^i{}_j v^j \tag{6.52}$$

and hence the components of the vector $\mathbf{v}' \equiv A(\mathbf{v})$ are given by:

$$v^{i\prime} = A^i{}_j v^j \tag{6.53}$$

The association:

$$A \mapsto A^i{}_j \tag{6.54}$$

is an isomorphism of $\mathrm{Hom}(V,V)$ onto the ring $M_n(\mathbb{C})$ of $n \times n$ matrices with complex coefficients. For simplicity, from now on in this chapter, we no longer care about upper and lower indices of matrices writing them all at the same level.

Definition 6.3.1. A matrix A_{ij} such that $A_{ij} = 0$ if $i > j$ is named **upper triangular.** A matrix such that $A_{ij} = 0$ if $i < j$ is named **lower triangular**. Finally, a matrix that is simultaneously upper and lower triangular is named **diagonal.**

We recall the concept of eigenvalue:

Definition 6.3.2. Let $A \in \mathrm{Hom}(V,V)$. A complex number $\lambda \in \mathbb{C}$ is named an **eigenvalue** of A if $\exists \mathbf{v} \in V$ such that:

$$A\mathbf{v} = \lambda\mathbf{v} \tag{6.55}$$

Definition 6.3.3. Let λ be an eigenvalue of the endomorphism $A \in \mathrm{Hom}(V,V)$, the set of vectors $\mathbf{v} \in V$ such that $A\mathbf{v} = \lambda\mathbf{v}$ is named the **eigenspace** $V_\lambda \subset V$ pertaining to the eigenvalue λ. It is obvious that it is a vector subspace.

As is known from elementary courses in Geometry and Algebra, the possible eigenvalues of A are the roots of the secular equation:

$$\det(\lambda \mathbf{1} - \mathscr{A}) = 0 \tag{6.56}$$

where $\mathbf{1}$ is the unit matrix and $\mathscr{A}$ is the matrix representing the endomorphism A in an arbitrary basis.

Definition 6.3.4. An endomorphism $N \in \mathrm{Hom}(V, V)$ is named nilpotent if there exists an integer: $\exists k \in \mathbb{N}$ such that:

$$N^k = 0 \tag{6.57}$$

Lemma 6.3.1. *A nilpotent endomorphism has always the unique eigenvalue* $0 \in \mathbb{C}$.

Proof of Lemma 6.3.1. Let λ be an eigenvalue and let $\mathbf{v} \in V_\lambda$ be an eigenvector. We have:

$$N^r \mathbf{v} = \lambda^r \mathbf{v} \tag{6.58}$$

Choosing $r = k$ we obtain $\lambda^k = 0$ which necessarily implies $\lambda = 0$. □

Lemma 6.3.2. *Let* $N \in \mathrm{Hom}(V, V)$ *be a nilpotent endomorphism. In this case one can choose a basis* $\{\mathbf{e}_i\}$ *of* V *such that in this basis the matrix* N_{ij} *satisfies the condition* $N_{ij} = 0$ *for* $i \geq j$.

Proof of Lemma 6.3.2. Let $\mathbf{e}_1$ be a null eigenvector of N, namely $N\mathbf{e}_1 = 0$, and let E_1 be the subspace of V generated by $\mathbf{e}_1$. From N we induce an endomorphism N_1 acting on the space V/E_1, namely, the vector space of equivalence classes of vectors in V modulo the relation:

$$\mathbf{v} \sim \mathbf{w} \quad \Leftrightarrow \quad \mathbf{v} - \mathbf{w} = m\mathbf{e}_1, \quad m \in \mathbb{C} \tag{6.59}$$

Also, the new endomorphism $N_1 : V/E_1 \to V/E_1$ is nilpotent. If $\dim V/E_1 \neq 0$, then we can find another vector $\mathbf{e}_2 \in V$ such that $(\mathbf{e}_2 + E_1) \in V/E_1$ is an eigenvector of N_1. Continuing iteratively this process we obtain a basis $\mathbf{e}_1, \dots, \mathbf{e}_n$ of V such that:

$$N\mathbf{e}_1 = 0; \quad N\mathbf{e}_p = 0 \bmod(\mathbf{e}_1, \dots, \mathbf{e}_{p-1}); \quad 2 \leq p \leq n \tag{6.60}$$

where $(\mathbf{e}_1, \dots, \mathbf{e}_{p-1})$ denotes the subspace of V generated by the vectors $\mathbf{e}_1, \dots, \mathbf{e}_{p-1}$. In this basis the matrix representing N is triangular. Similarly, if N_{ij} is triangular with $N_{ij} = 0$ for $i \geq j$, then the corresponding endomorphism is nilpotent. □

Definition 6.3.5. Let $\mathscr{S} \subset \mathrm{Hom}(V, V)$ be a subset of the ring of endomorphisms and $W \subset V$ a vector subspace. The subspace W is named **invariant** with respect to $\mathscr{S}$ if $\forall S \in \mathscr{S}$ we have $SW \subset W$. The space V is named **irreducible** if it does not contain invariant subspaces.

Definition 6.3.6. A subset $\mathscr{S} \subset \mathrm{Hom}(V, V)$ is named **semisimple** if every invariant subspace $W \subset V$ admits an orthogonal complement which is also invariant. In that case we can write:

$$V = \bigoplus_{i=1}^{p} W_i \tag{6.61}$$

where each subspace W_i is invariant.

A fundamental and central result in Linear Algebra, essential for the further development of Lie algebra theory is the *Jordan's decomposition theorem* that we quote without proof.

Theorem 6.3.1. *Let $L \in \mathrm{Hom}(V, V)$ be an endomorphism of a finite dimensional vector space V. Then there exists and is unique the following Jordan decomposition:*

$$L = S_L + N_L \tag{6.62}$$

where S_L is semisimple and N_L is nilpotent. Furthermore, both S_L and N_L can be expressed as polynomials in L.

6.4 Types of Lie algebras and Levi's decomposition

In the previous section we have discussed the notion of semisimplicity and of nilpotency for endomorphisms of vector spaces, namely for matrices. Such notions can now be extended to entire Lie algebras. This is not surprising since Lie algebras admit linear representations where each of their elements is replaced by a matrix. In the present section we discuss *solvable*, *nilpotent* and *semisimple* Lie algebras. Solvable and nilpotent Lie algebras are those for which all linear representations are provided by triangular matrices. Semisimple Lie algebras are those which do not admit any invariant subalgebra (or ideal) which is solvable. The main result in this section will be Levi's theorem which is the counterpart for algebras of Jordan's decomposition theorem 6.3.1 holding true for matrices.

Consider a Lie algebra $\mathbb{G}$ and define:

$$\mathscr{D}\mathbb{G} = [\mathbb{G}, \mathbb{G}] \tag{6.63}$$

the set of all elements $\mathbf{X} \in \mathbb{G}$ that can be written as the Lie bracket of two other elements $\mathbf{X} = [\mathbf{X}_1, \mathbf{X}_2]$. Clearly, $\mathscr{D}\mathbb{G}$ is an ideal in $\mathbb{G}$.

Definition 6.4.1. The sequence $\mathscr{D}^n\mathbb{G} = [\mathscr{D}^{n-1}\mathbb{G}, \mathscr{D}^{n-1}\mathbb{G}]$ of ideals:

$$G \supset \mathscr{D}\mathbb{G} \supset \mathscr{D}^2\mathbb{G} \supset \cdots \supset \mathscr{D}^n\mathbb{G} \tag{6.64}$$

is named the **derivative series** of the Lie algebra $\mathbb{G}$.

6.4.1 Solvable Lie algebras

Definition 6.4.2. A Lie algebra $\mathbb{G}$ is named **solvable** if there exists an integer $n \in \mathbb{N}$ such that

$$\mathscr{D}^n\mathbb{G} = \{0\} \tag{6.65}$$

Lemma 6.4.1. *A subalgebra $\mathbb{K} \subset \mathbb{G}$ of a solvable Lie algebra is also solvable.*

Proof of Lemma 6.4.1. Indeed, let G be solvable and $\mathbb{K} \subset \mathbb{G}$ be a subalgebra. Clearly, $\mathscr{D}\mathbb{K} \subset \mathscr{D}\mathbb{G}$ and hence at every level n we have $\mathscr{D}^n\mathbb{K} \subset \mathscr{D}^n\mathbb{G}$, so that the lemma follows. □

Definition 6.4.3. A Lie algebra $\mathbb{G}$ has the **chain property** if and only if, for each ideal $\mathbb{H} \subset \mathbb{G}$, there exists an ideal $\mathbb{H}_1 \subset \mathbb{H}$ of the considered ideal which has codimension one in $\mathbb{H}$.

The above definition can be illustrated in the following way. Let $\mathbb{H}$ be the considered ideal in $\mathbb{G}$. If G has the chain property, then $\mathbb{H}$ can be written in the following way:

$$\mathbb{H} = \mathbb{H}_1 \oplus \lambda\mathbf{X} \tag{6.66}$$

where $\mathbb{H}_1$ is a subspace of dimension:

$$\dim \mathbb{H}_1 = \dim \mathbb{H} - 1 \tag{6.67}$$

and $\mathbf{X} \in \mathbb{H}$, $(\mathbf{X} \notin \mathbb{H}_1)$ is an element that belongs to $\mathbb{H}$ but it does not to $\mathbb{H}_1$. Furthermore, we have:

$$\forall \mathbf{Z} \in \mathbb{H}_1, \quad [\mathbf{Z}, \mathbf{X}] \in \mathbb{H}_1 \tag{6.68}$$

From this definition we obtain the following:

Lemma 6.4.2. *A Lie algebra $\mathbb{G}$ is solvable if and only if it admits the chain property.*

Proof of Lemma 6.4.2. Let G be solvable and let us put $\dim \mathbb{G} = n$ and $\dim \mathscr{D}\mathbb{G} = m$. By hypothesis of solvability we have that $\mathscr{D}\mathbb{G} \neq \mathbb{G}$, so that $n - m = p > 0$. Let us choose $p - 1$ linear independent elements $\{\mathbf{X}_1, \dots, \mathbf{X}_{p-1}\} \in \mathbb{G}$, such that $\mathbf{X}_i \notin \mathscr{D}\mathbb{G}$, and define the subspace:

$$H_1 = \mathscr{D}\mathbb{G} + \lambda_1\mathbf{X}_1 + \cdots + \lambda_{p-1}\mathbf{X}_{p-1}, \quad (\lambda_i \in \mathbb{C}) \tag{6.69}$$

By construction H_1 has codimension one and is an ideal. This construction can be repeated for each ideal $\mathbb{H} \subset \mathbb{G}$ since it is solvable. Hence $\mathbb{G}$ admits the chain property.

Conversely, let $\mathbb{G}$ be a Lie algebra admitting the chain property. Then we find a sequence of ideals:

$$\mathbb{G} = \mathbb{G}^0 \supset \mathbb{G}^1 \supset \mathbb{G}^2 \supset \cdots \supset \mathbb{G}^n = \{0\} \tag{6.70}$$

such that $\mathbb{G}^r$ is an ideal in $\mathbb{G}^{r-1}$ of codimension one so that $\mathscr{D}^{r-1}\mathbb{G} \subset \mathbb{G}^r$. Hence $\mathbb{G}$ is solvable. □

We can now state the most relevant property of solvable Lie algebras. This is encoded in the following Levi's theorem and in its corollary.

Theorem 6.4.1. *Let $\mathbb{G}$ be a solvable Lie algebra and let V be a finite dimensional vector space over the field $\mathbb{F} = \mathbb{R}$, or $\mathbb{C}$, algebraically closed. Furthermore, let:*

$$\pi : \mathbb{G} \to \mathrm{Hom}(V, V) \tag{6.71}$$

be a homomorphism of $\mathbb{G}$ on the algebra of linear endomorphisms of V. Then there exists a vector $\mathbf{v} \in V$ such that it is a simultaneous eigenvector for all elements $\pi(\mathbf{X})$, $(\forall \mathbf{X} \in \mathbb{G})$.

Proof of Theorem 6.4.1. The proof is constructed by induction. If $\dim \mathbb{G} = 1$, then there is just one endomorphism $\pi(g)$ and it necessarily admits an eigenvector. Suppose next that the theorem is true for each solvable algebra $\mathbb{K}$ of dimension

$$\dim \mathbb{K} < \dim \mathbb{G} \tag{6.72}$$

Consider an ideal $\mathbb{H} \subset \mathbb{G}$ of codimension one:

$$\dim \mathbb{G} = \dim \mathbb{H} + 1 \tag{6.73}$$

Such an ideal exists because the Lie algebra is solvable and, therefore, admits the chain property. Write:

$$\mathbb{G} = \mathbb{H} + \lambda \mathbf{X} \quad (\lambda \in \mathbb{F}) \tag{6.74}$$

where $\mathbf{X}$ is an element of $\mathbb{G}$ not contained in $\mathbb{H}$. By the induction hypothesis there exists a vector $\mathbf{e}_0 \in V$ such that:

$$\forall \mathbf{H} \in \mathbb{H} : \quad \pi(\mathbf{H})\mathbf{e}_0 = \lambda(\mathbf{H})\mathbf{e}_0 \tag{6.75}$$

where $\lambda(\mathbf{H}) \in \mathbb{F}$ is an eigenvalue depending on the considered element $\mathbf{H}$. Define next the following vectors:

$$\mathbf{e}_p = [\pi(\mathbf{X})]^p \mathbf{e}_0 \quad p = 1, 2, \ldots \tag{6.76}$$

The subspace $W \subset V$ spanned by the vectors $\mathbf{e}_p$ $(p \geq 0)$ is clearly invariant with respect to $\pi(\mathbf{X})$. We can also show what follows:

$$\pi(\mathbf{H})\mathbf{e}_p = \lambda(\mathbf{H})\mathbf{e}_p \bmod(\mathbf{e}_0, \dots, \mathbf{e}_{p-1}); \quad (\forall \mathbf{H} \in \mathbb{H}) \tag{6.77}$$

Indeed, eq. (6.77) is true for $p = 0$ and assuming it true for p we get:

$$\begin{aligned} \pi(\mathbf{H})\mathbf{e}_{p+1} &= \pi(\mathbf{H})\pi(\mathbf{X})\mathbf{e}_p = \pi([\mathbf{H}, X])\mathbf{e}_p + \pi(\mathbf{X})\pi(\mathbf{H})\mathbf{e}_p \\ &= \lambda([\mathbf{H}, \mathbf{X}])\mathbf{e}_p + \lambda(\mathbf{H})\mathbf{e}_{p+1} + \bmod(\mathbf{e}_0, \dots, \mathbf{e}_{p-1}) \end{aligned} \tag{6.78}$$

(Note that $[\mathbf{H}, \mathbf{X}] \in \mathbb{H}$.) Hence we find:

$$\pi(\mathbf{H})\mathbf{e}_{p+1} = \lambda(\mathbf{H})\mathbf{e}_{p+1} + \bmod(\mathbf{e}_0, \dots, \mathbf{e}_p) \tag{6.79}$$

It follows that the subspace W is invariant with respect to $\pi(\mathbb{G})$ and that:

$$\mathrm{Tr}_W \pi(\mathbf{H}) = \lambda(\mathbf{H}) \dim W \tag{6.80}$$

On the other hand, we have:

$$\mathrm{Tr}_W(\pi([\mathbf{H}, \mathbf{X}])) = 0 \quad \Rightarrow \quad \lambda([\mathbf{H}, \mathbf{X}]) = 0 \tag{6.81}$$

Repeating the argument by induction, from the relation:

$$\pi(\mathbf{H})\mathbf{e}_{p+1} = \pi([\mathbf{H}, \mathbf{X}])\mathbf{e}_p + \pi(\mathbf{X})\pi(\mathbf{H})\mathbf{e}_p \tag{6.82}$$

and the original definition of the eigenvalue $\lambda(\mathbf{H})$ in eq. (6.75) we conclude that:

$$\pi(\mathbf{H})\mathbf{e}_p = \lambda(\mathbf{H})\mathbf{e}_p \quad (p \geq 0) \tag{6.83}$$

This shows that $\forall \mathbf{H} \in \mathbb{H}$ we have $\pi(\mathbf{H}) = \lambda(\mathbf{H})\mathbf{1}$ on the vector subspace W. Choosing a vector $\mathbf{e}'_p \in W$ that is an eigenvector of $\pi(\mathbf{X})$ we find that it is a simultaneous eigenvector for all elements $\pi(\mathbf{X})$, $(\forall \mathbf{X} \in \mathbb{G})$. □

Corollary 6.4.1. *Let* $\mathbb{G}$ *be a solvable Lie algebra and* π *a linear representation of* $\mathbb{G}$ *on a finite dimensional vector space* V. *Then there exists a basis* $\{\mathbf{e}_1, \dots, \mathbf{e}_n\}$ *where every* $\pi(\mathbf{X})$, $(\forall \mathbf{X} \in \mathbb{G})$ *is a triangular matrix.*

Proof of Corollary 6.4.1. Let $\mathbf{e}_1$ be the simultaneous eigenvector of all $\pi(\mathbf{X})$. The representation π induces a new linear representation on the quotient vector space V/E_1 where $E_1 \equiv \lambda \mathbf{e}_1$. Hence, applying Theorem 6.4.1 to this new representation we conclude that there is a common eigenvector $\mathbf{e}_2 + \lambda \mathbf{e}_1$ for all $\pi(\mathbf{X})$. Continuing in this way we obtain a basis such that:

$$\pi(\mathbf{X})\mathbf{e}_i \equiv 0 \bmod(\mathbf{e}_1, \mathbf{e}_2, \dots, \mathbf{e}_i) \tag{6.84}$$

so that $\pi(\mathbf{X})$ is indeed upper triangular. □

We are now ready to introduce the concept of semisimple Lie algebras and discuss their general properties.

6.4.2 Semisimple Lie algebras

We introduce some more definitions.

Definition 6.4.4. Let $\mathbb{G}$ be a Lie algebra. An ideal $\mathbb{H} \subset \mathbb{G}$ is named **maximal** if there is no other ideal $\mathbb{H}' \subset \mathbb{G}$ such that $\mathbb{H}' \supset \mathbb{H}$ except for $\mathbb{H}$ itself.

Definition 6.4.5. The maximal solvable ideal of a Lie algebra $\mathbb{G}$ is named the radical of $\mathbb{G}$ and is denoted $\operatorname{Rad}\mathbb{G}$.

Definition 6.4.6. A Lie algebra $\mathbb{G}$ is named semisimple if and only if $\operatorname{Rad}\mathbb{G} = 0$.

As an immediate consequence of the definition we have:

Theorem 6.4.2. *A Lie algebra $\mathbb{G}$ is semisimple if and only if it does not have any non-trivial abelian ideal.*

Proof of Theorem 6.4.2. We have to show the equivalence of the following two propositions:

a: $\mathbb{G}$ has a solvable ideal

b: $\mathbb{G}$ has an abelian ideal

i) Let us show that b $\Rightarrow$ a. Let $\mathscr{I} \subset \mathbb{G}$ be the abelian ideal. By definition we have $[\mathscr{I}, \mathscr{I}] = \mathscr{D}\mathscr{I} = 0$. Hence $\mathscr{I}$ itself is a solvable ideal and this proves a.

ii) Let us now show that a $\Rightarrow$ b. To this effect let $\mathscr{I} \subset \mathbb{G}$ be the solvable ideal. By definition, since $\mathscr{I}$ is non-trivial, $\exists k \in \mathbb{N}$ such that $\mathscr{D}^{k-1}\mathscr{I} \neq 0$ and $\mathscr{D}^{k}\mathscr{I} = 0$. Then the $\mathscr{D}^{k-1}\mathscr{I}$ is abelian and being the derivative of an ideal is an ideal. Hence b is true and this concludes the proof of the theorem. □

6.4.3 Levi's decomposition of Lie algebras

We want to prove that any Lie algebra can be seen as the semidirect product of its radical with a semisimple Lie algebra. Such a decomposition is named the Levi decomposition and is what we want to illustrate in the present section. To this effect we need to introduce some preliminary notions. The first is the notion of **Lie algebra cohomology**. It is a further very relevant example of an algebraic construction that realizes the paradigm introduced in Section 6.2.3. The second is the equally important notion of semidirect product.

6.4.3.1 Lie algebra cohomology

Let $\mathbb{G}$ be a Lie algebra and let $\rho : \mathbb{G} \to \mathrm{End}(V)$ be a representation of $\mathbb{G}$ on a complex, finite dimensional vector space V. Let $V^s(\mathbb{G},\rho)$ be the vector space of all antisymmetric linear maps

$$\theta : \underbrace{\mathbb{G} \times \mathbb{G} \times \cdots \times \mathbb{G}}_{s \text{ times}} \to V \tag{6.85}$$

The spaces $V^s(\mathbb{G},\rho)$ are the spaces of s-**cochains**. We can next define a coboundary operator d in the following way. Let $\theta \in V^s(\mathbb{G},\rho)$ be an s-cochain, the value of the $(s+1)$-cochain $d\theta$ on any set of $s+1$ elements $X_1, \dots, X_{s+1}$ of the Lie algebra $\mathbb{G}$ is given by the following expression:

$$\begin{aligned} d\theta(\mathbf{X}_1, \mathbf{X}_2, \dots, \mathbf{X}_{s+1}) = & \sum_{i=1}^{s+1} (-)^{i+1} \rho(\mathbf{X}_i) \theta(\mathbf{X}_1, \dots, \widehat{\mathbf{X}}_i, \dots, \mathbf{X}_{s+1}) \\ & - \sum_{r=1}^{s+1} \sum_{q<r} (-1)^{r+q} \theta(\mathbf{X}_1, \dots, \widehat{\mathbf{X}}_q, \dots, \widehat{\mathbf{X}}_r, \dots, \mathbf{X}_{s+1}, [\mathbf{X}_q, \mathbf{X}_r]) \end{aligned} \tag{6.86}$$

where the hat on top of an X-element means that it is omitted. It is straightforward to verify that by applying a second time the coboundary operator d we obtain identically zero, namely that:

$$d^2 = 0 \tag{6.87}$$

In particular if we consider the case of 1-cochains $\theta^{[1]}$ that are maps $\mathbb{G} \to V$ from the Lie algebra to the vector space V, by applying the general definition (6.86) we obtain:

$$\forall \mathbf{X}, \mathbf{Y} \in \mathbb{G} : \quad d\theta^{[1]}(\mathbf{X}, \mathbf{Y}) = \rho(\mathbf{X})\theta(\mathbf{Y}) - \rho(\mathbf{Y})\theta(\mathbf{X}) - \theta([\mathbf{X}, \mathbf{Y}]) \tag{6.88}$$

Given the coboundary operator d we have the usual definitions of an *elliptic complex*:

1. The space $C^{(n)}(\mathbb{G},\rho)$ of n-cocycles is the vector space of all n-chains $\theta^{[n]}$ that are closed $d\theta^{[n]} = 0$.
2. The space $B^{(n)}(\mathbb{G},\rho)$ of n-coboundaries is the vector space of all n-chains $\theta^{[n]}$ that are exact, namely, that can be written as $\theta^{[n]} = d\phi^{[n-1]}$ for some $(n-1)$-chain $\phi^{[n-1]}$.
3. The n-th cohomology group $H^{[n]}(\mathbb{G},\rho)$ of the Lie algebra $\mathbb{G}$ relative to the linear representation ρ is the quotient:

$$H^{[n]}(\mathbb{G},\rho) = \frac{C^{(n)}(\mathbb{G},\rho)}{B^{(n)}(\mathbb{G},\rho)} \tag{6.89}$$

   namely, it is the vector space whose equivalence classes are the n-cocycles modulo the n-coboundaries.

We have a useful general cohomological property of semisimple Lie algebras that follows from the above definitions but whose proof we omit for brevity.

Theorem 6.4.3. *Let $\mathbb{G}$ be a semisimple Lie algebra and ρ a linear representation of $\mathbb{G}$ on a finite dimensional vector space V. Then the first two cohomology groups are trivial:*

$$H^{[1]}(\mathbb{G},\rho) = H^{[2]}(\mathbb{G},\rho) = 0 \tag{6.90}$$

6.4.3.2 Semidirect product

We begin with two definitions:

Definition 6.4.7. Let $\mathbb{G}$ be a Lie algebra and let $\sigma : \mathbb{G} \to \mathbb{G}$ be an endomorphism of vector spaces. We say that σ is a **derivation** of the algebra if the following property holds true:

$$\forall \mathbf{X}, \mathbf{Y} \in \mathbb{G} : \quad \sigma([\mathbf{X}, \mathbf{Y}]) = [\sigma(\mathbf{X}), \mathbf{Y}] + [\mathbf{X}, \sigma(\mathbf{Y})] \tag{6.91}$$

Definition 6.4.8. Let $\mathbb{Q}$ and $\mathbb{M}$ be two Lie algebras and let σ be a linear representation of $\mathbb{M}$ on $\mathbb{Q}$ such that $\forall \mathbf{Y} \in \mathbb{M}$ the map $\sigma(\mathbf{Y})$ is a derivation of $\mathbb{Q}$. Next let $\mathbf{X}, \mathbf{X}'$ be elements of $\mathbb{Q}$ and $\mathbf{Y}, \mathbf{Y}'$ be elements of $\mathbb{M}$. We define the Lie bracket of the ordered pair $(\mathbf{X}, \mathbf{Y})$ with the ordered pair $(\mathbf{X}', \mathbf{Y}')$ in the following way:

$$[(\mathbf{X}, \mathbf{Y}), (\mathbf{X}', \mathbf{Y}')] = ([\mathbf{X}, \mathbf{X}'] + \sigma(\mathbf{Y})\mathbf{X}' - \sigma(\mathbf{Y}')\mathbf{X}, [\mathbf{Y}, \mathbf{Y}']) \tag{6.92}$$

With this definition of the Lie bracket, $\mathbb{Q} \times_\sigma \mathbb{M}$ becomes a Lie algebra and it is named the semidirect product of $\mathbb{Q}$ with $\mathbb{M}$ relative to the representation σ.

It is a straightforward exercise to check that the definition of the Lie bracket (6.92) is consistent and satisfies Jacobi identity.

Let us now consider a Lie algebra $\mathbb{G}$ and let $\mathscr{Q} \subset \mathbb{G}$ be an ideal and $\mathbb{M} \subset \mathbb{G}$ a subalgebra such that, as vector spaces, we have the following orthogonal decomposition:

$$\mathbb{G} = \mathscr{Q} \oplus \mathbb{M} \quad \Rightarrow \quad \mathscr{Q} \cap \mathbb{M} = 0 \tag{6.93}$$

Obviously $\mathbb{G}$ can be regarded as the semidirect product of $\mathbb{Q}$ with $\mathbb{M}$. It suffices to use as derivation σ the internal derivation provided by the Lie bracket of $\mathbb{G}$:

$$\forall \mathbf{Y} \in \mathbb{M}, \ \forall \mathbf{X} \in \mathscr{Q} : \quad \sigma(\mathbf{Y})\mathbf{X} \equiv -[\mathbf{X}, \mathbf{Y}] \tag{6.94}$$

Definition 6.4.9. Let $\mathbb{G}$ be a Lie algebra. We say that $\mathbb{G}$ is **decomposed according to Levi** if there exists a subalgebra $\mathbb{L} \subset \mathbb{G}$ such that:

$$\mathbb{G} = \mathbb{L} \times_{[,]} \operatorname{Rad} \mathbb{G} \tag{6.95}$$

Obviously, since $\mathbb{G}/\operatorname{Rad}\mathbb{G}$ is semisimple and since $\mathbb{L} \sim \mathbb{G}/\operatorname{Rad}\mathbb{G}$, also $\mathbb{L}$ is semisimple. It is named a **Levi subalgebra**.

Relying on this definition we can state the following fundamental theorem:

Theorem 6.4.4. *Let G be a Lie algebra and denote $\mathcal{Q} \equiv \text{Rad}\,\mathbb{G}$. Every $\mathbb{G}$ admits Levi subalgebras. Furthermore, if $\mathbb{L} \subset \mathbb{G}$ is a Levi subalgebra of $\mathbb{G}$, it is also a Levi subalgebra of $\mathcal{D}\mathbb{G}$ and*

$$\mathcal{D}\mathbb{G} = [\mathcal{Q}, \mathbb{G}] \oplus \mathbb{L} \tag{6.96}$$

is a Levi decomposition of $\mathcal{D}\mathbb{G}$.

In order to prove Theorem 6.4.4, we need the following lemma:

Lemma 6.4.3. *Let $\mathbb{H}$ be a Lie algebra and $\mathcal{Q}$ its radical. If $\mathcal{A} \subset \mathbb{H}$ is an ideal such that $\mathbb{H}/\mathcal{A}$ is semisimple, then $\mathcal{Q} \sqsubseteq \mathcal{A}$. Furthermore, if π is a homomorphism of $\mathbb{H}$ onto an algebra $\mathbb{H}'$, then $\pi(\mathcal{Q})$ is the radical of $\mathbb{H}'$.*

Proof of Lemma 6.4.3. Consider the natural map:

$$\tau : \mathbb{H} \to \mathbb{H}/\mathcal{A} \tag{6.97}$$

that to each element $h \in \mathbb{H}$ associates the equivalence class $h + \mathcal{A}$. If $\mathcal{Q} \nsubseteq \mathcal{A}$, we have that $\tau(\mathcal{Q})$ is a non-zero and solvable ideal of $\mathbb{H}/\mathcal{A}$. Indeed, under the homomorphism τ the ideal $\mathcal{Q}$ flows into an ideal $\tau(\mathcal{Q})$. Furthermore, under the homomorphism we have $\mathcal{D}\tau(\mathcal{Q}) = \tau(\mathcal{D}\mathbb{G})$ so that if $\mathcal{Q}$ is solvable, the same is true also for $\tau(\mathcal{Q})$. The existence of a solvable ideal is in contradiction with the assumption that $\mathbb{H}/\mathcal{A}$ is semisimple. Hence $\mathcal{Q} \subseteq \mathcal{A}$, necessarily.

Let us come to the second part of the lemma and let $\mathcal{Q}' = \text{Rad}\,\mathbb{H}'$. The homomorphism π induces a homomorphism of $\mathbb{H}/\mathcal{Q}$ in $\mathbb{H}'/\pi(\mathcal{Q})$, hence since $\mathbb{H}/\mathcal{Q}$ is semisimple, also $\mathbb{H}/\pi(\mathcal{Q})$ is semisimple. Therefore, relying on the previous result, $\mathcal{Q}' \subset \pi(\mathcal{Q})$. On the other hand, $\pi(\mathcal{Q})$ is a solvable ideal of $\mathbb{H}$. This implies $\pi(\mathcal{Q}) \subset \mathcal{Q}'$. We conclude $\pi(\mathcal{Q}) = \mathcal{Q}'$ and the lemma is proved. □

Let us now come to the proof of the main theorem 6.4.4.

Proof of Theorem 6.4.4. The proof of Theorem 6.4.4 is by induction on the dimension of the radical $\dim \mathcal{Q}$. If $\dim \mathcal{Q} = 0$, then $\mathbb{G}$ is semisimple and it is by itself a Levi subalgebra. Let us then assume that $\dim \mathcal{Q} \geq 1$ and that Levi subalgebras do exist for any Lie algebra $\mathbb{G}'$ such that $\dim \text{Rad}\,\mathbb{G}' < \dim \text{Rad}\,\mathbb{G}$. Consider two cases:

1st Case The radical $\mathcal{Q}$ is non-abelian, namely ($\mathcal{D}\mathcal{Q} \neq 0$).

As we know $\mathcal{D}\mathcal{Q}$ is by itself an ideal. Hence consider the Lie algebra $\mathbb{G}' \equiv \mathbb{G}/\mathcal{D}\mathcal{Q}$ and let π be the natural map:

$$\pi : \mathbb{G} \to \mathbb{G}' \equiv \mathbb{G}/\mathcal{D}\mathcal{Q} \tag{6.98}$$

Relying on Lemma 6.4.3, we have: $\mathscr{Q}' = \mathrm{Rad}\,\mathbb{G}' = \pi[\mathscr{Q}]$. Hence:

$$\mathscr{Q}' = \mathscr{Q}/\mathscr{D}\mathscr{Q} \quad \Rightarrow \quad \dim \mathscr{Q}' < \dim \mathscr{Q} \tag{6.99}$$

By induction hypothesis $\mathbb{G}'$ admits a Levi subalgebra $\mathscr{M}'$ and we have, as vector spaces:

$$\mathbb{G}' = \mathscr{Q}' \oplus \mathscr{M}' \tag{6.100}$$

Define $\mathscr{M}_0 = \pi^{-1}(\mathscr{M}')$. We obtain:

$$\mathbb{G} = \pi^{-1}(\mathbb{G}') = \mathscr{Q} \oplus \mathscr{M}_0 \tag{6.101}$$

Furthermore, it is true that $\mathscr{D}\mathscr{Q} = \mathscr{Q} \cap \mathscr{M}_0$ (indeed, the common elements of $\mathscr{Q}$ and $\mathscr{M}_0$ must be contained in the kernel of π, namely $\pi^{-1}(0)$). Hence $\mathscr{D}\mathscr{Q}$ is a solvable ideal of $\mathscr{M}_0$ and since $\mathscr{M}_0/\mathscr{D}\mathscr{Q} \sim \mathscr{M}'$ = semisimple algebra, then $\mathscr{D}\mathscr{Q} \subset \mathrm{Rad}\,\mathscr{M}_0$. Yet in force of Lemma 6.4.3 we also have $\mathrm{Rad}\,\mathscr{M}_0 \subset \mathscr{D}\mathscr{Q}$ which implies $\mathrm{Rad}\,\mathscr{M}_0 = \mathscr{D}\mathscr{Q}$. Since $\mathscr{Q}$ is solvable by definition, we have: $\dim \mathscr{D}\mathscr{Q} < \dim \mathscr{Q}$. Then by induction hypothesis we conclude that $\mathscr{M}_0$ admits a Levi decomposition:

$$\mathscr{M}_0 = \mathbb{L} \oplus \mathscr{D}\mathscr{Q}; \quad \mathbb{L} \cap \mathscr{D}\mathscr{Q} = 0 \tag{6.102}$$

from which we conclude:

$$\mathbb{G} = \mathbb{L} \oplus \mathscr{Q} \tag{6.103}$$

and the theorem is proved in this case.

2nd Case The radical $\mathscr{Q}$ is abelian, namely ($\mathscr{D}\mathscr{Q} = 0$).

To prove the theorem in this case we have to use Theorem 6.4.3 stating that the second cohomology group of a semisimple Lie algebra vanishes. Define $\mathbb{G}_1 = \mathbb{G}/\mathscr{Q}$ (so that $\mathbb{G}_1$ is semisimple) and let π be the natural map of $\mathbb{G}$ onto $\mathbb{G}_1$. Let μ be any linear map of $\mathbb{G}_1$ onto $\mathbb{G}$ such that $\pi \circ \mu = \mathrm{id}$. For each $\mathbf{X}_1 \in \mathbb{G}_1$ define $\rho(\mathbf{X}_1)$ the endomorphism $\mathrm{ad}(\mathbf{X})|_{\mathscr{Q}}$ where $\mathbf{X}$ is such that $\pi(\mathbf{X}) = \mathbf{X}_1$. Since $\mathscr{Q}$ is abelian, this is a well-posed definition. Indeed, $\mathbf{X}_1 = \mathbf{X} + \mathscr{Q}$ and $\forall q \in \mathscr{Q}$:

$$\rho(\mathbf{X}_1)q = [\mathbf{X} + \mathscr{Q}, q] = [\mathbf{X}, q] \tag{6.104}$$

The map $\mathbf{X}_1 \mapsto \rho(\mathbf{X}_1)$ is a linear representation of the semisimple Lie algebra $\mathbb{G}_1$ on the vector space $\mathscr{Q}$. Obviously $\rho(\mathbf{X}_1) = \mathrm{ad}\,\mu(\mathbf{X}_1)|_{\mathbb{G}}$. Next define:

$$\forall \mathbf{X}, \mathbf{Y} \in \mathbb{G}_1 : \quad \theta(\mathbf{X}, \mathbf{Y}) \equiv [\mu(\mathbf{X}), \mu(\mathbf{Y})] - \mu([\mathbf{X}, \mathbf{Y}]) \tag{6.105}$$

Since π is a homomorphism and $\pi \circ \mu = \mathrm{id}$, then we have that $\pi(\theta(\mathbf{X}, \mathbf{Y})) = 0 \Rightarrow \theta(\mathbf{X}, \mathbf{Y}) \in \mathscr{Q}$. This guarantees that $\theta \in V^2(\mathbb{G}_1, \rho)$ is a 2-cochain of the Lie algebra $\mathbb{G}_1$ relative to the representation ρ. By direct calculation and use of Jacobi identity we can immediately verify that θ is actually a 2-cycle, namely $d\theta = 0$. Since the second cohomology group vanishes for semisimple Lie algebra $H^2(\mathbb{G}_1,\rho) = 0$, it follows that there exists a linear map

$$\exists \nu : \quad \mathbb{G}_1 \mapsto \mathcal{Q} \tag{6.106}$$

such that $d\nu = \theta$, namely:

$$\begin{aligned}[\mu(\mathbf{X}),\mu(\mathbf{Y})] - \mu([\mathbf{X},\mathbf{Y}]) = [\mu(\mathbf{X}),\mu(\mathbf{Y})] - [\mu(\mathbf{X}),\nu(\mathbf{Y})] \\ - [\mu(\mathbf{Y}),\nu(\mathbf{X})] - \nu([\mathbf{X},\mathbf{Y}])\end{aligned} \tag{6.107}$$

Since $\nu(\mathbf{X}) \in \mathcal{Q}$, we have $[\nu(\mathbf{X}),\nu(\mathbf{Y})] = 0$. Hence defining:

$$\lambda(\mathbf{X}) = \mu(\mathbf{X}) - \nu(\mathbf{X}) \tag{6.108}$$

we see that $\lambda : \mathbb{G}_1 \mapsto \mathbb{G}_1$ is a homomorphism. It is also evident that $\pi \circ \lambda = \text{id}$. So we have found a map $\lambda : \mathbb{G}_1 \mapsto \mathbb{G}$ which is a homomorphism of algebras. It follows that $\lambda(\mathbb{G}_1) \subset \mathbb{G}$ is a subalgebra. Furthermore, by construction:

$$\mathbb{G} = \mathcal{Q} \oplus \lambda(\mathbb{G}_1); \quad \mathcal{Q} \cap \lambda(\mathbb{G}_1) = 0 \tag{6.109}$$

Hence $\lambda(\mathbb{G}_1)$ is a Levi subalgebra and we have completed the induction argument also in this case.

The theorem is proved. □

6.4.4 An illustrative example: the Galilean group

The invariance group of classical non-relativistic mechanics is the *Galilean group* which consists of the following transformations on the space-time manifold whose points are labeled by the three space coordinates x^i and by the instant of time t:

$$\begin{pmatrix} x^i \\ t \end{pmatrix} \mapsto \begin{pmatrix} x^{i\prime} \\ t' \end{pmatrix} \tag{6.110}$$

where

$$\begin{cases} x^{i\prime} = R^i{}_j x^j + v^i t + c^i \\ t' = t + T \end{cases} \tag{6.111}$$

and

$$\begin{aligned} R^i{}_j &= \text{rotation matrix } RR^T = 1 \\ x^i &\mapsto x^i + c^i \quad \text{is a translation} \\ x^i &\mapsto x^i + v^i t \quad \text{corresponds to a special Galilean transformation} \\ t &\mapsto t + T \quad \text{corresponds to a time translation} \end{aligned} \tag{6.112}$$

The total number of parameters is 10 just as for the relativistic Poincaré group. Let us write the corresponding Lie algebra. For the rotations we have the *angular momentum* generators:

$$J_{ij} = x_i \partial_j - x_j \partial_i \quad \rightarrow \quad J_i = \epsilon_{ijk} x_j \partial_k \tag{6.113}$$

for the space translations we have the *momentum generators*

$$P_i = \partial_i \tag{6.114}$$

while the *Galilean boosts* are generated by:

$$K_i = t \partial_i \tag{6.115}$$

Finally the *Hamiltonian* generates time translations:

$$H = \partial_t \tag{6.116}$$

By explicit evaluation of the commutators we find that the Galilean Lie algebra has the following structure:

$$\begin{aligned}
&[J_i, J_j] = \epsilon_{ijk} J_k; && [J_i, P_j] = -\epsilon_{ijk} P_k \\
&[J_i, K_j] = -\epsilon_{ijk} K_k; && [J_i, H] = 0 \\
&[P_i, H] = 0; && [P_i, P_j] = 0 \\
&[K_i, H] = -P_i; && [K_i, K_j] = 0 \\
&[P_i, K_j] = 0
\end{aligned} \tag{6.117}$$

We can ask the question whether the Galilean algebra $\mathbb{G}$ is *semisimple*. The answer is no. Indeed, P_i $(i = 1, 2, 3)$ generates an abelian ideal since we easily verify that $[P, X] \subset P$, $\forall \mathbf{X} \in \mathbb{G}$, so that P is an ideal. Next we inquire whether $\mathbb{G}$ is *solvable*. The derivative algebra $D\mathbb{G}$ is made by J_i, P_i, K_i. We easily verify, however, that $D^2\mathbb{G} = D\mathbb{G}$ so that $\mathbb{G}$ is not solvable. On the other hand, if we consider the subalgebra $S^{(0)}$ generated by $\{P, K, H\}$, we see that:

$$DS^{(0)} = S^{(1)} = \{P\}; \quad DS^{(1)} = \{0\} \tag{6.118}$$

so that $S^{(0)}$ is solvable. The algebra generated by J_i is instead semisimple. Hence the Galilean algebra is, according to Levi's theorem, the direct product of a semisimple algebra with a solvable one.

6.5 The adjoint representation and Cartan's criteria

Let us now introduce the concept of *adjoint representation* of a Lie algebra $\mathbb{G}$. Given such an algebra, to each element $\mathbf{X} \in \mathbb{G}$ we can associate a *linear endomorphism*:

$$\mathrm{ad}_{\mathbf{X}} : \mathbb{G} \rightarrow \mathbb{G} \tag{6.119}$$

defined by:

$$\forall \mathbf{Y} \in \mathbb{G}: \quad \mathrm{ad}_{\mathbf{X}}(\mathbf{Y}) \equiv [\mathbf{X}, \mathbf{Y}] \tag{6.120}$$

If we choose a basis $\{T_A\}$, we immediately get:

$$(\mathrm{ad}_{\mathbf{X}})_A{}^B = X^M f_{MA}{}^B \tag{6.121}$$

where $f_{MA}{}^B$ are the Lie algebra structure constants, defined by:

$$[T_A, T_B] = f_{AB}{}^C T_C \tag{6.122}$$

Then we can introduce the bilinear symmetric Killing form of the Lie algebra:

$$\kappa : \mathbb{G} \otimes \mathbb{G} \to \mathbb{F} \tag{6.123}$$

defined by:

$$\forall \mathbf{X}, \mathbf{Y} \in \mathbb{G}: \quad \kappa(\mathbf{X}, \mathbf{Y}) = \mathrm{Tr}(\mathrm{ad}_{\mathbf{X}} \mathrm{ad}_{\mathbf{Y}}) \tag{6.124}$$

where $\mathbb{F}$ is the field over which the Lie algebra is constructed, namely, $\mathbb{F} = \mathbb{C}$ for complex Lie algebras and $\mathbb{F} = \mathbb{R}$ for real Lie algebras.

In a basis we obtain:

$$\begin{aligned} \kappa(\mathbf{X}, \mathbf{Y}) &= (\mathrm{ad}_{\mathbf{X}})_A{}^B (\mathrm{ad}_{\mathbf{Y}})_B{}^A = \mathbf{X}^M \mathbf{Y}^N f_{MA}{}^B f_{NB}{}^A \\ &= \mathbf{X}^M \mathbf{Y}^N g_{MN}^{(\mathrm{Killing})} \end{aligned} \tag{6.125}$$

where the symmetric tensor $g_{MN}^{(\mathrm{Killing})} = f_{MA}{}^B f_{NB}{}^A$ is named the Killing metric.

6.5.1 Cartan's criteria

Whether a Lie algebra is solvable or semisimple is fully encoded in the properties of the Killing form, which therefore provides a very useful global tool to test the structure of the Lie algebra. That this is the case is established by two simple but very important theorems that go under the name of Cartan's criteria.

The first Cartan's criterion establishes a test of solvability and is provided by the following theorem.

Theorem 6.5.1. *A Lie algebra $\mathbb{G}$ is solvable if and only if*

$$\forall \mathbf{X}, \mathbf{Y}, \mathbf{Z} \in \mathbb{G} \quad \kappa(\mathbf{X}, [\mathbf{Y}, \mathbf{Z}]) = 0 \tag{6.126}$$

For brevity we omit the proof of this theorem.

The second Cartan's criterion, which uses the first in its own proof, is a test of semisimplicity. It is given by the following theorem.

Theorem 6.5.2. *A Lie algebra $\mathbb{G}$ is semisimple if and only if the Killing form $\kappa(,)$ on $\mathbb{G}$ is non-degenerate.*

Proof of Theorem 6.5.2. We recall that a bilinear form $\kappa(,)$ on a vector space $\mathbb{G}$ is degenerate if $\exists \mathbf{X} \in \mathbb{G}$ such that $\forall \mathbf{Y} \in \mathbb{G}$ we have $\kappa(\mathbf{X}, \mathbf{Y}) = 0$. In a basis $\mathbf{X}_i$ this implies that the determinant of the matrix $\kappa_{ij} = \kappa(\mathbf{X}_i, \mathbf{X}_j)$ vanishes $\det \kappa_{ij} = 0$.

To prove the theorem we have to show that the following statements are both true:

a) If $\mathbb{G}$ is semisimple then κ is non-degenerate.
b) If κ is non-degenerate then $\mathbb{G}$ is semisimple.

Let us begin with case a) and let us assume that κ is degenerate, namely, the set:

$$B = \{\mathbf{X} : \kappa(\mathbf{X}, \mathbf{Y}) = 0, \ \forall \mathbf{Y} \in \mathbb{G}\} \tag{6.127}$$

contains non-trivial elements besides $\mathbf{0}$. We can immediately verify that B is an ideal of $\mathbb{G}$. Indeed, $\forall \mathbf{X} \in B$ and $\forall \mathbf{Z} \in \mathbb{G}$ we have $[\mathbf{X}, \mathbf{Y}] \in B$ since $\kappa([\mathbf{X}, \mathbf{Z}], \mathbf{Y}) = 0 \ \forall \mathbf{Y} \in \mathbb{G}$. This follows from the properties of the Killing form that imply $\kappa([\mathbf{X}, \mathbf{Z}], \mathbf{Y}) = \kappa(\mathbf{X}, [\mathbf{X}, \mathbf{Y}]) = 0$. Next we can show that:

$$\forall \mathbf{X}, \mathbf{X}' \in B: \quad \kappa(\mathbf{X}, \mathbf{X}') = \kappa_B(\mathbf{X}, \mathbf{X}') \tag{6.128}$$

where $\kappa_B(,)$ denotes the restriction of the Killing form to ideal B. Indeed, given $\mathbf{Z} \in \mathbb{G}$, we have:

$$\mathrm{ad}_{\mathbf{X}} \mathrm{ad}_{\mathbf{X}'} \mathbf{Z} = [\mathbf{X}, [\mathbf{X}', \mathbf{Z}]] \in B \quad \text{since } [\mathbf{X}', \mathbf{Z}] \in B \tag{6.129}$$

This means that the image of the linear map $\mathrm{ad}_{\mathbf{X}} \mathrm{ad}_{\mathbf{X}'}$ is contained in the ideal B which implies that the only contribution to the trace comes from its restriction to the subspace B. By our definition of the ideal B we have $\kappa(\mathbf{X}, \mathbf{X}') = 0$ for all $\mathbf{X}, \mathbf{X}' \in B$, which by the above argument implies also $\kappa_B(\mathbf{X}, \mathbf{X}') = 0$. Hence the algebra $\mathbb{G}$ admits an ideal B whose Killing form is identically vanishing. By the first Cartan criterion 6.5.1 it follows that the ideal B is solvable. Yet this contradicts the assumption that the Lie algebra $\mathbb{G}$ was semisimple, so B necessarily contains only the zero element $\mathbf{0}$ and the Killing form is non-degenerate.

Let us turn to case b). Assume that the Lie algebra $\mathbb{G}$ is not semisimple and let us show that this implies that the Killing form is degenerate. If $\mathbb{G}$ is not semisimple there is a non-trivial solvable ideal $\mathscr{Q}$. By definition, $\exists k \in \mathbb{N}$ such that:

$$\mathscr{A} \equiv \mathscr{D}^k \mathscr{Q} \neq 0; \quad \mathscr{D}^{k+1} \mathscr{Q} = 0 \tag{6.130}$$

The subalgebra $\mathscr{A}$ is a non-trivial abelian ideal. As a next step we show that:

$$\forall \mathbf{X} \in \mathscr{A}, \ \forall \mathbf{Y} \in \mathbb{G}: \quad \kappa(\mathbf{X}, \mathbf{Y}) = \kappa_{\mathscr{A}}(\mathbf{X}, \mathbf{Y}) \tag{6.131}$$

Indeed, given $\mathbf{Z} \in \mathbb{G}$ we have $\mathrm{ad}_{\mathbf{X}} \circ \mathrm{ad}_{\mathbf{Y}}(\mathbf{Z}) = [\mathbf{X}, [\mathbf{Y}, \mathbf{Z}]] \in \mathscr{A}$ since $\mathscr{A}$ is an ideal. Hence the image of $\mathrm{ad}_{\mathbf{X}} \circ \mathrm{ad}_{\mathbf{Y}}$ as $\mathbf{Z}$ varies in $\mathbb{G}$ takes values only in $\mathscr{A}$ and therefore its trace takes contributions only from $\mathscr{A}$. This suffices to prove that eq. (6.131) is true. Next we observe that:

$$\forall \mathbf{X} \in \mathscr{A}, \ \forall \mathbf{Y} \in \mathbb{G} \quad \text{we have } \kappa_{\mathscr{A}}(\mathbf{X}, \mathbf{Y}) = 0 \tag{6.132}$$

Indeed, given $\mathbf{X}' \in \mathscr{A}$ we have $\mathrm{ad}_{\mathbf{X}} \circ \mathrm{ad}_{\mathbf{Y}}(\mathbf{X}') = [\mathbf{X}, [\mathbf{Y}, \mathbf{X}']] = 0$ since both $\mathbf{X} \in \mathscr{A}$, $[\mathbf{Y}, \mathbf{X}'] \in \mathscr{A}$ and $\mathscr{A}$ is abelian. Hence there is no contribution to the trace. On the other hand, in force of eq. (6.131) we conclude that $\kappa(\mathbf{X}, \mathbf{Y}) = 0$ for all $\mathbf{Y} \in \mathbb{G}$ and all $\mathbf{X} \in \mathscr{A}$. This means that the Killing form is degenerate unless $\mathscr{A}$ is empty. So there cannot be any non-trivial solvable ideal and the algebra $\mathbb{G}$ has to be semisimple. This concludes the proof of the theorem. □

6.6 Bibliographical note

Main bibliographical sources for this chapter are:
1. the textbook [92]
2. the textbook [138]
3. the textbook [84]
4. the textbook [77]
5. the textbook [117]
6. volume 2 of [51].

7 Root systems and their classification

Tout ce qu'on invente est vrai, soi-en sure. La poésie est une chose aussi précise que la géométrie.
Gustave Flaubert, letter to Louise Colet

7.1 Cartan subalgebras

We consider a semisimple Lie algebra $\mathbb{G}$ and we introduce the fundamental concept of Cartan subalgebra that will be the primary instrument to set up the reduction of the Lie algebra to a canonical form and its identification in terms of a **root system**.

Definition 7.1.1. A Cartan subalgebra $\mathscr{H} \subset \mathbb{G}$ is a subalgebra that satisfies the following two defining properties:

i) $\mathscr{H}$ is a **maximal abelian subalgebra**
ii) $\forall H \in \mathscr{H}$ the map $\mathrm{ad}(H)$ is a **semisimple endomorphism**

First we prove that every semisimple Lie algebra $\mathbb{G}$ has a Cartan subalgebra (frequently abbreviated as CSA). Then we show that if $\mathscr{H}_1$ and $\mathscr{H}_2$ are two CSAs then they are isomorphic.

Let $H \in \mathbb{G}$ be an element of the semisimple Lie algebra and let $\lambda_0, \lambda_1, \dots, \lambda_r$ be the eigenvalues of $\mathrm{ad}(H)$: define

$$g(H, \lambda_i) = \{X \in \mathbb{G} / \mathrm{ad}(H)X = \lambda_i X\} \tag{7.1}$$

the subspace of $\mathbb{G}$ pertaining to the eigenvalue λ_i. We have:

$$\mathbb{G} = \bigoplus_{i=0}^{r} g(H, \lambda_i) \tag{7.2}$$

Definition 7.1.2. An element $H_0 \in \mathbb{G}$ is named **regular** if

$$\dim g(H_0, 0) = \min_{X \in \mathbb{G}} (\dim g(X, 0)) \tag{7.3}$$

We have:

Theorem 7.1.1. *If H_0 is a regular element then $g(H_0, 0)$ is a Cartan subalgebra.*

Proof of Theorem 7.1.1. We have to show that:

a) $g(H_0, 0)$ is a subalgebra
b) $g(H_0, 0)$ is a maximal abelian subalgebra
c) if $H \in g(H_0, 0)$ then $\mathrm{ad}(H)$ is semisimple as an endomorphism

https://doi.org/10.1515/9783110551204-007

We begin by observing that

$$[g(Z,\lambda),g(Z,\mu)] \subset g(Z,\lambda+\mu) \tag{7.4}$$

which immediately follows from Jacobi identities. This implies that $\mathscr{H} = g(H_0,0)$ is a subalgebra. Next we prove that $\mathscr{H}$ is abelian. To this effect let us denote by $0 = \lambda_0,\lambda_1,\lambda_2,\dots,\lambda_r$ the different eigenvalues of $\mathrm{ad}(H_0)$ and set:

$$\mathbb{G}' = \bigoplus_{i=1}^{r} g(H,\lambda_i) \tag{7.5}$$

From eq. (7.4) it follows that $[\mathscr{H},\mathbb{G}'] \subset \mathbb{G}'$. $\forall H \in \mathscr{H}$ let us denote $\mathrm{ad}'(H)$ the restriction of $\mathrm{ad}(H)$ to the subspace $\mathbb{G}'$ and name $d(H) = \det[\mathrm{ad}'(H)]$ the determinant of such an endomorphism. By definition $d(H)$ is a polynomial function on the finite dimensional vector space (algebra) $\mathscr{H}$; furthermore, by definition of the subspace $\mathbb{G}'$, the map $\mathrm{ad}'(H_0)$ has only non-vanishing eigenvalues, so that $d(H_0) \neq 0$. If a polynomial function vanishes on an open set then it is identically zero. Since $d(H_0) \neq 0$ it follows that $d(H)$ is not identically zero and that its zeros are isolated. Calling S the set of elements of $\mathscr{H}$ for which $d(H) \neq 0$ we conclude that S is dense in $\mathscr{H}$. Let $H \in S \subset \mathscr{H}$: since $\det[\mathrm{ad}'(h)] \neq 0$ it follows that all the null eigenvectors of $\mathrm{ad}(H)$, if any, are contained in $\mathscr{H}$. Hence we have shown:

$$\forall H \in S: \quad g(H,0) \subset \mathscr{H} \tag{7.6}$$

Since the element H_0 is by hypothesis regular, we conclude that $g(H,0) = \mathscr{H}$. Hence it is proved that

$$\forall H \in S,\ \forall H_1 \in \mathscr{H}: \quad \mathrm{ad}(H)(H_1) = 0 \tag{7.7}$$

Hence the restriction of $\mathrm{ad}(H)$ to the subalgebra $\mathscr{H}$ is nilpotent since it vanishes. Since S is dense in $\mathscr{H}$, by continuity it follows that

$$\forall H \in \mathscr{H}: \quad \mathrm{ad}_{\mathscr{H}}(H) = 0 \tag{7.8}$$

namely that

$$\forall H_1,H_2 \in \mathscr{H}: \quad [H_1,H_2] = 0 \tag{7.9}$$

This concludes the proof that $\mathscr{H} \equiv g(H_0,0)$ is an abelian subalgebra. By definition it is also maximal. Indeed, if there existed an element $X \notin g(H_0,0)$ such that $[X,\mathscr{H}] = 0$, we would have a contradiction since, in particular $[X,H_0] = 0$, which implies $X \in g(H_0,0)$.

Let us now show that if λ is a non-vanishing eigenvalue of $\mathrm{ad}(H_0)$, then every endomorphism $\mathrm{ad}(H)$ with $H \in \mathscr{H}$ maps the subspace $g(H_0,\lambda)$ onto itself. Hence, denoting by $\mathrm{ad}_\lambda(H)$ the restriction of $\mathrm{ad}(H)$ to this subspace we have that $\mathrm{ad}_\lambda(H)$ is a *representation* of $\mathscr{H}$ on $g(H_0,\lambda)$. Since $\mathrm{ad}_\lambda(H)$ is a family of commuting endomorphisms (solvable algebra, in particular), we can put all of them simultaneously in a triangular form, by choosing some appropriate basis $\mathbf{e}_1, \dots, \mathbf{e}_s$ of $g(H_0, \lambda)$. In this basis the semisimple part of $\mathrm{ad}_\lambda(H)$ will be the diagonal part:

$$\mathrm{ad}_\lambda(H) = \begin{pmatrix} \alpha_1(H) & 0 & \dots & \dots & 0 \\ 0 & \alpha_2(H) & 0 & \dots & 0 \\ \dots & \dots & \dots & \dots & \dots \\ 0 & \dots & 0 & \alpha_{s-1}(H) & 0 \\ 0 & \dots & \dots & 0 & \alpha_s(H) \end{pmatrix} + \text{nilpotent matrix} \tag{7.10}$$

The diagonal elements $\alpha_i(H)$ are linear functions on $\mathscr{H}$ with the property that $\alpha_1(H_0) = \alpha_2(H_0) = \dots = \alpha_s(H_0) = \lambda$. Let $\beta(H)$ be any linear function on $\mathscr{H}$ that takes the value $\beta(H_0) = \lambda$ at H_0. Let V_β be the subspace of $g(H_0, \lambda)$ spanned by such basis vectors $\mathbf{e}_i$ that $\alpha_i(H) = \beta(H)$ $(\forall H \in \mathscr{H})$. By definition it follows that:

$$\forall X \in V_\beta \quad \Rightarrow \quad \exists k \in \mathbb{N} \quad / \quad (\mathrm{ad}(H) - \beta(H)\mathbf{1})^k X = 0 \tag{7.11}$$

Indeed, once we have subtracted the diagonal part, what remains is nilpotent. In general we have:

$$\mathbb{G} = \sum_i V_{\beta_i} \quad \text{for suitable } \beta_i \tag{7.12}$$

hence if $\kappa(,)$ is the Killing form and we can write:

$$\forall H, H' \in \mathscr{H}: \quad \kappa(H, H') = \sum_i \beta_i(H)\beta_i(H') \dim V_{\beta_i} \tag{7.13}$$

We decompose $\mathrm{ad}(H)$ à la Jordan:

$$\mathrm{ad}(H) = \underbrace{S(H)}_{\text{semisimple}} + \underbrace{N(H)}_{\text{nilpotent}} \tag{7.14}$$

and we recall that $S(H)$ is polynomial in $\mathrm{ad}(H)$. By construction the endomorphism $S(H)$ leaves each V_β subspace invariant and:

$$S(H)X = \beta(H)X; \quad \forall X \in V_\beta \tag{7.15}$$

Furthermore, since $[V_\alpha, V_\beta] \subset V_{\alpha+\beta}$ it follows that S is a derivation of the algebra. But for a semisimple Lie algebra every derivation is internal, hence $\exists Z \in \mathbb{G}$ such that $\mathrm{ad}(Z) = S(H)$. Since $S(H)$ commutes with all elements of $\mathscr{H}$ it follows that $Z \in \mathscr{H}$. In other words, $Z = H$ and this shows that $\mathrm{ad}(H)$ coincides with its semisimple part. □

7.2 Root systems

Let $\mathscr{H} \subset \mathbb{G}$ be a Cartan subalgebra of the semisimple Lie algebra $\mathbb{G}$. Consider an element $\alpha \in \mathscr{H}^\star$, namely a linear functional:

$$\begin{aligned} &\alpha : \mathscr{H} \to \mathbb{C} \\ &\forall H_1, H_2 \in \mathscr{H};\ \forall \lambda, \mu \in \mathbb{C} : \quad \alpha(\lambda H_1 + \mu H_2) = \lambda\alpha(H_1) + \mu\alpha(H_2) \end{aligned} \tag{7.16}$$

Let us define the linear subspace $\mathbb{G}^\alpha \subset \mathbb{G}$:

$$\mathbb{G}^\alpha : \{X \in \mathbb{G} \setminus [H, X] = \alpha(H)X,\ \forall H \in \mathscr{H}\} \tag{7.17}$$

If $\mathbb{G}^\alpha \neq \varnothing$ is not empty then we say that $\alpha \in \mathscr{H}^\star$ is a **root** and $\mathbb{G}^\alpha$ is named the **corresponding subspace** of root α. Since, by definition, $\mathscr{H}$ is maximal abelian, then we have $\mathbb{G}^0 = \mathscr{H}$. On the other hand, from Jacobi identity we immediately obtain:

$$[\mathbb{G}^\alpha, \mathbb{G}^\beta] \subset \mathbb{G}^{\alpha+\beta} \quad \forall \alpha, \beta \in \mathscr{H}^\star \tag{7.18}$$

Let us next denote by Φ the set of all non-vanishing roots and with $\kappa(,)$ the Killing form. We have:

Theorem 7.2.1. *The following statements are true:*

i) $\mathbb{G} = \mathscr{H} \bigoplus \sum_{\alpha\in\Phi} \mathbb{G}^\alpha$ *(direct sum).*

ii) $\dim \mathbb{G}^\alpha = 1,\ \forall \alpha \in \Phi$.

iii) *Let $\alpha, \beta \in \Phi$ be two roots such that $\alpha + \beta \neq 0$, then the corresponding subspaces $\mathbb{G}^\alpha$ and $\mathbb{G}^\beta$ are mutually orthogonal with respect to the Killing form $\kappa(,)$.*

iv) *The restriction of the Killing form $\kappa(,)$ to $\mathscr{H} \bigotimes \mathscr{H}$ is non-degenerate and for each root $\forall \alpha \in \Phi$ there exists an element $H_\alpha \in \mathscr{H}$ of the Cartan subalgebra such that*

$$\kappa(H, H_\alpha) = \alpha(H) \quad \forall H \in \mathscr{H} \tag{7.19}$$

v) *If $\alpha \in \Phi$ is a root then also its negative is a root: $-\alpha \in \Phi$. Furthermore, we have:*

$$\begin{aligned} [\mathbb{G}^\alpha, \mathbb{G}^{-\alpha}] &= \text{const}\, H_\alpha \\ \alpha(H_\alpha) &\neq 0 \end{aligned} \tag{7.20}$$

Proof of Theorem 7.2.1. We begin with point **i)** in the above list and we show first that the sum is direct. If it were not, this would mean that there existed a linear relation:

$$H^\star + \sum_i X_{\alpha_i} = 0 \tag{7.21}$$

where $H^\star \in \mathscr{H}$ and $X_{\alpha_i} \in \mathbb{G}^{\alpha_i}$. We can choose an element $H \in \mathscr{H}$ such that $\alpha_i(H) \neq 0$ for all the roots α_i. Indeed, the subset $N \subset \mathscr{H}$ on which all the roots α_i are different and non-vanishing is the complement of the union of a finite number of hyperplanes

($\alpha(H) = 0 \Leftrightarrow$ hyperplane $\ni H$). Hence H with the required properties exists and, as a consequence, $H^\star$ and X_{α_i} belong to different eigenspaces of $\mathrm{ad}(H)$. As such they are linearly independent which contradicts the assumption of eq. (7.21). This shows that the sum of subspaces in statement **i)** is direct. On the other hand, since $\mathrm{ad}_{\mathbb{G}}(\mathscr{H})$ is a set of semisimple endomorphisms it follows that $\mathbb{G}$ can be decomposed into eigenspaces and the relation advocated in statement **i)** follows. Furthermore, if $\alpha(H_0) = 0$ for all roots $\alpha \in \Phi$ then $H_0 = 0$. Indeed, by hypothesis we have $[H_0, X] = 0$, $\forall X \in \mathbb{G}$ and since the Lie algebra $\mathbb{G}$ is semisimple, this implies $H_0 = 0$.

Let us next prove the statement **iii)**. To this effect we choose $X \in \mathbb{G}^\alpha$ and $Y \in \mathbb{G}^\beta$. With this choice the endomorphism $\mathrm{ad}(x).\mathrm{ad}(Y)$ maps the space $\mathbb{G}^\gamma$ to $\mathbb{G}^{\alpha+\beta+\gamma}$ and since $\alpha + \beta \neq 0$, we get:

$$\mathbb{G}^\gamma \cap \mathbb{G}^{\alpha+\beta+\gamma} = 0 \tag{7.22}$$

Therefore if we use a basis where every basis vector lies in some root subspace $\mathbb{G}^\gamma$, we immediately see that:

$$\kappa(X, Y) \equiv \mathrm{Tr}\left(\mathrm{ad}(X).\mathrm{ad}(Y)\right) = 0 \tag{7.23}$$

which is what we wanted to show.

Next let us prove statement **iv)**. If $H_0 \in \mathscr{H}$ satisfies the condition:

$$\kappa(H_0, H) = 0 \quad \forall H \in \mathscr{H} \tag{7.24}$$

then, as a consequence of statement **iii)** that we have already proved, it follows that:

$$\kappa(H_0, X) = 0 \quad \forall X \in \mathbb{G} \tag{7.25}$$

This would imply that the Killing form $\kappa(,)$ is degenerate in contradiction to Cartan's criterion for semisimple Lie algebras. Hence there are no vectors in $\mathscr{H}$ which are orthogonal to all vectors of $\mathscr{H}$, which proves the first part of statement **iv)**. Next choose a basis $\{H_i\}$ $(i = 1, \dots \mathrm{rank}\, \mathbb{G})$ of the Cartan subalgebra[1] $\mathscr{H}$ and set:

$$\kappa_{ij} = \kappa(H_i, H_j) \tag{7.26}$$

Writing $\forall H \in \mathscr{H}$ $H = h^i H_i$ we obtain $\alpha(H) = h^i \alpha_i$ where $\alpha_i \equiv \alpha(H_i)$. Statement **iv)** advocates that we should be able to find an element $H_\alpha = \alpha^i H_i$ such that $\kappa(H, H_\alpha) = \alpha(H)$. In the chosen basis this means $H^i \alpha^j \kappa_{ij} = h^i \alpha_i$ which implies:

$$\alpha^j \kappa_{ij} = \alpha_i \tag{7.27}$$

1 We recall that the dimension of the CSA of a Lie algebra $\mathbb{G}$ is the name of the rank of $\mathbb{G}$.

Since κ_{ij} is a non-degenerate matrix, we can always find its inverse and set

$$\alpha^j = (\kappa^{-1})^{ji}\alpha_i \tag{7.28}$$

which concludes the proof of statement **iv)**.

Let us come to the proof of statement **v)**. Let us assume that $-\alpha \notin \mathbf{\Phi}$. This would imply that $\mathbb{G}^{-\alpha} = \mathbf{0}$. In this case an element $X \in \mathbb{G}^{\alpha}$ being orthogonal to all the other subspaces $\mathbb{G}^{\beta}$ would imply that

$$\kappa(X_\alpha, Y) = 0 \quad \forall Y \in \mathbb{G} \tag{7.29}$$

In this case the Killing form would be degenerate which is impossible for a semisimple Lie algebra by Cartan criterion. So $-\alpha \in \mathbf{\Phi}$. Let now $H, X_\alpha, X_{-\alpha}$ be arbitrary elements respectively in $\mathscr{H}$, $\mathbb{G}^{\alpha}$ and $\mathbb{G}^{-\alpha}$. Then by properties of the Killing form we have:

$$\begin{aligned}\kappa([X_\alpha, X_{-\alpha}], H) &= \kappa(X_{-\alpha}, [H, X_\alpha]H) \\ &= \kappa(X_{-\alpha}, X_\alpha)\alpha(H) \\ &= \kappa(X_{-\alpha}, X_\alpha)\kappa(H_\alpha, H)\end{aligned} \tag{7.30}$$

so that we are forced to identify:

$$[X_\alpha, X_{-\alpha}] = \kappa(X_{-\alpha}, X_\alpha)H_\alpha = \kappa(X_{-\alpha}, X_\alpha)\alpha^i H_i \tag{7.31}$$

which concludes the proof of statement **v)**.

Finally let us prove statement **ii)**. Let us assume that $\dim \mathbb{G}^{\alpha} > 1$. In this case let us choose $X_\alpha \in \mathbb{G}^{\alpha}$ and $X_{-\alpha} \in \mathbb{G}^{-\alpha}$ such that:

$$\kappa(X_{-\alpha}, X_\alpha) = 1 \tag{7.32}$$

If $\dim \mathbb{G}^{\alpha} > 1$ it follows that there exists $D_\alpha \in \mathbb{G}^{\alpha}$ such that:

$$\kappa(D_\alpha, X_{-\alpha}) = 0 \tag{7.33}$$

Set $D_n = (\mathrm{ad}(X_\alpha))^n D_\alpha$ for $n = 0, 1, 2, \dots$. We have $D_n \in \mathbb{G}^{(n+1)\alpha}$ and hence

$$[H_\alpha, D_n] = \alpha(H)(n+1)D_n \tag{7.34}$$

Furthermore, by induction we can show that:

$$[X_{-\alpha}, D_n] = -n\frac{(n+1)}{2}\alpha(H_\alpha)D_{n-1} \tag{7.35}$$

For $n = 0$ we have $[X_{-\alpha}, D_\alpha] = \kappa(D_\alpha, X_{-\alpha})H_\alpha = 0$. On the other hand, if eq. (7.35) is true for n it follows that it is also true for $n+1$. Indeed:

$$\begin{aligned}[X_{-\alpha}, D_{n+1}] &= [X_{-\alpha}, [X_\alpha, D_n]] \\ &= -[X_\alpha, [D_n, X_{-\alpha}]] - [D_n, [X_{-\alpha}, X_\alpha]] \\ &= -n\frac{n+1}{2}\alpha(H_\alpha)D_n + (n+1)\alpha(H_\alpha)D_n \\ &= -(n+1)\frac{n+2}{2}\alpha(H_\alpha)D_n\end{aligned} \tag{7.36}$$

which shows what we claimed. Therefore if $D_0 = D_\alpha$ exists, also all the other D_n exist and are non-vanishing. This implies that there are infinite roots $(n+1)\alpha$ and correspondingly infinite orthogonal subspaces $\mathbb{G}^{(n+1)\alpha}$. This is manifestly absurd since the dimension of the semisimple Lie algebra $\mathbb{G}$ is finite. Hence D_α cannot exist and the dimension of the subspace $\mathbb{G}^\alpha = 1$ as claimed.

This concludes the proof of the theorem. □

7.2.1 Final form of the semisimple Lie algebra

Using the result provided by Theorem 7.2.1 we can now write a final general form of a semisimple Lie algebra in terms of Cartan generators H_i and step operators E^α associated with the roots α. To this effect we normalize the Cartan subalgebra (CSA) generators in the following way:

$$\begin{aligned} \kappa(H_i, H_j) &= \delta_{ij} \quad \Rightarrow \quad H_\alpha = \alpha_i H_i \\ \kappa(E^\alpha, E^{-\alpha}) &= 1 \\ \kappa(H_i, E^\alpha) &= 0 \end{aligned} \tag{7.37}$$

With this normalization the commutation relations of the *complex semisimple Lie algebra* take the following general form:

$$\begin{aligned} [H_i, H_j] &= 0 \\ [H_i, E^\alpha] &= \alpha_i E^\alpha \\ [E^\alpha, E^{-\alpha}] &= \alpha^i H_i \\ [E^\alpha, E^\beta] &= \begin{cases} N(\alpha,\beta) E^{\alpha+\beta} & \text{if } \alpha+\beta \in \Phi \\ 0 & \text{if } \alpha+\beta \notin \Phi \end{cases} \end{aligned} \tag{7.38}$$

where $N(\alpha, \beta)$ is a coefficient that has to be determined using Jacobi identities.

7.2.2 Properties of root systems

Let us now consider the properties of a root system associated with a semisimple Lie algebra. We have:

Theorem 7.2.2. *If $\alpha, \beta \in \Phi$ are two roots, then the following two statements are true:*
1. $2\frac{(\alpha,\beta)}{(\alpha,\alpha)} \in \mathbb{Z}$
2. $\sigma_\alpha(\beta) \equiv \beta - 2\alpha\frac{(\alpha,\beta)}{(\alpha,\alpha)} \in \Phi$ *is also a root*

The vector $\sigma_\alpha(\beta)$ defined above is named the reflection of β with respect to α and the second part of the thesis can be reformulated by saying that any root system Φ is invariant under reflection with respect to any of its elements.

Proof of Theorem 7.2.2. Let $\alpha, \beta \in \Phi$ be two roots, and let us define the non-negative integer $j \in \mathbb{N}$ by means of the following conditions:

$$\begin{aligned} \gamma &\equiv \beta + j\alpha \in \Phi \\ \gamma + \alpha &\notin \Phi \end{aligned} \tag{7.39}$$

In other words, j is the maximal integer n for which $\beta + n\alpha$ is a root.

We know that $-\alpha$ is a root and hence we can conclude that:

$$\begin{aligned} [E^{-\alpha}, E^{\gamma}] &= \widehat{E}^{\gamma-\alpha} \\ [E^{-\alpha}, \widehat{E}^{\gamma-\alpha}] &= \widehat{E}^{\gamma-2\alpha} \\ &\dots \dots \dots \end{aligned} \tag{7.40}$$

where $\widehat{E}^{\gamma-n\alpha}$ denotes some element in the one-dimensional subspace pertaining to the root $\gamma - n\alpha$. Since the number of roots is necessarily finite it follows that there exists some positive integer $g \in \mathbb{N}$ such that:

$$[E^{-\alpha}, \widehat{E}^{\gamma-g\alpha}] = \widehat{E}^{\gamma-(g+1)\alpha} = 0 \tag{7.41}$$

In general, due to the one-dimensionality of each root space we can set:

$$[E^{\alpha}, \widehat{E}^{\gamma-(n+1)\alpha}] = \mu_{n+1} \widehat{E}^{\gamma-n\alpha} \tag{7.42}$$

where μ_{n+1} is some normalization factor. From Jacobi identities we immediately obtain a recursion relation satisfied by these normalization factors. Indeed:

$$\begin{aligned} [E^{\alpha}[E^{-\alpha}, \widehat{E}^{\gamma-n\alpha}]] &= -[E^{-\alpha}[\widehat{E}^{\gamma-n\alpha}, E^{\alpha}]] - [E^{\gamma-n\alpha}[E^{\alpha}, E^{-\alpha}]] \\ &= \mu_n \widehat{E}^{\gamma-n\alpha} + \alpha^i [H_i, \widehat{E}^{\gamma-n\alpha}] \\ &= (\mu_n + (\gamma, \alpha) - n(\alpha, \alpha)) \widehat{E}^{\gamma-n\alpha} \end{aligned} \tag{7.43}$$

which implies the recursion relation:

$$\mu_{n+1} = \mu_n + (\gamma, \alpha) - n(\alpha, \alpha) \tag{7.44}$$

Since by hypothesis $\gamma + \alpha$ is not a root we have

$$[E^{\alpha}, E^{\gamma}] = \mu_0 E^{\gamma+\alpha} = 0 \quad \text{namely } \mu_0 = 0 \tag{7.45}$$

This allows to solve the recursion relation explicitly obtaining:

$$\mu_n = n(\alpha, \gamma) - \frac{n(n-1)}{2}(\alpha, \alpha) \tag{7.46}$$

Since, at the other end of the chain, we have assumed that $\gamma - (g+1)\alpha$ is not a root, we conclude that $\mu_{g+1} = 0$ and hence:

$$(g+1)\left\{(\alpha, \gamma) - \frac{g}{2}(\alpha, \alpha)\right\} = 0 \tag{7.47}$$

This implies that $2\frac{(\gamma,\alpha)}{(\alpha,\alpha)} = g \in \mathbb{N}$. Hence for each pair of roots $\alpha\beta$ there exists a non-negative integer $j \geq 0$ such that $\gamma = \beta + j\alpha$ is a root and

$$(\alpha,\beta) = (\alpha,\gamma) - j(\alpha,\alpha) = \left(\frac{g}{2} - j\right)(\alpha,\alpha) \tag{7.48}$$

namely:

$$2\frac{(\alpha,\beta)}{(\alpha,\alpha)} = g - 2j \in \mathbb{Z} \quad \text{(positive or negative)} \tag{7.49}$$

This concludes the first part of our proof. Let us now consider the string of roots that we have constructed to make the above argument:

$$\begin{aligned} \beta_0 &= \gamma = \beta + j\alpha \\ \beta_1 &= \gamma - \alpha = \beta + (j-1)\alpha \\ \beta_2 &= \gamma - 2\alpha = \beta + (j-2)\alpha \\ &\dots\ \dots\ \dots \qquad (7.50) \\ \beta_g &= \gamma - g\alpha = \beta + (j-g)\alpha \qquad (7.51) \end{aligned}$$

Since $2\frac{(\gamma,\alpha)}{(\alpha,\alpha)} = g$, it is evident by means of the replacement:

$$\beta \mapsto \beta - 2\alpha\frac{(\beta,\alpha)}{(\alpha,\alpha)} \tag{7.52}$$

that the string (7.51) is simply reflected onto itself $\beta_g \mapsto \beta_0, \beta_{g-1} \mapsto \beta_1, \dots$. So not only we proved that if β and α are roots then the reflection of β with respect to α is a root but also that the entire string of α through β is invariant under such reflection. □

Collecting together the properties of the roots that we have so far, we can axiomatize the notion of root system in the following way.

Definition 7.2.1. Let $\mathbb{E}$ be a Euclidean space of dimension ℓ. A subset $\Phi \subset \mathbb{E}$ is named a **root system** if:
1. Φ is finite, spans $\mathbb{E}$ and does not contain $\mathbf{0}$.
2. If $\alpha \in \Phi$ the only multiples of α contained in Φ are $\pm\alpha$.
3. $\forall \alpha, \beta \in \Phi$ we have $2\frac{(\alpha,\beta)}{(\alpha,\alpha)} \in \mathbb{Z}$.
4. $\forall \alpha, \beta \in \Phi$ we have $\sigma_\alpha(\beta) \equiv \beta - 2\alpha\frac{(\alpha,\beta)}{(\alpha,\alpha)} \in \Phi$.

7.2.2.1 Angles between the roots

It is convenient to introduce the following notation of a *hook product*

$$\langle \beta, \alpha \rangle \equiv 2\frac{(\beta,\alpha)}{(\alpha,\alpha)} \tag{7.53}$$

From Theorem 7.2.2 we have learned that $\langle\beta,\alpha\rangle \in \mathbb{Z}$, but at the same time also $\langle\alpha,\beta\rangle \in \mathbb{Z}$. Hence we conclude that

$$\langle\beta,\alpha\rangle\langle\alpha,\beta\rangle = 4\cos^2\theta_{\alpha\beta} \in \mathbb{Z} \tag{7.54}$$

where $\theta_{\alpha\beta}$ is the angle between the two roots.

This implies that the angles between the roots are quantized and the available cases are listed in the following table.

$\langle\alpha,\beta\rangle$	$\langle\beta,\alpha\rangle$	θ	$\frac{\|\beta\|^2}{\|\alpha\|^2}$
0	0	$\frac{\pi}{2}$	undetermined
1	1	$\frac{\pi}{3}$	1
−1	−1	$\frac{2\pi}{3}$	1
1	2	$\frac{\pi}{4}$	2
−1	−2	$\frac{3\pi}{4}$	2
1	3	$\frac{\pi}{6}$	3
−1	−3	$\frac{5\pi}{6}$	3

(7.55)

As one sees, also the ratio between the squared lengths of two roots that are not orthogonal to each other is quantized and we have three possibilities: 1, 2, 3. As we will see in the sequel, this ratios are very much relevant and distinguish the Lie algebras in two categories, the *simply laced* and the *non-simply laced* algebras.

A very useful lemma which is just a consequence of the above argument is the following one:

Lemma 7.2.1. *Let $\alpha,\beta \in \Phi$ be two non-proportional roots. If $(\alpha,\beta) > 0$ then $\alpha-\beta \in \Phi$ is a root. If $(\alpha,\beta) < 0$ then $\alpha+\beta \in \Phi$ is a root.*

Proof of Lemma 7.2.1. Looking at table (7.55) we see that if $(\alpha,\beta) > 0$, then one or the other of the hook products $\langle\alpha,\beta\rangle$ or $\langle\beta,\alpha\rangle$ is equal to 1. In the first case $\alpha-\beta = \sigma_\beta(\alpha)$ and then it is a root. In the second case $\beta-\alpha = \sigma_\alpha(\beta)$ is a root but also its negative, namely, $\alpha-\beta$ is a root. To prove the lemma in the case $(\alpha,\beta) < 0$ it suffices to apply the same argument to $\beta' = -\beta$. □

7.3 Simple roots, the Weyl group and the Cartan matrix

The next steps in our discussion are all rather simple constructions of Euclidean Geometry whose consequences are however quite far-reaching. We begin with the notion of simple roots.

Definition 7.3.1. Given a root system $\Phi \subset \mathbb{E}^\ell$ in a Euclidean space of dimension ℓ, a set Δ of exactly ℓ roots is named a simple root basis if:

1. Δ is a basis for the entire $\mathbb{E}^\ell$.
2. Every root $\alpha \in \Phi$ can be written as a linear combination of the elements α_i whose coefficients are either all positive or all negative integers

$$\alpha = \sum_{i=1}^{\ell} k^i \alpha_i; \quad k^i \in \begin{cases} \mathbb{Z}_+ \\ \text{or } \mathbb{Z}_- \end{cases} \tag{7.56}$$

The vectors α_i comprised in Δ are named the simple roots of Φ.

What might look at first sight surprising is the following:

Theorem 7.3.1. *Every root system Φ admits a basis Δ of simple roots.*

Proof of Theorem 7.3.1. The proof is a constructive one since it indicates a precise algorithm how to construct the simple roots.

Let us consider any vector $\gamma \in \mathbb{E}^\ell$ which does not lie in any of the hyperplanes Π_α orthogonal to the roots $\alpha \in \Phi$. Since the roots are in a finite number, such vectors certainly exist. The hyperplane Π_γ separates the set of roots in two subsets:

$$\Phi = \Phi_+ \cup \Phi_- \tag{7.57}$$

The set of **positive roots** Φ_+ comprises all those roots α for which $(\gamma, \alpha) > 0$. The set of **negative roots** Φ_- comprises all those for which the reverse is true. A positive root $\alpha \in \Phi_+$ is named **decomposable** if it can be written as the sum of two other positive roots: $\alpha = \beta_1 + \beta_2$ with $\beta_{1,2} \in \Phi_+$. Otherwise it is named **indecomposable.**

The search for simple root basis is provided by the set of indecomposable positive roots that we name Δ_γ. To demonstrate such a statement we proceed in the several steps listed below.

Step 1) We begin by proving that every element $\alpha \in \Phi_+$ is a linear combination with coefficients in $\mathbb{Z}_+$ of the roots in Δ_γ. We proceed by *reductio ad absurdum*. Suppose that there are roots $\alpha \in \Phi_+$ which do not admit such a decomposition. Choose among them that one for which (γ, α) is minimal. Since by hypothesis $\alpha \notin \Delta_\gamma$ it follows that there exists $\beta_{1,2} \in \Phi_+$ such that $\alpha = \beta_1 + \beta_2$ and hence $(\gamma, \alpha) = (\gamma, \beta_1) + (\gamma, \beta_2)$. But each of the two contributions in the above sum are positive and we are at a contradiction: if $\beta_{1,2}$ both admit a decomposition in terms of the simple roots with positive integral coefficients, then also α does which contradicts the initial hypothesis. If, on the other hand, one of the two does not, then for that one, say β_1, the scalar product with γ is smaller than the same scalar product for α. This contradicts the hypothesis that for α it was minimal. Henceforth exceptions do not exist and the statement is true.

Step 2) If $\alpha, \beta \in \Delta_\gamma$, then, necessarily $(\alpha, \beta) \leq 0$. Otherwise in force of Lemma 7.2.1, $\alpha - \beta \in \Phi$. In that case either $\alpha - \beta$ or $\beta - \alpha$ are positive roots of Φ_+. In the first case, $\alpha = \beta + (\alpha - \beta)$ is shown to be decomposable. In the second case it is $\beta = (\beta - \alpha) + \alpha$

that is shown to be decomposable. In either case the hypothesis is contradicted. Hence $(\alpha, \beta) \leq 0$ is true.

Step 3) The vectors in Δ_γ are linearly independent. Let us prove this by *reductio ad absurdum*. Suppose that there is a vanishing linear combination of the vectors in Δ_γ, namely $0 = \sum_i \alpha_i$. Separating the coefficients into two subsets, $p_i > 0$ the positive ones and $n_i < 0$ the negative ones, the same condition can be written as $\epsilon = \eta$ where $\epsilon = \sum_i p_i \alpha_i$ and $\eta = -\sum_i n_i \alpha_j$. Because of what we proved in step 2) we have $(\epsilon, \epsilon) = \sum_{ij} p_i p_j (\alpha_i, \alpha_j) \leq 0$. Hence $\epsilon = 0$ and $p_i = 0$. Similarly for η. Therefore the conjectured linear combination does not exist and all the vectors in Δ_γ are linearly independent.

Step 4) Combining step 3 with step 1 we arrive at the conclusion that Δ_γ spans $\mathbf{\Phi}$ and then also $\mathbb{E}$. Furthermore, for each root the decomposition involves either non-negative or non-positive integers. Hence Δ_γ is indeed a simple root basis and the theorem is proved. □

From now on we can associate to every complex simple Lie algebra its root system $\mathbf{\Phi}$ and to every root system its simple root basis Δ. Furthermore, each root system singles out a well-defined finite group, named the Weyl group, that is obtained combining the reflections with respect to all the roots.

Definition 7.3.2. Let $\mathbf{\Phi}$ be a root system in dimension ℓ. The Weyl group of $\mathbf{\Phi}$, denoted $\mathscr{W}(\mathbf{\Phi})$, is the finite group generated by the reflections σ_α, $\forall \alpha \in \mathbf{\Phi}$.

Since for any two vectors $\mathbf{v}, \mathbf{w} \in \mathbb{E}$ we have:

$$(\sigma_\alpha(\mathbf{v}), \sigma_\alpha(\mathbf{w})) = (\mathbf{v}, \mathbf{w}) \tag{7.58}$$

it follows that the Weyl group, which is finite, is always a finite subgroup of the rotation group in ℓ dimensions:

$$\mathscr{W}(\mathbf{\Phi}) \subset \mathrm{SO}(\ell) \tag{7.59}$$

7.4 Classification of the irreducible root systems

Having established that all possible irreducible root systems $\mathbf{\Phi}$ are uniquely determined (up to isomorphisms) by the Cartan matrix:

$$C_{ij} = \langle \alpha_i, \alpha_j \rangle \equiv 2\frac{(\alpha_i, \alpha_j)}{(\alpha_j, \alpha_j)} \tag{7.60}$$

we can classify *all the complex simple Lie algebras* by classifying all possible Cartan matrices.

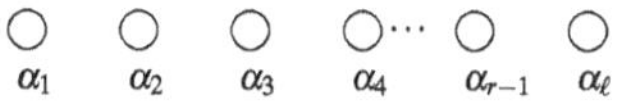

Figure 7.1: The simple roots α_i are represented by circles.

7.4.1 Dynkin diagrams

Each Cartan matrix can be given a graphical representation in the following way. To each simple root α_i we associate a circle ○ as in Figure 7.1 and then we link the i-th circle with the j-th circle by means of a line which is *simple*, *double* or *triple*, depending on whether

$$\langle\alpha_i,\alpha_j\rangle\langle\alpha_j,\alpha_i\rangle = 4\cos^2\theta_{ij} = \begin{cases}1\\2\\3\end{cases} \tag{7.61}$$

having denoted θ_{ij} the angle between the two simple roots α_i and α_j. The corresponding graph is named a **Coxeter graph.**

If we consider the simplest case of two-dimensional Cartan matrices we have the four possible Coxeter graphs depicted in Figure 7.2. Given a Coxeter graph if it is *simply laced*, namely, if there are only simple lines, then all the simple roots appearing in such a graph have the same *length* and the corresponding Cartan matrix is completely identified. On the other hand, if the Coxeter graph involves double or triple lines, then, in order to identify the corresponding Cartan matrix, we need to specify which of the two roots sitting at the end points of each multiple line is the *long* root and which is the *short* one. This can be done by associating an arrow to each multiple line. By convention we decide that this *arrow points* in the direction of the *short root*. A Coxeter graph equipped with the necessary arrows is named a **Dynkin diagram**. Applying this convention to the case of the Coxeter graphs of Figure 7.2 we obtain the result displayed in Figure 7.3. The one-to-one correspondence between the Dynkin diagram and the associated Cartan matrix is illustrated by considering in some detail the case $\mathfrak{b}_2$ in Figure 7.3. By definition of the Cartan matrix we have:

$$\begin{aligned}2\frac{(\alpha_1,\alpha_2)}{(\alpha_2,\alpha_2)} &= 2\frac{|\alpha_1|}{|\alpha_2|}\cos\theta = -2\\ 2\frac{(\alpha_2,\alpha_1)}{(\alpha_1,\alpha_1)} &= 2\frac{|\alpha_2|}{|\alpha_1|}\cos\theta = -1\end{aligned} \tag{7.62}$$

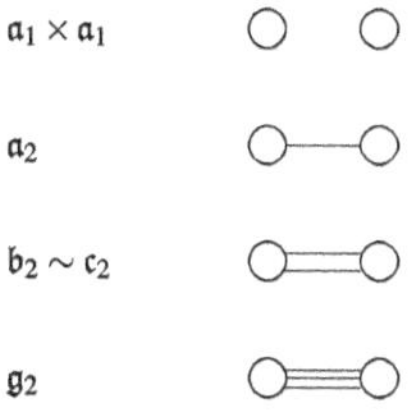

Figure 7.2: The four possible Coxeter graphs with two vertices.

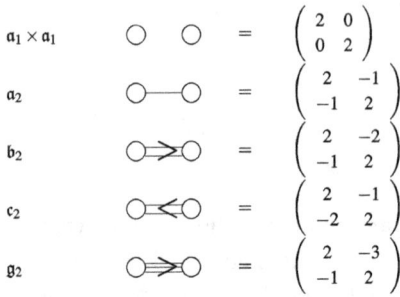

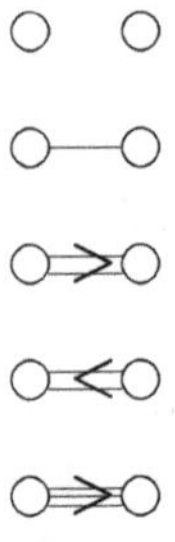

$$\mathfrak{a}_1 \times \mathfrak{a}_1 \quad = \begin{pmatrix} 2 & 0 \\ 0 & 2 \end{pmatrix}$$

$$\mathfrak{a}_2 \quad = \begin{pmatrix} 2 & -1 \\ -1 & 2 \end{pmatrix}$$

$$\mathfrak{b}_2 \quad = \begin{pmatrix} 2 & -2 \\ -1 & 2 \end{pmatrix}$$

$$\mathfrak{c}_2 \quad = \begin{pmatrix} 2 & -1 \\ -2 & 2 \end{pmatrix}$$

$$\mathfrak{g}_2 \quad = \begin{pmatrix} 2 & -3 \\ -1 & 2 \end{pmatrix}$$

Figure 7.3: The distinct Cartan matrices in two dimensions (and therefore the simple algebras in rank two) correspond to the Dynkin diagrams displayed above. We have distinguished a $\mathfrak{b}_2$ and a $\mathfrak{c}_2$ matrix since they are the limiting case for $\ell = 2$ of two series of Cartan matrices the $\mathfrak{b}_\ell$ and the $\mathfrak{c}_\ell$ series that for $\ell > 2$ are truly different. However, $\mathfrak{b}_2$ is the transposed of $\mathfrak{c}_2$ so that they correspond to isomorphic algebras obtained one from the other by renaming the two simple roots $\alpha_1 \leftrightarrow \alpha_2$.

so that we conclude:

$$|\alpha_1|^2 = 2|\alpha_2|^2 \tag{7.63}$$

which shows that α_1 is a long root, while α_2 is a short one. Hence the arrow in the Dynkin diagram pointing towards the short root α_2 tells us that the matrix elements C_{12} is -2 while the matrix element C_{21} is -1. It happens the opposite in the example $\mathfrak{c}_2$.

7.4.2 The classification theorem

Having clarified the notation of Dynkin diagrams the basic classification theorem of *complex simple Lie algebras* is the following:

Theorem 7.4.1. *If Φ is an irreducible system of roots of rank ℓ then its Dynkin diagram is either one of those shown in Figure 7.4 or for special values of ℓ is one of those shown in Figure 7.5. There are no other irreducible root systems besides these ones.*

Proof of Theorem 7.4.1. Let us consider a Euclidean space E and in E let us consider set of vectors:

$$\mathscr{U} = \{\epsilon_1, \epsilon_2, \dots, \epsilon_\ell\} \tag{7.64}$$

that satisfy the following three conditions:

$$\begin{aligned} (\epsilon_i, \epsilon_i) &= 1 \\ (\epsilon_i, \epsilon_j) &\le 0 \quad i \neq j \\ 4(\epsilon_i, \epsilon_j)^2 &= 0, 1, 2, 3 \quad i \neq j \end{aligned} \tag{7.65}$$

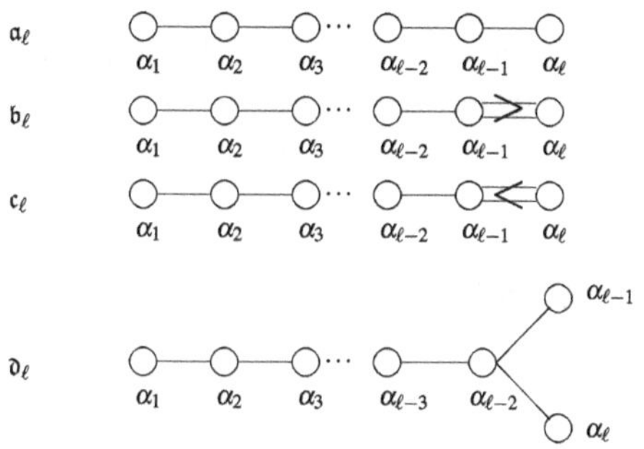

Figure 7.4: The Dynkin diagrams of the four infinite families of classical simple algebras.

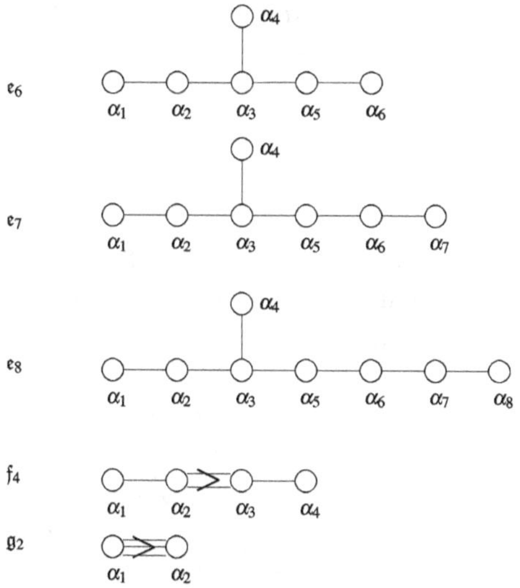

Figure 7.5: The Dynkin diagrams of the five exceptional algebras.

Such a system of vectors is named *admissible*. It is clear that each admissible system of vectors singles out a Coxeter graph Γ. Indeed, the vectors ϵ_i correspond to the simple roots α_i divided by their norm:

$$\epsilon_i = \frac{\alpha_i}{\sqrt{|\alpha_i|^2}} \tag{7.66}$$

Our task is that of classifying all connected Coxeter graphs.

We proceed through a series of steps.

Step 1 We note that by deleting a subset of vectors ϵ_i in an admissible system those that are left still form an admissible system whose Coxeter graph is obtained from the original one by deleting the corresponding vertices and all the lines that end in these vertices.

Step 2 The number of pairs of vertices that are connected by at least one line is strictly less than the number of vectors ϵ_i namely strictly less than ℓ. Indeed let us set $\epsilon = \sum_{i=1}^{\ell} \epsilon_i$ and observe what follows. Since all the ϵ_i are independent we have $\epsilon \neq 0$. Hence

$$0 < (\epsilon, \epsilon) = \ell + 2\sum_{i<j} (\epsilon_i, \epsilon_j) \tag{7.67}$$

If the i-th vertex is joined to the j-th vertex we have $4(\epsilon_i, \epsilon_j)^2 = 1, 2, 3$. Hence we can conclude that, in this case:

$$2(\epsilon_i, \epsilon_j) \leq -1 \tag{7.68}$$

On the other hand, if the i-th vertex is not joined to the j-th vertex we have $2(\epsilon_i, \epsilon_j) = 0$. Naming N_J the number of pairs of vertices joined by at least one line we conclude that:

$$0 < (\epsilon, \epsilon) < \ell - N_J \quad \Rightarrow \quad N_J \leq \ell - 1 \tag{7.69}$$

which is what we have asserted.

Step 3 The Coxeter graph Γ cannot contain any loop. Indeed, if a loop existed this would constitute a subgraph Γ' for which the number of pairs joined by a line N_J would be larger than the number of vertices and this we have shown to be impossible.

Step 4 The number of lines that end up in any vertex can be at most three. Indeed, let $\epsilon \in \mathscr{U}$ and let us denote $\eta_1, \eta_2, \ldots, \eta_k$ the vectors connected to ϵ by some link. In other words, we have $(\epsilon, \eta_i) < 0 (\forall \eta_i)$. Since there are no loops in the graph it follows that no η_i can be connected to any other η_j, namely $(\eta_i, \eta_j) = 0\ \forall i \neq j$. Since $\mathscr{U}$ is a set of linearly independent vectors there must exist a unit vector η_0 in the vector span of $\epsilon, \eta_1, \ldots, \eta_k$ which is orthogonal to $\eta_1, \ldots, \eta_k$. Obviously the projection of such a vector η_0 on ϵ is non-vanishing, namely $(\epsilon, \eta_0) \neq 0$. The set $\eta_0, \eta_1, \ldots, \eta_k$ makes an orthogonal basis for the linear span of the vectors $\epsilon, \eta_1, \ldots, \eta_k$ and we can write:

$$\begin{aligned} \epsilon &= \sum_{i=0}^{k} (\epsilon, \eta_i) \eta_i \\ 1 = (\epsilon, \epsilon) &= \sum_{i=0}^{k} (\epsilon, \eta_i)^2 \end{aligned} \tag{7.70}$$

This reasoning implies that $\sum_{i=1}^{k} (\epsilon, \eta_i)^2 < 1$ and hence

$$4\sum_{i=1}^{k} (\epsilon, \eta_i)^2 < 4 \tag{7.71}$$

On the other hand, $4(\epsilon, \eta_i)^2$ is precisely the number of lines that link η_i to ϵ so that eq. (7.71) is precisely the statement we wanted to prove in **Step 4.**

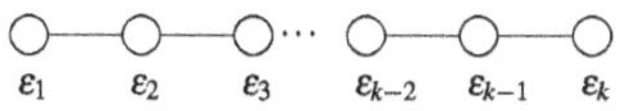

Figure 7.6: A simple line Coxeter graph.

Step 5 The only connected Coxeter graph that contains a triple line is the $\mathfrak{g}_2$ graph of Figure 7.2. This immediately follows from **Step 4**.

Step 6 Let $\{\epsilon_1, \dots, \epsilon_k\} \subset \mathcal{U}$ be a subset of vectors corresponding to a simple line as in Figure 7.6. Then the subset $\mathcal{U}' \equiv \{\mathcal{U} - \{\epsilon_1, \dots, \epsilon_k\}\} \cup \{\epsilon\}$ where $\epsilon \equiv \sum_{i=1}^{k} \epsilon_i$ is still an admissible system. Graphically the operation of making the transition from the admissible system $\mathcal{U}$ to the admissible system $\mathcal{U}'$ corresponds to collapsing the entire simple line to a single vertex. That this statement is true can be proved in the following way. That the vectors composing $\mathcal{U}'$ are linearly independent is obvious. By hypothesis of a simple chain we have:

$$2(\epsilon_i, \epsilon_{i+1}) = -1 \quad 1 \le i \le k-1 \tag{7.72}$$

so that

$$(\epsilon, \epsilon) = k + 2\sum_{i<j}(\epsilon_i, \epsilon_j) = k - (k-1) = 1 \tag{7.73}$$

and hence ϵ is a unit vector. Furthermore, each $\eta \in \mathcal{U} - \{\epsilon_1, \dots, \epsilon_k\}$ can be joined at most to one of the vectors $\epsilon_1, \dots, \epsilon_k$. Otherwise we would generate a loop. Hence we either have $(\eta, \epsilon) = 0$ or we have $(\eta, \epsilon_i) \neq 0$ for some value of i. In any case we conclude $4(\eta, \epsilon_i)^2 = 0, 1, 2, 3$ which is what makes $\mathcal{U}'$ an admissible system.

Step 7 A Coxeter graph cannot contain subgraphs of the form displayed in Figure 7.7. Indeed, in all these three cases, by using the property shown in **Step 6** and collapsing a simple chain we obtain a graph that contains a vertex where four lines converge. This was shown to be forbidden in **Step 4**.

Step 8 Relying on the properties we have so far proven we are left with four types of possible Coxeter graphs, namely i) the simple chains of length ℓ corresponding to

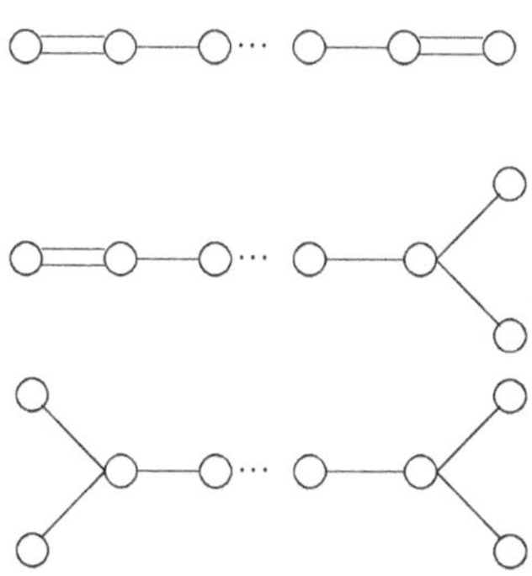

Figure 7.7: Prohibited subgraphs.

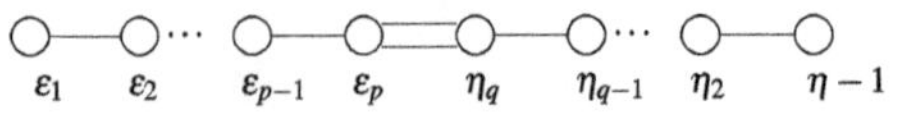

Figure 7.8: Coxeter graph with a double link that is preceded by a simple chain of length p and followed by a simple chain of length q.

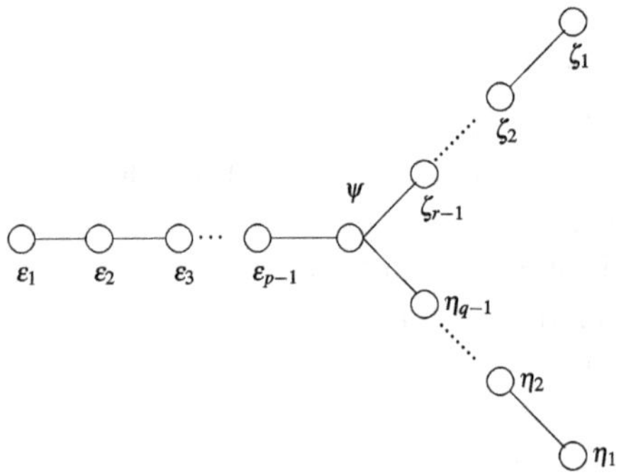

Figure 7.9: Coxeter graph with a node. The unit vector in the node is named ψ while the unit vectors along the three simple lines departing from the node are respectively named $\epsilon_1, \dots, \epsilon_{p-1}, \eta_1, \dots, \eta_{q-1}, \zeta_1, \dots, \zeta_{r-1}$. The graph is characterized by the three integer numbers p, q, r that denote the lengths of the three simple lines departing from the node.

the $\mathfrak{a}_\ell$ Dynkin diagrams of Figure 7.4, ii) the $\mathfrak{g}_2$ graph of Figure 7.2, iii) the graphs of Figure 7.8 with a double line, and finally iv) the graphs of Figure 7.9 with a node.

Step 9 If we consider the graphs of the type shown in Figure 7.8, there are only two solutions, namely:

$$\begin{array}{llll} p = 2; & q = 2 & \Rightarrow \mathfrak{f}_4 & \text{Dynkin diagram} \\ p = \ell \in \mathbb{N}; & q = 1 & \Rightarrow \mathfrak{b}_\ell \quad \text{or} \quad \mathfrak{c}_\ell & \text{Dynkin diagrams} \end{array} \tag{7.74}$$

Indeed, let us set $\epsilon = \sum_{i=1}^{p} i\epsilon_i$ and $\eta = \sum_{i=1}^{q} i\eta_i$. By hypothesis of simple chains we have $2(\epsilon_i, \epsilon_{i+1}) = -1$, $2(\eta_i, \eta_{i+1}) = -1$ and all the other pairs of vectors are mutually orthogonal. In this way we obtain:

$$\begin{aligned} (\epsilon, \epsilon) &= \sum_{i=1}^{p} i^2 - \sum_{i=1}^{p-1} i(i-1) = p\frac{p-1}{2} \\ (\eta, \eta) &= \sum_{i=1}^{q} i^2 - \sum_{i=1}^{q-1} i(i-1) = q\frac{q-1}{2} \end{aligned} \tag{7.75}$$

and since by hypothesis of double line we have: $4(\epsilon_p, \eta_q)^2 = 2$, it follows that

$$(\epsilon, \eta)^2 = p^2q^2(\epsilon_p, \eta_q)^2 = \frac{1}{2}p^2q^2 \tag{7.76}$$

On the other hand, from the triangular Schwarz inequality of Euclidean geometry we have:

$$\begin{gathered}(\epsilon,\eta) < (\epsilon,\epsilon)(\eta,\eta) \\ \Downarrow \\ (p-1)(q-1) < 2\end{gathered} \tag{7.77}$$

which for positive integers p,q admits only the two solutions advocated in eq. (7.74). The first solution leads to the Dynkin diagram of the exceptional Lie algebra $\mathfrak{f}_4$, while the second solution leads to the two infinite series of classical Lie algebras $\mathfrak{b}_\ell$ and $\mathfrak{c}_\ell$.

Step 10 Let us finally consider the Coxeter graphs of the type shown in Figure 7.9. We claim that the only possible solutions are:

$$(p,q,r) = \begin{cases} (\ell,1,1) & \Rightarrow \quad \mathfrak{a}_\ell \quad \text{Dynkin diagrams} \quad \ell \in \mathbb{N} \\ (\ell-2,2,2) & \Rightarrow \quad \mathfrak{d}_\ell \quad \text{Dynkin diagrams} \quad 4 \le \ell \in \mathbb{N} \\ (3,3,2) & \Rightarrow \quad \mathfrak{e}_6 \quad \text{Dynkin diagram} \\ (4,3,2) & \Rightarrow \quad \mathfrak{e}_7 \quad \text{Dynkin diagram} \\ (5,3,2) & \Rightarrow \quad \mathfrak{e}_8 \quad \text{Dynkin diagram} \end{cases} \tag{7.78}$$

To prove this statement we follow a strategy similar to that used in the proof of **Step 9** and we define the following three vectors:

$$\epsilon = \sum_{i=1}^{p-1} i\epsilon_i; \quad \eta = \sum_{i=1}^{q-1} i\eta_i; \quad \sum_{i=1}^{r-1} i\zeta_i \tag{7.79}$$

Clearly ϵ,η,ζ are mutually orthogonal and ψ, the vector in the node, is not in the subspace generated by ϵ,η,ζ. Hence if in the linear span of $\{\psi,\epsilon,\eta,\zeta\}$ we construct a vector y that is orthogonal to $\{\epsilon,\eta,\zeta\}$, we obtain that $(y,\psi) \neq 0$. Normalizing this vector to 1 we can write:

$$\psi = (\psi,y)y + \frac{(\psi,\epsilon)}{\sqrt{(\epsilon,\epsilon)}}\epsilon + \frac{(\psi,\eta)}{\sqrt{(\eta,\eta)}}\eta + \frac{(\psi,\zeta)}{\sqrt{(\zeta,\zeta)}}\zeta \tag{7.80}$$

and we obtain:

$$(\psi,\psi) = 1 = (\psi,y)^2 + \frac{(\psi,\epsilon)^2}{(\epsilon,\epsilon)} + \frac{(\psi,\eta)^2}{(\eta,\eta)} + \frac{(\psi,\zeta)^2}{(\zeta,\zeta)} \tag{7.81}$$

that implies the inequality:

$$1 > \frac{(\psi,\epsilon)^2}{(\epsilon,\epsilon)} + \frac{(\psi,\eta)^2}{(\eta,\eta)} + \frac{(\psi,\zeta)^2}{(\zeta,\zeta)} \tag{7.82}$$

By definition of the Coxeter graph in Figure 7.9 we have:

$$\begin{aligned}(\psi,\epsilon) = (p-1)(\epsilon_{p-1},\psi) \quad &\Rightarrow \quad (\psi,\epsilon)^2 = \frac{(p-1)^2}{4} \\ & \quad (\epsilon,\epsilon) = \frac{p(p-1)}{2}\end{aligned} \tag{7.83}$$

and similarly for the scalar products associated with the other chains. Inserting these results into the inequality (7.82) we obtain the Diophantine inequality:

$$\frac{1}{p} + \frac{1}{q} + \frac{1}{r} > 1 \tag{7.84}$$

whose independent solutions are those displayed in eq. (7.78). To this effect it is sufficient to note that eq. (7.84) has an obvious permutational symmetry in the three numbers p, q, r. To avoid double counting of solutions we break this symmetry by setting $p \geq q \geq r$ and then we see that the only possibilities are those listed in eq. (7.78). □

Having concluded the proof of the classification theorem we can look back and compare the just obtained results with those summarized in Section 4.2.4. The anticipated correspondence between finite rotation subgroups and simply laced Lie algebras should now be clear.

7.5 Identification of the classical Lie algebras

In the previous sections we have classified the allowed Dynkin diagrams and hence the allowed simple root systems. We have not shown that all of them do indeed exist. This is what we do in the present section by explicit construction. Furthermore, we identify the classical or exceptional complex Lie algebra that corresponds to each of the constructed root systems.

7.5.1 The $\mathfrak{a}_\ell$ root system and the corresponding Lie algebra

The Dynkin diagram is that recalled in Figure 7.10. We want to perform the explicit construction of a root system that admits a basis corresponding to such a diagram.

To this effect let us consider the $(\ell+1)$-dimensional Euclidean space $\mathbb{R}^{\ell+1}$ and let $\epsilon_1, \dots, \epsilon_{\ell+1}$ denote the unit vectors along the $\ell+1$ axes:

$$\epsilon_1 = \begin{pmatrix} 1 \\ 0 \\ \dots \\ \dots \\ 0 \end{pmatrix}, \quad \epsilon_2 = \begin{pmatrix} 0 \\ 1 \\ 0 \\ \dots \\ 0 \end{pmatrix} \quad \dots \quad \epsilon_{\ell+1} = \begin{pmatrix} 0 \\ 0 \\ \dots \\ \dots \\ 1 \end{pmatrix} \tag{7.85}$$

$\mathfrak{a}_\ell$

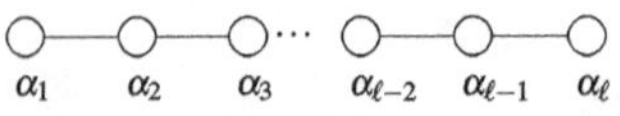

Figure 7.10: The Dynkin diagram of $\mathfrak{a}_\ell$ type.

Define the vector $v = \epsilon_1 + \epsilon_2 + \cdots + \epsilon_{\ell+1}$:

$$v = \begin{pmatrix} 1 \\ 1 \\ 1 \\ \dots \\ 1 \end{pmatrix} \tag{7.86}$$

and define $\mathbb{I} \subset \mathbb{R}^{\ell+1}$ to be the $(\ell+1)$-dimensional cubic lattice immersed in $\mathbb{R}^{\ell+1}$:

$$\mathbb{I} = \{x \in \mathbb{R}^{\ell+1} / x = n^i \epsilon_i,\ n^i \in \mathbb{Z}\} \tag{7.87}$$

In the cubic lattice $\mathbb{I}$ consider the sublattice:

$$\mathbb{I}' = \mathbb{I} \cap E \tag{7.88}$$

where E is the hyperplane of vectors orthogonal to the vector v:

$$E = \{y \in \mathbb{R}^{\ell+1} / (v, y) = 0\} \tag{7.89}$$

Finally in the sublattice $\mathbb{I}'$ consider the finite set of vectors whose norm is $\sqrt{2}$:

$$\mathbf{\Phi} = \{\alpha \in \mathbb{I}' / (\alpha, \alpha) = 2\} \tag{7.90}$$

Theorem 7.5.1. *The above defined set* $\mathbf{\Phi}$ *is a root system and it corresponds to the* $\mathfrak{a}_\ell$ *Dynkin diagram.*

Proof of Theorem 7.5.1. To prove this proposition let us first summarize the properties of $\mathbf{\Phi}$. We have:

$$\alpha \in \mathbf{\Phi} \quad \Rightarrow \quad \begin{cases} 1) & \alpha = n^i \epsilon_i & & n^i \in \mathbb{Z} \\ 2) & (\alpha, v) = 0 & \Leftrightarrow & \sum_{i=1}^{\ell+1} n^i = 0 \\ 3) & (\alpha, \alpha) = 2 & \Leftrightarrow & \sum_{i=1}^{\ell+1} (n^i)^2 = 2 \end{cases} \tag{7.91}$$

These diophantine equations have the following solutions:

$$\alpha = \epsilon_i - \epsilon_j \quad (i \neq j) \tag{7.92}$$

The number of such solutions is equal to twice the number of pairs (ij) in $\ell+1$-dimensional space:

$$\#\alpha = 2\frac{1}{2}(\ell+1)(\ell+1-1) = \ell(\ell+1) = (\ell+1)^2 - 1 - \ell \tag{7.93}$$

We verify that this finite set of vectors is a root system. First we check that for all pairs $\alpha, \beta \in \Phi$ their hook product is an integer. Indeed, we have:

$$\begin{aligned} \langle \alpha, \beta \rangle &= 2\frac{(\alpha, \beta)}{(\beta, \beta)} = (\alpha, \beta) \\ &= (\epsilon_i - \epsilon_j, \epsilon_k - \epsilon_\ell) = \delta_{ik} - \delta_{jk} - \delta_{i\ell} + \delta_{j\ell} \in \mathbb{Z} \end{aligned} \tag{7.94}$$

Secondly, we check that the reflection of any candidate root $\beta \in \Phi$ with respect to any other candidate root $\alpha \in \Phi$ belongs to the same set Φ:

$$\begin{aligned} \sigma_\alpha(\beta) &= \beta - (\alpha, \beta) \\ &= \epsilon_k - \epsilon_\ell - (\delta_{ik} - \delta_{jk} - \delta_{i\ell} + \delta_{j\ell})(\epsilon_i - \epsilon_j) \end{aligned} \tag{7.95}$$

If (k, ℓ) are both different from (i, j) then $\sigma_\alpha(\beta) = \beta \in \Phi$, so the statement is true. If $k = i$ then, necessarily $k \neq j$ and $i \neq \ell$, so that:

$$\sigma_\alpha(\beta) = \epsilon_k - \epsilon_\ell - (\delta_{ik} + \delta_{jl})(\epsilon_i - \epsilon_j) \tag{7.96}$$

If $j \neq \ell$ then:

$$\sigma_\alpha(\beta) = \epsilon_k - \epsilon_\ell - (1)(\epsilon_i - \epsilon_j) = \epsilon_j - \epsilon_\ell \in \Phi \tag{7.97}$$

If $j = \ell$ then

$$\sigma_\alpha(\beta) = \epsilon_k - \epsilon_\ell - 2(\epsilon_k - \epsilon_\ell) = \epsilon_\ell - \epsilon_k \in \Phi \tag{7.98}$$

which exhausts all possible cases.

Hence, in $\mathbb{R}^{\ell+1}$ we have constructed a root system of $\ell(\ell + 1)$ roots.

Consider the roots:

$$\alpha_i = \epsilon_i - \epsilon_{i+1}, \quad (i = 1, \dots, \ell) \tag{7.99}$$

These roots are clearly linearly independent and given a root $\alpha \in \Phi$ it can be expressed as a linear combination of the α_i. Subdivide the set of roots into a positive and negative set according to the following rule:

$$\begin{aligned} &\Phi = \Phi_+ \cup \Phi_- \\ &\alpha \in \Phi_+ : \quad \{\alpha = \epsilon_i - \epsilon_j,\ i < j\} \\ &\alpha \in \Phi_- : \quad \{\alpha = \epsilon_j - \epsilon_i,\ i < j\} \end{aligned} \tag{7.100}$$

Clearly, positive roots can be written as follows:

$$\alpha = \epsilon_i - \epsilon_j = \alpha_i + \alpha_{i+1} + \dots + \alpha_{j-1} \tag{7.101}$$

and, consequently, have integer positive components in the $\{\alpha, \dots, \alpha_\ell\}$ basis. Negative roots have negative integer components. It follows that $\{\alpha, \dots, \alpha_\ell\}$ form a basis of simple roots.

Let us compute the Cartan matrix:

$$\begin{aligned}(\alpha_i,\alpha_j) &= (\epsilon_i-\epsilon_j,\epsilon_j-\epsilon_{j+1}), \quad i<j\\ &= \delta_{ij}-\delta_{i+1,j}-\delta_{i,j+1}+\delta_{i+1,j+1}\end{aligned} \tag{7.102}$$

If	$i=j$	$(\alpha_i,\alpha_i)=2$
If	$j=i+1$	$(\alpha_i,\alpha_{i+1})=-1$
If	$j=i-1$	$(\alpha_i,\alpha_{i-1})=-1$

Hence we have precisely the Dynkin diagram of Figure 7.10.

This concludes the proof of our theorem. □

Theorem 7.5.2. *The root system $\mathfrak{a}_\ell$ corresponds to the complex Lie algebra $\mathfrak{sl}(\ell+1,\mathbb{C})$ of traceless matrices in $\ell+1$ dimensions.*

Proof of Theorem 7.5.2. Note that the dimension of the $\mathfrak{sl}(\ell+1,\mathbb{C})$ Lie algebra is:

$$\dim \mathfrak{sl}(\ell+1,\mathbb{C}) = (\ell+1)^2-1 \tag{7.103}$$

since on the $(\ell+1)\times(\ell+1)$ matrix A we just impose one scalar condition, namely $\operatorname{Tr} A = 0$. This agrees with the number of roots in the system Φ:

$$\operatorname{card}\Phi = \ell(\ell+1) = (\ell+1)^2-\ell-1 \tag{7.104}$$

if the rank of $\mathfrak{sl}(\ell+1,\mathbb{C})$ is precisely ℓ:

$$\operatorname{card}\Phi = \dim \mathbb{G} - \dim \mathscr{H} \tag{7.105}$$

$\mathscr{H}$ being the Cartan subalgebra.

This is indeed the case. Let e_{ij} denote the $(\ell+1)\times(\ell+1)$ matrix whose entries are all zero except for the ij-th entry which is one

$$e_{ij} = \begin{pmatrix} 0 & 0 & \dots & \dots & \dots & 0 & \\ 0 & 0 & \dots & \dots & \dots & 0 & \\ \dots & \dots & 0 & 0 & 0 & \dots & \\ \dots & \dots & 0 & 1 & 0 & \dots & i\text{-th} \\ \dots & \dots & 0 & 0 & 0 & \dots & \\ 0 & 0 & \dots & \dots & \dots & 0 & \\ & & & j\text{-th} & & & \end{pmatrix} \tag{7.106}$$

and define:

$$\begin{aligned} H_i &= e_{ii} - e_{\ell+1,\ell+1} \quad (i=1,\dots,\ell)\\ \mathscr{E}_{ij} &= e_{ij} \quad (i\neq j)\end{aligned} \tag{7.107}$$

Since:

$$e_{ij} \cdot e_{km} = \delta_{jk} e_{im} \tag{7.108}$$

we have:

$$\begin{aligned}
[H_i, H_j] &= 0 \\
[H_i, \mathscr{E}_{rs}] &= \delta_{ir} e_{is} - \delta_{si} e_{ri} - \delta_{\ell+1,r} e_{\ell+1,s} + \delta_{s,\ell+1} e_{r,\ell+1} \\
&= (\delta_{ir} - \delta_{si} - \delta_{\ell+1,r} + \delta_{s,\ell+1}) \mathscr{E}_{rs}
\end{aligned} \tag{7.109}$$

Now observe that a basis for the space E of vectors orthogonal to $v = (1, 1, \dots, 1)$ is provided by:

$$u_i = \epsilon_i - \epsilon_{\ell+1} \quad (i = 1, \dots, \ell) \tag{7.110}$$

Indeed, this is a system of ℓ linearly independent vectors in an ℓ-dimensional space. Hence we can identify:

$$(\delta_{ir} - \delta_{si} - \delta_{\ell+1,r} + \delta_{s,\ell+1}) = (\epsilon_r - \epsilon_s, u_i); \quad (r, s = 1, \dots, \ell + 1), \ (i = 1, \dots, \ell) \tag{7.111}$$

This implies that to every Cartan subalgebra element $h = \omega^r H_r$ we can associate the vector $\omega \equiv \omega^r u_r \in E$ and to every root $\epsilon_r - \epsilon_s$ we can associate the linear functional:

$$[\epsilon_r - \epsilon_s](\omega) = (\epsilon_r - \epsilon_s, \omega) \tag{7.112}$$

With such identifications the $\mathfrak{sl}(\ell + 1, \mathbb{C})$ is cast into the canonical Weyl form of eqs (7.38) and our theorem is proved. □

7.5.2 The $\mathfrak{d}_\ell$ root system and the corresponding Lie algebra

The Dynkin diagram is that recalled in Figure 7.11. We want to perform the explicit construction of a root system that admits a basis corresponding to such a diagram.

To this effect let us consider $E = \mathbb{R}^\ell$ and the set

$$\Phi \equiv \{\alpha \in \mathbb{I}_\ell \mid (\alpha, \alpha) = 2\} \tag{7.113}$$

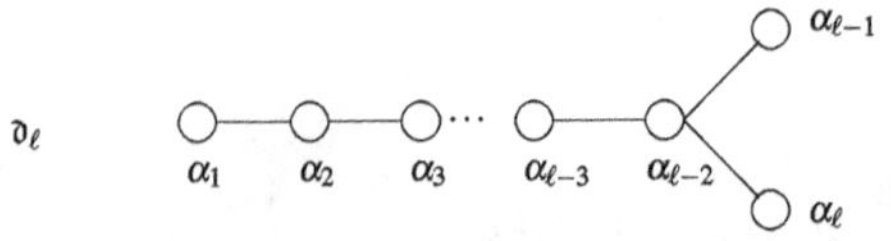

Figure 7.11: The Dynkin diagram of $\mathfrak{d}_\ell$ type.

namely, the set of all length $\sqrt{2}$ vectors in the cubic lattice:

$$\alpha = n^i \epsilon_i; \quad \sum_{i=1}^{\ell} (n^i)^2 = 2 \tag{7.114}$$

The solution of the above constraint is given by all the vectors of the following type:

$$\alpha = \pm(\epsilon_i \pm \epsilon_j) \quad i \neq j \tag{7.115}$$

These vectors are easily counted. We have:

$$\text{card}\, \mathbf{\Phi} = 2\ell(\ell - 1) \tag{7.116}$$

In a completely similar way to the previous case one verifies that $\mathbf{\Phi}$ is a root system. The simple roots can be chosen as follows:

$$\begin{aligned} \alpha_1 &= \epsilon_1 - \epsilon_2 \\ \alpha_2 &= \epsilon_2 - \epsilon_3 \\ \vdots &\;\vdots\; \vdots \\ \alpha_{\ell-1} &= \epsilon_{\ell-1} - \epsilon_\ell \\ \alpha_\ell &= \epsilon_{\ell-1} + \epsilon_\ell \end{aligned} \tag{7.117}$$

which clearly corresponds to the chosen Dynkin diagram in Figure 7.11.

Next we identify the corresponding Lie algebra with $\mathfrak{so}(2\ell, \mathbb{C})$. The procedure is the following. Consider in 2ℓ dimensions the following symmetric matrix:

$$\mathfrak{Q} \equiv \left(\begin{array}{c|c} \mathbf{0}_{\ell\times\ell} & \mathbf{1}_{\ell\times\ell} \\ \hline \mathbf{1}_{\ell\times\ell} & \mathbf{0}_{\ell\times\ell} \end{array} \right) \tag{7.118}$$

The Lie algebra $\mathfrak{so}(2\ell, \mathbb{C})$ can be identified as the set of $2\ell \times 2\ell$-matrices satisfying the $\mathfrak{Q}$-antisymmetricity constraint:

$$\mathfrak{so}(2\ell, \mathbb{C}) = \left\{ A = \left(\begin{array}{c|c} a_{11} & a_{12} \\ \hline a_{21} & a_{21} \end{array} \right) \middle| A^T \mathfrak{Q} + \mathfrak{Q} A = 0 \right\} \tag{7.119}$$

The constraint is easily solved by setting:

$$a_{22} = -a_{11}^T; \quad a_{21} = -a_{21}^T; \quad a_{12} = -a_{12}^T \tag{7.120}$$

We leave to the reader the easy task of explicitly verifying that a Cartan–Weyl basis spanning the space of all matrices of type (7.119) is the following one:

$$\begin{aligned} H_i &= e_{ii} - e_{\ell+i,\ell+i} \\ \mathscr{E}^{\epsilon_i - \epsilon_j} &= e_{ji} - e_{i+\ell,j+\ell} \\ \mathscr{E}^{\epsilon_i + \epsilon_j} &= e_{i+\ell,j} - e_{j+\ell,i} \\ \mathscr{E}^{-\epsilon_i - \epsilon_j} &= e_{j,i+\ell} - e_{i,j+\ell} \end{aligned} \tag{7.121}$$

The diagonal matrices H_i are the Cartan generators. The commutator of each of the step operators $\mathcal{E}^\alpha$ with each of the Cartan operators yields the appropriate answer according to eq. (7.38). This concludes the proof that the root systems of type $\mathfrak{d}_\ell$ are associated with the Lie algebras $\mathfrak{so}(2\ell, \mathbb{C})$ of the orthogonal groups in even dimensions.

7.5.3 The $\mathfrak{b}_\ell$ root system and the corresponding Lie algebra

The Dynkin diagram is that recalled in Figure 7.12. We want to perform the explicit construction of a root system that admits a basis corresponding to such a diagram.

$\mathfrak{b}_\ell$

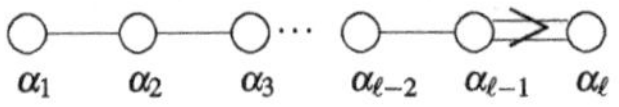

Figure 7.12: The Dynkin diagram of $\mathfrak{b}_\ell$ type.

To this effect let us consider $E = \mathbb{R}^\ell$: within this Euclidean space the candidate root system is defined by the following set of vectors

$$\Phi \equiv \underbrace{\{\pm\epsilon_i \pm \epsilon_j\}}_{2\ell(\ell-1)} \cup \underbrace{\{\pm\epsilon_i\}}_{2\ell} \tag{7.122}$$

where in the underbrace we have spelled out the number of vectors for each of the two types. Hence we have:

$$\operatorname{card}\Phi = 2\ell^2 \tag{7.123}$$

In a completely similar way to the previous cases one can verify that Φ is a root system. The simple roots can be chosen as follows:

$$\begin{aligned} \alpha_1 &= \epsilon_1 - \epsilon_2 \\ \alpha_2 &= \epsilon_2 - \epsilon_3 \\ \vdots &\;\vdots\; \vdots \\ \alpha_{\ell-1} &= \epsilon_{\ell-1} - \epsilon_\ell \\ \alpha_\ell &= \epsilon_\ell \end{aligned} \tag{7.124}$$

which clearly corresponds to the chosen Dynkin diagram in Figure 7.12.

Next we identify the corresponding Lie algebra with $\mathfrak{so}(2\ell+1, \mathbb{C})$. The procedure is similar to that used in the previous case. Consider in $(2\ell+1)$ dimensions the following symmetric matrix:

$$\mathfrak{Q} \equiv \left(\begin{array}{c|c|c} 1 & \mathbf{0}_{1\times\ell} & \mathbf{0}_{1\times\ell} \\ \hline \mathbf{0}_{\ell\times 1} & \mathbf{0}_{\ell\times\ell} & \mathbf{1}_{\ell\times\ell} \\ \hline \mathbf{0}_{\ell\times 1} & \mathbf{1}_{\ell\times\ell} & \mathbf{0}_{\ell\times\ell} \end{array}\right) \tag{7.125}$$

The Lie algebra $\mathfrak{so}(2\ell+1,\mathbb{C})$ can be identified as the set of $(2\ell+1)\times(2\ell+1)$-matrices satisfying the $\mathfrak{Q}$-antisymmetricity constraint:

$$A^T\mathfrak{Q}+\mathfrak{Q}A=0 \tag{7.126}$$

whose solution in terms of independent subblocks is displayed below:

$$A=\begin{pmatrix} 0 & v_1 & v_2 \\ -v_2^T & a & b \\ -v_1^T & -b^T & -a^T \end{pmatrix} \tag{7.127}$$

We leave to the reader the easy task of explicitly verifying that a Cartan–Weyl basis spanning the space of all matrices of type (7.127) is the following one:

$$\begin{aligned} H_i &= e_{i+1,i+1}-e_{\ell+i+1,\ell+i+1} \\ \mathcal{E}^{\epsilon_i-\epsilon_j} &= e_{j+1,i+1}-e_{i+\ell+1,j+\ell+1} \\ \mathcal{E}^{\epsilon_i+\epsilon_j} &= e_{i+\ell+1,j+1}-e_{j+\ell+1,i+1} \\ \mathcal{E}^{-\epsilon_i-\epsilon_j} &= e_{j+1,i+\ell+1}-e_{i+1,j+\ell+1} \\ \mathcal{E}^{\epsilon_i} &= e_{1,i+1}-e_{i+\ell+1,1} \\ \mathcal{E}^{-\epsilon_i} &= e_{i+1,1}-e_{1,i+\ell+1} \end{aligned} \tag{7.128}$$

The diagonal matrices H_i are the Cartan generators. The commutator of each of the step operators $\mathcal{E}^\alpha$ with each of the Cartan operators yields the appropriate answer according to eq. (7.38). This concludes the proof that the root systems of type $\mathfrak{b}_\ell$ are associated with the Lie algebras $\mathfrak{so}(2\ell+1,\mathbb{C})$ of the orthogonal groups in odd dimensions.

7.5.4 The $\mathfrak{c}_\ell$ root system and the corresponding Lie algebra

The Dynkin diagram is that recalled in Figure 7.13. We want to perform the explicit construction of a root system that admits a basis corresponding to such a diagram.

To this effect let us consider $E=\mathbb{R}^\ell$: within this Euclidean space the candidate root system is defined by the following set of vectors

$$\Phi \equiv \underbrace{\{\pm\epsilon_i\pm\epsilon_j\}}_{2\ell(\ell-1)}\cup\underbrace{\{\pm 2\epsilon_i\}}_{2\ell} \tag{7.129}$$

$\mathfrak{c}_\ell$

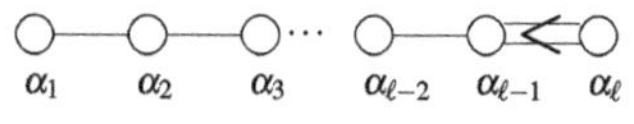

Figure 7.13: The Dynkin diagram of $\mathfrak{c}_\ell$ type.

where in the underbrace we have spelled out the number of vectors for each of the two types. Hence we have:

$$\mathrm{card}\,\Phi = 2\ell^2 \tag{7.130}$$

In a completely similar way to the previous cases one can verify that Φ is a root system. The simple roots can be chosen as follows:

$$\begin{aligned} \alpha_1 &= \epsilon_1 - \epsilon_2 \\ \alpha_2 &= \epsilon_2 - \epsilon_3 \\ \vdots &\ \vdots\ \vdots \\ \alpha_{\ell-1} &= \epsilon_{\ell-1} - \epsilon_\ell \\ \alpha_\ell &= 2\epsilon_\ell \end{aligned} \tag{7.131}$$

which clearly corresponds to the chosen Dynkin diagram in Figure 7.13.

Next we identify the corresponding Lie algebra with $\mathfrak{sp}(2\ell,\mathbb{C})$. The procedure is similar to that used in the previous case. Consider in 2ℓ dimensions the following antisymmetric matrix:

$$\mathfrak{Q} \equiv \left(\begin{array}{c|c} \mathbf{0}_{\ell\times\ell} & \mathbf{1}_{\ell\times\ell} \\ \hline -\mathbf{1}_{\ell\times\ell} & \mathbf{0}_{\ell\times\ell} \end{array}\right) \tag{7.132}$$

The Lie algebra $\mathfrak{sp}(2\ell,\mathbb{C})$ can be identified as the set of $2\ell \times 2\ell$-matrices satisfying the $\mathfrak{Q}$-antisymmetricity constraint:

$$A^T\mathfrak{Q} + \mathfrak{Q}A = 0 \tag{7.133}$$

whose solution in terms of independent subblocks is displayed below:

$$\begin{aligned} A &= \left(\begin{array}{c|c} a_{11} & a_{12} \\ \hline a_{21} & a_{21} \end{array}\right) \\ a_{22} &= -a_{11}^T; \quad a_{21} = a_{21}^T; \quad a_{12} = a_{12}^T \end{aligned} \tag{7.134}$$

We leave to the reader the easy task of explicitly verifying that a Cartan–Weyl basis spanning the space of all matrices of type (7.127) is the following one:

$$\begin{aligned} H_i &= e_{i,i} - e_{\ell+i,\ell+i} \\ \mathcal{E}^{\epsilon_i-\epsilon_j} &= e_{i,j} - e_{j+\ell,i+\ell} \quad i<j \\ \mathcal{E}^{-\epsilon_i+\epsilon_j} &= e_{j,i} - e_{i+\ell,j+\ell} \quad i<j \\ \mathcal{E}^{\epsilon_i+\epsilon_j} &= e_{i,j+\ell} + e_{j,i+\ell} \\ \mathcal{E}^{-\epsilon_i-\epsilon_j} &= e_{i+\ell,j} + e_{j+\ell,i} \\ \mathcal{E}^{2\epsilon_i} &= e_{i,i+\ell} \\ \mathcal{E}^{-2\epsilon_i} &= e_{i+\ell,i} \end{aligned} \tag{7.135}$$

The diagonal matrices H_i are the Cartan generators. The commutator of each of the step operators $\mathscr{E}^\alpha$ with each of the Cartan operators yields the appropriate answer according with eq. (7.38). This concludes the proof that the root systems of type $\mathfrak{c}_\ell$ are associated with the Lie algebras $\mathfrak{sp}(2\ell, \mathbb{C})$ of the symplectic groups in even dimensions.

7.5.5 The exceptional Lie algebras

We should now construct, as Cartan did, the explicit form of the exceptional Lie algebras. We postpone this to Chapter 9. Each exceptional Lie algebra will be studied in relation with a particular homogeneous manifold of interest in physics.

7.6 Bibliographical note

Main bibliographical sources for the present chapter are:
- unpublished lecture notes by one of the authors (P.F.) dating back to the time when he usually gave the Lie algebra course at SISSA (beginning of the 1990s)
- the textbook [88]
- the textbook [92]
- the textbook [29]
- the monograph [84]

Furthermore, let us mention that a detailed historical account of how Lie algebras were discovered by Lie and were independently classified by Killing (who also discovered exceptional Lie algebras) and finally were systematized by Cartan, stressing also the key role played in such developments by Klein and Engels, is presented in the forthcoming book [53].

8 Lie algebra representation theory

My work always tried to unite the truth with the beautiful, but when I had to choose one or the other, I usually chose the beautiful...
Hermann Weyl

8.1 Linear representations of a Lie algebra

The general concept of linear representations of groups was introduced in Section 3.3 and it equally applies to discrete groups as to Lie groups. The reader is invited to revisit Definitions 3.3.1, 3.3.2, 3.3.3, 3.3.4 which apply without any change to continuous groups.

What is new in the case of Lie groups is that linear group representations can be constructed first as Lie algebra linear representations, using next the exponential map (5.193), described in Section 5.9, in order to obtain the matrix representation of finite group elements from that of Lie algebra elements.

In full analogy to the above quoted definitions that apply to groups we have the following ones:

Definition 8.1.1. Let $\mathbb{G}$ be a finite dimensional Lie algebra and let V be a vector space of dimension n. Any homomorphism:

$$D : \mathbb{G} \to \mathrm{Hom}(V, V) \tag{8.1}$$

such that:

$$\forall A, B \in \mathbb{G} : \quad D([A, B]) = [D(A), D(B)] \tag{8.2}$$

is named a linear representation of dimension n of the Lie algebra $\mathbb{G}$. In the left-hand side of eq. (8.2), the symbol $[\,,\,]$ denotes the abstract Lie bracket, while in the right-hand side of the same equation it denotes the commutator of two linear maps applied in sequence.

If the vectors $\{\mathbf{e}_i\}$ form a basis of the vector space V, then each Lie algebra element $\gamma \in \mathbb{G}$ is mapped onto a $n \times n$ matrix $D_{ij}(\gamma)$ such that:

$$D(\gamma).\mathbf{e}_i = \mathbf{e}_j D_{ji}(\gamma) \tag{8.3}$$

and we see that the homomorphism D can also be rephrased as the following one:[1]

$$D : G \to \mathfrak{gl}(n, \mathbb{F}) \quad \mathbb{F} = \begin{cases} \mathbb{R} \\ \mathbb{C} \end{cases} \tag{8.4}$$

1 Let us recall that by $\mathfrak{gl}(n, \mathbb{F})$ one refers to the Lie algebra of arbitrary $n \times n$ matrices.

https://doi.org/10.1515/9783110551204-008

where the field $\mathbb{F}$ is that of the real or complex numbers, depending on whether V is a real or complex vector space. Correspondingly we say that D is a *real* or *complex representation.*

Definition 8.1.2. Let $D : G \to \mathrm{Hom}(V, V)$ be a linear representation of a Lie algebra $\mathbb{G}$. A vector subspace $W \subset V$ is said to be **invariant** iff:

$$\forall \gamma \in \mathbb{G},\ \forall \mathbf{w} \in W : \quad D(\gamma).\mathbf{w} \in W \tag{8.5}$$

Definition 8.1.3. A linear representation $D : G \to \mathrm{Hom}(V, V)$ of a Lie algebra $\mathbb{G}$ is named **irreducible** iff the only invariant subspaces of V are $\mathbf{0}$ and V itself.

In other words, a representation is irreducible if it does not admit any proper invariant subspace.

Definition 8.1.4. A linear representation $D : \mathbb{G} \to \mathrm{Hom}(V, V)$ that admits at least one proper invariant subspace $W \subset V$ is named **reducible.** A reducible representation $D : G \to \mathrm{Hom}(V, V)$ is named **fully reducible** iff the orthogonal complement $W^\perp$ of any invariant subspace W is also invariant.

8.1.1 Weights of a representation and the weight lattice

The classification of complex Lie algebras in terms of Cartan subalgebras and root systems provides a very powerful weaponry for the explicit construction of linear representations of Lie algebras. In this context the contributions of Herman Weyl have been extremely important and the role of the Weyl group is quite essential.

Given the root system Φ of rank ℓ and its basis of simple roots α_i ($i = 1, \ldots, \ell$) we can define the root lattice:

$$\Lambda_{\text{root}} \subset \mathbb{R}^\ell \quad / \quad \mathbf{v} \in \Lambda_{\text{root}} \quad \Leftrightarrow \quad \mathbf{v} = n^i \alpha_i, \quad n^i \in \mathbb{Z} \tag{8.6}$$

As we explained in Section 4.3.1, given any lattice Λ we can define its dual $\Lambda^\star$ by means of the definition (4.73). Concretely, given a basis of Λ, one introduces a dual basis as in eq. (4.74) and takes the linear combinations of these dual basis vectors with integer valued coefficients. In view of this general construction a naturally arising question concerns the interpretation of the dual of the root lattice. This latter is named the *weight lattice* and it is of the utmost relevance for *representation theory*:

$$\Lambda_{\text{weight}} \equiv \Lambda^\star_{\text{root}} \tag{8.7}$$

Explicitly one defines the basis of simple weights dual to the basis of simple roots through the following condition:

$$\langle \lambda^i, \alpha_j \rangle \equiv 2\frac{(\lambda^i, \alpha_j)}{(\alpha_j, \alpha_j)} = \delta^i_j \tag{8.8}$$

and sets:

$$\Lambda_{\text{weight}} \subset \mathbb{R}^{\ell} \quad / \quad \mathbf{w} \in \Lambda_{\text{weight}} \quad \Leftrightarrow \quad \mathbf{w} = n_i \lambda^i, \quad n_i \in \mathbb{Z} \tag{8.9}$$

As we presently show, linear representations Γ of the Lie algebra $\mathbb{G}$ associated with the considered root system Φ are essentially encoded in certain precisely defined subsets $\Pi_\Gamma \subset \Lambda_{\text{weight}}$ named the set of *weights of the representation.*

8.1.1.1 The maximally split and the maximal compact real sections of a simple Lie algebra $\mathbb{G}(\mathbb{C})$

So far, in our discussion of Lie algebras $\mathbb{G}$ we have assumed that, as vector spaces, they are defined over the complex numbers $\mathbb{C}$. An important topic that we will address later is that of real sections $\mathbb{G}_r \subset \mathbb{G}(\mathbb{C})$. Given the simple Lie algebra generators in the canonical Cartan–Weyl form, $T_A = \{H_i, E^\alpha, E^{-\alpha}\}$, the question is which restrictions on the imaginary and the real parts of the coefficients c^A of Lie algebra elements $c^A T_A$ can be introduced that are consistent with the Lie bracket and produce a real Lie algebra $\mathbb{G}_r$. Furthermore, one would like to know how many such real sections exist up to isomorphism.

Here we just introduce two real sections that are simply and universally defined for all simple Lie algebras:

a) **The maximally split real section $\mathbb{G}_{\mathbf{max}}$.** This is defined by assuming that the allowed coefficients c^A are all real. In any linear representation of $\mathbb{G}_{\mathbf{max}}$ the matrices representing

$$T_A \equiv \{H_i, E^\alpha, E^{-\alpha}\} \tag{8.10}$$

are all *real.* From the representations of $\mathbb{G}_{\mathbf{max}}$, by taking linear combinations of the generators with complex coefficients one obtains all the linear representations of the complex Lie algebra $\mathbb{G}(\mathbb{C})$.

b) **The maximally compact real section $\mathbb{G}_c$.** This real section, whose exponentiation produces a compact Lie group, is obtained by allowing linear combinations with real coefficients of the set of generators:

$$\mathfrak{T}_A \equiv \{iH_i, i(E^\alpha + E^{-\alpha}), (E^\alpha - E^{-\alpha})\} \tag{8.11}$$

In all linear representations of $\mathbb{G}_\mathbf{c}$ the matrices representing the generators $\mathfrak{T}_A$ are *anti-Hermitian.*

One easily obtains the Hermitian matrix representation of the generators $\mathfrak{T}_A$ from the real representation of the generators T_A and vice versa. It also follows that the matrices representing $E^{-\alpha}$ are the transposed of those representing E^α.

A very useful instrument in the explicit construction of matrix representations that has also important consequences for later developments is provided by the notion of Borel subalgebra. Starting from the Cartan–Weyl basis, if one considers the subset of generators:

$$\mathrm{Bor}[\mathbb{G}] = \mathrm{span}\{H_i, E^\alpha\}; \quad \alpha > 0 \tag{8.12}$$

we see that it corresponds to a *solvable subalgebra of* $\mathbb{G}$. Hence according to Corollary 6.4.1, every representation of Bor[$\mathbb{G}$] can be put into an upper triangular form. This gives a powerful construction criterion for the fundamental representation. We just construct an upper triangular representation of the Bor[$\mathbb{G}$] subalgebra and then we promote it to a representation of the full Lie algebra $\mathbb{G}$, by setting:

$$E^{-\alpha} = (E^\alpha)^T \tag{8.13}$$

Furthermore, in view of the above discussion, the representations of the real sections $\mathbb{G}_{\mathbf{max}}$ and $\mathbb{G}_{\mathbf{c}}$ can be considered together and on that we rely in the following.

Let us assume that

$$\Gamma : \mathbb{G}_c \quad \Longrightarrow \quad \mathrm{Hom}(V, V) \tag{8.14}$$

is a linear representation of the compact section which, by the above argument, can be turned into a representation of the maximally split real section. From linearity we conclude that:

$$\Gamma(H_i) = -i\Gamma(iH_i); \quad (i = 1, \dots, \ell) \tag{8.15}$$

are a set of ℓ *commuting Hermitian matrices*. Therefore we can choose a basis in the vector space V where all the $\Gamma(H_i)$ are diagonal matrices. This is equivalent to the statement that the vector space V of dimension n can be decomposed into a direct sum of subspaces:

$$V = \bigoplus_{\lambda \in \Pi_\Gamma} V_\lambda; \quad (\dim V_\lambda \geq 1) \tag{8.16}$$

where $\Pi_\Gamma \subset \mathcal{H}^\star$ is a subset of linear functionals on the Cartan subalgebra $\mathcal{H}$ that is characteristic of the considered representation Γ:

$$\lambda : \mathcal{H} \to \mathbb{C} \tag{8.17}$$

and $\forall \mathbf{v} \in V_\lambda$ we have:

$$\Gamma(h)\mathbf{v} = \lambda(h)\mathbf{v}; \quad \forall h \in \mathcal{H} \tag{8.18}$$

The set Π_Γ is necessarily a finite set if Γ is a finite dimensional representation. It is the *set of weights of the representation* as we already anticipated above. The dimension of V_λ is named the weight multiplicity and is denoted by $m(\lambda)$:

$$m(\lambda) = \dim V_\lambda \tag{8.19}$$

It follows that:

$$n = \dim V = \sum_{\lambda \in \Pi_\Gamma} m(\lambda) \tag{8.20}$$

Since the Cartan subalgebra is endowed with a natural scalar product induced by the Killing metric $\kappa(,)$ restricted to $\mathscr{H}$, the elements of $\mathscr{H}^\star$ can be represented by r-dimensional vectors that belong to the same Euclidean space which includes the roots. Let $\{H_i\}$ be the basis of $\mathscr{H}$ in which $\kappa(H_i, H_j) = \delta_{ij}$ and let us define the components of the λ vector by:

$$\lambda_i = \lambda(H_i) \tag{8.21}$$

Furthermore, let us utilize the bracket notation for the vectors belonging to V_λ:

$$\begin{aligned} &|\lambda, p\rangle; \quad p = 1, \dots, m(\lambda) \\ \langle \mu, r | \lambda, p\rangle &= \delta_{\lambda\mu}\delta_{rp} \\ \Gamma(h)|\lambda, p\rangle &= \lambda(h)|\lambda, p\rangle \end{aligned} \tag{8.22}$$

then we have the following:

Lemma 8.1.1. *Let $\lambda \in \Pi_\Gamma$ be a weight and let $\alpha \in \Phi$ be a root, then $\alpha + \lambda \in \Pi_\Gamma$ is another weight if there exists a vector $|\lambda\rangle \in V_\lambda$ such that:*

$$\Gamma(E^\alpha)|\lambda\rangle \neq 0 \tag{8.23}$$

Proof of Lemma 8.1.1. To prove the statement it suffices to set:

$$|\lambda + \alpha\rangle = \Gamma(E^\alpha)|\lambda\rangle \tag{8.24}$$

and verify that:

$$\begin{aligned} \Gamma(h)|\lambda + \alpha\rangle &= [\Gamma(h), \Gamma(E^\alpha)]|\lambda\rangle + \Gamma(E^\alpha)\Gamma(h)|\lambda\rangle \\ &= (\alpha(h) + \lambda(h))|\lambda + \alpha\rangle \end{aligned} \tag{8.25}$$

which shows that $(\alpha + \lambda)$ is a weight. □

8.1.1.2 The subalgebra $A_1 \subset G$ associated with a positive root and the properties of the weights

To each positive root $\alpha \in \Phi_+$ of a root system we can associate a subalgebra $\mathbb{G} \supset A_1 \sim \mathfrak{su}(2)$ which plays an important role in determining the properties of weights of a representation. Given the canonical Cartan–Weyl form of the simple Lie algebra (7.38) let us define:

$$\mathscr{H}_\alpha = \frac{2}{(\alpha, \alpha)}\alpha^i H_i; \quad \mathscr{E}_\alpha^\pm = \sqrt{\frac{2}{(\alpha, \alpha)}} E^{\pm\alpha} \tag{8.26}$$

We obtain the commutation relations:

$$\begin{aligned}
[\mathcal{H}_\alpha, \mathcal{H}_\alpha] &= 0 \\
[\mathcal{H}_\alpha, \mathcal{E}^\pm_\alpha] &= \pm 2\mathcal{E}^\pm_\alpha \\
[\mathcal{E}^+_\alpha, \mathcal{E}^-_\alpha] &= \mathcal{H}_\alpha
\end{aligned} \tag{8.27}$$

This algebra is isomorphic to the angular momentum algebra

$$\begin{aligned}
[J_3, J_3] &= 0 \\
[J_3, J^\pm] &= \pm J^\pm \\
[J^+, J^-] &= 2J_3
\end{aligned} \tag{8.28}$$

upon the identifications:

$$\mathcal{H}_\alpha = 2J_3; \quad \mathcal{E}^\pm_\alpha = J^\pm \tag{8.29}$$

Since any irreducible $\mathbb{G}$ representation necessarily decomposes into a sum of irreducible representations of each $A_1(\alpha)$ subalgebra and since in each irreducible representation of A_1 the eigenvalues of J_3 are necessarily either integer or half-integer,[2] it follows that for any root α the eigenvalues of H_α must be integer. This provides the relation between the weights of the representation defined above and the weight lattice introduced in eqs (8.9) and (8.8). Indeed, from the outlined reasoning we conclude:

$$\begin{aligned}
H_\alpha|\lambda, i\rangle &= \lambda(H_\alpha)|\lambda, i\rangle \\
\lambda(H_\alpha) = 2\frac{(\lambda,\alpha)}{(\alpha,\alpha)} &\equiv \langle\lambda,\alpha\rangle \in \mathbb{Z} \\
&\Downarrow \\
\lambda &\in \Lambda_{\mathbf{weight}}
\end{aligned} \tag{8.30}$$

So the weights of a representation Π_Γ constitute a finite subset of vectors lying in the weight lattice. We have the following very simple but very important theorem:

Theorem 8.1.1. *Given any finite dimensional representation Γ of a simple Lie algebra $\mathbb{G}$, the set of its weights Π_Γ is invariant with respect to the Weyl group defined in Section 7.3 Definition 7.3.2. Explicitly we have:*

$$\begin{aligned}
&\forall\alpha \in \mathbf{\Phi}: \quad \lambda \in \Pi_\Gamma \quad \Rightarrow \quad \sigma_\alpha(\lambda) \in \Pi_\Gamma \\
&m(\sigma_\alpha(\lambda)) = m(\lambda)
\end{aligned} \tag{8.31}$$

2 Here we use the properties of angular momenta that are known to any student of physics, mathematics or engineering from the course of elementary quantum mechanics. A representation of spin j contains $(2j+1)$ states with third component $m = -j, -j+1, \dots, j-1, j$ where j is integer or half-integer.

Proof of Theorem 8.1.1. Let us note that:

$$\sigma_\alpha\lambda(h) = \lambda(h) - \langle\lambda,\alpha\rangle\alpha(h) = \lambda(h) - \lambda(H_\alpha)\alpha(h) \tag{8.32}$$

and let us define:

$$\beta = \begin{cases} \alpha & \text{if } \lambda(H_\alpha) \geq 0 \\ -\alpha & \text{if } \lambda(H_\alpha) < 0 \end{cases} \tag{8.33}$$

With such a notation we always have $\lambda(H_\beta) \geq 0$ and $\forall h \in \mathscr{H}_{CSA}$, $\sigma_\alpha\lambda(h) = \sigma_\beta\lambda(h)$. The number $\lambda(H_\beta) = R \in \mathbb{Z}_+$ is by construction a non-negative integer. If $R = 0$ the theorem is proved. Hence consider the case $R > 0$.

We have:

$$\sigma_\alpha\lambda(h) = \lambda(h) - R\beta(h) \tag{8.34}$$

The linear functional $\sigma_\alpha\lambda(h)$ is a weight of the representation with multiplicity $m(\sigma_\beta\lambda) = m(\lambda)$ if

$$[\Gamma(E^{-\beta})]^R|\lambda,p\rangle \neq 0; \quad p = 1,\dots,m(\lambda) \tag{8.35}$$

Indeed, if A, B are two linear operators, we have:

$$[A,B] = xB \quad \Rightarrow \quad [A,B^k] = kxB; \quad \forall k \in \mathbb{Z}_+ \tag{8.36}$$

so that

$$\Gamma(h)[\Gamma(E^{-\beta})]^R|\lambda,p\rangle = (\lambda(h) - R\beta(h))[\Gamma(E^{-\beta})]^R|\lambda,p\rangle \tag{8.37}$$

and

$$[\Gamma(E^{-\beta})]^R|\lambda,p\rangle \equiv |\lambda - R\beta,p\rangle \in V_{\lambda-R\beta} \tag{8.38}$$

unless $[\Gamma(E^{-\beta})]^R|\lambda,p\rangle = 0$. Let us now consider the representation of the $A_1(\beta)$ Lie algebra induced by Γ, where $|\lambda,p\rangle$ is an eigenvector of J_3 with eigenvalue $\frac{1}{2}R$, since

$$J_3|\lambda,p\rangle = \frac{1}{2}\Gamma(H_\beta)|\lambda,p\rangle = \frac{1}{2}R|\lambda,p\rangle \tag{8.39}$$

By general properties of the angular momentum representations, in the same representation there must be also the eigenstate of angular momentum $J_3 = -\frac{1}{2}R$ and this is precisely the state $[\Gamma(E^{-\beta})]^R|\lambda,p\rangle$. Hence:

$$[\Gamma(E^{-\beta})]^R|\lambda,p\rangle \neq 0 \tag{8.40}$$

and the theorem is proved. □

Next let us generalize the concept of α-string through β and consider the weights of the form $\lambda + k\alpha$ where λ is a weight, α is a root and k is an integer. We have another theorem:

Theorem 8.1.2. *Let α be a non-vanishing root and let $\lambda \in \Pi_\Gamma$ be a weight of a representation Γ of a semisimple Lie algebra $\mathbb{G}$. Then there exist two integers p, q, such that $\lambda + k\alpha \in \Pi_\Gamma$ is a weight of the representation for all the integers $-p \leq k \leq q$. Furthermore, we have:*

$$p - q = \langle \lambda, \alpha \rangle \equiv 2\frac{(\lambda, \alpha)}{(\alpha, \alpha)} \tag{8.41}$$

Proof of Theorem 8.1.2. Using the above definition (8.33) for β we conclude that for $k > 0$ $\lambda + k\beta$ is a weight if and only if:

$$0 \neq (\Gamma(E^\beta))^k |\lambda\rangle \tag{8.42}$$

for some $|\lambda\rangle \in V_\lambda$. On the other hand, this happens only if $\lambda(H_\beta) + k\beta(H_\beta)$ is an eigenvalue of H_β in the restriction of the given representation to the subalgebra $A_1(\beta)$. Similarly in the case $k < 0$ one argues for $(\Gamma(E^{-\beta}))^{-k}|\lambda\rangle$. Therefore, let us suppose that the largest irreducible representation of $A_1(\beta)$ contained in Γ is of spin j and is of dimension $2j + 1$. In this case the possible eigenvalues of $\frac{1}{2}H_\beta$ are extended from j to $-j$ with an integer jump from one to the next one. Hence k takes all values in the interval determined by:

$$\begin{aligned} \lambda(H_\beta) + q &= j \\ \lambda(H_\beta) - p &= -j \end{aligned} \tag{8.43}$$

and we have:

$$p - q = \lambda(H_\beta) = \langle \lambda, \beta \rangle \tag{8.44}$$

as we wanted to show. □

As we see from the above results, all properties of linear representations of higher dimensional algebras follow from the properties of the representations of the angular momentum algebra (8.28) that every student of physics learns in elementary courses of quantum mechanics.

8.1.1.3 Irreducible representations and maximal weights

One important notion that applies to the weights of a representation is their partial ordering that relies on the following:

Definition 8.1.5. Given two weights λ, μ of an irreducible representation Γ of a simple Lie algebra $\mathbb{G}$ we say that λ is larger than μ, if and only if their difference is a sum of positive roots

$$\lambda \succ \mu \quad \Leftrightarrow \quad \lambda - \mu = \sum_{\alpha_k \in \Phi_+} \alpha_k \tag{8.45}$$

The main theorem of representation theory whose proof we omit is the following:

Theorem 8.1.3. *If the irreducible representation Γ of a simple Lie algebra $\mathbb{G}$ is finite dimensional then the set of weights Π_Γ admits a maximal weight Λ such that any other weight $\lambda \in \Pi_\Gamma$ is $\lambda \prec \Lambda$. Consequently, we have:*

$$\Gamma(E^\alpha)|\Lambda\rangle = 0 \tag{8.46}$$

Furthermore, we have the following:

Corollary 8.1.1. *The highest weights of all existing irreducible representations belong to the* ***Weyl chamber*** $\mathfrak{W}$ *which is the convex hull defined by the following conditions:*

$$\mathbf{v} \in \mathfrak{W} \quad \Leftrightarrow \quad 2\frac{(\mathbf{v}, \alpha_i)}{(\alpha_i, \alpha_i)} > 0 \quad i = 1, \dots, \ell \tag{8.47}$$

where α_i are the simple roots.

Theorem 8.1.3 and its corollary are much more than an existence statement. They provide an actual construction algorithm for the weights of the representation. Starting from any dominant weight Λ in the Weyl chamber and imposing the condition (8.46) we can begin the construction of the representation by acting on $|\Lambda\rangle$ with the step-down operators $\Gamma(E^{-\alpha})$ obtaining in this way the states $|\Lambda - k\alpha\rangle$ of the α-strings through Λ. The last value of k is determined, for each string, by $\langle \Lambda, \alpha \rangle$. Next, by acting once again with the operators $\Gamma(E^{-\alpha})$ one derives all the states and the weights of the representation.

8.1.1.4 Weyl's character formula

Weyl brought Lie group theory to perfection by providing the analog of characters of finite group theory.

In the case of finite groups, characters are associated with the conjugacy classes of group elements. In the case of Lie groups, every group element $\gamma \in G$ is conjugate to some element $h \in T \subset G$ where, by definition, the maximal torus $T \equiv \exp[\mathscr{H}]$ is the exponential of the Cartan subalgebra. This statement can be easily understood if we consider any linear representation of G. Given a group element γ we can always rotate it to a basis where it is diagonal: in that basis it lies in T. Then the character of a representation can be defined as the following linear functional on the Cartan subalgebra:

$$\forall h \in \mathscr{H}: \quad \chi_\Gamma(h) \equiv \mathrm{Tr}_\Gamma(\exp[h]) = \sum_{\lambda \in \Pi_\Gamma} m(\lambda) \exp[\lambda(h)] \tag{8.48}$$

where the sum is extended to all the weights of the representation with their multiplicities.

Weyl demonstrated that the character can be written by means of a universal formula whose only input is the maximal weight of the representation. His result is the following:

$$\chi_\Gamma = \frac{\sum_{w \in W} \epsilon(w) \exp[\Lambda_\Gamma + \rho]}{\exp[\rho] \prod_{\alpha \in \Phi_+} (1 - \exp[-\alpha])} \tag{8.49}$$

where:
1. $\rho \equiv \frac{1}{2}(\sum_{\alpha \in \Phi_+} \alpha)$ is half of the sum of positive roots.
2. W denotes the Weyl group and $\epsilon(w)$ denotes the determinant of the action of the Weyl group element w on the Cartan subalgebra $\mathscr{H} \subset \mathbb{G}$.
3. Λ_Γ is the highest weight of the irreducible representation.

Weyl's formula (8.49) is to be interpreted in the linear functional sense. Weights and roots are all linear functionals on the Cartan subalgebra $\mathscr{H}$ and the character defined by (8.49) is also a functional on $\mathscr{H}$.

From eq. (8.49), considering the limiting character $h \to 0$, Weyl obtained the formula that gives the dimension of the irreducible representation with highest weight Λ_Γ:

$$\dim\Gamma = \frac{\prod_{\alpha>0}(\alpha, \Lambda_\Gamma + \rho)}{\prod_{\alpha>0}(\alpha, \rho)} \tag{8.50}$$

Although Weyl's formulae are not always friendly for explicit calculations since one has to cancel vanishing factors between the numerator and the denominator, yet they have an outstanding conceptual relevance. They explicitly show that complete information about any linear representation is encoded in its highest weight vector and what are the possible dominant weights is completely fixed by the weight lattice, ultimately by the Cartan matrix.

Next we turn to illustrative examples.

8.2 Discussion of tensor products and examples

In this section we illustrate the concepts introduced in the previous ones by means of some examples that will be worked out in detail. For pedagogical effectiveness we will mainly consider Lie algebras of rank $r = 2$ where the roots and the weights are easily visualized in an Euclidean plane so that they can be intuitively grasped.

One of our goals is that of comparing the universal Weyl–Dynkin approach to representations with the natural approach in terms of tensors with fixed symmetry patterns.

The underlying question is a very important and conceptual one. The following digression tries to illustrate the issue at stake.

8.2.1 Tensor products and irreps

Every simple Lie algebra $\mathbb{G}$ possesses a faithful representation of smallest dimension that we can dub the *fundamental representation* $\Gamma^{\mathbf{fun}}$ and a simple way of constructing higher dimensional representations is that of taking tensor products of that representation. Assume that the fundamental representation is of dimension n:

$$\Gamma^{\mathbf{fun}} : \mathbb{G} \to \mathrm{Hom}(V_n, V_n) \tag{8.51}$$

If we take the tensor product:

$$V_{m|n} \equiv \underbrace{V_n \otimes V_n \otimes \cdots \otimes V_n}_{m \text{ times}} \tag{8.52}$$

eq. (8.51) canonically induces a new representation of dimension n^m:

$$\Gamma_{m|n} : \mathbb{G} \to \mathrm{Hom}(V_{m|n}, V_{m|n}) \tag{8.53}$$

If $\mathbf{e}_i$, $(i =, 1, \ldots, n)$ are a set of basis vectors of V_n, then a natural basis for $V_{m|n}$ is provided by:

$$|i_1 \ldots i_m\rangle \equiv \mathbf{e}_{i_1} \otimes \cdots \otimes \mathbf{e}_{i_m} \tag{8.54}$$

and the matrices of the induced representation $\Gamma_{m|n}$ are constructed as follows:

$$\begin{aligned} \forall \mathfrak{g} \in \mathbb{G} \quad & \langle j_1 \ldots j_m | \Gamma_{m|n}(\mathfrak{g}) | i_1 \ldots i_m \rangle \\ &= \sum_{p=1}^{m} \delta_{j_1,i_1} \delta_{j_2,i_2} \ldots \delta_{j_{p-1},i_{p-2}} \Gamma^{\mathbf{fun}}_{j_p,i_p}(\mathfrak{g}) \delta_{j_{p+1},i_{p+2}} \ldots \delta_{j_m,i_m} \end{aligned} \tag{8.55}$$

Typically the representation $\Gamma_{m|n}$ is reducible since one can find several invariant subspaces, one way of finding invariant subspaces being the construction of irreducible tensors with respect to permutations.

A generic element of the vector space $V_{m|n}$ is of the form:

$$V_{m|n} \ni \mathbf{w} = w^{i_1 \ldots i_m} |i_1 \ldots i_m\rangle \tag{8.56}$$

Since the definition of eq. (8.55) is invariant under any permutation of the m indices, we can impose that the coefficients $w^{i_1 \ldots i_m}$ transform into an irreducible representation of the symmetric group $\mathscr{S}_m$ operating on the set of indices: the corresponding subspace of $V_{m|n}$ will be an invariant subspace under the Lie algebra $\mathbb{G}$. The simplest example is provided by the singlet representation of the symmetric group, namely, by the full symmetrization of the indices. Let us assume that the coefficients $w^{i_1 \dots i_m}$ have the following property:

$$\forall P \in \mathscr{S}_m : \quad w^{P(i_1) \dots P(i_m)} = w^{i_1 \dots i_m} \tag{8.57}$$

the subspace $SV_{m|n} \subset V_{m|n}$ made by those vectors whose components are antisymmetric tensors, namely fulfill property (8.57), form an invariant subspace. What is the dimension of such a subspace? Answering this question amounts to determining how many different m-tuples of n-numbers are there up to permutations. We easily find:

$$\dim SV_{m|n} = \frac{n \cdot (n+1) \cdot (n+2) \dots (n+m)}{m!} \tag{8.58}$$

Another simple example corresponds to the alternating representation of $\mathscr{S}_m$ or, if you prefer such a wording, to complete antisymmetrization. Suppose that the coefficients $w^{i_1 \dots i_m}$ have the following property:

$$\forall P \in \mathscr{S}_m : \quad w^{P(i_1) \dots P(i_m)} = (-)^{\delta_P} w^{i_1 \dots i_m} \tag{8.59}$$

where δ_P denotes the parity of the permutation. The subspace $AV_{m|n} \subset V_{m|n}$ made by those vectors whose components are antisymmetric tensors, namely have property (8.59), form an invariant subspace. What is the dimension of such a subspace? Answering this question amounts to determine how many different m-tuples of n-numbers are there, without repetitions. We easily find:

$$\dim AV_{m|n} = \frac{n \cdot (n-1) \cdot (n-2) \dots (n-m)}{m!} \tag{8.60}$$

In Section 3.2.4 and Subsection 3.2.4.1 we discussed the conjugacy classes of the symmetric group $\mathscr{S}_m$ and we show that they are in one-to-one correspondence with the partitions of m into integers, which can be represented by Young tableaux (see eq. (3.13)). On the other hand, we know from the general principles of finite group theory that the irreducible representations of a finite group G are in one-to-one correspondence with its conjugacy classes. It follows that the irreducible representations of the symmetric group S_m are in one-to-one correspondence with the Young tableaux one can construct out of m-boxes and that to each Young tableau there corresponds an irreducible tensor spanning an invariant subspace under the full considered Lie algebra $\mathbb{G}$.

Let us consider a Young tableau like that in the following picture:

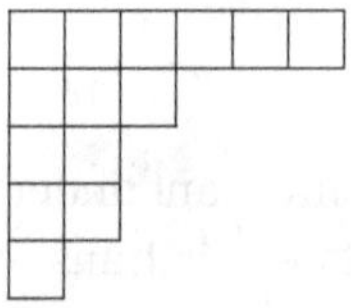

(8.61)

and let us fill its m-boxes with the numbers $1, 2, \ldots, m$ in any preferred order. For instance we can arrange the 14 symbols at our disposal in the example of eq. (8.61) as follows:

$$\begin{array}{|c|c|c|c|c|c|}\hline 1 & 2 & 3 & 4 & 5 & 6 \\ \hline 7 & 8 & 9 \\ \cline{1-3} 10 & 11 \\ \cline{1-2} 12 & 13 \\ \cline{1-2} 14 \\ \cline{1-1} \end{array} \tag{8.62}$$

Once the tableau has been fixed, we consider two types of permutations. *Horizontal permutations* p are permutations which interchange only symbols in the same row. *Vertical permutations* q interchange only symbols in the same column. Then we construct the following index-operators:

$$P = \sum_p p \quad \text{symmetrizer} \tag{8.63}$$

$$Q = \sum_q \delta_q q \quad \text{antisymmetrizer} \tag{8.64}$$

where the sum is extended respectively to all the horizontal and to all the vertical permutations encoded in the tableau. Finally we define the Young operator:

$$Y = QP \tag{8.65}$$

that can be applied to any tensor with m-indices once these latter have been divided into blocks corresponding to the rows of the tableau. The resulting tensor forms an irreducible representation of the symmetric group and certainly belongs to an invariant subspace in the tensor product of m fundamental representations $\Gamma^{\mathbf{fun}}$ of any Lie algebra $\mathbb{G}$. In particular the symmetric tensor introduced in eq. (8.57) corresponds to the following Young tableau composed of a single row:

$$\begin{array}{|c|c|c|c|c|c|}\hline 1 & 2 & \cdot & \cdot & \cdot & m \\ \hline \end{array} \tag{8.66}$$

while the antisymmetric tensor discussed in eq. (8.59) corresponds to the Young tableau composed of a single column:

$$\begin{array}{|c|}\hline 1 \\ \hline 2 \\ \hline \cdot \\ \hline \cdot \\ \hline m \\ \hline \end{array} \tag{8.67}$$

The question is whether these irreducible tensors always correspond to irreducible representations of $\mathbb{G}$ and whether all irreps of $\mathbb{G}$ can be obtained in this way.

The answer to these questions is not universal and it depends on the type of considered Lie algebras. Let us first discuss the case of classical Lie algebras.

8.2.1.1 Classical Lie algebras

$\mathfrak{a}_\ell$) For the algebras $\mathfrak{a}_\ell$ that correspond to the matrix algebras $\mathfrak{sl}(\ell+1,\mathbb{C})$ the answer is positive to both the above posed questions. There is only one fundamental representation, namely, the defining one in $\ell+1$ dimensions and the set of all possible irreducible representations is in one-to-one correspondence with the set of irreducible m-tensors obtained by means of the Young tableau discussed above. Every *highest weight* module or *irrep* can be uniquely identified with a corresponding irreducible tensor associated with a Young tableau.

C_ℓ) For the algebras $\mathfrak{c}_\ell$ that correspond to the matrix algebras $\mathfrak{sp}(2\ell,\mathbb{C})$ the answer is more elaborated. There is only one fundamental representation, which is the defining one in 2ℓ dimensions, but the candidate irreducible tensors obtained acting with a Young operator on the multiple tensor product of that fundamental irrep are not guaranteed to be fully irreducible with respect to $\mathfrak{sp}(2\ell,\mathbb{C})$. The reason is that differently from the previous case there exists a symplectic invariant antisymmetric tensor $\mathbb{C}_{\alpha\beta}$ by means of which any antisymmetric pair of indices of a multiple tensors can be contracted giving rise to an invariant symplectic trace. To obtain an irreducible tensor corresponding to a truly irreducible representation of the Lie algebra $\mathfrak{sp}(2\ell,\mathbb{C})$, one has to enforce the additional condition that all the symplectic traces vanish.

$\mathfrak{b}_\ell, D_\ell$) The algebras $\mathfrak{b}_\ell$ correspond to the orthogonal Lie algebras in odd dimensions $\mathfrak{so}(2\ell+1,\mathbb{C})$, while the algebras D_ℓ correspond to the same in even dimensions $\mathfrak{so}(2\ell,\mathbb{C})$. In both cases we do not have a single fundamental representation, rather we have two of them. One is the defining representation $\Gamma^{\mathbf{vec}}$, that we name the vector representation, the other is the spinor representation $\Gamma^{\mathbf{spin}}$ which is a peculiarity of the orthogonal Lie algebras. For $\mathfrak{so}(n,\mathbb{C})$ with $n \le 6$ the dimension of the spinor representation is smaller or equal to the dimension of the vector representation, yet for $n > 6$ the dimension of the spinor representation becomes larger than n and grows exponentially with it. While the vector representation is typically contained in the square of the spinor one, there is no way to construct it generically as an irreducible symmetric or antisymmetric tensor in spinors. Hence the general truth is that for orthogonal groups there are two types of representations:
1. the bosonic ones that are obtained, as in the symplectic case, applying Young tableaux operators to tensor products of the vector representation and setting to zero all symmetric traces taken with the symmetric invariant matrix $\eta_{\alpha\beta}$;
2. the spinor ones that are obtained taking the tensor product of any bosonic irreducible one with the spinor representation and setting to zero the gamma-traces that can be constructed with the gamma-matrices.

In conclusion for classical algebras the irreducible highest weight representations can be identified with irreducible tensors or tensor-spinors but one has to be careful with standard traces, symplectic traces or gamma-traces.

8.2.1.2 Exceptional Lie algebras

Since the exceptional Lie algebras have no classical definition in terms of matrix algebras, the irreducible representations cannot be identified with irreducible tensors starting from a fundamental representation. The matrix representations have to be constructed *ad hoc* case by case.

8.2.1.3 The value of the highest weight module construction

From the point of view of the highest weight module construction all representations are on the same footing and there is no essential distinction between classical and exceptional Lie algebras. This shows that although more abstract, this approach is the deepest and most algorithmic one which encodes the essence of the underlying mathematical structure. In Weyl's spirit, it provides that symbolic system which is the real mathematical truth behind the screen.[3]

Let us then turn to our examples.

8.2.2 The Lie algebra $\mathfrak{a}_2$, its Weyl group and examples of its representations

In order to illustrate the general ideas discussed in previous sections, we begin with an analysis of the rank $r = 2$ Lie algebra $\mathfrak{sl}(3, \mathbb{R})$ and of its fundamental representation. This Lie algebra is the maximally split real section of the $\mathfrak{a}_2$ Lie algebra, encoded in the Dynkin diagram of Figure 8.1.

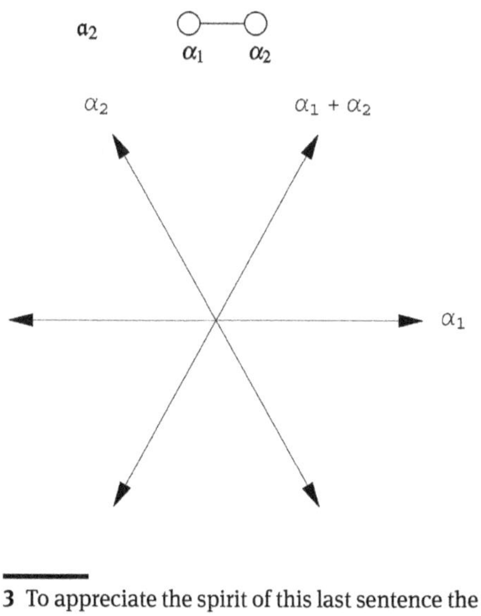

Figure 8.1: The $\mathfrak{a}_2$ Dynkin diagram and root system.

3 To appreciate the spirit of this last sentence the reader is referred to Weyl's lecture *on the mathematical way of thinking* fully analyzed and discussed in the forthcoming book [53].

As discussed in previous sections, from the representations of the maximally split real section we easily obtain those of the compact real section which, in this case, is $\mathfrak{su}(3)$.

The root system has rank two and it is composed of the six vectors displayed below and pictured in Figure 8.1:

$$\Phi_{a_2} = \begin{cases} \alpha_1 = (\sqrt{2}, 0) \\ \alpha_2 = (-\frac{1}{\sqrt{2}}, \sqrt{\frac{3}{2}}) \\ \alpha_1 + \alpha_2 = (\frac{1}{\sqrt{2}}, \sqrt{\frac{3}{2}}) \\ -\alpha_1 = (-\sqrt{2}, 0) \\ -\alpha_2 = (\frac{1}{\sqrt{2}}, -\sqrt{\frac{3}{2}}) \\ -\alpha_1 - \alpha_2 = (-\frac{1}{\sqrt{2}}, -\sqrt{\frac{3}{2}}) \end{cases} \tag{8.68}$$

The simple roots are α_1 and α_2.

A complete set of generators for the Lie algebra $\mathfrak{sl}(3, \mathbb{R})$ is provided by the following 3×3 matrices:

$$\begin{aligned} H_1 &= \begin{pmatrix} \frac{1}{\sqrt{2}} & 0 & 0 \\ 0 & -\frac{1}{\sqrt{2}} & 0 \\ 0 & 0 & 0 \end{pmatrix}; \quad H_2 = \begin{pmatrix} \frac{1}{\sqrt{6}} & 0 & 0 \\ 0 & \frac{1}{\sqrt{6}} & 0 \\ 0 & 0 & -\sqrt{\frac{2}{3}} \end{pmatrix} \\ E^{\alpha_1} &= \begin{pmatrix} 0 & 1 & 0 \\ 0 & 0 & 0 \\ 0 & 0 & 0 \end{pmatrix}; \quad E^{\alpha_2} = \begin{pmatrix} 0 & 0 & 0 \\ 0 & 0 & 1 \\ 0 & 0 & 0 \end{pmatrix} \\ E^{\alpha_1+\alpha_2} &= \begin{pmatrix} 0 & 0 & 1 \\ 0 & 0 & 0 \\ 0 & 0 & 0 \end{pmatrix} \\ E^{-\alpha_1} &= (E^{\alpha_1})^T; \quad E^{-\alpha_2} = (E^{\alpha_2})^T; \quad E^{-\alpha_1-\alpha_2} = (E^{\alpha_1+\alpha_2})^T \end{aligned} \tag{8.69}$$

where $H_{1,2}$ are the two Cartan generators and E^{α} are the step operators associated to the corresponding roots.

As a useful exercise, let us calculate the eight generators of the compact subalgebra $\mathfrak{su}(3)$. We have:

$$\begin{aligned} h_1 = iH_1 &= \begin{pmatrix} \frac{i}{\sqrt{2}} & 0 & 0 \\ 0 & -\frac{i}{\sqrt{2}} & 0 \\ 0 & 0 & 0 \end{pmatrix} \\ h_2 = iH_2 &= \begin{pmatrix} \frac{i}{\sqrt{6}} & 0 & 0 \\ 0 & \frac{i}{\sqrt{6}} & 0 \\ 0 & 0 & -i\sqrt{\frac{2}{3}} \end{pmatrix} \end{aligned} \tag{8.70}$$

$$
\begin{aligned}
r_1 &= E^{\alpha_1} - E^{-\alpha_1} = \begin{pmatrix} 0 & 1 & 0 \\ -1 & 0 & 0 \\ 0 & 0 & 0 \end{pmatrix} \\
r_2 &= E^{\alpha_2} - E^{-\alpha_2} = \begin{pmatrix} 0 & 0 & 0 \\ 0 & 0 & 1 \\ 0 & -1 & 0 \end{pmatrix} \\
r_3 &= E^{\alpha_1+\alpha_2} - E^{-\alpha_1-\alpha_2} = \begin{pmatrix} 0 & 0 & 1 \\ 0 & 0 & 0 \\ -1 & 0 & 0 \end{pmatrix}
\end{aligned} \tag{8.71}
$$

and

$$
\begin{aligned}
\ell_1 &= i(E^{\alpha_1} + E^{-\alpha_1}) = \begin{pmatrix} 0 & i & 0 \\ i & 0 & 0 \\ 0 & 0 & 0 \end{pmatrix} \\
\ell_2 &= i(E^{\alpha_2} + E^{-\alpha_2}) = \begin{pmatrix} 0 & 0 & 0 \\ 0 & 0 & i \\ 0 & i & 0 \end{pmatrix} \\
\ell_3 &= i(E^{\alpha_1+\alpha_2} + E^{-\alpha_1-\alpha_2}) = \begin{pmatrix} 0 & 0 & i \\ 0 & 0 & 0 \\ i & 0 & 0 \end{pmatrix}
\end{aligned} \tag{8.72}
$$

Any linear combination with real coefficients of $\{h_1, h_2, r_1, r_2, r_3, \ell_1, \ell_2, \ell_3\}$ is an anti-Hermitian matrix and an element of the compact Lie algebra $\mathfrak{su}(3)$ in its fundamental representation.[4]

The Borel Lie algebra is composed by the following five operators:

$$
\operatorname{Bor}[\mathfrak{sl}(3,\mathbb{R})] = \operatorname{span}\{H_1, H_2, E^{\alpha_1}, E^{\alpha_2}, E^{\alpha_1+\alpha_2}\} \tag{8.73}
$$

and it is clearly represented by upper triangular matrices.

Let us now remark that the two real sections $\mathfrak{sl}(3,\mathbb{R})$ and $\mathfrak{su}(3)$ share a common maximal compact subalgebra:

$$
\mathfrak{so}(3) \equiv \mathbb{H} \subset \mathbb{G} \equiv \mathfrak{sl}(3,\mathbb{R}) \tag{8.74}
$$

defined by the following generators:

$$
\begin{aligned}
\mathbb{H} &= \operatorname{span}\{J_1, J_2, J_3\} \\
&\equiv \left\{ \frac{1}{\sqrt{2}}(E^{\alpha_1} - E^{-\alpha_1}), \frac{1}{\sqrt{2}}(E^{\alpha_2} - E^{-\alpha_2}), \frac{1}{\sqrt{2}}(E^{\alpha_1+\alpha_2} - E^{-\alpha_1-\alpha_2}) \right\}
\end{aligned} \tag{8.75}
$$

4 In the physical literature of the 1960s and 1970s the eight generators $\{h_1, h_2, r_1, r_2, r_3, \ell_1, \ell_2, \ell_3\}$ were named the Gell–Mann lambda matrices.

This is a general phenomenon. The maximally split real section $\mathbb{G}_{\mathbf{max}}$ always contains a maximal compact subalgebra:

$$\mathbb{H}_{\mathbf{compact}} = \text{span}(E^{\alpha} - E^{-\alpha}); \quad \alpha > 0 \tag{8.76}$$

The orthogonal splitting

$$\mathbb{G}_{\mathbf{max}} = \mathbb{H}_{\mathbf{compact}} \oplus \mathbb{K} \tag{8.77}$$

where:

$$\mathbb{K} = \text{span}\{H_i, (E^{\alpha} - E^{-\alpha})\} \tag{8.78}$$

plays an important role in our subsequent discussion of symmetric spaces and in other geometrical issues.

8.2.2.1 The Weyl group of $\mathfrak{a}_2$

Let us now consider a generic element of the Cartan subalgebra, namely, a diagonal matrix of the form:

$$\text{CSA} \ni \mathscr{C}(\{\lambda_1, \lambda_2\}) = \begin{pmatrix} \lambda_1 & 0 & 0 \\ 0 & \lambda_2 & 0 \\ 0 & 0 & -\lambda_1 - \lambda_2 \end{pmatrix} \tag{8.79}$$

By definition the Weyl group maps the Cartan subalgebra onto itself, so that we have:

$$\forall w \in \mathscr{W} \subset \text{SO}(3): \quad w^T \mathscr{C}(\{\lambda_1, \lambda_2\}) w = \mathscr{C}(w\{\lambda_1, \lambda_2\}) \in \text{CSA} \tag{8.80}$$

where $\mathscr{C}(w\{\lambda_1, \lambda_2\})$ denotes the diagonal matrix of type (8.79) with eigenvalues λ_1', λ_2', $-\lambda_1' - \lambda_2'$ obtained from the action of the Weyl group on the original ones. In the case of the Lie algebras A_n the Weyl group is the symmetric group $\mathscr{S}_{n+1}$ and its action on the eigenvalues $\lambda_1, \lambda_2, \ldots, \lambda_n, \lambda_{n+1} = -\sum_{i=1}^{n} \lambda_i$ is just that of permutations on these $n+1$ eigenvalues. For $\mathfrak{a}_2$ we have $\mathscr{S}_3$ whose order is six. The six group elements can be enumerated in the following way:

$$\begin{aligned}
w_1 &= \begin{pmatrix} 1 & 0 & 0 \\ 0 & 1 & 0 \\ 0 & 0 & 1 \end{pmatrix}; \quad (\lambda_1, \lambda_2, \lambda_3) \mapsto (\lambda_1, \lambda_2, \lambda_3) \\
w_2 &= \begin{pmatrix} 0 & 1 & 0 \\ 1 & 0 & 0 \\ 0 & 0 & 1 \end{pmatrix}; \quad (\lambda_1, \lambda_2, \lambda_3) \mapsto (\lambda_2, \lambda_1, \lambda_3) \\
w_3 &= \begin{pmatrix} 0 & 0 & 1 \\ 0 & 1 & 0 \\ 1 & 0 & 0 \end{pmatrix}; \quad (\lambda_1, \lambda_2, \lambda_3) \mapsto (\lambda_3, \lambda_2, \lambda_1)
\end{aligned} \tag{8.81}$$

$$W_4 = \begin{pmatrix} 1 & 0 & 0 \\ 0 & 0 & 1 \\ 0 & 1 & 0 \end{pmatrix}; \quad (\lambda_1, \lambda_2, \lambda_3) \mapsto (\lambda_1, \lambda_3, \lambda_2)$$

$$W_5 = \begin{pmatrix} 0 & 0 & 1 \\ 1 & 0 & 0 \\ 0 & 1 & 0 \end{pmatrix}; \quad (\lambda_1, \lambda_2, \lambda_3) \mapsto (\lambda_2, \lambda_3, \lambda_1) \tag{8.82}$$

$$W_6 = \begin{pmatrix} 0 & 1 & 0 \\ 0 & 0 & 1 \\ 1 & 0 & 0 \end{pmatrix}; \quad (\lambda_1, \lambda_2, \lambda_3) \mapsto (\lambda_3, \lambda_1, \lambda_2)$$

8.2.2.2 The weight lattice and the weights of some irreps

Next we discuss some of the irreducible representations of $\mathfrak{a}_2$ in relation to their weights.

The weight lattice of $\mathfrak{a}_2$ is defined below:

$$\Lambda_{\mathbf{weight}}[\mathfrak{a}_2] \ni n_1\lambda^1 + n_2\lambda^2; \quad n_{1,2} \in \mathbb{Z} \tag{8.83}$$

where the two fundamental weights have the following explicit analytic form:

$$\lambda^1 = \left\{\frac{1}{\sqrt{2}}, \frac{1}{\sqrt{6}}\right\} \tag{8.84}$$

$$\lambda^2 = \left\{0, \sqrt{\frac{2}{3}}\right\} \tag{8.85}$$

and are shown in Figure 8.2. Given the fundamental weights, one immediately determines the Weyl-chamber which is displayed in Figure 8.3.

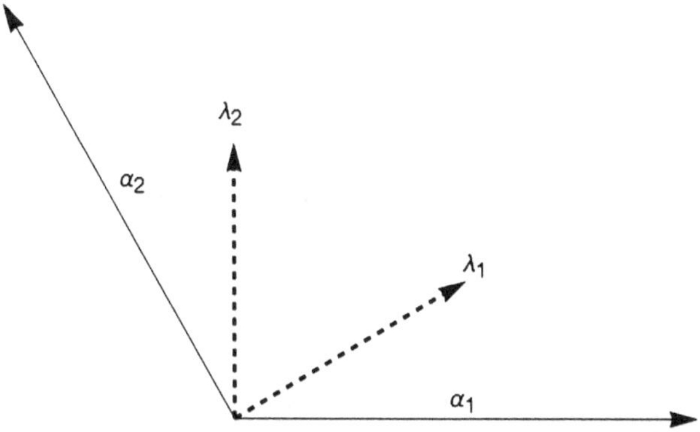

Figure 8.2: The fundamental weights of the $\mathfrak{a}_2$ Lie algebra, compared to the simple roots.

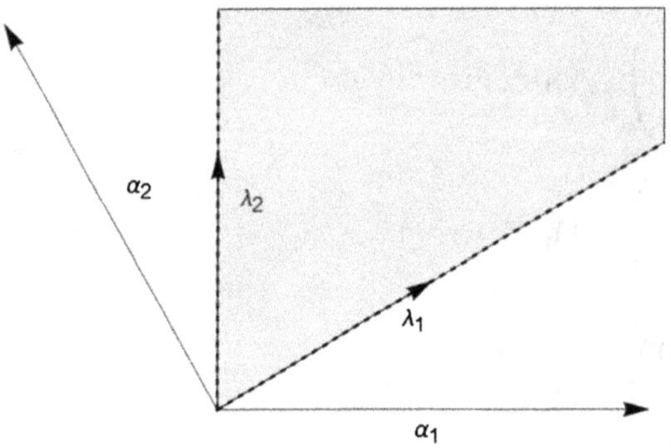

Figure 8.3: The Weyl chamber of the $\mathfrak{a}_2$ Lie algebra is, in the above picture, the infinite shaded region.

8.2.2.3 The weights of the fundamental representation

The fundamental representation of the $\mathfrak{sl}(3,\mathbb{R})$ Lie algebra or equivalently of the $\mathfrak{su}(3)$ algebra is three-dimensional and we have already constructed the matrix realization of its generators in eqs (8.69). Let us work out its weights. These are easily seen from the explicit form of the Cartan matrices displayed in eq. (8.69). Indeed, setting:

$$\begin{aligned} \Pi^{\mathbf{triplet}} &= \{\Lambda^1,\Lambda^2,\Lambda^3\} \\ \Lambda^1 &= \lambda^1 && \Rightarrow \quad (\mathbf{1},\mathbf{0}) \\ \Lambda^2 &= -\lambda^1+\lambda^2 && \Rightarrow \quad (\mathbf{-1},\mathbf{1}) \\ \Lambda^3 &= -\lambda^2 && \Rightarrow \quad (\mathbf{0},\mathbf{-1}) \end{aligned} \tag{8.86}$$

where the bold faced components on the right are those with respect to the fundamental weight basis (the Dynkin labels), we have:

$$\begin{aligned} |\Lambda^1\rangle &= \{1,0,0\}; \quad h^iH_i|\Lambda^1\rangle = h^i\Lambda^1_i|\Lambda^1\rangle \\ |\Lambda^2\rangle &= \{0,1,0\}; \quad h^iH_i|\Lambda^2\rangle = h^i\Lambda^2_i|\Lambda^2\rangle \\ |\Lambda^3\rangle &= \{0,0,1\}; \quad h^iH_i|\Lambda^3\rangle = h^i\Lambda^3_i|\Lambda^3\rangle \end{aligned} \tag{8.87}$$

as the reader can easily verify looking at eqs (8.69).

The three weights of the fundamental representation are displayed in Figure 8.4. As mentioned in the figure caption, the highest weight of this representation is clearly singled out: it is

$$\lambda_{\max} = \Lambda_1 = \lambda^1 \tag{8.88}$$

and from this simple information, using the structure of the root system, we might construct all the matrices of the representation, which are those displayed in eq. (8.69).

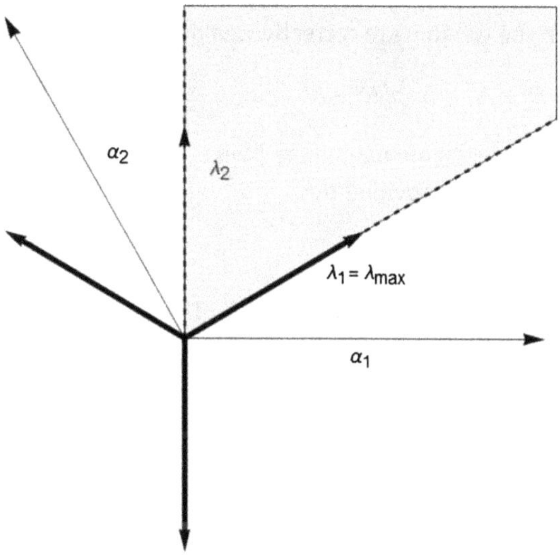

Figure 8.4: The three weights of the fundamental representation of the $\mathfrak{a}_2$ Lie algebra correspond to the three thicker arrows. The highest weight is clearly λ_1, the only positive one lying in the Weyl chamber, actually on its boundary.

8.2.2.4 The weights of the adjoint representation: the octet

The next representation that we consider is the adjoint representation, namely, the octet. This representation is already encoded in the canonical Cartan–Weyl structure of the Lie algebra and it is 8-dimensional. It is interesting to re-examine it from the point of view of the weights that we can immediately predict. The eight weights are indeed the weight $\{0,0\}$ with multiplicity $m(\{0,0\}) = 2$ corresponding to the two Cartan generators and the six roots with multiplicity $m = 1$. These latter can be expressed in terms of the fundamental weights as follows. Indeed, setting:

$$\Pi^{\mathbf{octet}} = \{\Lambda^1, \Lambda^2, \Lambda^3, \Lambda^4, \Lambda^5, \Lambda^6, \Lambda^7\} \tag{8.89}$$

we have

$$\begin{aligned}
\Lambda^1 = \alpha_1 + \alpha_2 = \lambda^1 + \lambda^2 \quad &\Rightarrow \quad \Lambda^1 = (\mathbf{1},\mathbf{1}) \\
\Lambda^2 = \alpha_1 = 2\lambda^1 - \lambda^2 \quad &\Rightarrow \quad \Lambda^2 = (\mathbf{2},\mathbf{-1}) \\
\Lambda^3 = \alpha_2 = -\lambda^1 + 2\lambda^2 \quad &\Rightarrow \quad \Lambda^3 = (\mathbf{-1},\mathbf{2}) \\
\Lambda^4 = \{0,0\} = \{0,0\} \quad &\Rightarrow \quad \Lambda^4 = (\mathbf{0},\mathbf{0}) \\
\Lambda^5 = -\alpha_2 = \lambda^1 - 2\lambda^2 \quad &\Rightarrow \quad \Lambda^5 = (\mathbf{1},\mathbf{-2}) \\
\Lambda^6 = -\alpha_1 = -2\lambda^1 + \lambda^2 \quad &\Rightarrow \quad \Lambda^6 = (\mathbf{-2},\mathbf{1}) \\
\Lambda^7 = -\alpha_1 - \alpha_2 = -\lambda^1 - \lambda^2 \quad &\Rightarrow \quad \Lambda^7 = (\mathbf{-1},\mathbf{-1})
\end{aligned} \tag{8.90}$$

where the boldface components are the integer ones in the basis of the fundamental weights (the Dynkin labels of the weight). Recalling that the height of a weight is the sum of such components we see that the weights are correctly weakly ordered since:

$$\Lambda^1 \succ \Lambda^2 \succeq \Lambda^3 \succ \Lambda^4 \succ \Lambda^5 \succeq \Lambda^6 \succ \Lambda^7 \quad (8.91)$$

The positive components of the highest weight are named the *Dynkin labels of the representation.* In this case the highest weight is provided by:

$$\lambda_{max} = \alpha_1 + \alpha_2 = \lambda^1 + \lambda^2 \quad (8.92)$$

and the Dynkin labels of the adjoint (octet) representation are $(1,1)$. The weights of the octet representation are displayed in Figure 8.5.

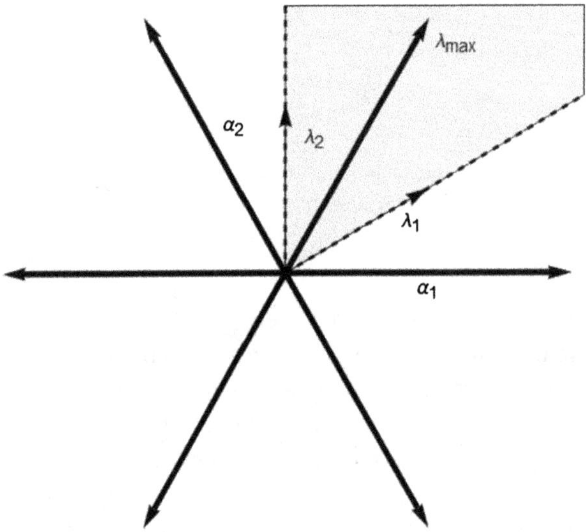

Figure 8.5: The six non-vanishing weights of the adjoint representation of the $\mathfrak{a}_2$ Lie algebra correspond to the six thicker arrows. The highest weight is clearly $\lambda^1 + \lambda^2$, the only positive one lying in the Weyl chamber.

8.2.2.5 The weights of a 15-dimensional representation

As a final example we consider the weights of the representation whose maximal weight is the following:

$$\lambda_{\mathbf{max}} = 2\lambda^1 + \lambda^2; \quad \Rightarrow \quad \lambda_{\mathbf{max}} = (\mathbf{2},\mathbf{1}) \quad (8.93)$$

Utilizing the Weyl group symmetry of the weight set and the α-strings through λ technique we can easily work out the full set of weights together with their multiplicities. The result is the following one:

$$\Pi_\Gamma = \{\Lambda_1, \Lambda_2, \Lambda_3, \Lambda_4, \Lambda_5, \Lambda_6, \Lambda_7, \Lambda_8, \Lambda_9, \Lambda_{10}, \Lambda_{11}, \Lambda_{12}\} \quad (8.94)$$

where the list of ordered weights with their multiplicities is displayed below:

Name				Orth. comp.		Dynk. lab.	mult.
Λ_1	=	$2\lambda^1+\lambda^2$	=	$\{\sqrt{2}, 2\sqrt{\frac{2}{3}}\}$	=	$(2,1)$	1
Λ_2	=	$3\lambda^1-\lambda^2$	=	$\{\frac{3}{\sqrt{2}}, \frac{1}{\sqrt{6}}\}$	=	$(3,-1)$	1
Λ_3	=	$2\lambda^2$	=	$\{0, 2\sqrt{\frac{2}{3}}\}$	=	$(0,2)$	1
Λ_4	=	λ^1	=	$\{\frac{1}{\sqrt{2}}, \frac{1}{\sqrt{6}}\}$	=	$(1,0)$	2
Λ_5	=	$-2\lambda^1+3\lambda^2$	=	$\{-\sqrt{2}, 2\sqrt{\frac{2}{3}}\}$	=	$(-2,3)$	1
Λ_6	=	$2\lambda^1-2\lambda^2$	=	$\{\sqrt{2}, -\sqrt{\frac{2}{3}}\}$	=	$(2,-2)$	1
Λ_7	=	$-\lambda^1+\lambda^2$	=	$\{-\frac{1}{\sqrt{2}}, \frac{1}{\sqrt{6}}\}$	=	$(-1,1)$	2
Λ_8	=	$-\lambda^2$	=	$\{0, -\sqrt{\frac{2}{3}}\}$	=	$(0,-1)$	2
Λ_9	=	$-3\lambda^1+2\lambda^2$	=	$\{-\frac{3}{\sqrt{2}}, \frac{1}{\sqrt{6}}\}$	=	$(-3,2)$	1
Λ_{10}	=	$-3\lambda^2+\lambda^1$	=	$\{\frac{1}{\sqrt{2}}, -\frac{5}{\sqrt{6}}\}$	=	$(1,-3)$	1
Λ_{11}	=	$-2\lambda^1$	=	$\{-\sqrt{2}, -\sqrt{\frac{2}{3}}\}$	=	$(-2,0)$	1
Λ_{12}	=	$-\lambda^1-2\lambda^2$	=	$\{-\frac{1}{\sqrt{2}}, -\frac{5}{\sqrt{6}}\}$	=	$(-1,-2)$	1

(8.95)

The weight set is displayed in Figure 8.6. Counting the multiplicity 2 of the three shortest weights it appears that this representation has dimension 15. The question is how can we understand this representation in the tensorial way.

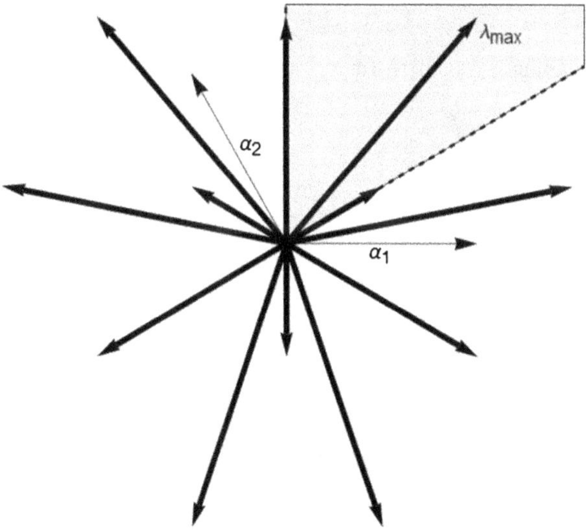

Figure 8.6: The 12 weights of the irreducible representation of the $\mathfrak{a}_2$ Lie algebra corresponding to the maximal weight $(2,1)$ are given by the 12 thicker arrows. The highest weight is clearly $2\lambda^1+\lambda^2$. It is the largest of the three lying in the Weyl chamber.

The answer is the following. Those displayed in eq. (8.95) and in Figure 8.6 are the weights of the representation corresponding to the following irreducible tensor:

$$\mathbf{15} \equiv \begin{array}{|c|c|c|}\hline \; & \; & \; \\ \hline \; & \multicolumn{2}{c}{} \\ \cline{1-1} \end{array} \tag{8.96}$$

For pedagogical reasons we dwell upon the details of this identifications. First let us introduce four tri-vectors transforming in the fundamental representation of $\mathfrak{sl}(3,\mathbb{C})$:

$$\mathbf{v} = \begin{pmatrix} v_1 \\ v_2 \\ v_3 \end{pmatrix}; \quad \mathbf{w} = \begin{pmatrix} w_1 \\ w_2 \\ w_3 \end{pmatrix}; \quad \mathbf{z} = \begin{pmatrix} z_1 \\ z_2 \\ z_3 \end{pmatrix}; \quad \mathbf{t} = \begin{pmatrix} t_1 \\ t_2 \\ t_3 \end{pmatrix} \tag{8.97}$$

Next construct the irreducible tensor in the following way:

$$\begin{array}{|c|c|c|}\hline \mathbf{v} & \mathbf{w} & \mathbf{z} \\ \hline \mathbf{t} & \multicolumn{2}{c}{} \\ \cline{1-1} \end{array} = \begin{array}{|c|c|c|}\hline \mathbf{v} & \mathbf{w} & \mathbf{z} \\ \hline \end{array} \otimes \begin{array}{|c|}\hline \mathbf{t} \\ \hline \end{array} - \begin{array}{|c|c|c|c|}\hline \mathbf{v} & \mathbf{w} & \mathbf{z} & \mathbf{t} \\ \hline \end{array} \tag{8.98}$$

where the Young tableau is, as explained above, a shorthand for the application to the product $v_i w_j z_k t_k$ of the row symmetrization and column antisymmetrization. The possible components of the tensor are obtained by replacing, in each of the boxes, one of the three components of each vector **v**, **w**, **z** and **t**. In this way the tensor components are quartic polynomials in the components $v_i w_j z_k t_k$. These polynomials can be expressed in a basis of 15 independent ones that correspond to a basis of the irreducible representation. A possible basis is provided by the polynomials corresponding to the tensor components displayed in Table 8.1. The simplest way to verify that this is indeed the proper correspondence relies on finite group transformations. Given a Cartan Lie algebra element:

$$\mathfrak{h} = h_1 H_1 + h_2 H_2; \quad \mathbf{h} \equiv \{h_1, h_2\} \tag{8.99}$$

the corresponding finite group element has the form:

$$\mathfrak{T} = \exp[\mathfrak{h}] = \begin{pmatrix} \frac{h_1}{\sqrt{2}} + \frac{h_2}{\sqrt{6}} & 0 & 0 \\ 0 & \frac{h_2}{\sqrt{6}} - \frac{h_1}{\sqrt{2}} & 0 \\ 0 & 0 & -\sqrt{\frac{2}{3}} h_2 \end{pmatrix} \tag{8.100}$$

and under this group transformation we have:

$$\mathbf{v}' = \mathfrak{T}\mathbf{v}; \quad \mathbf{w}' = \mathfrak{T}\mathbf{w}; \quad \mathbf{z}' = \mathfrak{T}\mathbf{z}; \quad \mathbf{t}' = \mathfrak{T}\mathbf{t} \tag{8.101}$$

Replacing the unprimed vectors with the primed ones in the expression for the tensor:

$$\begin{array}{|c|c|c|}\hline a & b & c \\ \hline d & \multicolumn{2}{c}{} \\ \cline{1-1} \end{array} \tag{8.102}$$

we find that each of its components mentioned in Table 8.1 transforms by multiplication with an exponential factor:

$$\exp[\Lambda \cdot \mathbf{h}] \tag{8.103}$$

where Λ is the corresponding weight vector also mentioned in Table 8.1. This proves the correspondence between weights and tensor components.

Weight	Tensor comp.	Weight mult.
Λ_1	1 1 1 / 2	1
Λ_2	1 1 1 / 3	1
Λ_3	1 1 2 / 2	1
Λ_4	{1 1 2 / 3 ; 1 1 3 / 2}	2
Λ_5	1 2 2 / 2	1
Λ_6	1 1 3 / 3	1
Λ_7	{1 2 2 / 3 ; 1 2 3 / 2}	2
Λ_8	{1 2 3 / 3 ; 1 3 3 / 2}	2
Λ_9	2 2 2 / 3	1
Λ_{10}	1 3 3 / 3	1
Λ_{11}	2 2 3 / 3	1
Λ_{12}	2 3 3 / 3	1

Table 8.1: Components of the irreducible tensor corresponding to the weights of the 15-dimensional representation corresponding to the highest weight $(\mathbf{2},\mathbf{1})$.

8.2.3 The Lie algebra $\mathfrak{sp}(4,\mathbb{R}) \simeq \mathfrak{so}(2,3)$, its fundamental representation and its Weyl group

The next example that we consider is that of the Lie algebras $\mathfrak{c}_2$ and $\mathfrak{b}_2$ that are accidentally isomorphic. This isomorphism is very convenient to our pedagogical purposes since it allows us to discuss in a simple way an instance of spin representation.

When we restrict our attention to maximally split real sections, the isomorphism between the complex $\mathfrak{c}_2$ and $\mathfrak{b}_2$ Lie algebras translates into the isomorphism between the following two real Lie algebras:

$$\mathfrak{so}(2,3) \simeq \mathfrak{sp}(4,\mathbb{R}) \tag{8.104}$$

On the contrary, if we focus on maximally compact real sections, the implied isomorphism between compact algebras is the following one:

$$\mathfrak{so}(5) \simeq \mathfrak{usp}(4) \tag{8.105}$$

The Lie algebra $\mathfrak{usp}(4)$ is by definition composed by all those 4×4 matrices that are simultaneously anti-Hermitian and symplectic, in the sense that they preserve an antisymmetric real matrix which squares to $-\mathbf{1}$.

Considering the $\mathfrak{b}_2$ and $\mathfrak{c}_2$ formulations at the same time we discover that the fundamental representation of $\mathfrak{sp}(4, \mathbb{C})$ is the spinor representation of $\mathfrak{so}(5, \mathbb{C})$. In this way we can easily work out all the weights of the spinor representations to be compared with those of the vector representation. The structure of the spinor weights is rather general for all orthogonal Lie algebras.

8.2.3.1 The root system

The $\mathfrak{b}_2$ and $\mathfrak{c}_2$ Dynkin diagrams are displayed in Figure 8.7. Because of the isomorphism we can rely on either formulation in terms of 4×4 symplectic matrices or 5×5 pseudo-orthogonal matrices to obtain a fundamental representation of the Lie algebra. The first will prove to be the spinor representation of the second, as we anticipated.

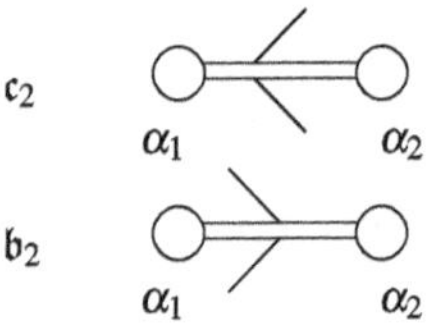

Figure 8.7: The Dynkin diagram of the $\mathfrak{c}_2 \sim \mathfrak{b}_2$ Lie algebra. The exchange of the α_1 with the α_2 root maps one Dynkin diagram onto the other and this is sufficient to prove the isomorphism.

In the symplectic $\mathfrak{sp}(4)$ interpretation, the $\mathfrak{c}_2$ root system can be realized by the following eight two-dimensional vectors:

$$\Phi_{\mathfrak{c}_2} = \{\pm\epsilon^i \pm \epsilon^j, \pm 2\epsilon^i\} \tag{8.106}$$

where ϵ^i $(i = 1, 2)$ denotes a basis of orthonormal unit vectors. In the pseudo-orthogonal $\mathfrak{so}(2, 3)$ interpretation of the same algebra the $\mathfrak{b}_2$ root system is instead realized by the following eight vectors:

$$\Phi_{\mathfrak{b}_2} = \{\pm\epsilon^i \pm \epsilon^j, \pm\epsilon^i\} \tag{8.107}$$

The two root systems are displayed in Figure 8.8. Because of the isomorphism between the two Lie algebras, we can use only one of the two root systems and consider the other as belonging to the weight lattice of the first. We choose to utilize the $\mathfrak{c}_2$ realization and we introduce the simple root basis:

$$\begin{aligned} \alpha_1 &= \{1, -1\} \\ \alpha_2 &= \{0, 2\} \end{aligned} \tag{8.108}$$

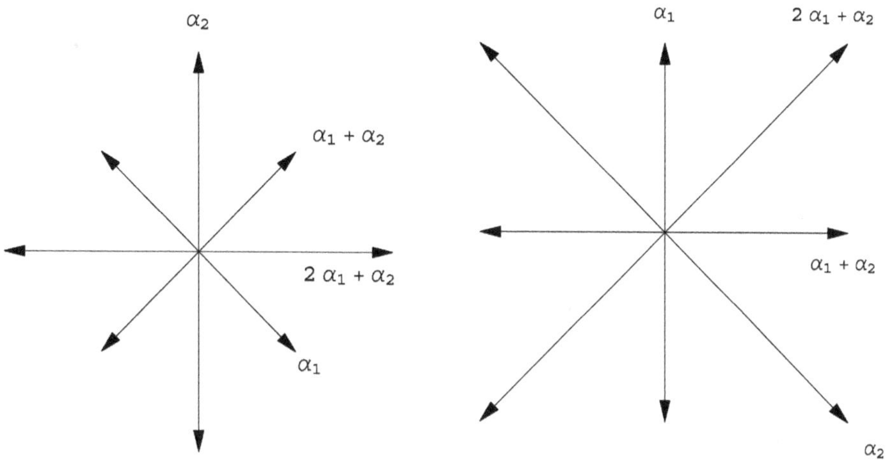

Figure 8.8: The $\mathfrak{c}_2$ and $\mathfrak{b}_2$ root systems. They are related by the exchange of the long with the short roots and vice versa.

leading to the Cartan matrix:

$$\mathbb{C} = \begin{pmatrix} 2 & -1 \\ -2 & 2 \end{pmatrix} \tag{8.109}$$

The fundamental weights satisfying the necessary relation:

$$2\frac{(\lambda^i, \alpha_j)}{(\alpha_j, \alpha_j)} = \delta^i_j \tag{8.110}$$

are easily determined and one finds:

$$\begin{aligned} \lambda^1 &= \{1, 0\} \\ \lambda^2 &= \{1, 1\} \end{aligned} \tag{8.111}$$

The complete set of positive roots is written below in the simple root and in the fundamental weight basis:

$$\begin{aligned} \alpha_1 &= 2\lambda_1 - \lambda_2 \\ \alpha_2 &= -2\lambda_1 + 2\lambda_2 \\ \alpha_1 + \alpha_2 &= \lambda_2 \\ 2\alpha_1 + \alpha_2 &= 2\lambda_1 \end{aligned} \tag{8.112}$$

From eq. (8.112) we also read off the Dynkin labels of the adjoint representation that are those of the highest root, namely $(2, 0)$. The simple roots and the simple weights are displayed in Figure 8.9.

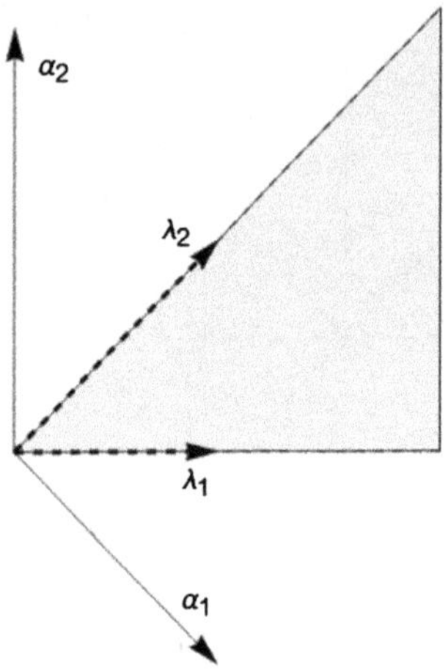

Figure 8.9: The simple roots and the simple weights of the $\mathfrak{c}_2$ Lie algebra. The shaded region is the Weyl chamber.

8.2.3.2 The Weyl group $\mathfrak{sp}(4,\mathbb{R})$

Abstractly the Weyl group Weyl($\mathfrak{c}_2$) of the Lie algebra $\mathfrak{sp}(4,\mathbb{R})$ is given by $(\mathbb{Z}_2 \times \mathbb{Z}_2) \ltimes S_2$ which is just the dihedral group Dih_4. The eight elements $w_i \in \mathrm{Weyl}(\mathfrak{c}_2)$ of the Weyl group can be described by their action on the two Cartan fields h_1, h_2:

$$\begin{aligned} w_1 &: (h_1,h_2) \to (h_1,h_2) \\ w_2 &: (h_1,h_2) \to (-h_1,-h_2) \\ w_3 &: (h_1,h_2) \to (-h_1,h_2) \\ w_4 &: (h_1,h_2) \to (h_1,-h_2) \\ w_5 &: (h_1,h_2) \to (h_2,h_1) \\ w_6 &: (h_1,h_2) \to (h_2,-h_1) \\ w_7 &: (h_1,h_2) \to (-h_2,h_1) \\ w_8 &: (h_1,h_2) \to (-h_2,-h_1) \end{aligned} \tag{8.113}$$

We leave to the reader the explicit verification that the reflections along the four positive roots generate the order 8 group explicitly displayed in eq. (8.113).

8.2.3.3 Construction of the $\mathfrak{sp}(4,\mathbb{R})$ Lie algebra

The most compact way of presenting our basis is the following. According to our general strategy let us begin with the Borel Lie algebra Bor[$\mathfrak{sp}(4,\mathbb{R})$]. Abstractly the most general element of this algebra is given by

$$\mathcal{T} = h_1\mathcal{H}_1 + h_2\mathcal{H}_2 + e_1E^{\alpha_1} + e_2E^{\alpha_2} + e_3E^{\alpha_1+\alpha_2} + e_4E^{2\alpha_1+\alpha_2} \tag{8.114}$$

If we write the explicit form of $\mathcal{T}$ as a 4×4 upper triangular symplectic matrix

$$\mathcal{T}_{\text{sym}} = \begin{pmatrix} h_1 & e_1 & e_3 & -\sqrt{2}e_4 \\ 0 & h_2 & \sqrt{2}e_2 & e_3 \\ 0 & 0 & -h_2 & -e_1 \\ 0 & 0 & 0 & -h_1 \end{pmatrix} \in \mathfrak{sp}(4, \mathbb{R}) \tag{8.115}$$

which satisfies the condition

$$\mathcal{T}_{\text{sym}}^T \begin{pmatrix} \mathbf{0}_2 & \sigma_1 \\ -\sigma_1 & \mathbf{0}_2 \end{pmatrix} + \begin{pmatrix} \mathbf{0}_2 & \sigma_1 \\ -\sigma_1 & \mathbf{0}_2 \end{pmatrix} \mathcal{T}_{\text{sym}} = 0; \quad \sigma_1 = \begin{pmatrix} 0 & 1 \\ 1 & 0 \end{pmatrix} \tag{8.116}$$

all the generators of the solvable algebra are defined in the 4-dimensional symplectic representation. Moreover, also the generators associated with negative roots are defined by transposition:

$$\forall \alpha \in \Phi; \quad E^{-\alpha} = [E^{\alpha}]^T \tag{8.117}$$

By writing the same Lie algebra element (8.114) as a 5×5 matrix

$$\mathcal{T}_{\text{so}} = \begin{pmatrix} h_1 + h_2 & -\sqrt{2}e_2 & -\sqrt{2}e_3 & -\sqrt{2}e_4 & 0 \\ 0 & h_1 - h_2 & -\sqrt{2}e_1 & 0 & \sqrt{2}e_4 \\ 0 & 0 & 0 & \sqrt{2}e_1 & \sqrt{2}e_3 \\ 0 & 0 & 0 & h_2 - h_1 & \sqrt{2}e_2 \\ 0 & 0 & 0 & 0 & -h_1 - h_2 \end{pmatrix} \in \mathfrak{so}(2,3) \tag{8.118}$$

which satisfies the condition

$$\mathcal{T}_{\text{so}}^T \left(\begin{array}{cc|c|cc} 0 & 0 & 0 & 0 & 1 \\ 0 & 0 & 0 & 1 & 0 \\ \hline 0 & 0 & 1 & 0 & 0 \\ \hline 0 & 1 & 0 & 0 & 0 \\ 1 & 0 & 0 & 0 & 0 \end{array}\right) + \left(\begin{array}{cc|c|cc} 0 & 0 & 0 & 0 & 1 \\ 0 & 0 & 0 & 1 & 0 \\ \hline 0 & 0 & 1 & 0 & 0 \\ \hline 0 & 1 & 0 & 0 & 0 \\ 1 & 0 & 0 & 0 & 0 \end{array}\right) \mathcal{T}_{\text{so}} \equiv \mathcal{T}_{\text{so}}^T \eta + \eta \mathcal{T}_{\text{so}} = 0 \tag{8.119}$$

we define the same generators also in the 5-dimensional pseudo-orthogonal representation. The choice of the invariant metric displayed in eq. (8.119) is that which guarantees the upper triangular structure of the solvable Lie algebra generators.[5] This allows to define the generators associated with negative roots in the same way as in eq. (8.117) and the 5-dimensional representation is fully constructed.

5 The eigenvalues of the matrix η are $(-1, -1, 1, 1, 1,)$. This shows that by means of a change of basis the matrices fulfilling eq. (8.119) span the $\mathfrak{so}(2,3)$ Lie algebra, according to its conventional definition.

8.2.3.4 Weights of the vector, spinor and adjoint representations

Relying on the results of the previous section we have an explicit realization of three linear representations of the same complex Lie algebra, a 4-dimensional one, a 5-dimensional one and the 10-dimensional adjoint representation. It is interesting to construct the weights of these three representations and compare them.

8.2.3.5 Weights of the 4-dimensional spinor representation

The weights of the fundamental representation of $\mathfrak{sp}(4,\mathbb{C})$ which corresponds to the spinor representation of $\mathfrak{so}(3,2,\mathbb{C}) \sim \mathfrak{so}(5,\mathbb{C})$ are easily read off from the explicit form of the matrix representation (8.115). They are listed below:

Dynk. lab.	Orth. comp.	mult.
$\{1,0\}$	$\{1,0\}$	1
$\{-1,1\}$	$\{0,1\}$	1
$\{1,-1\}$	$\{0,-1\}$	1
$\{-1,0\}$	$\{-1,0\}$	1

(8.120)

where the first column provides their components in the fundamental weight basis, while the second column gives their expression in an orthonormal basis. The weights are graphically shown in Figure 8.10.

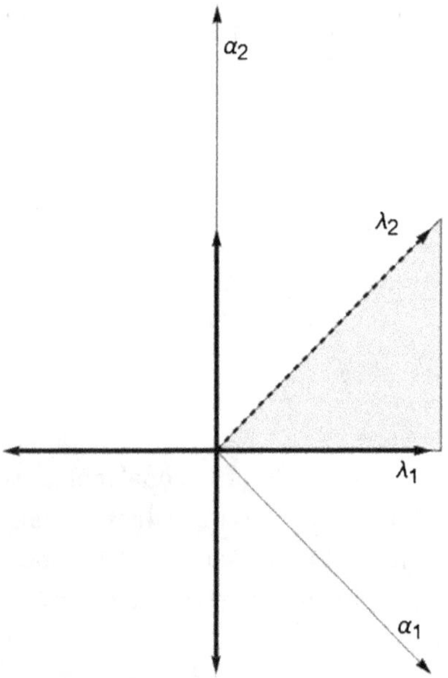

Figure 8.10: The weights of the spinor representation of $\mathfrak{so}(5)$ which is the fundamental representation of $\mathfrak{sp}(4)$. The shaded region is the Weyl chamber. One sees that the fundamental weight λ^1 is just the highest weight of the spinor representation.

8.2.3.6 Weights of the 5-dimensional vector representation

The weights of the fundamental vector representation of $\mathfrak{so}(5,\mathbb{C})$ are easily read off from the explicit form of the matrix representation (8.118). They are listed below:

Dynk. lab.	Orth. comp.	mult.
$\{0,1\}$	$\{1,1\}$	1
$\{2,-1\}$	$\{1,-1\}$	1
$\{0,0\}$	$\{0,0\}$	1
$\{-2,1\}$	$\{-1,1\}$	1
$\{0,-1\}$	$\{-1,-1\}$	1

(8.121)

where the first column provides their components in the fundamental weight basis, while the second column gives their expression in an orthonormal basis. The weights are graphically shown in Figure 8.11.

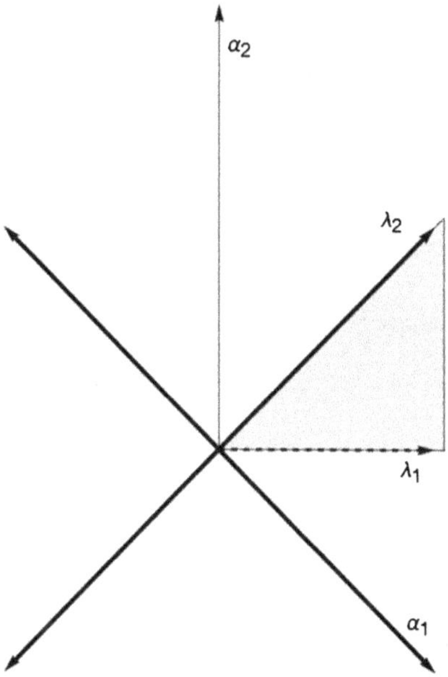

Figure 8.11: The weights of the vector defining representation of $\mathfrak{so}(5)$. The shaded region is the Weyl chamber. One sees that fundamental weight λ^2 is just the highest weight of the vector.

8.2.3.7 Weights of the 10-dimensional adjoint representation

The weights of the adjoint representation of $\mathfrak{so}(5,\mathbb{C}) \sim \mathfrak{sp}(4,\mathbb{C})$ are easily read from the transcription of the roots in terms of fundamental weights (8.112). They are listed below:

Dynk. lab.	Orth. comp.	mult.
{2, 0}	{2, 0}	1
{0, 1}	{1, 1}	1
{2, −1}	{1, −1}	1
{−2, 2}	{0, 2}	1
{0, 0}	{0, 0}	2
{2, −2}	{0, −2}	1
{−2, 1}	{−1, 1}	1
{0, −1}	{−1, −1}	1
{−2, 0}	{−2, 0}	1

(8.122)

where the first column provides their components in the fundamental weight basis, while the second column gives their expression in an orthonormal basis. The weights are graphically shown in Figure 8.12.

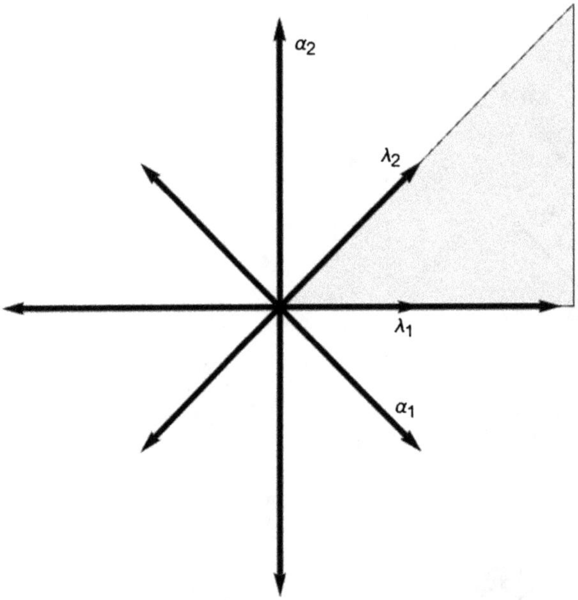

Figure 8.12: The weights of the adjoint representation of $\mathfrak{so}(5)$. The shaded region is the Weyl chamber. One sees that the weight $2\lambda^1 = 2\alpha_1 + \alpha_2$ is the highest weight of the adjoint representation.

8.2.3.8 Spinor interpretation of the 4-dimensional representation

We have repeatedly claimed that the 4-dimensional representation is the spinor representation with respect to the $\mathfrak{so}(5, \mathbb{C})$ Lie algebra. How do we prove this claim? It suffices to construct a well adapted set of gamma matrices.

Let us recall eqs (8.119), (8.116) and let us set

$$\eta = \left(\begin{array}{cc|c|c|c} 0 & 0 & 0 & 0 & 1 \\ 0 & 0 & 0 & 1 & 0 \\ \hline 0 & 0 & 1 & 0 & 0 \\ \hline 0 & 1 & 0 & 0 & 0 \\ 1 & 0 & 0 & 0 & 0 \end{array}\right) ; \quad \mathbf{C} = \begin{pmatrix} \mathbf{0}_2 & \sigma_1 \\ -\sigma_1 & \mathbf{0}_2 \end{pmatrix} \tag{8.123}$$

The $\mathfrak{so}(5,\mathbb{C}) \sim \mathfrak{so}(2,3,\mathbb{C})$ algebra is spanned by the 4×4 matrices $\mathscr{O}$ that satisfy the equation:

$$\mathscr{O}^T \eta + \eta \mathscr{O} = 0 \tag{8.124}$$

while the isomorphic $\mathfrak{sp}(4,\mathbb{C})$ algebra is spanned by the 5×5 matrices $\mathscr{S}$ that satisfy the equation:

$$\mathscr{S}^T \mathbf{C} + \mathbf{C}\mathscr{S} = 0 \tag{8.125}$$

In order to discuss the spinor representation we have to introduce a set of gamma-matrices Γ_a which satisfy the appropriate Clifford algebra:

$$\{\Gamma_a, \Gamma_b\} = 2\eta_{ab} \tag{8.126}$$

Furthermore, we remark that the gamma-matrices should be antisymmetric with respect to the symplectic matrix $\mathbf{C}$:

$$\Gamma_a^T \mathbf{C} - \mathbf{C}\Gamma_a = 0 \tag{8.127}$$

and $\mathbf{C}$-traceless:

$$\mathrm{Tr}(\Gamma_a \mathbf{C}) = 0 \tag{8.128}$$

The reasoning is the following: the space of general 4×4 matrices is 16-dimensional and it corresponds to the tensor product of two fundamental representations of $\mathfrak{sp}(4)$. The symmetric part (with respect to the matrix $\mathbf{C}$) spans the adjoint representation and has dimension 10. The antisymmetric part, which has dimension 6, further splits into a $\mathbf{C}$-trace which is by definition a singlet and into a 5-dimensional irreducible representation that must be, out of necessity, the fundamental representation of $\mathfrak{so}(5)$. The gamma-matrices are just the projection operators onto this 5-dimensional representation. In other words if $\mathfrak{s}_{1,2}$ are two spinors, we obtain a vector by setting:

$$\mathbf{v}^a = \mathfrak{s}_1^T \mathbf{C}\Gamma^a \mathfrak{s}_2 \tag{8.129}$$

We are supposed to solve the three constraints (8.126), (8.127) and (8.128). An ansatz for the solution is inspired by the knowledge of the weights. Since the directions $(1,2,3,4,5)$ in vector spaces are in one-to-one correspondence with the weights of the 5-dimensional representation and since the directions $(1,2,3,4)$ in spinor space are in one-to-one correspondence with the weights of the 4-dimensional representation, it suffices to check which spinor weights add up to which vector weight and we immediately know which entries of each gamma-matrix are different from zero. In this way we find the following solution:

$$\Gamma_1 = \begin{pmatrix} 0 & 0 & 1 & 0 \\ 0 & 0 & 0 & -1 \\ 0 & 0 & 0 & 0 \\ 0 & 0 & 0 & 0 \end{pmatrix}; \quad \Gamma_2 = \begin{pmatrix} 0 & -1 & 0 & 0 \\ 0 & 0 & 0 & 0 \\ 0 & 0 & 0 & -1 \\ 0 & 0 & 0 & 0 \end{pmatrix}$$
$$\Gamma_3 = \begin{pmatrix} 1 & 0 & 0 & 0 \\ 0 & -1 & 0 & 0 \\ 0 & 0 & -1 & 0 \\ 0 & 0 & 0 & 1 \end{pmatrix}; \quad \Gamma_4 = \begin{pmatrix} 0 & 0 & 0 & 0 \\ -2 & 0 & 0 & 0 \\ 0 & 0 & 0 & 0 \\ 0 & 0 & -2 & 0 \end{pmatrix} \tag{8.130}$$
$$\Gamma_5 = \begin{pmatrix} 0 & 0 & 0 & 0 \\ 0 & 0 & 0 & 0 \\ 2 & 0 & 0 & 0 \\ 0 & -2 & 0 & 0 \end{pmatrix}$$

Introducing the generators of the $\mathfrak{so}(5)$ Lie algebra in the spinor representation:

$$\mathfrak{t}_{ab} = -\frac{1}{4}[\Gamma_a, \Gamma_b] \tag{8.131}$$

we verify that all of them satisfy:

$$\mathfrak{t}_{ab}^T C + C\mathfrak{t}_{ab} = 0 \tag{8.132}$$

namely, they are elements of the $\mathfrak{sp}(4)$ Lie algebra and constitute a basis for it. This is as much as it is needed to conclude that the fundamental of $\mathfrak{sp}(4)$ is just the spinor representation of $\mathfrak{so}(5)$.

8.2.3.9 Maximal compact subgroup of the maximally split real section

In view of our previous remarks about the maximal compact subalgebra and its orthogonal complement let us see what is their structure in the present case.

The orthonormal generators of the $\mathbb{K}$ subspace are as follows:

$$\begin{aligned} \mathrm{K}_1 &= \mathscr{H}_1 \\ \mathrm{K}_2 &= \mathscr{H}_2 \\ \mathrm{K}_3 &= \frac{1}{\sqrt{2}}(E^{\alpha_1} + (E^{\alpha_1})^T) \\ \mathrm{K}_4 &= \frac{1}{\sqrt{2}}(E^{\alpha_2} + (E^{\alpha_2})^T) \\ \mathrm{K}_5 &= \frac{1}{\sqrt{2}}(E^{\alpha_1+\alpha_2} + (E^{\alpha_1+\alpha_2})^T) \\ \mathrm{K}_6 &= \frac{1}{\sqrt{2}}(E^{\alpha_1+2\alpha_2} + (E^{\alpha_1+2\alpha_2})^T) \end{aligned} \tag{8.133}$$

and those of the maximal compact subalgebra $\mathbb{H} = \mathfrak{u}(2)$ are as follows:

$$\begin{aligned} J_1 &= \frac{1}{\sqrt{2}}(E^{\alpha_1} - (E^{\alpha_1})^T), & J_2 &= \frac{1}{\sqrt{2}}(E^{\alpha_2} - (E^{\alpha_2})^T) \\ J_3 &= \frac{1}{\sqrt{2}}(E^{\alpha_1+\alpha_2} - (E^{\alpha_1+\alpha_2})^T), & J_4 &= \frac{1}{\sqrt{2}}(E^{\alpha_1+2\alpha_2} - (E^{\alpha_1+2\alpha_2})^T) \end{aligned} \tag{8.134}$$

Those above span a Lie algebra $\mathfrak{u}(2) = \mathfrak{u}(1) \oplus \mathfrak{su}(2)$.

8.3 Conclusions for this chapter

As the examples provided in the present chapter and those about exceptional Lie algebras to be discussed in the next one should clearly illustrate, although deterministic and implicitly defined by the Dynkin diagram, the actual construction of large Lie algebras is far from being a trivial matter and involves a series of strategies and long calculations that are best done by means of computer codes. One deals with large matrices that is difficult to display on paper and the best approach is to save the constructions in electronic libraries that can be utilized in subsequent calculations. It is not surprising that it took such a giant of mathematics as Elie Cartan to explicitly construct the fundamental representations of the exceptional Lie algebras, especially at a time when computers were not available.

From another point of view, the existing mathematical literature usually presents the construction of Lie algebra representations in a very compact format that is not of too friendly use to physicists and differential geometers concerned with their application to several problems of theoretical physics. As we stressed it is not only a question of convenience but also a conceptual one. There are in the architecture of Lie algebras and of their representations deep and significant aspects that are easily lost if you are not looking at them in the proper way, motivated by those questions that are posed, in particular, by the various special geometries implied by supersymmetry. The explicit construction of the exceptional and non-exceptional Lie algebras in the light of supergravity is one of the motivations that pushed one of us (P.F.) to write the more advanced book [54] and its historical twin [53]. The aim is that of presenting the *Mathematics of Symmetry* in a conceptually unified yet practical and computational way, at many stages different from the conventional approaches of most textbooks [88, 84, 138, 92].

8.4 Bibliographical note

In addition to the already quoted textbooks [88, 84, 138, 92], the most relevant bibliographical source for the first part of this chapter is the textbook by Cornwell [29]. The examples analyzed in full-fledged fashion in later sections are partially based on private notes of one of the authors (P.F.), partially have their basis in a research paper written in collaboration with A. Sorin [55].

9 Exceptional Lie algebras

9.1 The exceptional Lie algebra $\mathfrak{g}_2$

It was Killing who, through his own classification of the root systems, first discovered the possible existence of the exceptional Lie algebras: yet their concrete existence was proved only later by Cartan who was able to construct the fundamental representation of all of them. In this section we study the smallest of the five exceptional algebras, namely $\mathfrak{g}_2$, and we explicitly exhibit its fundamental representation which is 7-dimensional.

Our presentation is aimed not only at showing that $\mathfrak{g}_2$ exists but also at enlightening some features of its structure that will turn out to be general within a certain algebraic scheme that encompasses an entire set of classical and exceptional Lie algebras relevant for the *special geometries* implied by supergravity and superstrings.

Before the advent of supergravity, exceptional Lie algebras were viewed by physicists as some mathematical extravagance good only for a Dickensian *Old Curiosity Shop*. Supergravity quite surprisingly shows that all exceptional Lie algebras have a distinct and essential role to play in the connected web of gravitational theories that one obtains through dimensional reduction and coupling of matter multiplets in diverse dimensions. Furthermore, there is an inner algebraic structure of the exceptional algebras, shared with other classical algebras, that appears to be specially prepared to fit the geometrical yields of supersymmetry. This provides a new structural viewpoint motivated by physics that, in Weyl's spirit, encodes a deep truth, at the same time physical and mathematical, the distinction being somewhat irrelevant. For these aspects we refer the reader to the forthcoming book [54]: here we just focus on the construction issue with two motivations; the first is to derive concrete results utilizable in several contexts, while the second, more educational, is to illustrate in full detail all the steps needed to arrive at such results. Hence let us turn to the specific topic of the present section.

The complex Lie algebra $\mathfrak{g}_2(\mathbb{C})$ has rank two and it is defined by the 2×2 Cartan matrix encoded in the following Dynkin diagram:

$\mathfrak{g}_2$  $= \begin{pmatrix} 2 & -1 \\ -3 & 2 \end{pmatrix}$

The $\mathfrak{g}_2$ root system Φ consists of the following six positive roots plus their negatives:

$$\begin{aligned} &\alpha_1 = \alpha_1 = (1,0); && \alpha_2 = \alpha_2 = \frac{\sqrt{3}}{2}(-\sqrt{3},1) \\ &\alpha_3 = \alpha_1 + \alpha_2 = \frac{1}{2}(-1,\sqrt{3}); && \alpha_4 = 2\alpha_1 + \alpha_2 = \frac{1}{2}(1,\sqrt{3}) \\ &\alpha_5 = 3\alpha_1 + \alpha_2 = \frac{\sqrt{3}}{2}(\sqrt{3},1); && \alpha_6 = 3\alpha_1 + 2\alpha_2 = (0,\sqrt{3}) \end{aligned} \tag{9.1}$$

https://doi.org/10.1515/9783110551204-009

The two fundamental weights are easily derived and have the following form:

$$\lambda^1 = \{1, \sqrt{3}\}$$
$$\lambda^2 = \left\{0, \frac{2}{\sqrt{3}}\right\} \tag{9.2}$$

Simple roots, fundamental weights and the Weyl chamber are displayed in Figure 9.1. Figure 9.2 instead displays the entire root system. The fundamental representation of the Lie algebra is identified as the one which admits the fundamental weight λ^1 as highest weight. Using the Weyl group symmetry and the α through λ string technique one derives all the weights of the 7-dimensional fundamental representation that are

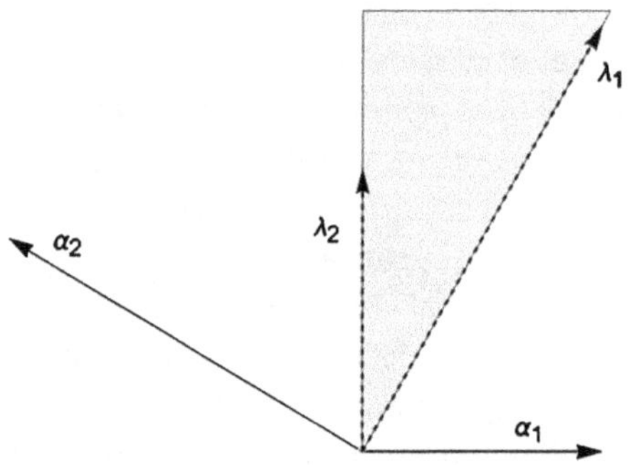

Figure 9.1: The simple roots and the fundamental weights of the $\mathfrak{g}_2$ Lie algebra. The shaded region is the Weyl chamber.

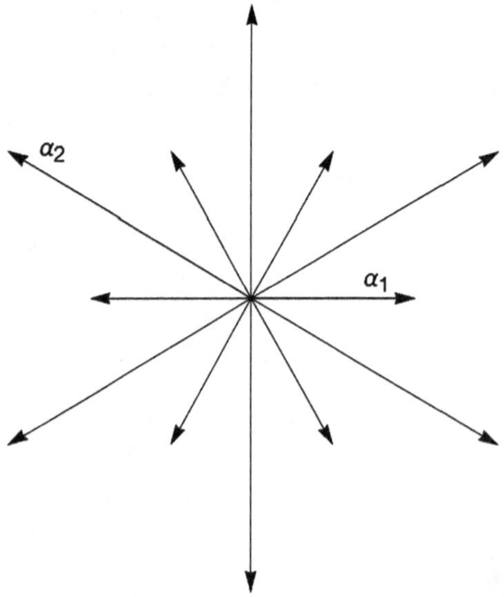

Figure 9.2: The complete root system of the $\mathfrak{g}_2$ Lie algebra.

the following ones:

Name		Dynk. lab.		Orth. comp.	mult.
Λ_1	=	$\{1,0\}$	⇒	$\{1,\sqrt{3}\}$	1
Λ_2	=	$\{-1,1\}$	⇒	$\{-1,-\frac{1}{\sqrt{3}}\}$	1
Λ_3	=	$\{2,-1\}$	⇒	$\{2,\frac{4}{\sqrt{3}}\}$	1
Λ_4	=	$\{0,0\}$	⇒	$\{0,0\}$	1
Λ_5	=	$\{-2,1\}$	⇒	$\{-2,-\frac{4}{\sqrt{3}}\}$	1
Λ_6	=	$\{1,-1\}$	⇒	$\{1,\frac{1}{\sqrt{3}}\}$	1
Λ_7	=	$\{-1,0\}$	⇒	$\{-1,-\sqrt{3}\}$	1

(9.3)

The six non-vanishing weights are displayed in Figure 9.3.

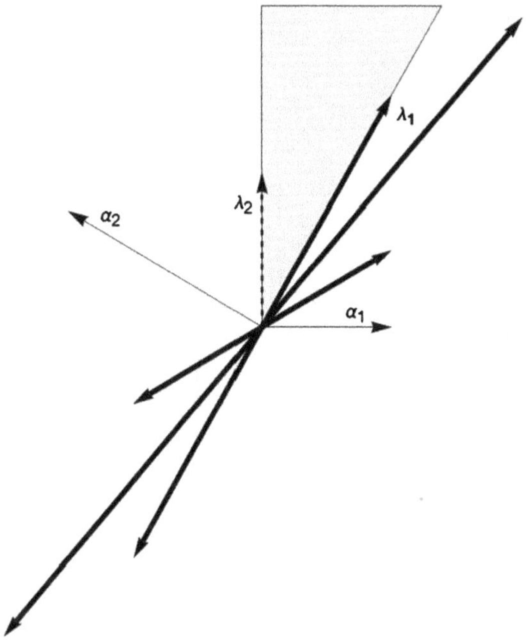

Figure 9.3: The six non-vanishing weights of the fundamental representation of the $\mathfrak{g}_2$ Lie algebra. The fundamental weight λ^1 is the highest weight of this representation.

Given this information we are ready to derive the fundamental representation of the algebra. According to our general strategy we are supposed to construct 7×7 upper triangular matrices spanning the Borel subalgebra of the maximally split real section $g_{2(2)}$ of $\mathfrak{g}_2(\mathbb{C})$:

$$\text{Bor}[\mathfrak{g}_2] = \text{span}\{H_1, H_2, E^{\alpha_1}, E^{\alpha_2}, \dots, E^{\alpha_6}\} \tag{9.4}$$

As for all maximally split algebras, the Cartan generators H_i and the step operators E^α associated with each positive root α can be chosen completely real in all representations.

In the fundamental 7-dimensional representation the explicit form of the $\mathfrak{g}_{2(2)}$-generators with the above properties is presented hereby. Naming $\{H_1, H_2\}$ the Cartan generators along the two orthonormal directions and adopting the standard Cartan–Weyl normalizations:

$$[E_\alpha, E_{-\alpha}] = \alpha^i H_i, \quad [H_i, E_\alpha] = \alpha^i E_\alpha \tag{9.5}$$

we have:

$$H_1 = \begin{pmatrix} \frac{1}{2} & 0 & 0 & 0 & 0 & 0 & 0 \\ 0 & -\frac{1}{2} & 0 & 0 & 0 & 0 & 0 \\ 0 & 0 & 1 & 0 & 0 & 0 & 0 \\ 0 & 0 & 0 & 0 & 0 & 0 & 0 \\ 0 & 0 & 0 & 0 & -1 & 0 & 0 \\ 0 & 0 & 0 & 0 & 0 & \frac{1}{2} & 0 \\ 0 & 0 & 0 & 0 & 0 & 0 & -\frac{1}{2} \end{pmatrix}$$

$$H_2 = \begin{pmatrix} \frac{\sqrt{3}}{2} & 0 & 0 & 0 & 0 & 0 & 0 \\ 0 & \frac{\sqrt{3}}{2} & 0 & 0 & 0 & 0 & 0 \\ 0 & 0 & 0 & 0 & 0 & 0 & 0 \\ 0 & 0 & 0 & 0 & 0 & 0 & 0 \\ 0 & 0 & 0 & 0 & 0 & 0 & 0 \\ 0 & 0 & 0 & 0 & 0 & -\frac{\sqrt{3}}{2} & 0 \\ 0 & 0 & 0 & 0 & 0 & 0 & -\frac{\sqrt{3}}{2} \end{pmatrix}$$

$$E_{\alpha_1} = \begin{pmatrix} 0 & \frac{1}{\sqrt{2}} & 0 & 0 & 0 & 0 & 0 \\ 0 & 0 & 0 & 0 & 0 & 0 & 0 \\ 0 & 0 & 0 & 1 & 0 & 0 & 0 \\ 0 & 0 & 0 & 0 & 1 & 0 & 0 \\ 0 & 0 & 0 & 0 & 0 & 0 & 0 \\ 0 & 0 & 0 & 0 & 0 & 0 & \frac{1}{\sqrt{2}} \\ 0 & 0 & 0 & 0 & 0 & 0 & 0 \end{pmatrix}$$

$$E_{\alpha_2} = \begin{pmatrix} 0 & 0 & 0 & 0 & 0 & 0 & 0 \\ 0 & 0 & \sqrt{\frac{3}{2}} & 0 & 0 & 0 & 0 \\ 0 & 0 & 0 & 0 & 0 & 0 & 0 \\ 0 & 0 & 0 & 0 & 0 & 0 & 0 \\ 0 & 0 & 0 & 0 & 0 & -\sqrt{\frac{3}{2}} & 0 \\ 0 & 0 & 0 & 0 & 0 & 0 & 0 \\ 0 & 0 & 0 & 0 & 0 & 0 & 0 \end{pmatrix}$$

$$E_{\alpha_1+\alpha_2}=\begin{pmatrix}0&0&\frac{1}{\sqrt{2}}&0&0&0&0\\0&0&0&-1&0&0&0\\0&0&0&0&0&0&0\\0&0&0&0&0&-1&0\\0&0&0&0&0&0&\frac{1}{\sqrt{2}}\\0&0&0&0&0&0&0\\0&0&0&0&0&0&0\end{pmatrix} \tag{9.6}$$

$$E_{2\alpha_1+\alpha_2}=\begin{pmatrix}0&0&0&-1&0&0&0\\0&0&0&0&\frac{1}{\sqrt{2}}&0&0\\0&0&0&0&0&-\frac{1}{\sqrt{2}}&0\\0&0&0&0&0&0&1\\0&0&0&0&0&0&0\\0&0&0&0&0&0&0\\0&0&0&0&0&0&0\end{pmatrix}$$

$$E_{3\alpha_1+\alpha_2}=\begin{pmatrix}0&0&0&0&-\sqrt{\frac{3}{2}}&0&0\\0&0&0&0&0&0&0\\0&0&0&0&0&0&-\sqrt{\frac{3}{2}}\\0&0&0&0&0&0&0\\0&0&0&0&0&0&0\\0&0&0&0&0&0&0\\0&0&0&0&0&0&0\end{pmatrix}$$

$$E_{3\alpha_1+2\alpha_2}=\begin{pmatrix}0&0&0&0&0&-\sqrt{\frac{3}{2}}&0\\0&0&0&0&0&0&-\sqrt{\frac{3}{2}}\\0&0&0&0&0&0&0\\0&0&0&0&0&0&0\\0&0&0&0&0&0&0\\0&0&0&0&0&0&0\\0&0&0&0&0&0&0\end{pmatrix}$$

9.1.1 A golden splitting for quaternionic algebras

In [54] one of us (P.F.) addresses the study of *special Kähler geometry* and of *quaternionic geometry* that are both implied by $\mathcal{N}=2$ supersymmetry, the first applying to the scalars of vector multiplets, the second to the scalar of hypermultiplets. Furthermore, a very interesting relation between such geometries, that is named the *c*-map, is discussed in the same monograph:

$$\mathfrak{c}\text{-map} : \mathcal{SK}_{2n} \to \mathcal{Q}_{4n+4} \tag{9.7}$$

where $\mathcal{SK}_{2n}$ denotes a special Kähler manifold of $2n$ real dimension while $\mathcal{Q}_{4n+4}$ denotes a quaternionic Kähler manifold of $4n+4$ real dimension. For the definition and properties of such manifolds we refer the reader to [54]. What is relevant to us here is that among the special Kähler and quaternionic manifolds there are also classes of homogeneous symmetric spaces G/H leading to a split of the Lie algebra:

$$\mathbb{G} = \mathbb{H} \oplus \mathbb{K} \tag{9.8}$$

into a subalgebra $\mathbb{H}$ and an orthogonal subspace $\mathbb{K}$. We refer the reader to the later Chapter 11 for the notion of coset manifolds and symmetric spaces: here we just focus on the fact that the existence of a *c*-map between two symmetric spaces implies the existence of a well-defined relation between two Lie algebras that we can respectively dub the *special Kählerian algebra* $\mathbb{U}_{\mathcal{S}\mathcal{K}}$ and the *quaternionic algebra* $\mathbb{U}_{\mathcal{Q}}$. This relation has been brought to attention by supersymmetric theories but it is intrinsic to Lie algebra structural theory and therefore it is relevant to our present scope, independently from supergravity and from other physical applications. The announced relation is provided by the following decomposition of the adjoint representation of the quaternionic algebra $\mathbb{U}_{\mathcal{Q}}$ with respect to its special Kähler subalgebra:

$$\mathrm{adj}(\mathbb{U}_{\mathcal{Q}}) = \mathrm{adj}(\mathbb{U}_{\mathcal{S}\mathcal{K}}) \oplus \mathrm{adj}(\mathfrak{sl}(2,\mathbb{R})_E) \oplus \mathbf{W}_{(2,\mathbf{W})} \tag{9.9}$$

where $\mathbf{W}$ is a **symplectic** representation of $\mathbb{U}_{\mathcal{S}\mathcal{K}}$ in which the symplectic section of Special Geometry[1] transforms. Denoting the generators of $\mathbb{U}_{\mathcal{S}\mathcal{K}}$ by T^a, the generators of $\mathfrak{sl}(2,\mathbb{R})_E$, which is named the Ehlers subalgebra, by L^x and denoting by $\mathbf{W}^{i\alpha}$ the generators in $\mathbf{W}_{(2,\mathbf{W})}$, the commutation relations that correspond to the decomposition (9.9) have the following general form:

$$\begin{aligned}
[T^a, T^b] &= f^{ab}{}_c T^c \\
[L^x_E, L^y_E] &= f^{xy}{}_z L^z \\
[T^a, \mathbf{W}^{i\alpha}] &= (\Lambda^a)^\alpha{}_\beta \mathbf{W}^{i\beta} \\
[L^x_E, \mathbf{W}^{i\alpha}] &= (\lambda^x)^i{}_j \mathbf{W}^{j\alpha} \\
[\mathbf{W}^{i\alpha}, \mathbf{W}^{j\beta}] &= \epsilon^{ij} (K_a)^{\alpha\beta} T^a + \mathbb{C}^{\alpha\beta} k^{ij}_x L^x_E
\end{aligned} \tag{9.10}$$

where the 2×2 matrices $(\lambda^x)^i_j$ are the canonical generators of $\mathfrak{sl}(2,\mathbb{R})$ in the fundamental, defining representation:

$$\lambda^3 = \begin{pmatrix} \frac{1}{2} & 0 \\ 0 & -\frac{1}{2} \end{pmatrix}; \quad \lambda^1 = \begin{pmatrix} 0 & \frac{1}{2} \\ \frac{1}{2} & 0 \end{pmatrix}; \quad \lambda^2 = \begin{pmatrix} 0 & \frac{1}{2} \\ -\frac{1}{2} & 0 \end{pmatrix} \tag{9.11}$$

while Λ^a are the generators of $\mathbb{U}_{\mathcal{S}\mathcal{K}}$ in the symplectic representation $\mathbf{W}$. By

$$\mathbb{C}^{\alpha\beta} \equiv \left(\begin{array}{c|c} \mathbf{0}_{(n+1)\times(n+1)} & \mathbf{1}_{(n+1)\times(n+1)} \\ \hline -\mathbf{1}_{(n+1)\times(n+1)} & \mathbf{0}_{(n+1)\times(n+1)} \end{array} \right) \tag{9.12}$$

1 This concept is explained in [54]: the Lie algebra student can ignore it at this level.

we denote the antisymmetric symplectic metric in $2n+2$ dimensions, n being the complex dimension of the special Kähler manifold $\frac{\mathbb{U}_{\mathscr{S}\mathscr{K}}}{\mathbb{H}_{\mathscr{S}\mathscr{K}}}$. The symplectic character of the representation $\mathbf{W}$ is asserted by the identity:

$$\Lambda^a\mathbb{C} + \mathbb{C}(\Lambda^a)^T = 0 \tag{9.13}$$

The fundamental doublet representation of $\mathfrak{sl}(2,\mathbb{R})_E$ is also symplectic and we have denoted by $\epsilon^{ij} = \begin{pmatrix} 0 & 1 \\ -1 & 0 \end{pmatrix}$ the 2-dimensional symplectic metric, so that:

$$\lambda^x\epsilon + \epsilon(\lambda^x)^T = 0 \tag{9.14}$$

The matrices $(K_a)^{\alpha\beta} = (K_a)^{\beta\alpha}$ and $(k_x)^{ij} = (k_y)^{ji}$ are just symmetric matrices in one-to-one correspondence with the generators of $\mathbb{U}_{\mathscr{Q}}$ and $\mathfrak{sl}(2,\mathbb{R})$, respectively. Implementing Jacobi identities we find the following relations:

$$K_a\Lambda^c + \Lambda^c K_a = f^{bc}{}_a K_b, \quad k_x\lambda^y + \lambda^y k_x = f^{yz}{}_x k_z$$

which admit the unique solution:

$$K_a = c_1\mathbf{g}_{ab}\Lambda^b\mathbb{C}; \quad k_x = c_2\mathbf{g}_{xy}\lambda^y\epsilon \tag{9.15}$$

where $\mathbf{g}_{ab}$, $\mathbf{g}_{xy}$ are the Cartan–Killing metrics on the algebras $\mathbb{U}_{\mathscr{S}\mathscr{K}}$ and $\mathfrak{sl}(2,\mathbb{R})$, respectively, and c_1 and c_2 are two arbitrary constants. These latter can always be reabsorbed into the normalization of the generators $\mathbf{W}^{i\alpha}$ and correspondingly set to one. Hence the algebra (9.10) can always be put into the following elegant form:

$$\begin{aligned}
&[T^a, T^b] = f^{ab}{}_c T^c \\
&[L^x, L^y] = f^{xy}{}_z L^z \\
&[T^a, \mathbf{W}^{i\alpha}] = (\Lambda^a)^\alpha{}_\beta \mathbf{W}^{i\beta} \\
&[L^x, \mathbf{W}^{i\alpha}] = (\lambda^x)^i{}_j \mathbf{W}^{j\alpha} \\
&[\mathbf{W}^{i\alpha}, \mathbf{W}^{j\beta}] = \epsilon^{ij}(\Lambda_a)^{\alpha\beta} T^a + \mathbb{C}^{\alpha\beta}\lambda_x^{ij} L^x
\end{aligned} \tag{9.16}$$

where we have used the convention that symplectic indices are raised and lowered with the symplectic metric, while adjoint representation indices are raised and lowered with the Cartan–Killing metric.

We name (9.16) the golden splitting of quaternionic Lie algebras and it is obviously an intrinsic property of certain Lie algebras that might have been discovered by Killing, Cartan or Weyl if they had searched for it, independently of any supersymmetry or dimensional reduction of supergravity theories. It is the algebraic basis of the c-map and it has far-reaching geometrical consequences.

As we emphasized above, starting from eq. (9.16) we can embark on the programme of classifying all pairs of Lie algebras $(\mathbb{U}_{\mathscr{Q}}, \mathbb{U}_{\mathscr{S}\mathscr{K}})$ whose structure fits into such a presentation with the additional necessary constraint that the dimension $2n+2$ of the representation $\mathbf{W}$ should be consistent with

$$2n = \dim[\mathbb{U}_{\mathscr{S}\mathscr{K}}] - \dim[\mathbb{H}_{\mathscr{S}\mathscr{K}}] \tag{9.17}$$

the subalgebra $\mathbb{H}_{\mathscr{S}\mathscr{K}} \subset \mathbb{U}_{\mathscr{S}\mathscr{K}}$ being compact.

The result of such a scanning leads to the classification of all the homogeneous special Kähler manifolds and of their quaternionic images through the c-map, which is presented in [54].

Here we illustrate the first example of the golden splitting with the case of the $\mathfrak{g}_2$ Lie algebra.

9.1.2 The golden splitting of the quaternionic algebra $\mathfrak{g}_2$

The Lie algebra $\mathfrak{g}_2$ is quaternionic since it contains two $\mathfrak{a}_1 \sim \mathfrak{sl}(2,\mathbb{C})$ subalgebras with respect to which the adjoint representation decomposes as follows:

$$\mathrm{adj}[\mathfrak{g}] = (\mathrm{adj}[\mathfrak{sl}(2,\mathbb{C})_E], \mathbf{1}) \oplus (\mathbf{1}, \mathrm{adj}[\mathfrak{sl}(2,\mathbb{C})]) \oplus (\mathbf{2},\mathbf{4}) \tag{9.18}$$

where $\mathbf{4}$, which is the present instance of the symplectic $\mathbf{W}$, denotes the $J=\frac{3}{2}$ irreducible representation of the Lie algebra $\mathfrak{so}(3,\mathbb{C}) \sim \mathfrak{sl}(2,\mathbb{C})$.

To show this we begin to analyze the $\mathbf{W}$-representation proving that it is symplectic. To this effect we find it convenient to restrict our attention to the maximally split real section of the algebra.

9.1.2.1 The $J=\frac{3}{2}$-representation of $\mathfrak{sl}(2,\mathbb{R})$

The group $SL(2,\mathbb{R})$ is locally isomorphic to $SO(1,2)$ and the fundamental representation of the first corresponds to the spin $J=\frac{1}{2}$ of the latter. The spin $J=\frac{3}{2}$ representation is obviously 4-dimensional and, in the $SL(2,\mathbb{R})$ language, it corresponds to a symmetric three-index tensor t_{abc}. Let us explicitly construct the 4×4 matrices of such a representation. This is easily done by choosing an order for the four independent components of the symmetric tensor t_{abc}. For instance we can identify the four axes of the representation with $t_{111}, t_{112}, t_{122}, t_{222}$. So doing, the image of the group element:

$$\mathfrak{A} = \begin{pmatrix} a & b \\ c & d \end{pmatrix}; \quad ad - bc = 1 \tag{9.19}$$

in the cubic symmetric tensor product representation is the following 4×4 matrix:

$$\mathscr{D}_3(\mathfrak{A}) = \begin{pmatrix} a^3 & 3a^2b & 3ab^2 & b^3 \\ a^2c & da^2+2bca & cb^2+2adb & b^2d \\ ac^2 & bc^2+2adc & ad^2+2bcd & bd^2 \\ c^3 & 3c^2d & 3cd^2 & d^3 \end{pmatrix} \tag{9.20}$$

By explicit evaluation we can easily check that:

$$\mathscr{D}_3^T(\mathfrak{A})\widehat{\mathbb{C}}_4\mathscr{D}_3(A) = \widehat{\mathbb{C}}_4 \quad \text{where } \widehat{\mathbb{C}}_4 = \begin{pmatrix} 0 & 0 & 0 & 1 \\ 0 & 0 & -3 & 0 \\ 0 & 3 & 0 & 0 \\ -1 & 0 & 0 & 0 \end{pmatrix} \tag{9.21}$$

Since $\widehat{\mathbb{C}}_4$ is antisymmetric, eq. (9.21) is already a clear indication that the triple symmetric representation defines a symplectic embedding. To make this manifest it suffices to change basis. Consider the matrix:

$$S = \begin{pmatrix} 0 & 1 & 0 & 0 \\ -\frac{1}{\sqrt{3}} & 0 & 0 & 0 \\ 0 & 0 & \frac{1}{\sqrt{3}} & 0 \\ 0 & 0 & 0 & 1 \end{pmatrix} \tag{9.22}$$

and define:

$$\Lambda(\mathfrak{A}) = S^{-1}D_3(\mathfrak{A})S \tag{9.23}$$

We can easily check that:

$$\Lambda^T(\mathfrak{A})\mathbb{C}_4\Lambda(\mathfrak{A}) = \mathbb{C}_4 \quad \text{where } \mathbb{C}_4 = \begin{pmatrix} 0 & 0 & 1 & 0 \\ 0 & 0 & 0 & 1 \\ -1 & 0 & 0 & 0 \\ 0 & -1 & 0 & 0 \end{pmatrix} \tag{9.24}$$

So we have indeed constructed a standard symplectic embedding $\mathrm{SL}(2,\mathbb{R}) \mapsto \mathrm{Sp}(4,\mathbb{R})$ whose explicit form is the following:

$$\mathfrak{A} = \begin{pmatrix} a & b \\ c & d \end{pmatrix} \mapsto \left(\begin{array}{cc|cc} da^2 + 2bca & -\sqrt{3}a^2c & -cb^2 - 2adb & -\sqrt{3}b^2d \\ -\sqrt{3}a^2b & a^3 & \sqrt{3}ab^2 & b^3 \\ \hline -bc^2 - 2adc & \sqrt{3}ac^2 & ad^2 + 2bcd & \sqrt{3}bd^2 \\ -\sqrt{3}c^2d & c^3 & \sqrt{3}cd^2 & d^3 \end{array}\right) \equiv \Lambda(\mathfrak{A}) \tag{9.25}$$

The 2×2 blocks A, B, C, D of the 4×4 symplectic matrix $\Lambda(\mathfrak{A})$ are easily readable from eq. (9.25).

9.1.2.2 Putting the $\mathfrak{g}_2$ Lie algebra in the quaternionic form

Explicitly the $\mathfrak{g}_2$ Lie algebra can be cast into the form (9.10) in the following way.

First we single out the two relevant $\mathfrak{sl}(2,\mathbb{C})$ subalgebras. The Ehlers algebra is associated with the highest root and we have:

$$L_0^E = \frac{1}{\sqrt{3}}H_2; \quad L_\pm^E = \sqrt{\frac{2}{3}}E^{\pm(3\alpha_1+2\alpha_2)} \tag{9.26}$$

while the special Kähler subalgebra $\mathbb{U}_{\mathscr{S}\mathscr{K}} =, \mathfrak{sl}(2,\mathbb{C})$ is associated with the first simple root orthogonal to the highest one and we have:

$$L_0 = H_1; \quad L_\pm = \sqrt{2}E^{\pm\alpha_1} \tag{9.27}$$

Then we can arrange the remaining eight generators in the tensor $W^{i\beta}$ as follows:

$$\begin{aligned} W^{1M} &= \sqrt{\frac{2}{3}}(E^{\alpha_1+\alpha_2}, E^{\alpha_2}, E^{2\alpha_1+\alpha_2}, E^{3\alpha_1+\alpha_2}) \\ W^{2M} &= \sqrt{\frac{2}{3}}(-E^{-2\alpha_1-\alpha_2}, -E^{-3\alpha_1-\alpha_2}, E^{-\alpha_1-\alpha_2}, E^{-\alpha_2}) \end{aligned} \tag{9.28}$$

Calculating the commutators of W^{iM} with the generators of the two $\mathfrak{sl}(2)$ algebras we find:

$$\begin{aligned} \left[L_0^E, \begin{pmatrix} W^1 \\ W^2 \end{pmatrix}\right] &= \left(\begin{array}{c|c} \frac{1}{2}\mathbf{1} & 0 \\ \hline 0 & -\frac{1}{2}\mathbf{1} \end{array}\right)\begin{pmatrix} W^1 \\ W^2 \end{pmatrix} \\ \left[L_+^E, \begin{pmatrix} W^1 \\ W^2 \end{pmatrix}\right] &= \left(\begin{array}{c|c} 0 & 0 \\ \hline -\mathbf{1} & 0 \end{array}\right)\begin{pmatrix} W^1 \\ W^2 \end{pmatrix} \\ \left[L_-^E, \begin{pmatrix} W^1 \\ W^2 \end{pmatrix}\right] &= \left(\begin{array}{c|c} 0 & -\mathbf{1} \\ \hline 0 & 0 \end{array}\right)\begin{pmatrix} W^1 \\ W^2 \end{pmatrix} \end{aligned} \tag{9.29}$$

and:

$$\begin{aligned} \left[L_0, \begin{pmatrix} W^1 \\ W^2 \end{pmatrix}\right] &= -\left(\begin{array}{c|c} U_0 & 0 \\ \hline 0 & U_0 \end{array}\right)\begin{pmatrix} W^1 \\ W^2 \end{pmatrix} \\ \left[L_\pm, \begin{pmatrix} W^1 \\ W^2 \end{pmatrix}\right] &= -\left(\begin{array}{c|c} U_\pm & 0 \\ \hline 0 & U_\pm \end{array}\right)\begin{pmatrix} W^1 \\ W^2 \end{pmatrix} \end{aligned} \tag{9.30}$$

where:

$$\begin{aligned} U_0 &= \begin{pmatrix} \frac{1}{2} & 0 & 0 & 0 \\ 0 & \frac{3}{2} & 0 & 0 \\ 0 & 0 & -\frac{1}{2} & 0 \\ 0 & 0 & 0 & -\frac{3}{2} \end{pmatrix} \\ U_+ &= \begin{pmatrix} 0 & 0 & -2 & 0 \\ -\sqrt{3} & 0 & 0 & 0 \\ 0 & 0 & 0 & \sqrt{3} \\ 0 & 0 & 0 & 0 \end{pmatrix} \\ U_- &= \begin{pmatrix} 0 & -\sqrt{3} & 0 & 0 \\ 0 & 0 & 0 & 0 \\ -2 & 0 & 0 & 0 \\ 0 & 0 & \sqrt{3} & 0 \end{pmatrix} \end{aligned} \tag{9.31}$$

which are the generators of $\mathfrak{sl}(2,\mathbb{C})$ in the symplectic embedding (9.25) as it can be easily verified by considering the embedding of a group element infinitesimally closed to the identity:

$$\begin{pmatrix} a & b \\ c & d \end{pmatrix} = \begin{pmatrix} 1+\frac{1}{2}\epsilon_0 & \epsilon_+ \\ \epsilon_- & 1-\frac{1}{2}\epsilon_0 \end{pmatrix} \tag{9.32}$$

and collecting the matrix coefficients of the first order terms in ϵ_0 and $\epsilon_\pm$.

9.1.3 Chevalley–Serre basis

We utilize the case of the $\mathfrak{g}_2$ algebra to illustrate another canonical presentation of the Lie algebra commutation relations that is named the presentation in terms of Chevalley–Serre triples. It is the analog for Lie algebras of the presentation of discrete groups through generators and relations and proves to be quite useful in several applications. Given a simple Lie algebra of rank r defined by its Cartan matrix C_{ij}, a Chevalley–Serre basis is given by r-triplets of generators:

$$(h_i, e_i, f_i); \quad i = 1, \dots, r \tag{9.33}$$

such that the following commutation relations are satisfied:

$$\begin{aligned} [h_i, h_j] &= 0 \\ [h_i, e_j] &= C_{ij} e_j \\ [h_i, f_j] &= -C_{ij} f_j \\ [e_i, f_j] &= \delta_{ij} h_i \\ \text{adj } [e_i]^{(C_{ji}+1)}(e_j) &= 0 \\ \text{adj } [f_i]^{(C_{ji}+1)}(f_j) &= 0 \end{aligned} \tag{9.34}$$

When such r-triplets are given, the entire algebra is defined. Indeed, all the other generators are constructed by commuting these ones modulo the relations (9.34). For simply-laced finite simple Lie algebras a Chevalley basis is easily constructed in terms of simple roots. Let α_i denote the simple roots, then it suffices to set:

$$(h_i, e_i, f_i) = (H_{\alpha_i}, E^{\alpha_i}, E^{-\alpha_i}) \tag{9.35}$$

where $H_{\alpha_i} \equiv \alpha_i \cdot H$ are the Cartan generator associated with the simple roots and $E^{\pm\alpha_i}$ are the step operators respectively associated with the simple roots and their negative.

9.1.3.1 The $\mathfrak{g}_2$ Lie algebra in terms of Chevalley triples

Let us rewrite the commutation relations of $\mathfrak{g}_{(2,2)}$ in terms of triples of Chevalley generators.

Since the algebra has rank two, there are two fundamental triples of Chevalley generators:

$$(\mathscr{H}_1, e_1, f_1); \quad (\mathscr{H}_2, e_2, f_2) \tag{9.36}$$

with the following commutation relations:

$$\begin{array}{llll} [\mathscr{H}_2, e_2] = 2e_2 & [\mathscr{H}_1, e_2] = -3e_2 & [\mathscr{H}_2, f_2] = -2f_2 & [\mathscr{H}_1, f_2] = 3f_2 \\ [\mathscr{H}_2, e_1] = -e_1 & [\mathscr{H}_1, e_1] = 2e_1 & [\mathscr{H}_2, f_1] = f_1 & [\mathscr{H}_1, f_1] = -2f_1 \\ [e_2, f_2] = \mathscr{H}_2 & [e_2, f_1] = 0 & [e_1, f_1] = \mathscr{H}_1 & [e_1, f_2] = 0 \end{array} \tag{9.37}$$

The remaining basis elements are defined as follows:

$$\begin{array}{llll} e_3 = [e_1, e_2] & e_4 = \frac{1}{2}[e_1, e_3] & e_5 = \frac{1}{3}[e_4, e_1] & e_6 = [e_2, e_5] \\ f_3 = [f_2, f_1] & f_4 = \frac{1}{2}[f_3, f_1] & f_5 = \frac{1}{3}[f_1, f_4] & f_6 = [f_5, f_2] \end{array} \tag{9.38}$$

and satisfy the following Serre relations:

$$[e_2, e_3] = [e_5, e_1] = [f_2, f_3] = [f_5, f_1] = 0 \tag{9.39}$$

The Chevalley form of the commutation relation is obtained from the standard Cartan–Weyl basis introducing the following identifications:

$$\begin{array}{ll} e_1 = \sqrt{2}E^{\alpha_1}; & e_2 = \sqrt{\frac{2}{3}}E^{\alpha_2} \\ e_3 = \sqrt{2}E^{\alpha_3}; & e_4 = \sqrt{2}E^{\alpha_4} \\ e_5 = \sqrt{\frac{2}{3}}E^{\alpha_5}; & e_6 = \sqrt{\frac{2}{3}}E^{\alpha_6} \\ f_1 = \sqrt{2}E^{-\alpha_1}; & f_2 = \sqrt{\frac{2}{3}}E^{-\alpha_2} \\ f_3 = \sqrt{2}E^{-\alpha_3}; & f_4 = \sqrt{2}E^{-\alpha_4} \\ f_5 = \sqrt{\frac{2}{3}}E^{-\alpha_5}; & f_6 = \sqrt{\frac{2}{3}}E^{-\alpha_6} \end{array} \tag{9.40}$$

and

$$\mathscr{H}_1 = 2\alpha_1 \cdot H; \quad \mathscr{H}_2 = \frac{2}{3}\alpha_2 \cdot H \tag{9.41}$$

All the steps of the construction described above are electronically implemented by the MATHEMATICA NoteBook described in Section B.3.

9.2 The Lie algebra $\mathfrak{f}_4$ and its fundamental representation

The next example that we consider is no longer of rank $r = 2$ and the root system cannot be easily visualized as in previous cases. We discuss the exceptional Lie algebra $\mathfrak{f}_4$

$\mathfrak{f}_4$

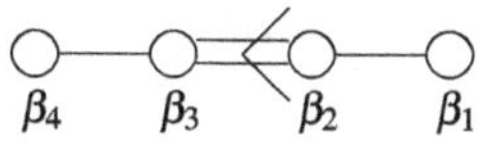

Figure 9.4: The Dynkin diagram of $\mathfrak{f}_4$ and the labeling of simple roots.

whose rank is four and whose structure is codified in the Dynkin diagram presented in Figure 9.4. We show how we can explicitly construct the fundamental and the adjoint representations of this exceptional, non-simply laced Lie algebra.

Another interesting issue associated with this example concerns the factors $N_{\alpha\beta}$ appearing in the general form (7.38) of the Lie algebra. We see in this non-simply laced case that the only effective way to derive them is necessarily based on the explicit construction of a representation, typically the fundamental one. Hence let us proceed to such a construction.

Calling $y_{1,2,3,4}$ a basis of orthonormal vectors:

$$y_i \cdot y_j = \delta_{ij} \tag{9.42}$$

a possible choice of simple roots β_i which reproduces the Cartan matrix encoded in the Dynkin diagram 9.4 is the following:

$$\begin{aligned}
\beta_1 &= -y_1 - y_2 - y_3 + y_4 \\
\beta_2 &= 2y_3 \\
\beta_3 &= y_2 - y_3 \\
\beta_4 &= y_1 - y_2
\end{aligned} \tag{9.43}$$

With this basis of simple roots the full root system composed of 48 vectors is given by:

$$\Phi_{\mathfrak{f}_4} \equiv \underbrace{\pm y_i \pm y_j;}_{24\text{ roots}} \ \underbrace{\pm y_i}_{8\text{ roots}} \ ; \ \underbrace{\pm y_1 \pm y_2 \pm y_3 \pm y_4}_{16\text{ roots}} \tag{9.44}$$

and one can list the positive roots by height as displayed in Table 9.1. Since the considered Lie algebra is not simply-laced, the 24 positive roots split into two subsets of 12 long roots α^ℓ and 12 short roots α^s. They are displayed in Tables 9.2 and 9.3, respectively.

Calling Φ_ℓ and Φ_s the two subsets, we have the following structure:

$$\begin{aligned}
\forall \alpha^\ell, \beta^\ell \in \Phi_\ell : \quad & \alpha^\ell + \beta^\ell = \begin{Bmatrix} \text{not a root or} \\ \gamma^\ell \in \Phi_\ell \end{Bmatrix} \\
\forall \alpha^\ell \in \Phi_\ell \text{ and } \forall \beta^s \in \Phi_s : \quad & \alpha^\ell + \beta^s = \begin{Bmatrix} \text{not a root or} \\ \gamma^s \in \Phi_s \end{Bmatrix} \\
\forall \alpha^s, \beta^s \in \Phi_s : \quad & \alpha^s + \beta^s = \begin{Bmatrix} \text{not a root or} \\ \gamma^s \in \Phi_s \text{ or} \\ \gamma^\ell \in \Phi_\ell \end{Bmatrix}
\end{aligned} \tag{9.45}$$

Table 9.1: List of the positive roots of the exceptional Lie algebra $\mathfrak{f}_4$. In this table the first column is the name of the root, the second column gives its decomposition in terms of simple roots, while the last column provides the component of the root vector in $\mathbb{R}^4$.

β_1	=	β_1	=	$\{-1,-1,-1,1\}$
β_2	=	β_2	=	$\{0,0,2,0\}$
β_3	=	β_3	=	$\{0,1,-1,0\}$
β_4	=	β_4	=	$\{1,-1,0,0\}$
β_5	=	$\beta_1+\beta_2$	=	$\{-1,-1,1,1\}$
β_6	=	$\beta_2+\beta_3$	=	$\{0,1,1,0\}$
β_7	=	$\beta_3+\beta_4$	=	$\{1,0,-1,0\}$
β_8	=	$\beta_1+\beta_2+\beta_3$	=	$\{-1,0,0,1\}$
β_9	=	$\beta_2+2\beta_3$	=	$\{0,2,0,0\}$
β_{10}	=	$\beta_2+\beta_3+\beta_4$	=	$\{1,0,1,0\}$
β_{11}	=	$\beta_1+\beta_2+2\beta_3$	=	$\{-1,1,-1,1\}$
β_{12}	=	$\beta_1+\beta_2+\beta_3+\beta_4$	=	$\{0,-1,0,1\}$
β_{13}	=	$\beta_2+2\beta_3+\beta_4$	=	$\{1,1,0,0\}$
β_{14}	=	$\beta_1+2\beta_2+2\beta_3$	=	$\{-1,1,1,1\}$
β_{15}	=	$\beta_1+\beta_2+2\beta_3+\beta_4$	=	$\{0,0,-1,1\}$
β_{16}	=	$\beta_2+2\beta_3+2\beta_4$	=	$\{2,0,0,0\}$
β_{17}	=	$\beta_1+2\beta_2+2\beta_3+\beta_4$	=	$\{0,0,1,1\}$
β_{18}	=	$\beta_1+\beta_2+2\beta_3+2\beta_4$	=	$\{1,-1,-1,1\}$
β_{19}	=	$\beta_1+2\beta_2+3\beta_3+\beta_4$	=	$\{0,1,0,1\}$
β_{20}	=	$\beta_1+2\beta_2+2\beta_3+2\beta_4$	=	$\{1,-1,1,1\}$
β_{21}	=	$\beta_1+2\beta_2+3\beta_3+2\beta_4$	=	$\{1,0,0,1\}$
β_{22}	=	$\beta_1+2\beta_2+4\beta_3+2\beta_4$	=	$\{1,1,-1,1\}$
β_{23}	=	$\beta_1+3\beta_2+4\beta_3+2\beta_4$	=	$\{1,1,1,1\}$
β_{24}	=	$2\beta_1+3\beta_2+4\beta_3+2\beta_4$	=	$\{0,0,0,2\}$

Table 9.2: The Φ_ℓ set of the 12 long positive roots in the $\mathfrak{f}_4$ root system.

	$\mathfrak{f}_4$ root labels	$\mathfrak{f}_4$ root in Eucl. basis	Root ordered by height
α_1^ℓ	$\{0,1,0,0\}$	$2y_3$	β_2
α_2^ℓ	$\{1,0,0,0\}$	$-y_1-y_2-y_3+y_4$	β_1
α_3^ℓ	$\{1,1,0,0\}$	$-y_1-y_2+y_3+y_4$	β_3
α_4^ℓ	$\{0,1,2,0\}$	$2y_2$	β_9
α_5^ℓ	$\{1,1,2,0\}$	$-y_1+y_2-y_3+y_4$	β_{11}
α_6^ℓ	$\{1,2,2,0\}$	$-y_1+y_2+y_3+y_4$	β_{14}
α_7^ℓ	$\{0,1,2,2\}$	$2y_1$	β_{16}
α_8^ℓ	$\{1,1,2,2\}$	$y_1-y_2-y_3+y_4$	β_{18}
α_9^ℓ	$\{1,2,2,2\}$	$y_1-y_2+y_3+y_4$	β_{20}
α_{10}^ℓ	$\{1,2,4,2\}$	$y_1+y_2-y_3+y_4$	β_{22}
α_{11}^ℓ	$\{1,3,4,2\}$	$y_1+y_2+y_3+y_4$	β_{23}
α_{12}^ℓ	$\{2,3,4,2\}$	$2y_4$	β_{24}

Table 9.3: The Φ_s set of 12 short positive roots in the $\mathfrak{f}_4$ root system.

	$\mathfrak{f}_4$ root labels	$\mathfrak{f}_4$ root in Eucl. basis	Root ordered by height
α_1^s	$\{0,0,0,1\}$	$y_1 - y_2$	β_4
α_2^s	$\{0,0,1,0\}$	$y_2 - y_3$	β_3
α_3^s	$\{0,1,1,0\}$	$y_2 + y_3$	β_6
α_4^s	$\{0,0,1,1\}$	$y_1 - y_3$	β_7
α_5^s	$\{1,1,1,0\}$	$-y_1 + y_4$	β_8
α_6^s	$\{0,1,1,1\}$	$y_1 + y_3$	β_{10}
α_7^s	$\{1,1,1,1\}$	$-y_2 + y_4$	β_{12}
α_8^s	$\{0,1,2,1\}$	$y_1 + y_2$	β_{13}
α_9^s	$\{1,1,2,1\}$	$-y_3 + y_4$	β_{15}
α_{10}^s	$\{1,2,2,1\}$	$y_3 + y_4$	β_{17}
α_{11}^s	$\{1,2,3,1\}$	$y_2 + y_4$	β_{19}
α_{12}^s	$\{1,2,3,2\}$	$y_1 + y_4$	β_{21}

The standard Cartan–Weyl form of the Lie algebra is as follows:

$$[\mathcal{H}_i, E^{\pm\beta}] = \pm\beta^i E^{\pm\beta_I} \tag{9.46}$$

$$[E^\beta, E^{-\beta}] = \beta \cdot \mathcal{H} \tag{9.47}$$

$$[E^\beta, E^\gamma] = \begin{cases} N_{\beta\gamma} E^{\beta+\gamma} & \text{if } \beta+\gamma \text{ is a root} \\ 0 & \text{if } \beta+\gamma \text{ is not a root} \end{cases} \tag{9.48}$$

where $N_{\beta\gamma}$ are numbers that can be worked constructing an explicit representation of the Lie algebra.

In the following three tables – (9.49), (9.50), (9.51) – we exhibit the values of $N_{\beta\gamma}$ for the $\mathfrak{f}_4$ Lie algebra.

α_1^ℓ	α_2^ℓ	α_3^ℓ	α_4^ℓ	α_5^ℓ	α_6^ℓ	α_7^ℓ	α_8^ℓ	α_9^ℓ	α_{10}^ℓ	α_{11}^ℓ	α_{12}^ℓ	
0	$-\sqrt{2}$	0	0	$-\sqrt{2}$	0	0	$-\sqrt{2}$	0	$\sqrt{2}$	0	0	α_1^ℓ
$\sqrt{2}$	0	0	$-\sqrt{2}$	0	0	$-\sqrt{2}$	0	0	0	$-\sqrt{2}$	0	α_2^ℓ
0	0	0	$-\sqrt{2}$	0	0	$-\sqrt{2}$	0	0	$-\sqrt{2}$	0	0	α_3^ℓ
0	$\sqrt{2}$	$\sqrt{2}$	0	0	0	0	$-\sqrt{2}$	$\sqrt{2}$	0	0	0	α_4^ℓ
$\sqrt{2}$	0	0	0	0	0	$\sqrt{2}$	0	$\sqrt{2}$	0	0	0	α_5^ℓ
0	0	0	0	0	0	$-\sqrt{2}$	$-\sqrt{2}$	0	0	0	0	α_6^ℓ
0	$\sqrt{2}$	$\sqrt{2}$	0	$-\sqrt{2}$	$\sqrt{2}$	0	0	0	0	0	0	α_7^ℓ
$\sqrt{2}$	0	0	$\sqrt{2}$	0	$\sqrt{2}$	0	0	0	0	0	0	α_8^ℓ
0	0	0	$-\sqrt{2}$	$-\sqrt{2}$	0	0	0	0	0	0	0	α_9^ℓ
$-\sqrt{2}$	0	$\sqrt{2}$	0	0	0	0	0	0	0	0	0	α_{10}^ℓ
0	$\sqrt{2}$	0	0	0	0	0	0	0	0	0	0	α_{11}^ℓ
0	0	0	0	0	0	0	0	0	0	0	0	α_{12}^ℓ

$N_{\alpha^\ell\beta^\ell}$

(9.49)

α_1^s	α_2^s	α_3^s	α_4^s	α_5^s	α_6^s	α_7^s	α_8^s	α_9^s	α_{10}^s	α_{11}^s	α_{12}^s	
0	$\sqrt{2}$	0	$-\sqrt{2}$	0	0	0	0	$\sqrt{2}$	0	0	0	α_1^ℓ
0	0	$-\sqrt{2}$	0	0	$-\sqrt{2}$	0	$\sqrt{2}$	0	0	0	0	α_2^ℓ
0	$-\sqrt{2}$	0	$\sqrt{2}$	0	0	0	$-\sqrt{2}$	0	0	0	0	α_3^ℓ
$\sqrt{2}$	0	0	0	0	0	$\sqrt{2}$	0	0	0	0	0	α_4^ℓ
$-\sqrt{2}$	0	0	0	0	$-\sqrt{2}$	0	0	0	0	0	0	α_5^ℓ
$\sqrt{2}$	0	0	$\sqrt{2}$	0	0	0	0	0	0	0	0	α_6^ℓ
0	0	0	0	$-\sqrt{2}$	0	0	0	0	0	0	0	α_7^ℓ
0	0	$\sqrt{2}$	0	0	0	0	0	0	0	0	0	α_8^ℓ
0	$\sqrt{2}$	0	0	0	0	0	0	0	0	0	0	α_9^ℓ
0	0	0	0	0	0	0	0	0	0	0	0	α_{10}^ℓ
0	0	0	0	0	0	0	0	0	0	0	0	α_{11}^ℓ
0	0	0	0	0	0	0	0	0	0	0	0	α_{12}^ℓ

$N_{\alpha^\ell \beta^s}$

(9.50)

α_1^s	α_2^s	α_3^s	α_4^s	α_5^s	α_6^s	α_7^s	α_8^s	α_9^s	α_{10}^s	α_{11}^s	α_{12}^s	
0	1	−1	0	−1	0	0	$\sqrt{2}$	$-\sqrt{2}$	$\sqrt{2}$	1	0	α_1^s
−1	0	$\sqrt{2}$	0	$\sqrt{2}$	1	−1	0	0	−1	0	$\sqrt{2}$	α_2^s
1	$-\sqrt{2}$	0	1	$-\sqrt{2}$	0	−1	0	1	0	0	$\sqrt{2}$	α_3^s
0	0	−1	0	1	$\sqrt{2}$	$\sqrt{2}$	0	0	1	$-\sqrt{2}$	0	α_4^s
1	$-\sqrt{2}$	$\sqrt{2}$	−1	0	1	0	1	0	0	0	$\sqrt{2}$	α_5^s
0	−1	0	$-\sqrt{2}$	−1	0	$\sqrt{2}$	0	1	0	$\sqrt{2}$	0	α_6^s
0	1	1	$-\sqrt{2}$	0	$-\sqrt{2}$	0	1	0	0	$\sqrt{2}$	0	α_7^s
$-\sqrt{2}$	0	0	0	−1	0	−1	0	$\sqrt{2}$	$\sqrt{2}$	0	0	α_8^s
$\sqrt{2}$	0	−1	0	0	−1	0	$-\sqrt{2}$	0	$-\sqrt{2}$	0	0	α_9^s
$-\sqrt{2}$	1	0	−1	0	0	0	$-\sqrt{2}$	$\sqrt{2}$	0	0	0	α_{10}^s
−1	0	0	$\sqrt{2}$	0	$-\sqrt{2}$	$-\sqrt{2}$	0	0	0	0	0	α_{11}^s
0	$-\sqrt{2}$	$-\sqrt{2}$	0	$-\sqrt{2}$	0	0	0	0	0	0	0	α_{12}^s

$N_{\alpha^s \beta^s}$

(9.51)

The ordering of long and short roots is that displayed in Tables 9.2 and 9.3. The explicit determination of the tensor $N_{\alpha\beta}$ was performed via the explicit construction of the fundamental 26-dimensional representation of this Lie algebra which we describe in the next subsection.

9.2.1 Explicit construction of the fundamental and adjoint representation of $\mathfrak{f}_4$

The semi-simple complex Lie algebra $\mathfrak{f}_4$ is defined by the Dynkin diagram in Figure 9.4 and a set of simple roots corresponding to such diagram was provided in eq. (9.43).

A complete list of the 24 positive roots was given in Table 9.1. The roots were further subdivided into the set of 12 long roots and 12 short roots respectively listed in Tables 9.2 and 9.3. The adjoint representation of $\mathfrak{f}_4$ is 52-dimensional, while its fundamental representation is 26-dimensional. This dimensionality is true for all real sections of the Lie algebra but the explicit structure of the representation is quite different in each real section. Here we are interested in the maximally split real section $\mathfrak{f}_4$. For such a section we have a maximal, regularly embedded, subgroup $\mathfrak{so}(5,4) \subset \mathfrak{f}_{4(4)}$. The decomposition of the representations with respect to this particular subgroup is the essential instrument for their actual construction. For the adjoint representation we have the decomposition:

$$\underbrace{\mathbf{52}}_{\text{adj}\,\mathfrak{f}_{4(4)}} \overset{\mathfrak{so}(5,4)}{\Longrightarrow} \underbrace{\mathbf{36}}_{\text{adj}\,\mathfrak{so}(5,4)} \oplus \underbrace{\mathbf{16}}_{\text{spinor of }\mathfrak{so}(5,4)} \tag{9.52}$$

while for the fundamental one we have:

$$\underbrace{\mathbf{26}}_{\text{fundamental}\,\mathfrak{f}_{4(4)}} \overset{\mathfrak{so}(5,4)}{\Longrightarrow} \underbrace{\mathbf{9}}_{\text{vector of }\mathfrak{so}(5,4)} \oplus \underbrace{\mathbf{16}}_{\text{spinor of }\mathfrak{so}(5,4)} \oplus \underbrace{\mathbf{1}}_{\text{singlet of }\mathfrak{so}(5,4)} \tag{9.53}$$

In view of this, we fix our conventions for the $\mathfrak{so}(5,4)$ invariant metric as follows

$$\eta_{AB} = \text{diag}\{+,+,+,+,+,-,-,-,-\} \tag{9.54}$$

and we perform an explicit construction of the 16×16 dimensional gamma matrices which satisfy the Clifford algebra

$$\{\Gamma_A, \Gamma_B\} = \eta_{AB}\mathbf{1} \tag{9.55}$$

and are *all completely real*. This construction is provided by the following tensor products:

$$\begin{aligned}
\Gamma_1 &= \sigma_1 \otimes \sigma_3 \otimes \mathbf{1} \otimes \mathbf{1} \\
\Gamma_2 &= \sigma_3 \otimes \sigma_3 \otimes \mathbf{1} \otimes \mathbf{1} \\
\Gamma_3 &= \mathbf{1} \otimes \sigma_1 \otimes \mathbf{1} \otimes \sigma_1 \\
\Gamma_4 &= \mathbf{1} \otimes \sigma_1 \otimes \sigma_1 \otimes \sigma_3 \\
\Gamma_5 &= \mathbf{1} \otimes \sigma_1 \otimes \sigma_3 \otimes \sigma_3 \\
\Gamma_6 &= \mathbf{1} \otimes i\sigma_2 \otimes \mathbf{1} \otimes \mathbf{1} \\
\Gamma_7 &= \mathbf{1} \otimes \sigma_1 \otimes i\sigma_2 \otimes \sigma_3 \\
\Gamma_8 &= \mathbf{1} \otimes \sigma_1 \otimes \mathbf{1} \otimes i\sigma_2 \\
\Gamma_9 &= i\sigma_2 \otimes \sigma_3 \otimes \mathbf{1} \otimes \mathbf{1}
\end{aligned} \tag{9.56}$$

where by σ_i we have denoted the standard Pauli matrices:

$$\sigma_1 = \begin{pmatrix} 0 & 1 \\ 1 & 0 \end{pmatrix}; \quad \sigma_2 = \begin{pmatrix} 0 & -i \\ i & 0 \end{pmatrix}; \quad \sigma_3 = \begin{pmatrix} 1 & 0 \\ 0 & -1 \end{pmatrix} \tag{9.57}$$

Moreover, we introduce the C_+ charge conjugation matrix, such that:

$$\begin{aligned} C_+ &= (C_+)^T; \quad C_+^2 = \mathbf{1} \\ C_+\Gamma_A C_+ &= (\Gamma_A)^T \end{aligned} \tag{9.58}$$

In the basis of eq. (9.56) the explicit form of C_+ is given by:

$$C_+ = i\sigma_2 \otimes \sigma_1 \otimes i\sigma_2 \otimes \sigma_1 \tag{9.59}$$

Then we define the usual generators $J_{AB} = -J_{BA}$ of the pseudo-orthogonal algebra $\mathfrak{so}(5,4)$ satisfying the commutation relations:

$$[J_{AB}, J_{CD}] = \eta_{BC}J_{AD} - \eta_{AC}J_{BD} - \eta_{BD}J_{AC} + \eta_{AD}J_{BC} \tag{9.60}$$

and we construct the spinor and the vector representations by respectively setting:

$$J^s_{CD} = \frac{1}{4}[\Gamma_C, \Gamma_D]; \quad (J^v_{CD})_A{}^B = \eta_{CA}\delta^B_D - \eta_{DA}\delta^B_C \tag{9.61}$$

In this way if v_A denotes the components of a vector, ξ those of a real spinor and $\epsilon^{AB} = -\epsilon^{BA}$ are the parameters of an infinitesimal $\mathfrak{so}(5,4)$ rotation, we can write the $\mathfrak{so}(5,4)$ transformation as follows:

$$\delta_{\mathfrak{so}(5,4)} v_A = 2\epsilon_{AB} v^B; \quad \delta_{\mathfrak{so}(5,4)}\xi = \frac{1}{2}\epsilon^{AB}\Gamma_{AB}\xi \tag{9.62}$$

where indices are raised and lowered with the metric (9.54). Furthermore, we introduce the conjugate spinors via the position:

$$\overline{\xi} \equiv \xi^T C_+ \tag{9.63}$$

With these preliminaries, we are now in a position to write the explicit form of the 26-dimensional fundamental representation of $\mathfrak{f}_4$ and in this way to construct also its structure constants and hence its adjoint representation, which is our main goal.

According to eq. (9.52) the parameters of an $\mathfrak{f}_4$ representation are given by an antisymmetric tensor ϵ_{AB} and a spinor q. On the other hand a *vector* in the 26-dimensional representation is specified by a collection of three objects, namely a scalar ϕ, a vector v_A and a spinor ξ. The representation is constructed if we specify the $\mathfrak{f}_{4(4)}$ transformation of these objects. This is done by writing:

$$\delta_{\mathfrak{f}_{4(4)}} \begin{pmatrix} \phi \\ v_A \\ \xi \end{pmatrix} \equiv [\epsilon^{AB}T_{AB} + \overline{q}Q] \begin{pmatrix} \phi \\ v_A \\ \xi \end{pmatrix} = \begin{pmatrix} \overline{q}\xi \\ 2\epsilon_{AB}v^B + a\overline{q}\Gamma_A\xi \\ \frac{1}{2}\epsilon^{AB}\Gamma_{AB}\xi - 3\phi q - \frac{1}{a}v^A\Gamma_A q \end{pmatrix} \tag{9.64}$$

where a is a numerical real arbitrary but non-null parameter. Equation (9.64) defines the generators T_{AB} and Q as 26×26 matrices and therefore completely specifies the fundamental representation of the Lie algebra $\mathfrak{f}_{4(4)}$. Explicitly we have:

$$T_{AB} = \left(\begin{array}{c|c|c} 0 & 0 & 0 \\ \hline 0 & J^{v}_{AB} & 0 \\ \hline 0 & 0 & J^{s}_{AB} \end{array}\right) \tag{9.65}$$

and

$$Q_\alpha = \left(\begin{array}{c|c|c} 0 & 0 & \delta^\beta_\alpha \\ \hline 0 & 0 & a(\Gamma_A)^\beta_\alpha \\ \hline -3\delta^\beta_\alpha & -\frac{1}{a}(\Gamma_B)^\beta_\alpha & 0 \end{array}\right) \tag{9.66}$$

and the Lie algebra commutation relations are evaluated to be the following ones:

$$\begin{aligned} [T_{AB}, T_{CD}] &= \eta_{BC}T_{AD} - \eta_{AC}T_{BD} - \eta_{BD}T_{AC} + \eta_{AD}T_{BC} \\ [T_{AB}, Q] &= \frac{1}{2}\Gamma_{AB}Q \\ [Q_\alpha, Q_\beta] &= -\frac{1}{12}(C_+\Gamma^{AB})_{\alpha\beta}T_{AB} \end{aligned} \tag{9.67}$$

Equation (9.67), together with eqs (9.56) and (9.58), provides an explicit numerical construction of the structure constants of the maximally split $\mathfrak{f}_{4(4)}$ Lie algebra. What we still have to do is to identify the relation between the tensorial basis of generators in eq. (9.67) and the Cartan–Weyl basis in terms of Cartan generators and step operators. To this effect let us enumerate the 52 generators of $\mathfrak{f}_4$ in the tensorial representation according to the following table:

$\Omega_1 = T_{12}$	$\Omega_2 = T_{13}$	$\Omega_3 = T_{14}$	$\Omega_4 = T_{15}$
$\Omega_5 = T_{16}$	$\Omega_6 = T_{17}$	$\Omega_7 = T_{18}$	$\Omega_8 = T_{19}$
$\Omega_9 = T_{23}$	$\Omega_{10} = T_{24}$	$\Omega_{11} = T_{25}$	$\Omega_{12} = T_{26}$
$\Omega_{13} = T_{27}$	$\Omega_{14} = T_{28}$	$\Omega_{15} = T_{29}$	$\Omega_{16} = T_{34}$
$\Omega_{17} = T_{35}$	$\Omega_{18} = T_{36}$	$\Omega_{19} = T_{37}$	$\Omega_{20} = T_{38}$
$\Omega_{21} = T_{39}$	$\Omega_{22} = T_{45}$	$\Omega_{23} = T_{46}$	$\Omega_{24} = T_{47}$
$\Omega_{25} = T_{48}$	$\Omega_{26} = T_{49}$	$\Omega_{27} = T_{56}$	$\Omega_{28} = T_{57}$
$\Omega_{29} = T_{58}$	$\Omega_{30} = T_{59}$	$\Omega_{31} = T_{67}$	$\Omega_{32} = T_{68}$
$\Omega_{33} = T_{69}$	$\Omega_{34} = T_{78}$	$\Omega_{35} = T_{79}$	$\Omega_{36} = T_{89}$
$\Omega_{37} = Q_1$	$\Omega_{38} = Q_2$	$\Omega_{39} = Q_3$	$\Omega_{40} = Q_4$
$\Omega_{41} = Q_5$	$\Omega_{42} = Q_6$	$\Omega_{43} = Q_7$	$\Omega_{44} = Q_8$
$\Omega_{45} = Q_9$	$\Omega_{46} = Q_{10}$	$\Omega_{47} = Q_{11}$	$\Omega_{48} = Q_{12}$
$\Omega_{49} = Q_{13}$	$\Omega_{50} = Q_{14}$	$\Omega_{51} = Q_{15}$	$\Omega_{52} = Q_{16}$

(9.68)

Then, as Cartan subalgebra we take the linear span of the following generators:

$$CSA \equiv \text{span}(\Omega_5, \Omega_{13}, \Omega_{20}, \Omega_{26}) \tag{9.69}$$

Table 9.4: Listing of the step operators corresponding to the positive roots of $\mathfrak{f}_4$.

Name	Dynkin lab.	Comp. root	Step operator $E^\beta =$
$\beta[1]$	$\{1,0,0,0\}$	$-y_1-y_2-y_3+y_4$	$(-\Omega_3-\Omega_8+\Omega_{23}-\Omega_{33})$
$\beta[2]$	$\{0,1,0,0\}$	$2y_3$	$(\Omega_{16}-\Omega_{21}+\Omega_{25}+\Omega_{36})$
$\beta[3]$	$\{0,0,1,0\}$	y_2-y_3	$(\Omega_{37}+\Omega_{39}+\Omega_{41}-\Omega_{43}+\Omega_{45}-\Omega_{47}+\Omega_{49}+\Omega_{51})$
$\beta[4]$	$\{0,0,0,1\}$	y_1-y_2	$(\Omega_{11}+\Omega_{28})$
$\beta[5]$	$\{1,1,0,0\}$	$-y_1-y_2+y_3+y_4$	$-\frac{1}{\sqrt{2}}(-\Omega_2+\Omega_7+\Omega_{18}+\Omega_{32})$
$\beta[6]$	$\{0,1,1,0\}$	y_2+y_3	$-\frac{1}{\sqrt{2}}(-\Omega_{38}-\Omega_{40}+\Omega_{42}-\Omega_{44}+\Omega_{46}-\Omega_{48}-\Omega_{50}-\Omega_{52})$
$\beta[7]$	$\{0,0,1,1\}$	y_1-y_3	$-(-\Omega_{37}-\Omega_{39}+\Omega_{41}-\Omega_{43}+\Omega_{45}-\Omega_{47}-\Omega_{49}-\Omega_{51})$
$\beta[8]$	$\{1,1,1,0\}$	$-y_1+y_4$	$-\frac{1}{2}(\Omega_{38}+\Omega_{40}-\Omega_{42}+\Omega_{44}+\Omega_{46}-\Omega_{48}-\Omega_{50}-\Omega_{52})$
$\beta[9]$	$\{0,1,2,0\}$	$2y_2$	$-\frac{1}{2}(\Omega_1+\Omega_6+\Omega_{12}-\Omega_{31})$
$\beta[10]$	$\{0,1,1,1\}$	y_1+y_3	$-\frac{1}{\sqrt{2}}(\Omega_{38}+\Omega_{40}+\Omega_{42}-\Omega_{44}+\Omega_{46}-\Omega_{48}+\Omega_{50}+\Omega_{52})$
$\beta[11]$	$\{1,1,2,0\}$	$-y_1+y_2-y_3+y_4$	$-\frac{1}{2\sqrt{2}}(\Omega_{10}+\Omega_{15}-\Omega_{24}+\Omega_{35})$
$\beta[12]$	$\{1,1,1,1\}$	$-y_2+y_4$	$-\frac{1}{2}(-\Omega_{38}-\Omega_{40}-\Omega_{42}+\Omega_{44}+\Omega_{46}-\Omega_{48}+\Omega_{50}+\Omega_{52})$
$\beta[13]$	$\{0,1,2,1\}$	y_1+y_2	$-\frac{1}{\sqrt{2}}(\Omega_4+\Omega_{27})$
$\beta[14]$	$\{1,2,2,0\}$	$-y_1+y_2+y_3+y_4$	$-\frac{1}{4}(-\Omega_9+\Omega_{14}+\Omega_{19}+\Omega_{34})$
$\beta[15]$	$\{1,1,2,1\}$	$-y_3+y_4$	$-\frac{1}{2}(\Omega_{22}-\Omega_{30})$
$\beta[16]$	$\{0,1,2,2\}$	$2y_1$	$-\frac{1}{2}(\Omega_1-\Omega_6+\Omega_{12}+\Omega_{31})$
$\beta[17]$	$\{1,2,2,1\}$	y_3+y_4	$-\frac{1}{2\sqrt{2}}(\Omega_{17}+\Omega_{29})$
$\beta[18]$	$\{1,1,2,2\}$	$y_1-y_2-y_3+y_4$	$-\frac{1}{2\sqrt{2}}(\Omega_{10}+\Omega_{15}+\Omega_{24}-\Omega_{35})$
$\beta[19]$	$\{1,2,3,1\}$	y_2+y_4	$-\frac{1}{2\sqrt{2}}(\Omega_{38}-\Omega_{40}+\Omega_{42}+\Omega_{44}+\Omega_{46}+\Omega_{48}+\Omega_{50}-\Omega_{52})$
$\beta[20]$	$\{1,2,2,2\}$	$y_1-y_2+y_3+y_4$	$-\frac{1}{4}(-\Omega_9+\Omega_{14}-\Omega_{19}-\Omega_{34})$
$\beta[21]$	$\{1,2,3,2\}$	y_1+y_4	$-\frac{1}{2\sqrt{2}}(-\Omega_{38}+\Omega_{40}+\Omega_{42}+\Omega_{44}+\Omega_{46}+\Omega_{48}-\Omega_{50}+\Omega_{52})$
$\beta[22]$	$\{1,2,4,2\}$	$y_1+y_2-y_3+y_4$	$-\frac{1}{4}(\Omega_3+\Omega_8+\Omega_{23}-\Omega_{33})$
$\beta[23]$	$\{1,3,4,2\}$	$y_1+y_2+y_3+y_4$	$-\frac{1}{4\sqrt{2}}(\Omega_2-\Omega_7+\Omega_{18}+\Omega_{32})$
$\beta[24]$	$\{2,3,4,2\}$	$2y_4$	$-\frac{1}{8}(\Omega_{16}+\Omega_{21}+\Omega_{25}-\Omega_{36})$

and furthermore we specify the following basis:

$$\begin{aligned} \mathscr{H}_1 &= \Omega_5+\Omega_{13}; \quad & \mathscr{H}_2 &= \Omega_5-\Omega_{13} \\ \mathscr{H}_3 &= \Omega_{20}+\Omega_{26}; \quad & \mathscr{H}_4 &= \Omega_{20}-\Omega_{26} \end{aligned} \tag{9.70}$$

With respect to this basis the step operators corresponding to the positive roots of $\mathfrak{f}_{4(4)}$ as ordered and displayed in Table 9.1 are those enumerated in Table 9.4. The step operators corresponding to negative roots are obtained from those associated with positive ones via the following relation:

$$E^{-\beta} = -\mathscr{C}E^{\beta}\mathscr{C} \tag{9.71}$$

where the 26×26 symmetric matrix $\mathscr{C}$ is defined in the following way:

$$\mathscr{C} = \left(\begin{array}{c|c|c} \mathbf{1} & 0 & 0 \\ \hline 0 & \eta & 0 \\ \hline 0 & 0 & C_+ \end{array}\right) \tag{9.72}$$

A further comment is necessary about the normalization of the step operators E^β which are displayed in Table 9.4. They have been fixed with the following criterion. Once we have constructed the algebra, via the generators (9.65), (9.66), we have the Lie structure constants encoded in eq. (9.67) and hence we can diagonalize the adjoint action of the Cartan generators (9.70) finding which linear combinations of the remaining generators correspond to which root. Each root space is one-dimensional and therefore we are left with the task of choosing an absolute normalization for what we want to call the step operators:

$$E^\beta = \lambda_\beta \quad \text{(linear combination of } \Omega\text{s)} \tag{9.73}$$

The values of λ_β are now determined by the following non-trivial conditions:

1. The differences $\mathbb{H}^i = (E^{\beta_i} - E^{-\beta_i})$ should close a subalgebra $\mathbb{H} \subset \mathfrak{f}_{4(4)}$, the maximal compact subalgebra $\mathfrak{su}(2)_R \oplus \mathfrak{usp}(6)$.
2. The sums $\mathbb{K}^i = \frac{1}{\sqrt{2}}(E^{\beta_i} + E^{-\beta_i})$ should span a 28-dimensional representation of $\mathbb{H}$, namely the aforementioned $\mathfrak{su}(2)_R \oplus \mathfrak{usp}(6)$.

We arbitrarily choose the first four λ_β associated with simple roots and then all the others are determined. The result is that displayed in Table 9.4. Using the Cartan generators defined by eqs (9.70) and the step operators enumerated in Table 9.4, one can calculate the structure constants of $\mathfrak{f}_4$ in the Cartan–Weyl basis, namely:

$$\begin{aligned} [\mathscr{H}_i, \mathscr{H}_j] &= 0 \\ [\mathscr{H}_i, E^\beta] &= \beta^i E^\beta \\ [E^\beta, E^{-\beta}] &= \beta \cdot \mathscr{H} \\ [E^{\beta_i}, E^{\beta_j}] &= N_{\beta_i,\beta_j} E^{\beta_i+\beta_j} \end{aligned} \tag{9.74}$$

In particular one obtains the explicit numerical value of the coefficients $\mathscr{N}_{\beta_i,\beta_j}$, which, as is well known, are the only ones not completely specified by the components of the root vectors in the root system. The result of this computation, following from eq. (9.67) is that encoded in eqs (9.49), (9.50), (9.51).

As a last point we can investigate the properties of the maximal compact subalgebra $\mathfrak{su}(2)_R \oplus \mathfrak{usp}(6) \subset \mathfrak{f}_{4(4)}$. As we know, a basis of generators for these subalgebras is provided by:

$$H_i = (E^{\beta_i} - E^{-\beta_i}); \quad (i = 1, \dots, 24) \tag{9.75}$$

but it is not a priori clear which are the generators of $\mathfrak{su}(2)_R$ and which of $\mathfrak{usp}(6)$. This distinction can be established by choosing a basis of Cartan generators of the compact algebra and diagonalizing their adjoint action. The generators of $\mathfrak{su}(2)_R$ are the following linear combinations:

$$\begin{aligned} J_X &= \frac{1}{4\sqrt{2}}(H_1 - H_{14} + H_{20} - H_{22}) \\ J_Y &= \frac{1}{4\sqrt{2}}(H_5 + H_{11} - H_{18} + H_{23}) \\ J_Z &= \frac{1}{4\sqrt{2}}(-H_2 + H_9 - H_{16} - H_{24}) \end{aligned} \tag{9.76}$$

satisfy the standard commutation relations:

$$[J_i, J_j] = \epsilon_{ijk} J_k \tag{9.77}$$

and commute with all the generators of $\mathfrak{usp}(6)$. These latter are defines as follows. The operators

$$\begin{aligned} \mathscr{H}_1^{(\mathrm{Usp}\,6)} &= -\frac{H_2}{2} - \frac{H_9}{2} + \frac{H_{16}}{2} - \frac{H_{24}}{2} \\ \mathscr{H}_2^{(\mathrm{Usp}\,6)} &= -\frac{H_2}{2} + \frac{H_9}{2} + \frac{H_{16}}{2} + \frac{H_{24}}{2} \\ \mathscr{H}_3^{(\mathrm{Usp}\,6)} &= \frac{H_2}{2} + \frac{H_9}{2} + \frac{H_{16}}{2} - \frac{H_{24}}{2} \end{aligned} \tag{9.78}$$

are the Cartan generators. On the other hand, the nine pairs of generators which are rotated one into the other by the Cartans with eigenvalues equal to the roots of the compact algebra are the following ones

$W_1 = H_{10}$	$Z_1 = H_7$
$W_2 = H_4$	$Z_2 = -H_{13}$
$W_3 = H_6$	$Z_3 = -H_3$
$W_4 = -H_1 + H_{14} + H_{20} - H_{22}$	$Z_4 = -H_5 - H_{11} - H_{18} + H_{23}$
$W_5 = H_{21}$	$Z_5 = -H_8$
$W_6 = H_1 + H_{14} + H_{20} + H_{22}$	$Z_6 = H_5 - H_{11} - H_{18} - H_{23}$
$W_7 = -H_1 - H_{14} + H_{20} + H_{22}$	$Z_7 = H_5 - H_{11} + H_{18} + H_{23}$
$W_8 = H_{17}$	$Z_8 = H_{15}$
$W_9 = H_{12}$	$Z_9 = H_{19}$

(9.79)

The construction of the $\mathfrak{f}_4$ Lie algebra presented in this section was published in [61]. All the steps described in the present section are electronically implemented by the MATHEMATICA NoteBook described in Section B.5.

9.3 The exceptional Lie algebra $\mathfrak{e}_7$

We consider the standard $\mathfrak{e}_7$ Dynkin diagram (see Figure 9.5) and we name α_i ($i = 1, \dots, 7$) the corresponding simple roots.

$\mathfrak{e}_7$

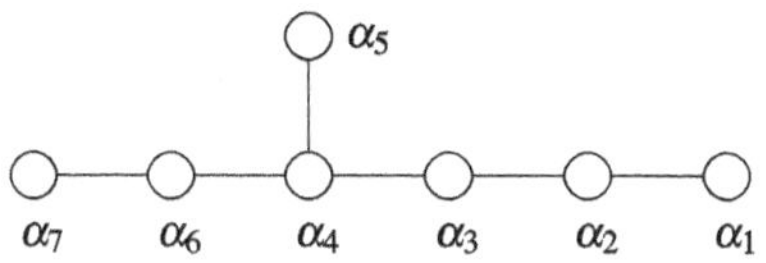

Figure 9.5: $\mathfrak{e}_7$ Dynkin diagram and root labeling.

An explicit representation of the simple roots in a Euclidean 7-dimensional space is the following one:

$$\begin{aligned}
\alpha_1 &= \{1,-1,0,0,0,0,0\}\\
\alpha_2 &= \{0,1,-1,0,0,0,0\}\\
\alpha_3 &= \{0,0,1,-1,0,0,0\}\\
\alpha_4 &= \{0,0,0,1,-1,0,0\}\\
\alpha_5 &= \{0,0,0,0,1,-1,0\}\\
\alpha_6 &= \{0,0,0,0,1,1,0\}\\
\alpha_7 &= \left\{-\frac{1}{2},-\frac{1}{2},-\frac{1}{2},-\frac{1}{2},-\frac{1}{2},-\frac{1}{2},\frac{1}{\sqrt{2}}\right\}
\end{aligned} \tag{9.80}$$

The chosen basis encodes the fundamental idea of the golden splitting, because of the properties that we spell out next. First observe that the first six roots are of the form:

$$\alpha_i = (\boldsymbol{\alpha}_i, 0) \quad (i = 1,\dots,6) \tag{9.81}$$

where the six vectors $\boldsymbol{\alpha}_i$ span the Dynkin diagram of the $\mathfrak{d}_6$ Lie algebra displayed in Figure 9.6. The seventh simple root α_7 is instead of the following form:

$$\begin{aligned}
\alpha_7 &= \left(\boldsymbol{w}_{\text{spin}}, \frac{1}{\sqrt{2}}\right)\\
\boldsymbol{w}_{\text{spin}} &= \left\{-\frac{1}{2},-\frac{1}{2},-\frac{1}{2},-\frac{1}{2},-\frac{1}{2},-\frac{1}{2}\right\}
\end{aligned} \tag{9.82}$$

where $\boldsymbol{w}_{\text{spin}}$ is the highest weight of the spinor representation of $\mathfrak{d}_6$. As we are going to see, utilizing this basis the highest root of the $\mathfrak{e}_7$ root system turns out to be of the form:

$$\psi \equiv \alpha_{63} = (\mathbf{0}, \sqrt{2}) \tag{9.83}$$

$\mathfrak{d}_6$

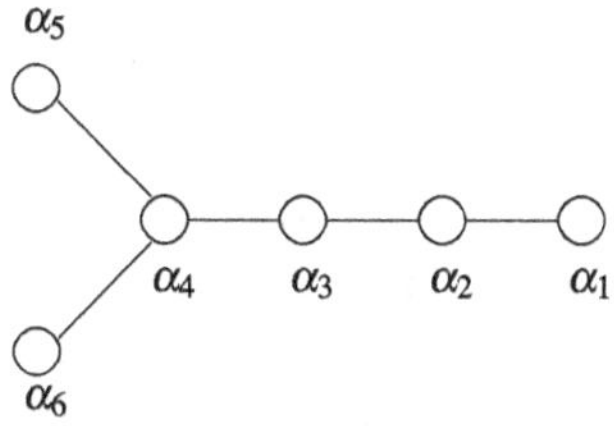

Figure 9.6: $\mathfrak{d}_6$ Dynkin diagram and root labeling.

where $\sqrt{2}$ is interpreted as the only root of an A_1 subalgebra, $\frac{1}{\sqrt{2}}$ being the highest weight of its fundamental doublet representation.

All this points to the following golden splitting of the $\mathfrak{e}_7$ Lie algebra:

$$\underbrace{\mathrm{adj}[\mathfrak{e}_7]}_{133} = \underbrace{(\mathrm{adj}[\mathfrak{so}(12,\mathbb{C})],1)}_{66} \oplus \underbrace{(1,\mathrm{adj}[\mathfrak{sl}(2,\mathbb{C})])}_{3} \oplus \underbrace{(32,2)}_{32\times 2} \tag{9.84}$$

$\mathfrak{sl}(2,\mathbb{C})$ being the complexification of the Ehlers algebra $\mathfrak{sl}(2,\mathbb{R})$, while $\mathfrak{so}(12,\mathbb{C})$ is the complexification of *the Kählerian subalgebra* $\mathscr{SK}$. The spinor representation of this latter, which is 32-dimensional, is the symplectic representation **W** mentioned in eq. (9.9).

Depending on the chosen real section of the full algebra $\mathfrak{e}_7$, the decomposition (9.84) takes different forms, whose complexification however brings always back to (9.84). For instance, we have:

$$\begin{aligned}
\mathrm{adj}[\mathfrak{e}_{7(7)}] &= (\mathrm{adj}[\mathfrak{so}(6,6)],1) \oplus (1,\mathrm{adj}[\mathfrak{sl}(2,\mathbb{R})]) \oplus (32,2)\\
\mathrm{adj}[\mathfrak{e}_{7(-5)}] &= (\mathrm{adj}[\mathfrak{so}^\star(12)],1) \oplus (1,\mathrm{adj}[\mathfrak{sl}(2,\mathbb{R})]) \oplus (32,2)\\
\mathrm{adj}[\mathfrak{e}_{7(-25)}] &= (\mathrm{adj}[\mathfrak{so}(10,2)],1) \oplus (1,\mathrm{adj}[\mathfrak{sl}(2,\mathbb{R})]) \oplus (32,2)\\
\mathrm{adj}[\mathfrak{e}_{7(7)}] &= (\mathrm{adj}[\mathfrak{so}^\star(12)],1) \oplus (1,\mathrm{adj}[\mathfrak{su}(2)]) \oplus (32,2)\\
\mathrm{adj}[\mathfrak{e}_{7(-5)}] &= (\mathrm{adj}[\mathfrak{so}(8,4)],1) \oplus (1,\mathrm{adj}[\mathfrak{su}(2)]) \oplus (32,2)\\
\mathrm{adj}[\mathfrak{e}_{7(-5)}] &= (\mathrm{adj}[\mathfrak{so}(12)],1) \oplus (1,\mathrm{adj}[\mathfrak{su}(2)]) \oplus (32,2)\\
\mathrm{adj}[\mathfrak{e}_{7(-25)}] &= (\mathrm{adj}[\mathfrak{so}^\star(12)],1) \oplus (1,\mathrm{adj}[\mathfrak{su}(2)]) \oplus (32,2)\\
\mathrm{adj}[\mathfrak{e}_{7(-133)}] &= (\mathrm{adj}[\mathfrak{so}(12)],1) \oplus (1,\mathrm{adj}[\mathfrak{su}(2)]) \oplus (32,2)
\end{aligned} \tag{9.85}$$

Having fixed this basis, each $\mathfrak{e}_7$ root is intrinsically identified by its Dynkin labels, namely, by its integer valued components in the simple root basis. We can construct all of them by induction on their height starting from the simple roots. Given the root of a given height we just verify which simple root can be added to each of them, the criterion being that the scalar product of the considered root with the considered simple one should be strictly negative. By means of this simple algorithm the 63 positive are easily found and they are displayed in Table 9.5.

Inspection of this table easily reveals the inner working of the decomposition (9.84). There are 30 roots whose 7th component is zero. These are the positive roots of the $\mathfrak{d}_6$ subalgebra. There are 32 roots whose last component is $\frac{1}{\sqrt{2}}$. These are the weights of the spinor representation of $\mathfrak{d}_6$. Finally there is only one root whose first six components are zero and the last is $\sqrt{2}$. This is the unique positive root of the $\mathfrak{sl}(2,\mathbb{C})$ subalgebra.

Having identified the roots, the next step we need is the construction of the real fundamental representation of the maximally non-compact real section $\mathfrak{e}_{7(7)}$. For this we need the corresponding weight vectors W.

As we have repeatedly stressed, a particularly relevant property of the maximally non-compact real sections of a simple complex Lie algebra is that all its irreducible

Table 9.5: Positive roots of $\mathfrak{e}_7$.

Name	Dynkin lab.	Eucl. comp.
α_1	$\{1,0,0,0,0,0,0\}$	$\{1,-1,0,0,0,0,0\}$
α_2	$\{0,1,0,0,0,0,0\}$	$\{0,1,-1,0,0,0,0\}$
α_3	$\{0,0,1,0,0,0,0\}$	$\{0,0,1,-1,0,0,0\}$
α_4	$\{0,0,0,1,0,0,0\}$	$\{0,0,0,1,-1,0,0\}$
α_5	$\{0,0,0,0,1,0,0\}$	$\{0,0,0,0,1,-1,0\}$
α_6	$\{0,0,0,0,0,1,0\}$	$\{0,0,0,0,1,1,0\}$
α_7	$\{0,0,0,0,0,0,1\}$	$\{-\frac{1}{2},-\frac{1}{2},-\frac{1}{2},-\frac{1}{2},-\frac{1}{2},-\frac{1}{2},\frac{1}{\sqrt{2}}\}$
α_8	$\{1,1,0,0,0,0,0\}$	$\{1,0,-1,0,0,0,0\}$
α_9	$\{0,1,1,0,0,0,0\}$	$\{0,1,0,-1,0,0,0\}$
α_{10}	$\{0,0,1,1,0,0,0\}$	$\{0,0,1,0,-1,0,0\}$
α_{11}	$\{0,0,0,1,1,0,0\}$	$\{0,0,0,1,0,-1,0\}$
α_{12}	$\{0,0,0,1,0,1,0\}$	$\{0,0,0,1,0,1,0\}$
α_{13}	$\{0,0,0,0,0,1,1\}$	$\{-\frac{1}{2},-\frac{1}{2},-\frac{1}{2},-\frac{1}{2},\frac{1}{2},\frac{1}{2},\frac{1}{\sqrt{2}}\}$
α_{14}	$\{1,1,1,0,0,0,0\}$	$\{1,0,0,-1,0,0,0\}$
α_{15}	$\{0,1,1,1,0,0,0\}$	$\{0,1,0,0,-1,0,0\}$
α_{16}	$\{0,0,1,1,1,0,0\}$	$\{0,0,1,0,0,-1,0\}$
α_{17}	$\{0,0,1,1,0,1,0\}$	$\{0,0,1,0,0,1,0\}$
α_{18}	$\{0,0,0,1,0,1,1\}$	$\{-\frac{1}{2},-\frac{1}{2},-\frac{1}{2},\frac{1}{2},-\frac{1}{2},\frac{1}{2},\frac{1}{\sqrt{2}}\}$
α_{19}	$\{0,0,0,1,1,1,0\}$	$\{0,0,0,1,1,0,0\}$
α_{20}	$\{1,1,1,1,0,0,0\}$	$\{1,0,0,0,-1,0,0\}$
α_{21}	$\{0,1,1,1,1,0,0\}$	$\{0,1,0,0,0,-1,0\}$
α_{22}	$\{0,1,1,1,0,1,0\}$	$\{0,1,0,0,0,1,0\}$
α_{23}	$\{0,0,1,1,0,1,1\}$	$\{-\frac{1}{2},-\frac{1}{2},\frac{1}{2},-\frac{1}{2},-\frac{1}{2},\frac{1}{2},\frac{1}{\sqrt{2}}\}$
α_{24}	$\{0,0,1,1,1,1,0\}$	$\{0,0,1,0,1,0,0\}$
α_{25}	$\{0,0,0,1,1,1,1\}$	$\{-\frac{1}{2},-\frac{1}{2},-\frac{1}{2},\frac{1}{2},\frac{1}{2},-\frac{1}{2},\frac{1}{\sqrt{2}}\}$
α_{26}	$\{1,1,1,1,1,0,0\}$	$\{1,0,0,0,0,-1,0\}$
α_{27}	$\{1,1,1,1,0,1,0\}$	$\{1,0,0,0,0,1,0\}$
α_{28}	$\{0,1,1,1,0,1,1\}$	$\{-\frac{1}{2},\frac{1}{2},-\frac{1}{2},-\frac{1}{2},-\frac{1}{2},\frac{1}{2},\frac{1}{\sqrt{2}}\}$
α_{29}	$\{0,1,1,1,1,1,0\}$	$\{0,1,0,0,1,0,0\}$
α_{30}	$\{0,0,1,1,1,1,1\}$	$\{-\frac{1}{2},-\frac{1}{2},\frac{1}{2},-\frac{1}{2},\frac{1}{2},-\frac{1}{2},\frac{1}{\sqrt{2}}\}$
α_{31}	$\{0,0,1,2,1,1,0\}$	$\{0,0,1,1,0,0,0\}$
α_{32}	$\{1,1,1,1,0,1,1\}$	$\{\frac{1}{2},-\frac{1}{2},-\frac{1}{2},-\frac{1}{2},-\frac{1}{2},\frac{1}{2},\frac{1}{\sqrt{2}}\}$
α_{33}	$\{1,1,1,1,1,1,0\}$	$\{1,0,0,0,1,0,0\}$
α_{34}	$\{0,1,1,1,1,1,1\}$	$\{-\frac{1}{2},\frac{1}{2},-\frac{1}{2},-\frac{1}{2},\frac{1}{2},-\frac{1}{2},\frac{1}{\sqrt{2}}\}$
α_{35}	$\{0,1,1,2,1,1,0\}$	$\{0,1,0,1,0,0,0\}$
α_{36}	$\{0,0,1,2,1,1,1\}$	$\{-\frac{1}{2},-\frac{1}{2},\frac{1}{2},\frac{1}{2},-\frac{1}{2},-\frac{1}{2},\frac{1}{\sqrt{2}}\}$
α_{37}	$\{1,1,1,1,1,1,1\}$	$\{\frac{1}{2},-\frac{1}{2},-\frac{1}{2},-\frac{1}{2},\frac{1}{2},-\frac{1}{2},\frac{1}{\sqrt{2}}\}$
α_{38}	$\{1,1,1,2,1,1,0\}$	$\{1,0,0,1,0,0,0\}$
α_{39}	$\{0,1,1,2,1,1,1\}$	$\{-\frac{1}{2},\frac{1}{2},-\frac{1}{2},\frac{1}{2},-\frac{1}{2},-\frac{1}{2},\frac{1}{\sqrt{2}}\}$
α_{40}	$\{0,1,2,2,1,1,0\}$	$\{0,1,1,0,0,0,0\}$
α_{41}	$\{0,0,1,2,1,2,1\}$	$\{-\frac{1}{2},-\frac{1}{2},\frac{1}{2},\frac{1}{2},\frac{1}{2},\frac{1}{2},\frac{1}{\sqrt{2}}\}$
α_{42}	$\{1,1,1,2,1,1,1\}$	$\{\frac{1}{2},-\frac{1}{2},-\frac{1}{2},\frac{1}{2},-\frac{1}{2},-\frac{1}{2},\frac{1}{\sqrt{2}}\}$
α_{43}	$\{1,1,2,2,1,1,0\}$	$\{1,0,1,0,0,0,0\}$
α_{44}	$\{0,1,1,2,1,2,1\}$	$\{-\frac{1}{2},\frac{1}{2},-\frac{1}{2},\frac{1}{2},\frac{1}{2},\frac{1}{2},\frac{1}{\sqrt{2}}\}$
α_{45}	$\{0,1,2,2,1,1,1\}$	$\{-\frac{1}{2},\frac{1}{2},\frac{1}{2},-\frac{1}{2},-\frac{1}{2},-\frac{1}{2},\frac{1}{\sqrt{2}}\}$
α_{46}	$\{1,1,1,2,1,2,1\}$	$\{\frac{1}{2},-\frac{1}{2},-\frac{1}{2},\frac{1}{2},\frac{1}{2},\frac{1}{2},\frac{1}{\sqrt{2}}\}$
α_{47}	$\{1,1,2,2,1,1,1\}$	$\{\frac{1}{2},-\frac{1}{2},\frac{1}{2},-\frac{1}{2},-\frac{1}{2},-\frac{1}{2},\frac{1}{\sqrt{2}}\}$
α_{48}	$\{1,2,2,2,1,1,0\}$	$\{1,1,0,0,0,0,0\}$
α_{49}	$\{0,1,2,2,1,2,1\}$	$\{-\frac{1}{2},\frac{1}{2},\frac{1}{2},-\frac{1}{2},\frac{1}{2},\frac{1}{2},\frac{1}{\sqrt{2}}\}$
α_{50}	$\{1,1,2,2,1,2,1\}$	$\{\frac{1}{2},-\frac{1}{2},\frac{1}{2},-\frac{1}{2},\frac{1}{2},\frac{1}{2},\frac{1}{\sqrt{2}}\}$
α_{51}	$\{1,2,2,2,1,1,1\}$	$\{\frac{1}{2},\frac{1}{2},-\frac{1}{2},-\frac{1}{2},-\frac{1}{2},-\frac{1}{2},\frac{1}{\sqrt{2}}\}$
α_{52}	$\{0,1,2,3,1,2,1\}$	$\{-\frac{1}{2},\frac{1}{2},\frac{1}{2},\frac{1}{2},-\frac{1}{2},\frac{1}{2},\frac{1}{\sqrt{2}}\}$
α_{53}	$\{1,1,2,3,1,2,1\}$	$\{\frac{1}{2},-\frac{1}{2},\frac{1}{2},\frac{1}{2},-\frac{1}{2},\frac{1}{2},\frac{1}{\sqrt{2}}\}$
α_{54}	$\{1,2,2,2,1,2,1\}$	$\{\frac{1}{2},\frac{1}{2},-\frac{1}{2},-\frac{1}{2},\frac{1}{2},\frac{1}{2},\frac{1}{\sqrt{2}}\}$
α_{55}	$\{0,1,2,3,2,2,1\}$	$\{-\frac{1}{2},\frac{1}{2},\frac{1}{2},\frac{1}{2},\frac{1}{2},-\frac{1}{2},\frac{1}{\sqrt{2}}\}$
α_{56}	$\{1,1,2,3,2,2,1\}$	$\{\frac{1}{2},-\frac{1}{2},\frac{1}{2},\frac{1}{2},\frac{1}{2},-\frac{1}{2},\frac{1}{\sqrt{2}}\}$
α_{57}	$\{1,2,2,3,1,2,1\}$	$\{\frac{1}{2},\frac{1}{2},-\frac{1}{2},\frac{1}{2},-\frac{1}{2},\frac{1}{2},\frac{1}{\sqrt{2}}\}$
α_{58}	$\{1,2,2,3,2,2,1\}$	$\{\frac{1}{2},\frac{1}{2},-\frac{1}{2},\frac{1}{2},\frac{1}{2},-\frac{1}{2},\frac{1}{\sqrt{2}}\}$
α_{59}	$\{1,2,3,3,1,2,1\}$	$\{\frac{1}{2},\frac{1}{2},\frac{1}{2},-\frac{1}{2},-\frac{1}{2},\frac{1}{2},\frac{1}{\sqrt{2}}\}$
α_{60}	$\{1,2,3,3,2,2,1\}$	$\{\frac{1}{2},\frac{1}{2},\frac{1}{2},-\frac{1}{2},\frac{1}{2},-\frac{1}{2},\frac{1}{\sqrt{2}}\}$
α_{61}	$\{1,2,3,4,2,2,1\}$	$\{\frac{1}{2},\frac{1}{2},\frac{1}{2},\frac{1}{2},-\frac{1}{2},-\frac{1}{2},\frac{1}{\sqrt{2}}\}$
α_{62}	$\{1,2,3,4,2,3,1\}$	$\{\frac{1}{2},\frac{1}{2},\frac{1}{2},\frac{1}{2},\frac{1}{2},\frac{1}{2},\frac{1}{\sqrt{2}}\}$
α_{63}	$\{1,2,3,4,2,3,2\}$	$\{0,0,0,0,0,0,\sqrt{2}\}$

representations are real. With the name $\mathfrak{e}_{7(7)}$ we just denote such maximally non-compact real section in the case of the complex Lie algebra $\mathfrak{e}_7$, hence all its irreducible representations Γ are real. This implies that if an element of the weight lattice $W \in \Lambda_w$ is a weight of a given irreducible representation $W \in \Gamma$ then also its negative is a weight of the same representation: $-W \in \Gamma$. Indeed, changing sign to the weights corresponds to complex conjugation.

According to standard Lie algebra lore, every irreducible representation of a simple Lie algebra $\mathbb{G}$ is identified by a unique *highest weight* W_{max} (see Section 8.1.1). Furthermore, all weights can be expressed as integral non-negative linear combinations of the *simple* weights $W_\ell(\ell = 1, \dots, r = \mathrm{rank}(\mathbb{G}))$, whose components are named the Dynkin labels of the weight. The simple weights W_i of $\mathbb{G}$ are the generators of the dual lattice to the root lattice and are defined by the condition:

$$\frac{2(W_i, \alpha_j)}{(\alpha_j, \alpha_j)} = \delta_{ij} \tag{9.86}$$

In the simply-laced $\mathfrak{e}_{7(7)}$ case, the previous equation simplifies as follows

$$(W_i, \alpha_j) = \delta_{ij} \tag{9.87}$$

where α_j are the simple roots. Using the Dynkin diagram of $\mathfrak{e}_{7(7)}$ (see Figure 9.5) from eq. (9.87) we can easily obtain the explicit expression of the simple weights that are listed below:

$$\begin{aligned}
W_1 &= \left\{1, 0, 0, 0, 0, 0, \frac{1}{\sqrt{2}}\right\} \\
W_2 &= \{1, 1, 0, 0, 0, 0, \sqrt{2}\} \\
W_3 &= \left\{1, 1, 1, 0, 0, 0, \frac{3}{\sqrt{2}}\right\} \\
W_4 &= \{1, 1, 1, 1, 0, 0, 2\sqrt{2}\} \\
W_5 &= \left\{\frac{1}{2}, \frac{1}{2}, \frac{1}{2}, \frac{1}{2}, \frac{1}{2}, -\frac{1}{2}, \sqrt{2}\right\} \\
W_6 &= \left\{\frac{1}{2}, \frac{1}{2}, \frac{1}{2}, \frac{1}{2}, \frac{1}{2}, \frac{1}{2}, \frac{3}{\sqrt{2}}\right\} \\
W_7 &= \{0, 0, 0, 0, 0, 0, \sqrt{2}\}
\end{aligned} \tag{9.88}$$

The Dynkin labels of the highest weight of an irreducible representation Γ give the Dynkin labels of the representation. Therefore the representation is usually denoted by $\Gamma[n_1, \dots, n_r]$. All the weights W belonging to the representation Γ can be described by r integer non-negative numbers q^ℓ defined by the following equation:

$$W_{max} - W = \sum_{\ell=1}^{r} q^\ell \alpha_\ell \tag{9.89}$$

where α_ℓ are the simple roots. According to this standard formalism the fundamental real representation of $\mathfrak{e}_{7(7)}$ is $\Gamma[1, 0, 0, 0, 0, 0, 0]$. Choosing this as highest weight $W_{max} = W_1$ and applying the inductive algorithm that follows from the two theorems, 8.1.1 and 8.1.2, we can construct the set of all the weights of the representation.

Table 9.6: Weights of the **56** representation of $\mathfrak{e}_{7(7)}$.

Weight name		q^ℓ vector	Weight name		q^ℓ vector
$W^{(1)}$	=	{2,3,4,5,3,3,1}	$W^{(2)}$	=	{2,2,2,2,1,1,1}
$W^{(3)}$	=	{1,2,2,2,1,1,1}	$W^{(4)}$	=	{1,1,2,2,1,1,1}
$W^{(5)}$	=	{1,1,1,2,1,1,1}	$W^{(6)}$	=	{1,1,1,1,1,1,1}
$W^{(7)}$	=	{2,3,3,3,1,2,1}	$W^{(8)}$	=	{2,2,3,3,1,2,1}
$W^{(9)}$	=	{2,2,2,3,1,2,1}	$W^{(10)}$	=	{2,2,2,2,1,2,1}
$W^{(11)}$	=	{1,2,2,2,1,2,1}	$W^{(12)}$	=	{1,1,2,2,1,2,1}
$W^{(13)}$	=	{1,1,1,2,1,2,1}	$W^{(14)}$	=	{1,2,2,3,1,2,1}
$W^{(15)}$	=	{1,2,3,3,1,2,1}	$W^{(16)}$	=	{1,1,2,3,1,2,1}
$W^{(17)}$	=	{2,2,2,2,1,1,0}	$W^{(18)}$	=	{1,2,2,2,1,1,0}
$W^{(19)}$	=	{1,1,2,2,1,1,0}	$W^{(20)}$	=	{1,1,1,2,1,1,0}
$W^{(21)}$	=	{1,1,1,1,1,1,0}	$W^{(22)}$	=	{1,1,1,1,1,0,0}
$W^{(23)}$	=	{3,4,5,6,3,4,2}	$W^{(24)}$	=	{2,4,5,6,3,4,2}
$W^{(25)}$	=	{2,3,5,6,3,4,2}	$W^{(26)}$	=	{2,3,4,6,3,4,2}
$W^{(27)}$	=	{2,3,4,5,3,4,2}	$W^{(28)}$	=	{2,3,4,5,3,3,2}
$W^{(29)}$	=	{1,1,1,1,0,1,1}	$W^{(30)}$	=	{1,2,3,4,2,3,1}
$W^{(31)}$	=	{2,2,3,4,2,3,1}	$W^{(32)}$	=	{2,3,3,4,2,3,1}
$W^{(33)}$	=	{2,3,4,4,2,3,1}	$W^{(34)}$	=	{2,3,4,5,2,3,1}
$W^{(35)}$	=	{1,1,2,3,2,2,1}	$W^{(36)}$	=	{1,2,2,3,2,2,1}
$W^{(37)}$	=	{1,2,3,3,2,2,1}	$W^{(38)}$	=	{1,2,3,4,2,2,1}
$W^{(39)}$	=	{2,2,3,4,2,2,1}	$W^{(40)}$	=	{2,3,3,4,2,2,1}
$W^{(41)}$	=	{2,3,4,4,2,2,1}	$W^{(42)}$	=	{2,2,3,3,2,2,1}
$W^{(43)}$	=	{2,2,2,3,2,2,1}	$W^{(44)}$	=	{2,3,3,3,2,2,1}
$W^{(45)}$	=	{1,2,3,4,2,3,2}	$W^{(46)}$	=	{2,2,3,4,2,3,2}
$W^{(47)}$	=	{2,3,3,4,2,3,2}	$W^{(48)}$	=	{2,3,4,4,2,3,2}
$W^{(49)}$	=	{2,3,4,5,2,3,2}	$W^{(50)}$	=	{2,3,4,5,2,4,2}
$W^{(51)}$	=	{0,0,0,0,0,0,0}	$W^{(52)}$	=	{1,0,0,0,0,0,0}
$W^{(53)}$	=	{1,1,0,0,0,0,0}	$W^{(54)}$	=	{1,1,1,0,0,0,0}
$W^{(55)}$	=	{1,1,1,1,0,0,0}	$W^{(56)}$	=	{1,1,1,1,0,1,0}

They are displayed in Table 9.6 where they are identified by the corresponding q^ℓ numbers.

As one sees, the highest weight corresponds in such a list to $W^{(51)}$ and the expression of its weights in terms of q^ℓ is given in Table 9.6, the highest weight being $W^{(51)}$.

We can now explain the specific ordering of the weights we have adopted.

First of all we have separated the 56 weights in two groups of 28 elements so that the first group:

$$\Lambda^{(n)} = W^{(n)} \quad n = 1, \dots, 28 \tag{9.90}$$

are the weights for the irreducible **28** dimensional representation of the subgroup $\mathrm{SL}(8,\mathbb{R}) \subset \mathfrak{e}_{7(7)}$. The remaining group of 28 weight vectors are the weights for the transposed representation of the same group that we name $\overline{\mathbf{28}}$.

Secondly, the 28 weights Λ have been arranged according to the decomposition with respect to the subalgebra $SO(6,6) \subset \mathfrak{e}_{7(7)}$. Such a decomposition is group-theoretical yet it is inspired by superstring theory. In maximal supergravity ($\mathcal{N} = 8$) the fundamental representation of $\mathfrak{e}_{7(7)}$ accommodates all the 28 vector field strengths and their magnetic duals. The decomposition with respect to the subalgebra $SO(6,6)$ traces back the origin of these fields from superstring theory compactified on a T^6 torus. The first 16 of these weights correspond to Ramond–Ramond vectors and are the weights of the spinor representation of $SO(6,6)$ while the last 12 are associated with Neveu–Schwarz fields and correspond to the weights of the vector representation of $SO(6,6)$.

9.3.1 The matrices of the fundamental 56 representation

Equipped with the weight vectors we can now proceed to the explicit construction of the $\mathbf{D_{56}}$ representation of $\mathfrak{e}_{7(7)}$. In this construction, the basis vectors are the 56 weights, according to the enumeration of Table 9.6. What we need are the 56×56 matrices associated with the seven Cartan generators H_{α_i} $(i = 1, \dots, 7)$ and with the 126 step operators E^α that are defined by:

$$\begin{aligned} [\mathbf{D_{56}}(H_{\alpha_i})]_{nm} &\equiv \langle W^{(n)} | H_{\alpha_i} | W^{(m)} \rangle \\ [\mathbf{D_{56}}(E^\alpha)]_{nm} &\equiv \langle W^{(n)} | E^\alpha | W^{(m)} \rangle \end{aligned} \tag{9.91}$$

Following [28] let us begin with the Cartan generators. As a basis of the Cartan subalgebra we use the generators H_{α_i} defined by the commutators:

$$[E^{\alpha_i}, E^{-\alpha_i}] \equiv H_{\alpha_i} \tag{9.92}$$

In the $\mathbf{D_{56}}$ representation the corresponding matrices are diagonal and of the form:

$$\langle W^{(p)} | H_{\alpha_i} | W^{(q)} \rangle = (W^{(p)}, \alpha_i)\delta_{pq}; \quad (p, q = 1, \dots, 56) \tag{9.93}$$

The scalar products

$$(\Lambda^{(n)} \cdot h, -\Lambda^{(m)} \cdot h) = (W^{(p)} \cdot h); \quad (n, m = 1, \dots, 28;\ p = 1, \dots, 56) \tag{9.94}$$

are to be understood in the following way:

$$W^{(p)} \cdot h = \sum_{i=1}^{7} (W^{(p)}, \alpha_i) h^i \tag{9.95}$$

Next we construct the matrices associated with the step operators. Here the first observation is that it suffices to consider the positive roots. Because of the reality of the representation, the matrix associated with the negative of a root is just the transposed of that associated with the root itself:

$$E^{-\alpha} = [E^{\alpha}]^T \leftrightarrow \langle W^{(n)}|E^{-\alpha}|W^{(m)}\rangle = \langle W^{(m)}|E^{\alpha}|W^{(n)}\rangle \tag{9.96}$$

The method followed in [28] to obtain the matrices for all the positive roots is that of constructing first the 56×56 matrices for the step operators E^{α_ℓ} ($\ell = 1, \dots, 7$) associated with the simple roots and then generating all the others through their commutators. The construction rules for the $\mathbf{D_{56}}$ representation of the six operators E^{α_ℓ} ($\ell \neq 5$) are:

$$\ell \neq 5 \quad \begin{cases} \langle W^{(n)}|E^{\alpha_\ell}|W^{(m)}\rangle = \delta_{W^{(n)},W^{(m)}+\alpha_\ell}; & n,m = 1,\dots,28 \\ \langle W^{(n+28)}|E^{\alpha_\ell}|W^{(m+28)}\rangle = -\delta_{W^{(n+28)},W^{(m+28)}+\alpha_\ell}; & n,m = 1,\dots,28 \end{cases} \tag{9.97}$$

The six simple roots α_ℓ with $\ell \neq 5$ belong also to the Dynkin diagram of the subgroup $\mathrm{SL}(8,\mathbb{R})$. Thus their shift operators have a block diagonal action on the **28** and $\overline{\mathbf{28}}$ subspaces of the $\mathbf{D_{56}}$ representation that are irreducible under the subgroup $\mathrm{SL}(8,\mathbb{R})$. From eq. (9.97) we conclude that:

$$\ell \neq 5 \quad \mathbf{D_{56}}(E^{\alpha_\ell}) = \begin{pmatrix} A[\alpha_\ell] & \mathbf{0} \\ \mathbf{0} & -A^T[\alpha_\ell] \end{pmatrix} \tag{9.98}$$

the 28×28 block $A[\alpha_\ell]$ being defined by the first line of eq. (9.97).

On the contrary, the operator E^{α_5}, corresponding to the only root of the $\mathfrak{e}_7$ Dynkin diagram that is not also part of the $\mathfrak{a}_7$ diagram, is represented by a matrix whose non-vanishing 28×28 blocks are off-diagonal. We have

$$\mathbf{D_{56}}(E^{\alpha_5}) = \begin{pmatrix} \mathbf{0} & B[\alpha_5] \\ C[\alpha_5] & \mathbf{0} \end{pmatrix} \tag{9.99}$$

where both $B[\alpha_5] = B^T[\alpha_5]$ and $C[\alpha_5] = C^T[\alpha_5]$ are symmetric 28×28 matrices. More explicitly, the matrix $\mathbf{D_{56}}(E^{\alpha_5})$ is given by:

$$\begin{aligned} \langle W^{(n)}|E^{\alpha_5}|W^{(m+28)}\rangle &= \langle W^{(m)}|E^{\alpha_5}|W^{(n+28)}\rangle \\ \langle W^{(n+28)}|E^{\alpha_5}|W^{(m)}\rangle &= \langle W^{(m+28)}|E^{\alpha_5}|W^{(n)}\rangle \end{aligned} \tag{9.100}$$

with

$$\begin{aligned} \langle W^{(7)}|E^{\alpha_5}|W^{(44)}\rangle &= -1 & \langle W^{(8)}|E^{\alpha_5}|W^{(42)}\rangle &= 1 \\ \langle W^{(9)}|E^{\alpha_5}|W^{(43)}\rangle &= -1 & \langle W^{(14)}|E^{\alpha_5}|W^{(36)}\rangle &= 1 \\ \langle W^{(15)}|E^{\alpha_5}|W^{(37)}\rangle &= -1 & \langle W^{(16)}|E^{\alpha_5}|W^{(35)}\rangle &= -1 \\ \langle W^{(29)}|E^{\alpha_5}|W^{(6)}\rangle &= -1 & \langle W^{(34)}|E^{\alpha_5}|W^{(1)}\rangle &= -1 \\ \langle W^{(49)}|E^{\alpha_5}|W^{(28)}\rangle &= 1 & \langle W^{(50)}|E^{\alpha_5}|W^{(27)}\rangle &= -1 \\ \langle W^{(55)}|E^{\alpha_5}|W^{(22)}\rangle &= -1 & \langle W^{(56)}|E^{\alpha_5}|W^{(21)}\rangle &= 1 \end{aligned} \tag{9.101}$$

In this way we have completed the construction of the E^{α_ℓ} operators associated with simple roots. For the matrices associated with higher roots we just proceed iteratively in the following way. As usual, we organize the roots by height:

$$\alpha = n^\ell \alpha_\ell \quad \rightarrow \quad \mathrm{ht}\,\alpha = \sum_{\ell=1}^{7} n^\ell \tag{9.102}$$

and for the roots $\alpha_i + \alpha_j$ of height ht = 2 we set:

$$\mathbf{D_{56}}(E^{\alpha_i+\alpha_j}) \equiv [\mathbf{D_{56}}(E^{\alpha_i}), \mathbf{D_{56}}(E^{\alpha_i})]; \quad i < j \tag{9.103}$$

Next for the roots of ht = 3 that can be written as $\alpha_i + \beta$ where α_i is simple and $\mathrm{ht}\,\beta = 2$ we write:

$$\mathbf{D_{56}}(E^{\alpha_i+\beta}) \equiv [\mathbf{D_{56}}(E^{\alpha_i}), \mathbf{D_{56}}(E^{\beta})] \tag{9.104}$$

Having obtained the matrices for the roots of ht = 3 one proceeds in a similar way for those of the next height and so on up to exhaustion of all the 63 positive roots.

This concludes our description of the algorithm by means of which a computer program can easily construct all the 133 matrices spanning the $\mathfrak{e}_{7(7)}$ Lie algebra in the $\mathbf{D_{56}}$ representation (see Section B.4 for the description of the MATHEMATICA Notebook that does the job). A fortiori, if we specify the embedding, we have the matrices generating the subgroup $\mathrm{SL}(8,\mathbb{R})$.

9.3.2 The $\mathfrak{sl}(8,\mathbb{R})$ subalgebra

The $\mathfrak{sl}(8,\mathbb{R})$ subalgebra is identified in $\mathfrak{e}_{7(7)}$ by specifying its simple roots β_i spanning the standard $\mathfrak{a}_7$ Dynkin diagram. The Cartan generators are the same for the $\mathfrak{e}_{7(7)}$ Lie algebra as for the $\mathfrak{sl}(8,\mathbb{R})$ subalgebra and if we give β_i, every other generator is defined. The basis we have chosen is the following one:

$$\begin{array}{ll} \beta_1 = \alpha_2 + 2\alpha_3 + 3\alpha_4 + 2\alpha_5 + 2\alpha_6 + \alpha_7; & \beta_2 = \alpha_1 \\ \beta_3 = \alpha_2; & \beta_4 = \alpha_3 \\ \beta_5 = \alpha_4; & \beta_6 = \alpha_6 \\ \beta_7 = \alpha_7 & \end{array} \tag{9.105}$$

The complete set of positive roots of $\mathfrak{a}_7$ is then composed of 28 elements that we name ρ_i $(i = 1, \ldots, 28)$ and that are enumerated according to our chosen order in Table 9.7.

Hence the 63 generators of the $\mathrm{SL}(8,\mathbb{R})$ Lie algebra are:

$$\begin{array}{ccc} \text{The 7 Cartan generators} & C_i = H_{\alpha_i} & i = 1, \ldots, 7 \\ \text{The 28 positive root generators} & E^{\rho_i} & i = 1, \ldots, 28 \\ \text{The 28 negative root generators} & E^{-\rho_i} & i = 1, \ldots, 28 \end{array} \tag{9.106}$$

Table 9.7: The choice of the order of the $\mathfrak{sl}(8,\mathbb{R})$ roots.

ρ_1	$\equiv$	β_2
ρ_2	$\equiv$	$\beta_2+\beta_3$
ρ_3	$\equiv$	$\beta_2+\beta_3+\beta_4$
ρ_4	$\equiv$	$\beta_2+\beta_3+\beta_4+\beta_5$
ρ_5	$\equiv$	$\beta_2+\beta_3+\beta_4+\beta_5+\beta_6$
ρ_6	$\equiv$	β_3
ρ_7	$\equiv$	$\beta_3+\beta_4$
ρ_8	$\equiv$	$\beta_3+\beta_4+\beta_5$
ρ_9	$\equiv$	$\beta_3+\beta_4+\beta_5+\beta_6$
ρ_{10}	$\equiv$	β_4
ρ_{11}	$\equiv$	$\beta_4+\beta_5$
ρ_{12}	$\equiv$	$\beta_4+\beta_5+\beta_6$
ρ_{13}	$\equiv$	β_5
ρ_{14}	$\equiv$	$\beta_5+\beta_6$
ρ_{15}	$\equiv$	β_6
ρ_{16}	$\equiv$	$\beta_1+\beta_2+\beta_3+\beta_4+\beta_5+\beta_6+\beta_7$
ρ_{17}	$\equiv$	$\beta_2+\beta_3+\beta_4+\beta_5+\beta_6+\beta_7$
ρ_{18}	$\equiv$	$\beta_3+\beta_4+\beta_5+\beta_6+\beta_7$
ρ_{19}	$\equiv$	$\beta_4+\beta_5+\beta_6+\beta_7$
ρ_{20}	$\equiv$	$\beta_5+\beta_6+\beta_7$
ρ_{21}	$\equiv$	$\beta_6+\beta_7$
ρ_{22}	$\equiv$	β_1
ρ_{23}	$\equiv$	$\beta_1+\beta_2$
ρ_{24}	$\equiv$	$\beta_1+\beta_2+\beta_3$
ρ_{25}	$\equiv$	$\beta_1+\beta_2+\beta_3+\beta_4$
ρ_{26}	$\equiv$	$\beta_1+\beta_2+\beta_3+\beta_4+\beta_5$
ρ_{27}	$\equiv$	$\beta_1+\beta_2+\beta_3+\beta_4+\beta_5+\beta_6$
ρ_{28}	$\equiv$	β_7

and since the 56×56 matrix representation of each $\mathfrak{e}_{7(7)}$ Cartan generator or step operator was constructed in the previous subsection, it is obvious that it is in particular given for the subset of those that belong to the $\mathfrak{sl}(8,\mathbb{R})$ subalgebra. The basis of this matrix representation is provided by the weights enumerated in Table 9.6.

In this way we have concluded our illustration of the construction of the fundamental representation of $\mathfrak{e}_{7(7)}$. For any other real section the corresponding representation can be obtained from the present real one by means of suitable linear combination of the generators with appropriate complex coefficients.

9.4 The exceptional Lie algebra $\mathfrak{e}_8$

The $\mathfrak{e}_8$ algebra is the largest exceptional Lie algebra with dimension 248 and rank 8. The root lattice, spanned by the integer linear combinations $n^i\alpha_i$, has the notable prop-

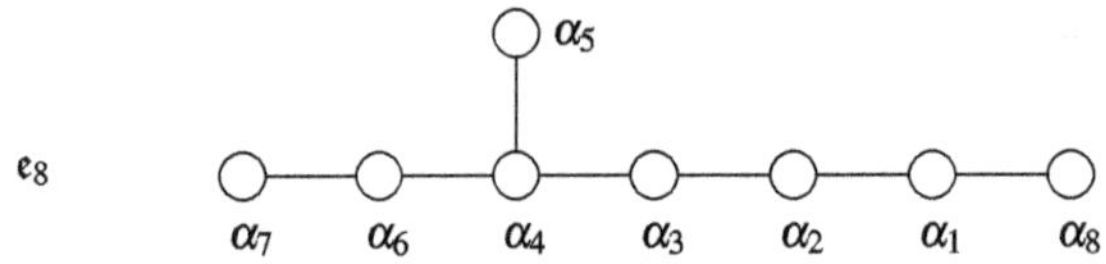

Figure 9.7: The Dynkin diagram of the $\mathfrak{e}_8$ Lie algebra.

erty of being self-dual, namely, the root lattice coincides with the root lattice:

$$\Lambda_{\text{weight}}[\mathfrak{e}_8] = \Lambda_{\text{root}}[\mathfrak{e}_8] \tag{9.107}$$

In line with this the other notable property of this remarkable algebra is the absence of a fundamental representation smaller than the adjoint one. Indeed, the smallest linear representation of $\mathfrak{e}_8$ is just the adjoint one whose highest weight is the highest positive root of the $\mathfrak{e}_8$ root system:

$$\Pi_{\text{highest}} = \psi \equiv \alpha_{120} \tag{9.108}$$

The $\mathfrak{e}_8$ Dynkin diagram is displayed in Figure 9.7 and our chosen explicit representation of the simple roots in $\mathbb{R}^8$ is the following one:

$$\begin{aligned}
\alpha_1 &= \{0,1,-1,0,0,0,0,0\} \\
\alpha_2 &= \{0,0,1,-1,0,0,0,0\} \\
\alpha_3 &= \{0,0,0,1,-1,0,0,0\} \\
\alpha_4 &= \{0,0,0,0,1,-1,0,0\} \\
\alpha_5 &= \{0,0,0,0,0,1,-1,0\} \\
\alpha_6 &= \{0,0,0,0,0,1,1,0\} \\
\alpha_7 &= \left\{-\frac{1}{2},-\frac{1}{2},-\frac{1}{2},-\frac{1}{2},-\frac{1}{2},-\frac{1}{2},-\frac{1}{2},-\frac{1}{2}\right\} \\
\alpha_8 &= \{1,-1,0,0,0,0,0,0\}
\end{aligned} \tag{9.109}$$

The Cartan matrix encoded in the Dynkin diagram displayed in Figure 9.7 has the following explicit appearance:

$$\mathcal{C}_{ij} \equiv \alpha_i \cdot \alpha_j = \begin{pmatrix}
2 & -1 & 0 & 0 & 0 & 0 & 0 & -1 \\
-1 & 2 & -1 & 0 & 0 & 0 & 0 & 0 \\
0 & -1 & 2 & -1 & 0 & 0 & 0 & 0 \\
0 & 0 & -1 & 2 & -1 & -1 & 0 & 0 \\
0 & 0 & 0 & -1 & 2 & 0 & 0 & 0 \\
0 & 0 & 0 & -1 & 0 & 2 & -1 & 0 \\
0 & 0 & 0 & 0 & 0 & -1 & 2 & 0 \\
-1 & 0 & 0 & 0 & 0 & 0 & 0 & 2
\end{pmatrix} \tag{9.110}$$

The remarkable property (9.107) simply follows from the fact that the determinant of the Cartan matrix is one, $\mathrm{Det}\,\mathscr{C} = 1$, so that the inverse $\mathscr{C}^{-1}$ has also integer-valued entries:

$$\mathscr{C}^{-1} = \begin{pmatrix} 6 & 8 & 10 & 12 & 6 & 8 & 4 & 3 \\ 8 & 12 & 15 & 18 & 9 & 12 & 6 & 4 \\ 10 & 15 & 20 & 24 & 12 & 16 & 8 & 5 \\ 12 & 18 & 24 & 30 & 15 & 20 & 10 & 6 \\ 6 & 9 & 12 & 15 & 8 & 10 & 5 & 3 \\ 8 & 12 & 16 & 20 & 10 & 14 & 7 & 4 \\ 4 & 6 & 8 & 10 & 5 & 7 & 4 & 2 \\ 3 & 4 & 5 & 6 & 3 & 4 & 2 & 2 \end{pmatrix} \tag{9.111}$$

This means that the fundamental weights:

$$\lambda^i \equiv (\mathscr{C}^{-1})^{ij} \alpha_j \tag{9.112}$$

are linear combinations with integer coefficients of the simple roots and hence lie in the root lattice.

In our chosen basis we have:

$$\begin{aligned} \lambda_1 &= \{1, 1, 0, 0, 0, 0, 0, -2\} \\ \lambda_2 &= \{1, 1, 1, 0, 0, 0, 0, -3\} \\ \lambda_3 &= \{1, 1, 1, 1, 0, 0, 0, -4\} \\ \lambda_4 &= \{1, 1, 1, 1, 1, 0, 0, -5\} \\ \lambda_5 &= \left\{\frac{1}{2}, \frac{1}{2}, \frac{1}{2}, \frac{1}{2}, \frac{1}{2}, \frac{1}{2}, -\frac{1}{2}, -\frac{5}{2}\right\} \\ \lambda_6 &= \left\{\frac{1}{2}, \frac{1}{2}, \frac{1}{2}, \frac{1}{2}, \frac{1}{2}, \frac{1}{2}, \frac{1}{2}, -\frac{7}{2}\right\} \\ \lambda_7 &= \{0, 0, 0, 0, 0, 0, 0, -2\} \\ \lambda_8 &= \{1, 0, 0, 0, 0, 0, 0, -1\} \end{aligned} \tag{9.113}$$

In these conventions the highest weight of the adjoint representation is $\lambda_8 = \alpha_{120}$ and hence it has Dynkin labels $\{0, 0, 0, 0, 0, 0, 0, 1\}$.

9.4.1 Construction of the adjoint representation

The explicit construction of the adjoint representation of $\mathfrak{e}_8$ in its maximally split real form $\mathfrak{e}_{8(8)}$ is done by means of the following steps:

1. Firstly, by means of an algorithm inductive on the height of the roots we construct all the 120 positive roots. It turns out that we have roots of all heights in the interval 1 to 29 and their explicit form is displayed in Section B.10.1.

2. Secondly, utilizing the following enumeration for the axes:

$$248 = \underbrace{1,\dots,8}_{\text{Cartan generators}}, \underbrace{1+8,\dots,8+120}_{\text{positive roots}}, \underbrace{1+128,\dots,120+128}_{\text{negative roots}} \tag{9.114}$$

we construct the 248×248 matrices representing the Cartan generators and fulfilling the following commutation relations:

$$[\mathcal{H}^i, \mathcal{H}^j] = 0 \tag{9.115}$$

$$[\mathcal{H}^i, \mathcal{E}^\alpha] = \alpha^i \mathcal{E}^\alpha; \quad \forall \alpha \in \text{root system} \tag{9.116}$$

3. Then we construct the 248×248 matrices that represent the step operators $\mathcal{E}^{\alpha_i}$ associated with the simple roots α_i $(i = 1, \dots, 8)$ in such a way that equation (9.116) is indeed confirmed

$$[\mathcal{H}^i, \mathcal{E}^{\alpha_j}] = \alpha^i_j \mathcal{E}^{\alpha_j}; \quad \forall \alpha_j > 0 \tag{9.117}$$

and setting:

$$\mathcal{E}^{-\alpha_j} \equiv (\mathcal{E}^{\alpha_j})^T \tag{9.118}$$

we also obtain:

$$\begin{aligned} [\mathcal{H}^i, \mathcal{E}^{-\alpha_j}] &= -\alpha^i_j \mathcal{E}^{-\alpha_j}; \quad \forall \alpha_j > 0 \\ [\mathcal{E}^{\alpha_j}, \mathcal{E}^{-\alpha_j}] &= \alpha^i_j \mathcal{H}_i; \quad \forall \alpha_j > 0 \end{aligned} \tag{9.119}$$

4. Finally, we obtain all the higher height step operators by defining:

$$\mathcal{E}^\beta \equiv [\mathcal{E}^{\alpha_i}, \mathcal{E}^\gamma] \tag{9.120}$$

where, by definition,

$$\beta = \alpha_i + \gamma \tag{9.121}$$

and the index i of the simple root α_i is the smallest for all pairs $\{\alpha_i, \gamma\}$ for which eq. (9.121) is true. This is an algorithmic way of fixing the normalization of the step operators and it is useful for computer-aided constructions.

9.4.1.1 The cocycle $N^{\alpha\beta}$

Before we proceed with the illustration of our explicit construction we have to pause and discuss a delicate point. The commutation relations of a Lie algebra in the Cartan–Weyl form are completely fixed, once also the cocycle $N^{\alpha,\beta}$ is given in the commutation rules of the step operators:

$$[\mathcal{E}^\alpha, \mathcal{E}^\beta] = N_{\alpha\beta} \mathcal{E}^{\alpha+\beta} \tag{9.122}$$

There is no way of representing $N^{\alpha,\beta}$ with an analytic formula, just because it is a cocycle. Since our Lie algebra is simply-laced, all the non-vanishing entries of $N^{\alpha,\beta}$ are either +1 or −1, yet there is no better way to specify it than to enumerate the pairs of roots for which it is positive and the pairs of roots for which it is negative. It suffices to enumerate the entries $N^{\alpha,\beta}$ for both α and β positive roots. Indeed, once we know $N^{\alpha,\beta}$ for $\alpha,\beta > 0$ we can extend its definition to the other cases through the following identities which follow from the Jacobi identities:

$$\begin{aligned} N^{-\alpha,-\beta} &= -N^{\alpha,\beta} \\ N^{\alpha,-\beta} &= N^{\alpha-\beta,\beta} \\ N^{\beta,\alpha,} &= -N^{\alpha,\beta} \end{aligned} \tag{9.123}$$

Our consistent choice of $N^{\alpha\beta}$ is encoded in an array named ***Nalfa*** which is generated in the library when the MATHEMATICA NoteBook described in Appendix B.2 is evaluated. The same NoteBook verifies also the identity:

$$N^{\alpha\beta}N^{\alpha+\beta,\gamma} + N^{\beta,\gamma}N^{\gamma+\beta,\alpha} + N^{\gamma,\alpha}N^{\alpha+\gamma,\beta} = 0 \tag{9.124}$$

which follows from Jacobi identities and which is the cocycle condition. Once this identity is verified, we are sure that the structure constants, as given in eq. (7.38) by the Cartan–Weyl basis, are consistent and we possess the adjoint representation.

9.4.1.2 The Cartan generators

First of all we set the form of the Cartan generators named $\mathscr{H}^i$ $(i = 1,\dots,8)$. The corresponding 248×248 matrices are diagonal and they are constructed as follows:

$$\begin{aligned} (\mathscr{H}^i)_m{}^{\ell} &= 0; & i &= 1,\dots,8;\ m,\ell = 1,\dots,8 \\ (\mathscr{H}^i)_{m+8}{}^{\ell+8} &= -\alpha^i_m\delta^{\ell}_m; & i &= 1,\dots,8;\ m,\ell = 1,\dots,120 \\ (\mathscr{H}^i)_{m+128}{}^{\ell+128} &= \alpha^i_m\delta^{\ell}_m; & i &= 1,\dots,8;\ m,\ell = 1,\dots,120 \\ (\mathscr{H}^i)_{m+128}{}^{\ell} &= 0; & i &= 1,\dots,8;\ m,\ell = 1,\dots,120 \\ (\mathscr{H}^i)_m{}^{\ell+128} &= 0; & i &= 1,\dots,8;\ m,\ell = 1,\dots,120 \end{aligned} \tag{9.125}$$

where by α^i_m we denote the ith component of the mth root.

9.4.1.3 The step operators associated with simple roots

The construction of the simple root step operators is done in three steps:

9.4.1.4 1st STEP

In the first step we fix the components of $\mathscr{E}^{\alpha_i}$ where one leg is in the Cartan subalgebra, the other along the simple roots:

$$\begin{aligned} (\mathscr{E}^{\alpha_i})_j{}^{i+8} &= -\alpha^j_i; & i &= 1,\dots,8;\ j = 1,\dots,8 \\ (\mathscr{E}^{\alpha_i})_{i+128}{}^{j} &= \alpha^j_i; & i &= 1,\dots,8;\ j = 1,\dots,8 \end{aligned} \tag{9.126}$$

Equation (9.126) is obligatory and unambiguous, since it is the transcription in the adjoint representation matrix of the commutation relation (9.117).

9.4.1.5 2nd STEP

In the second step we fix the components of $\mathcal{E}^{\alpha_i}$ where one leg is along the simple roots, while the other is along the roots of height $h = 2$.

$$\begin{aligned} (\mathcal{E}^{\alpha_i})_{k+8}{}^{p+8} &= \epsilon_{ik} \quad (\text{iff } \alpha_p = \alpha_i + \alpha_k) \quad i = 1,\dots,8;\ k = 1,\dots,8 \\ (\mathcal{E}^{\alpha_i})_{k+128}{}^{p+128} &= -\epsilon_{ik} \quad (\text{iff } \alpha_p = \alpha_i + \alpha_k) \quad i = 1,\dots,8;\ k = 1,\dots,8 \end{aligned} \tag{9.127}$$

where by definition, $\epsilon_{ik} = 1$ iff $i < k$, $\epsilon_{ik} = -1$ iff $i > k$, $\epsilon_{ik} = 0$ iff $i = k$.

9.4.1.6 3rd STEP

In the third step we fix the components of $\mathcal{E}^{\alpha_i}$ where one of the legs is along roots of height larger or equal than 3.

$$\begin{aligned} (\mathcal{E}^{\alpha_i})_{k+8}{}^{p+8} &= (-1)^{\frac{i(i+1)}{2}+h[\alpha_p]} \quad (\text{iff } \alpha_p = \alpha_i + \alpha_k) \\ (\mathcal{E}^{\alpha_i})_{k+128}{}^{p+128} &= -(-1)^{\frac{i(i+1)}{2}+h[\alpha_p]} \quad (\text{iff } \alpha_p = \alpha_i + \alpha_k) \\ &(i = 1,\dots,8 \quad k = 9,\dots,120) \end{aligned}$$

In the above equation $h[\alpha_p]$ denotes the height of the root α_p.

9.4.1.7 4th STEP

The fourth step is equivalent to determining by trial and error the $N_{\alpha\beta}$ cocycle. As we anticipated, it was through a series of educated guesses that we were able to determine a set of sign changes that was sufficient to guarantee the implementation of eqs (9.117), (9.118), (9.119) and the cocycle condition. This set of sign changes is displayed below:

$$\begin{aligned} &(\mathcal{E}^{\alpha_8})_{17,25} = -1 \\ &(\mathcal{E}^{\alpha_8})_{25+120,17+120} = 1 \\ &(\mathcal{E}^{\alpha_4})_{23,29} = -1 \\ &(\mathcal{E}^{\alpha_4})_{29+120,23+120} = 1 \\ &(\mathcal{E}^{\alpha_8})_{24,34} = 1 \\ &(\mathcal{E}^{\alpha_8})_{34+120,24+120} = -1 \\ &(\mathcal{E}^{\alpha_8})_{31,43} = -1 \\ &(\mathcal{E}^{\alpha_8})_{43+120,31+120} = 1 \\ &(\mathcal{E}^{\alpha_8})_{38,50} = 1 \\ &(\mathcal{E}^{\alpha_8})_{50+120,38+120} = -1 \\ &(\mathcal{E}^{\alpha_8})_{39,51} = 1 \\ &(\mathcal{E}^{\alpha_8})_{51+120,39+120} = -1 \\ &(\mathcal{E}^{\alpha_8})_{46,56} = -1 \\ &(\mathcal{E}^{\alpha_8})_{56+120,46+120} = 1 \end{aligned}$$

$$(\mathscr{E}^{\alpha_8})_{53,63} = 1$$
$$(\mathscr{E}^{\alpha_8})_{63+120,53+120} = -1$$
$$(\mathscr{E}^{\alpha_7})_{51,58} = -1$$
$$(\mathscr{E}^{\alpha_7})_{58+120,51+120} = 1$$
$$(\mathscr{E}^{\alpha_7})_{56,64} = 1$$
$$(\mathscr{E}^{\alpha_7})_{64+120,56+120} = -1$$
$$(\mathscr{E}^{\alpha_7})_{63,70} = -1$$
$$(\mathscr{E}^{\alpha_7})_{70+120,63+120} = 1$$
$$(\mathscr{E}^{\alpha_3})_{63,69} = -1$$
$$(\mathscr{E}^{\alpha_3})_{69+120,63+120} = 1$$

Implementing next the algorithm specified in eqs (9.120), (9.121) one obtains all the step operators and finally all the 248 generators of the $\mathfrak{e}_8$ Lie algebra given as 248×248 matrices. They are displayed in Section B.10.1.

9.4.2 Final comments on the $\mathfrak{e}_8$ root systems

From the above displayed table it is evident that the 120 positive roots subdivide into two subsets, one containing 56 elements, the other containing 64 elements. In the first set the Euclidean components of the roots are all integers, in the second set they are all half-integer. The set of 56 integer valued roots are the positive roots of the D_8 Lie subalgebra, namely, $\mathfrak{so}(16,\mathbb{C})$. The 64 half-integer valued roots are one-half of the 128 weights of the spinor representation of $\mathfrak{so}(16)$. Indeed, the $\mathfrak{e}_8$ Lie algebra has a regularly embedded $\mathfrak{so}(16)$ subalgebra with decomposition:

$$\mathrm{adj}[\mathfrak{e}_8] = \mathrm{adj}[\mathfrak{so}(16)] \oplus \mathrm{spinor}[\mathfrak{so}(16)] \tag{9.128}$$

9.4.2.1 The $\mathfrak{e}_{7(7)} \subset \mathfrak{e}_{8(8)}$ simple root basis

In addition to the decomposition (9.128), there is another one which corresponds to the golden splitting of the $\mathfrak{e}_{8(8)}$ Lie algebra, namely:

$$\mathrm{adj}[\mathfrak{e}_8] = (\mathrm{adj}[\mathfrak{e}_{7(7)}], \mathbf{1}) \oplus (\mathbf{1}, \mathrm{adj}[\mathfrak{sl}(2,\mathbb{R})]) \oplus (\mathbf{56}, \mathbf{2}) \tag{9.129}$$

To implement the decomposition (9.129) it is convenient to use a second basis for the root system which we presently describe. We choose $\alpha_1, \alpha_2, \dots, \alpha_7$ to be the simple roots of $\mathfrak{e}_{7(7)}$ extended by means of an 8th vanishing component. Then α_8 is uniquely determined in order to have the right scalar product with the other simple roots. Hence we set $\alpha_i = \beta_i$ where:

$$\begin{aligned}
\beta_1 &= \{0, 1, -1, 0, 0, 0, 0, 0\} \\
\beta_2 &= \{0, 0, 1, -1, 0, 0, 0, 0\} \\
\beta_3 &= \{0, 0, 0, 1, -1, 0, 0, 0\} \\
\beta_4 &= \{0, 0, 0, 0, 1, -1, 0, 0\} \\
\beta_5 &= \{0, 0, 0, 0, 0, 1, -1, 0\} \\
\beta_6 &= \{0, 0, 0, 0, 0, 1, 1, 0\} \\
\beta_7 &= \left\{0, -\frac{1}{2}, -\frac{1}{2}, -\frac{1}{2}, -\frac{1}{2}, -\frac{1}{2}, -\frac{1}{2}, \frac{1}{\sqrt{2}}\right\} \\
\beta_8 &= \left\{-\frac{1}{\sqrt{2}}, -1, 0, 0, 0, 0, 0, -\frac{1}{\sqrt{2}}\right\}
\end{aligned} \tag{9.130}$$

The MATHEMATICA NoteBook that implements all the construction algorithms explained above is briefly described in Section B.2.

9.5 Bibliographical note

The main sources for the constructions of the exceptional Lie algebras presented in this chapter are not in textbooks but rather in fairly recent research literature pertaining to the realm of supergravity and of its various applications to cosmology, black-hole physics and other fundamental issues. Indeed, the electronic implementation of these constructions by means of suitable MATHEMATICA codes was precisely motivated by supergravity applications. The versions of such MATHEMATICA NoteBooks presented in Section B are renewed and polished ones that have been made user-friendly for the purpose of distribution and use in connection with the present textbook. In particular we have:

1. The here presented construction of the $\mathfrak{g}_{2(2)}$ Lie algebra has its origin in [63, 62].
2. The here presented construction of the $\mathfrak{f}_{4(4)}$ Lie algebra has its origin in [61].
3. The here presented construction of the $\mathfrak{e}_{7(7)}$ Lie algebra has its origin in [28] and the other previous papers there quoted.
4. The here presented construction of the $\mathfrak{e}_{8(8)}$ Lie algebra has its origin in [59].

Some of the above constructions were further revised and refined in [64].

10 A primary on the theory of connections and metrics

We must admit with humility that, while number is purely a product of our minds, space has a reality outside our minds, so that we cannot completely prescribe its properties a priori...
Carl Friedrich Gauss

10.1 Introduction

The present chapter deals with the second stage in the development of modern differential geometry. Once the notions of manifold and of fiber-bundle have been introduced, the latter requiring Lie Groups as fundamental ingredients, one considers differential calculus on these spaces and rather than ordinary derivatives one utilizes *covariant derivatives*. In General Relativity we mainly use the covariant derivative on the tangent bundle but it is important to realize that one can define covariant derivatives on general fiber-bundles. Indeed, the covariant derivative is the physicist's name for the mathematical concept of *connection* that we are going to introduce and illustrate in this chapter. It is also important to stress that even restricting one's attention to the tangent bundle the connection used in General Relativity is a particular one, the so-called *Levi-Civita connection* that arises from a more fundamental object of the *metric*. As we are going to see soon, a manifold endowed with a *metric structure* is a space where one can measure lengths, specifically the length of all curves. A generic connection on the tangent bundle is named *an affine connection* and the Levi-Civita connection is a specific affine connection that is uniquely determined by the *metric structure*. As we shall presently illustrate, every connection has, associated with it, another object (actually a 2-form) that we name its curvature. The curvature of the Levi-Civita connection is what we name the Riemann curvature of a manifold and it is the main concern of General Relativity. It encodes the intuitive geometrical notion of curvature of a surface or hypersurface. The field equations of Einstein's theory are statements about the Riemann curvature of space-time that is related to its energy–matter content. We should be aware that the notion of curvature applies to generic connections on generic fiber-bundles, in particular on principal bundles. Physically these connections and curvatures are not less important than the Levi-Civita connection and the Riemann curvature. They constitute the main mathematical objects entering the theory of fundamental non-gravitational interactions, that are all *gauge theories*.

10.2 Connections on principal bundles: the mathematical definition

We come to the contemporary mathematical definition of a connection on a principal fiber-bundle. The adopted viewpoint is that of Ehresmann and what we are go-

https://doi.org/10.1515/9783110551204-010

ing to define is usually named an *Ehresmann connection* in the mathematical literature.

10.2.1 Ehresmann connections on a principal fiber-bundle

Let $P(\mathcal{M}, G)$ be a principal fiber-bundle with base-manifold $\mathcal{M}$ and structural group G. Let us moreover denote by π the projection:

$$\pi : P \to \mathcal{M} \tag{10.1}$$

Consider the action of the Lie group G on the total space P:

$$\forall g \in G \quad g : G \to G \tag{10.2}$$

By definition this action is vertical in the sense that

$$\forall u \in P,\ \forall g \in G : \quad \pi(g(u)) = \pi(u) \tag{10.3}$$

namely, it moves points only along the fibers. Given any element $X \in \mathbb{G}$ where we have denoted by $\mathbb{G}$ the Lie algebra of the structural group, we can consider the one-dimensional subgroup generated by it

$$g_X(t) = \exp[tX], \quad t \in \mathbb{R} \tag{10.4}$$

and consider the curve $\mathscr{C}_X(t,u)$ in the manifold P obtained by acting with $g_X(t)$ on some point $u \in P$:

$$\mathscr{C}_X(t,u) \equiv g_X(t)(u) \tag{10.5}$$

The vertical action of the structural group implies that:

$$\pi(\mathscr{C}_X(t,u)) = p \in \mathcal{M} \quad \text{if } \pi(u) = p \tag{10.6}$$

These items allow us to construct a linear map # which associates a vector field $\mathbf{X}^\#$ over P to every element X of the Lie algebra $\mathbb{G}$:

$$\begin{aligned} \# &: \mathbb{G} \ni X \to \mathbf{X}^\# \in \Gamma(TP,P) \\ \forall f \in C^\infty(P) \quad & \mathbf{X}^\# f(u) \equiv \frac{d}{dt} f(\mathscr{C}_X(t,u))\big|_{t=0} \end{aligned} \tag{10.7}$$

Focusing on any point $u \in P$ of the bundle, the map (10.7) reduces to a map:

$$\#_u : \mathbb{G} \to T_u P \tag{10.8}$$

from the Lie algebra to the tangent space at u. The map $\#_u$ is not surjective. We introduce the following:

Definition 10.2.1. At any point $p \in P(M,G)$ of a principal fiber-bundle we name vertical subspace V_uP of the tangent space T_uP the image of the map $\#_u$:

$$T_uP \supset V_u \equiv \text{Im}\,\#_u \tag{10.9}$$

Indeed, any vector $t \in \text{Im}\,\#_u$ lies in the tangent space to the fiber G_p, where $p \equiv \pi(u)$.

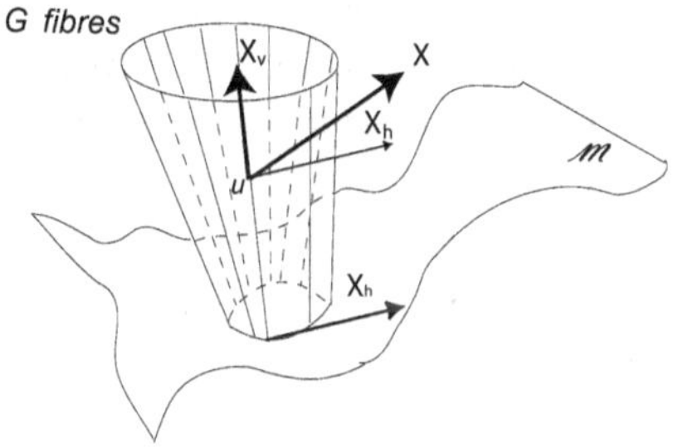

Figure 10.1: The tangent space to a principal bundle P splits at any point $u \in P$ into a vertical subspace along the fibers and a horizontal subspace parallel to the base manifold. This splitting is the intrinsic geometric meaning of a connection.

The meaning of Definition 10.2.1 becomes clearer if we consider both Figure 10.1 and a local trivialization of the bundle including $u \in P$. In such a local trivialization the bundle point u is identified by a pair:

$$u \overset{\text{loc. triv.}}{\longrightarrow} (p,f) \quad \text{where} \begin{cases} p \in \mathcal{M} \\ f \in G \end{cases} \tag{10.10}$$

and the curve $\mathscr{C}_X(t,u)$ takes the following appearance:

$$\mathscr{C}_X(t,u) \overset{\text{loc. triv.}}{\longrightarrow} (p, e^{tX}f), \quad p = \pi(u) \tag{10.11}$$

Correspondingly, naming α^i the group parameters, just as in the previous section, and x^μ the coordinates on the base space $\mathcal{M}$, a generic function $f \in C^\infty(P)$ is just a function $f(x,\alpha)$ of the m coordinates x and of the n parameters α. Comparing eq. (10.7) with eq. (5.154) we see that, in the local trivialization, the vertical tangent vector $\mathbf{X}^\#$ reduces to nothing else but to the right-invariant vector field on the group manifold associated with the same Lie algebra element X:

$$\mathbf{X}^\# \overset{\text{loc. triv.}}{\longrightarrow} \mathbf{X}_R = X_R^i(\alpha)\frac{\vec{\partial}}{\partial \alpha^i} \tag{10.12}$$

As we see from eq. (10.12), in a local trivialization a vertical vector contains no derivatives with respect to the base space coordinates. This means that its projection onto the base manifold is zero. Indeed, it follows from Definition 10.2.1 that:

$$\mathbf{X} \in V_uP \quad \Rightarrow \quad \pi_*\mathbf{X} = 0 \tag{10.13}$$

where π_* is the push-forward of the projection map.

Having defined the vertical subspace of the tangent space one would be interested in giving a definition also of the horizontal subspace H_uP, which, intuitively, must be somehow parallel to the tangent space $T_p\mathcal{M}$ to the base manifold at the projection point $p = \pi(u)$. The dimension of the vertical space is the same as the dimension of the fiber, namely $n = \dim G$. The dimension of the horizontal space H_uP must be the same as the dimension of the base manifold $m = \dim \mathcal{M}$. Indeed, H_uP should be the orthogonal complement of V_uP. Easy to say, but orthogonal with respect to what? This is precisely the point. Is there an a priori intrinsically defined way of defining the orthogonal complement to the vertical subspace $V_u \subset T_uP$? The answer is that there is not. Given a basis $\{\mathbf{v}^\mu\}$ of n vectors for the subspace V_uP, there are infinitely many ways of finding m extra vectors $\{\mathbf{h}^i\}$ which complete this basis to a basis of T_uP. The span of any such collection of m vectors $\{\mathbf{h}^i\}$ is a possible legitimate definition of the orthogonal complement H_uP. This arbitrariness is the root of the mathematical notion of a *connection*. Providing a fiber-bundle with a connection precisely means introducing a rule that uniquely defines the orthogonal complement H_uP.

Definition 10.2.2. Let $P(M, G)$ be a principal fiber-bundle. A **connection** on P is a rule which at any point $u \in P$ defines a unique splitting of the tangent space T_uP into the vertical subspace V_uP and into a horizontal complement H_uP satisfying the following properties:

(i) $T_uP = H_uP \oplus V_uP$.

(ii) Any smooth vector field $\mathbf{X} \in \Gamma(TP, P)$ separates into the sum of two smooth vector fields $\mathbf{X} = \mathbf{X}^H + \mathbf{X}^V$ such that at any point $u \in P$ we have $\mathbf{X}^H_u \in H_uP$ and $\mathbf{X}^V_u \in V_uP$.

(iii) $\forall \mathfrak{g} \in G$ we have that $H_{\mathfrak{g}u}P = L_{\mathfrak{g}*}H_uP$.

The third defining property of the connection states that all the horizontal spaces along the same fiber are related to each other by the push-forward of the left-translation on the group manifold.

This beautiful purely geometrical definition of the connection due to Ehresmann emphasizes that it is just an intrinsic attribute of the principal fiber-bundle. In a trivial bundle, which is a direct product of manifolds, the splitting between vertical and horizontal spaces is done once forever. The vertical space is the tangent space to the fiber, the horizontal space is the tangent space to the base. In a non-trivial bundle the splitting between vertical and horizontal directions has to be reconsidered at every next point and fixing this ambiguity is the task of the connection.

10.2.1.1 The connection one-form

The algorithmic way to implement the splitting rule advocated by Definition 10.2.2 is provided by introducing a connection one-form $\mathbf{A}$ which is just a Lie algebra-valued one-form on the bundle P satisfying two precise requirements.

Definition 10.2.3. Let $P(\mathcal{M}, G)$ be a principal fiber-bundle with structural Lie group G, whose Lie algebra we denote by $\mathbb{G}$. A connection one-form on this bundle is a Lie algebra-valued section of the cotangent bundle $\mathbf{A} \in \mathbb{G} \otimes \Gamma(T^*P, P)$ which satisfies the following defining properties:
(i) $\forall \mathbf{X} \in \mathbb{G} : \mathbf{A}(\mathbf{X}^\#) = \mathbf{X}$
(ii) $\forall \mathfrak{g} \in G : L_\mathfrak{g}^* \mathbf{A} = \mathrm{Adj}_{\mathfrak{g}^{-1}} \mathbf{A} \equiv \mathfrak{g}^{-1} \mathbf{A} \mathfrak{g}$

Given the connection one-form $\mathbf{A}$ the splitting between vertical and horizontal subspaces is performed through the following:

Definition 10.2.4. At any $u \in P$, the horizontal subspace $H_u P$ of the tangent space to the bundle is provided by the kernel of $\mathbf{A}$:

$$H_u P \equiv \{ \mathbf{X} \in T_u P \mid \mathbf{A}(\mathbf{X}) = 0 \} \tag{10.14}$$

The actual meaning of Definitions 10.2.3, 10.2.4 and their coherence with Definition 10.2.2 becomes clear if we consider a local trivialization of the bundle.

Just as before, let x^μ be the coordinates on the base-manifold and α^i the group parameters. With respect to this local patch, the connection one-form $\mathbf{A}$ has the following appearance:

$$\mathbf{A} = \widehat{\mathscr{A}} + \mathrm{A} \tag{10.15}$$

$$\widehat{\mathscr{A}} = \widehat{\mathscr{A}_\mu} dx^\mu = dx^\mu \, \mathscr{A}_\mu^I T_I \tag{10.16}$$

$$\mathrm{A} = \mathrm{A}_i \, d\alpha^i = d\alpha^i \, \mathrm{A}_i^I T_I \tag{10.17}$$

where, a priori, both the components $\widehat{\mathscr{A}_\mu}(x,\alpha)$ and $\mathrm{A}_i(x,\alpha)$ could depend on both x^μ and α^i. The second equality in eqs (10.17) and (10.16) is due to the fact that $\mathscr{A}$ and A are Lie algebra valued, hence they can be expanded along a basis of generators of $\mathbb{G}$. Actually the dependence on the group parameters or better on the group element $\mathfrak{g}(\alpha)$ is completely fixed by the defining axioms. Let us consider first the implications of property **(i)** in Definition 10.2.3. To this effect let $X \in \mathbb{G}$ be expanded along the same basis of generators, namely $X = X^A T_A$. As we already remarked in eq. (10.12), the image of X through the map # is $\mathbf{X}^\# = X^A \mathbf{T}_A^{(R)}$ where $\mathbf{T}_A^{(R)} = \overset{(r)}{\Sigma}{}_A^i(\alpha)\, \vec{\partial}/\partial\alpha^i$ are the right-invariant vector fields. Hence property **(i)** requires:

$$X^A \mathrm{A}_i^I \, \overset{(r)}{\Sigma}{}_A^i(\alpha) \, T_I = X^A T_A \tag{10.18}$$

which implies

$$\mathrm{A}_i^I \, \overset{(r)}{\Sigma}{}_A^i(\alpha) = \delta_A^I \tag{10.19}$$

Therefore $A_i^I = \overset{(r)}{\Sigma}{}_i^A(\alpha)$ and the vertical component of the one-form connection is nothing else but the right-invariant one-form:

$$A = \sigma_{(R)} \equiv d\mathfrak{g}\,\mathfrak{g}^{-1} \tag{10.20}$$

Consider now the implications of defining property **(ii)**. In the chosen local trivialization the left action of any element γ of the structural group G is $L_\gamma : (x, \mathfrak{g}(\alpha)) \to (x, \gamma \cdot \mathfrak{g}(\alpha))$ and its pullback acts on the right-invariant one-form $\sigma_{(R)}$ by adjoint transformations:

$$L_\gamma^* : \sigma_{(R)} \to d(\gamma \cdot \mathfrak{g})(\gamma \cdot \mathfrak{g})^{-1} = \gamma \cdot \sigma_{(R)} \cdot \gamma^{-1} \tag{10.21}$$

Property **(ii)** of Definition 10.2.3 requires that the same should be true for the complete connection one-form **A**. This fixes the $\mathfrak{g}$ dependence of $\widehat{\mathscr{A}}$, namely, we must have:

$$\widehat{\mathscr{A}}(x, \mathfrak{g}) = \mathfrak{g} \cdot \mathscr{A} \cdot \mathfrak{g}^{-1} \tag{10.22}$$

where

$$\mathscr{A} = \mathscr{A}_\mu^I(x)\, dx^\mu\, T_I \tag{10.23}$$

is a $\mathbb{G}$ Lie algebra-valued one-form on the base manifold $\mathscr{M}$. Consequently we can summarize the content of Definition 10.2.3 by stating that in any local trivialization $u = (x, \mathfrak{g})$ the connection one-form has the following structure:

$$\mathbf{A} = \mathfrak{g} \cdot \mathscr{A} \cdot \mathfrak{g}^{-1} + d\mathfrak{g} \cdot \mathfrak{g}^{-1} \tag{10.24}$$

10.2.1.2 Gauge transformations

From eq. (10.24) we easily work out the action on the one-form connection of the transition function from one local trivialization to another one. Let us recall eqs (5.68) and (5.69). In the case of a principal bundle the transition function $\mathfrak{t}_{\alpha\beta}(x)$ is just a map from the intersection of two open neighborhoods of the base space to the structural group:

$$\mathfrak{t}_{\alpha\beta} : \mathscr{M} \supset U_\alpha \cap U_\beta \to G \tag{10.25}$$

and we have:

$$\text{transition:} \quad (x, \mathfrak{g}_\alpha) \to (x, \mathfrak{g}_\beta) = (x, \mathfrak{g}_\alpha \cdot \mathfrak{t}_{\alpha\beta}(x)) \tag{10.26}$$

Correspondingly, combining this information with eq. (10.24) we conclude that:

$$\begin{aligned}
&\mathbf{A}_\alpha \to \mathbf{A}_\beta \\
&\mathbf{A}_\alpha = \mathfrak{g} \cdot \mathscr{A}_\alpha \cdot \mathfrak{g}^{-1} + d\mathfrak{g} \cdot \mathfrak{g}^{-1} \\
&\mathbf{A}_\beta = \mathfrak{g} \cdot \mathscr{A}_\beta \cdot \mathfrak{g}^{-1} + d\mathfrak{g} \cdot \mathfrak{g}^{-1} \\
&\mathscr{A}_\beta = \mathfrak{t}_{\alpha\beta}(x) \cdot \mathscr{A}_\alpha \cdot \mathfrak{t}_{\alpha\beta}^{-1}(x) + d\mathfrak{t}_{\alpha\beta}(x) \cdot \mathfrak{t}_{\alpha\beta}^{-1}(x)
\end{aligned} \tag{10.27}$$

The last of equations (10.27) is the most significant for the bridge from Mathematics to Physics. The one-form $\mathscr{A}$ defined over the base-manifold $\mathscr{M}$, which in particular can be space-time, is what physicists name *a gauge potential*, the prototype of which is the gauge-potential of electrodynamics; the transformation from one local trivialization to another one, encoded in the last of equations (10.27), is what physicists name a *gauge transformation*. The basic idea about gauge-transformations is that of *local symmetry*. Some physical system or rather some physical law is invariant with respect to the transformations of a group G but not only globally, rather also locally, namely, the transformation can be chosen differently from one point of space-time to the next one. The basic mathematical structure realizing this physical idea is that of fiber-bundle and the one-form connection on a fiber-bundle is the appropriate mathematical structure which encompasses all mediators of all physical interactions.

10.2.1.3 Horizontal vector fields and covariant derivatives

Let us come back to Definition 10.2.4 of horizontal vector fields. In a local trivialization a generic vector field on P is of the form:

$$\mathbf{X} = X^\mu \partial_\mu + X^i \partial_i \tag{10.28}$$

where we have used the short-writing $\partial_\mu = \partial/\partial x^\mu$ and $\partial_i = \partial/\partial \alpha^i$. The horizontality condition in eq. (10.14) becomes:

$$0 = X^\mu \mathscr{A}_\mu^A \mathfrak{g} \cdot T_A \cdot \mathfrak{g}^{-1} + X^i \partial_i \mathfrak{g} \cdot \mathfrak{g}^{-1} \tag{10.29}$$

Multiplying on the left by $\mathfrak{g}^{-1}$ and on the right by $\mathfrak{g}$, eq. (10.29) can be easily solved for X^i obtaining in this way the general form of a horizontal vector field on a principal fiber-bundle $P(\mathscr{M}, G)$ endowed with a connection one-form $\mathbf{A} \leftrightarrow \mathscr{A}$. We get:

$$\mathbf{X}^{(H)} = X^\mu (\vec{\partial}_\mu - \mathscr{A}_\mu^A \mathbf{T}_A^{(L)}) = X^\mu (\vec{\partial}_\mu - \mathscr{A}_\mu^A \overset{(\ell)}{\Sigma_A^i} \vec{\partial}_i) \tag{10.30}$$

To obtain the above result we made use of the following identities:

$$\mathfrak{g}^{-1} \cdot \partial_i \mathfrak{g} = \overset{(\ell)}{\Sigma_i^A}; \quad \mathbf{T}_A^{(L)} = \overset{(\ell)}{\Sigma_A^i} \partial_i; \quad \overset{(\ell)}{\Sigma_i^A} \overset{(\ell)}{\Sigma_B^i} = \delta_B^A \tag{10.31}$$

which follow from the definitions of left-invariant one-forms and vector fields discussed in previous subsections.

Equation (10.30) is equally relevant for the bridge between Mathematics and Physics as eq. (10.27). Indeed, from (10.30) we realize that the mathematical notion of horizontal vector fields coincides with the physical notion of *covariant derivative*. Expressions like that in the bracket of (10.30) firstly appeared in the early decades of the XXth century when Classical Electrodynamics was rewritten in the language of Special Relativity and the quadripotential $\mathscr{A}_\mu$ was introduced. Indeed, with the development first of Dirac equation and then of Quantum Electrodynamics, the covariant derivative of the electron wave-function made its entrance in the language of theoretical physics:

$$\nabla_\mu \psi = (\partial_\mu - i\mathscr{A}_\mu)\psi \tag{10.32}$$

The operator ∇_μ might be assimilated with the action of a horizontal vector field (10.30) if we just had a one-dimensional structural group U(1) with a single generator and if we were acting on functions over $P(\mathscr{M}, \mathrm{U}(1))$ that are eigenstates of the unique left-invariant vector field $\mathbf{T}^{(L)}$ with eigenvalue i. This is indeed the case of Electrodynamics. The group U(1) is composed by all unimodular complex numbers $\exp[i\alpha]$ and the left-invariant vector field is simply $\mathbf{T}^{(L)} = \vec{\partial}/\partial\alpha$. The wave-function $\psi(x)$ is actually a section of rank-1 complex vector bundle associated with the principal bundle $P(\mathscr{M}_4, \mathrm{U}(1))$. Indeed, $\psi(x)$ is a complex spinor whose overall phase factor $\exp[i\alpha(x)]$ takes values in the U(1)-fiber over $x \in \mathscr{M}_4$. Since $\mathbf{T}^{(L)} \exp[i\alpha] = i\exp[i\alpha]$, the covariant derivative (10.32) is of the form (10.30).

This shows that the one-form connection coincides, at least in the case of electrodynamics, with the physical notion of gauge-potential which, upon quantization, describes the photon-field, namely, the mediator of electromagnetic interactions. The 1954 invention of non-abelian gauge-theories by Yang and Mills started from the generalization of the covariant derivative (10.32) to the case of the $\mathfrak{su}(2)$ Lie algebra. Introducing three gauge potentials $\mathscr{A}_\mu^A$, corresponding to the three standard generators J_A of $\mathfrak{su}(2)$:

$$[\mathrm{J}_\mathrm{A}, \mathrm{J}_\mathrm{B}] = \epsilon_{ABC}\mathrm{J}_\mathrm{C}; \quad \mathrm{J}_\mathrm{A} = i\sigma^A; \quad (A, B, C = 1, 2, 3) \tag{10.33}$$

where σ^A denotes the 2×2 Pauli matrices:

$$\sigma^1 = \begin{pmatrix} 0 & 1 \\ 1 & 0 \end{pmatrix}; \quad \sigma^2 = \begin{pmatrix} 0 & -i \\ i & 0 \end{pmatrix}; \quad \sigma^3 = \begin{pmatrix} 1 & 0 \\ 0 & -1 \end{pmatrix} \tag{10.34}$$

they wrote:

$$\nabla_\mu \begin{pmatrix} \psi_1 \\ \psi_2 \end{pmatrix} = (\partial_\mu - i\mathscr{A}_\mu^A \mathrm{J}_\mathrm{A}) \begin{pmatrix} \psi_1 \\ \psi_2 \end{pmatrix} \tag{10.35}$$

From the $\mathfrak{su}(2) \times \mathfrak{u}(1)$ standard model of electro-weak interactions of Glashow, Salam and Weinberg we know nowadays that, in appropriate combinations with $\mathscr{A}_\mu$, the three gauge fields $\mathscr{A}_\mu^A$ describe the $W^\pm$ and Z_0 particles discovered at CERN in 1983 by the UA1 experiment of Carlo Rubbia. They have spin one and mediate the weak interactions.

The mentioned examples show that the Lie algebra-valued connection one-form defined by Ehresman on principal bundles is clearly related to the gauge-fields of particle physics since it enters the construction of covariant derivatives via the notion of horizontal vector fields. On the other hand, the same one-form must be related to gravitation as well, since also there one deals with covariant derivatives

$$\nabla_\mu t_{\lambda_1\dots\lambda_n} \equiv \partial_\mu t_{\lambda_1\dots\lambda_n} - \left\{ \begin{matrix} \rho \\ \mu\lambda_1 \end{matrix} \right\} t_{\rho\lambda_2\dots\lambda_n} - \left\{ \begin{matrix} \rho \\ \mu\lambda_2 \end{matrix} \right\} t_{\lambda_1\rho\dots\lambda_n} \cdots$$
$$- \left\{ \begin{matrix} \rho \\ \mu\lambda_n \end{matrix} \right\} t_{\lambda_1\dots\lambda_{n-1}\rho} \tag{10.36}$$

sustained by the Levi-Civita connection and the Christoffel symbols

$$\left\{ \begin{matrix} \lambda \\ \mu\nu \end{matrix} \right\} \equiv \frac{1}{2} g^{\lambda\sigma} (\partial_\mu g_{\nu\sigma} + \partial_\nu g_{\mu\sigma} - \partial_\sigma g_{\mu\nu}) \tag{10.37}$$

The key to understanding the unifying point of view offered by the notion of Ehresman connection on a principal bundle is obtained by recalling what we just emphasized in Chapter 5 when we introduced the very notion of fiber-bundles. Following Definition 5.6.4 let us recall that to each principal bundle $P(\mathscr{M}, G)$ one can associate as many *associated vector bundles* as there are linear representations of the structural group G, namely, infinitely many. It suffices to use as transition functions the corresponding linear representations of the transition functions of the principal bundle as displayed in eq. (5.80). In all such associated bundles the fibers are vector spaces of dimension r (the rank of the bundle) and the transition functions are $r \times r$ matrices. Every linear representation of a Lie group G of dimension r induces a linear representation of its Lie algebra $\mathbb{G}$, where the left(right)-invariant vector fields $\mathbf{T}_A^{(L/R)}$ are mapped into $r \times r$ matrices:

$$\mathbf{T}_A^{(L/R)} \to D(T_A) \tag{10.38}$$

satisfying the same commutation relations:

$$[D(T_A), D(T_B)] = C_{AB}{}^C D(T_C) \tag{10.39}$$

It is therefore tempting to assume that given a one-form connection $\mathbf{A} \leftrightarrow \mathscr{A}$ on a principal bundle one can define a one-form connection on every associated vector bundle by taking its $r \times r$ matrix representation $D(\mathbf{A}) \leftrightarrow D(\mathscr{A})$. In which sense this matrix-valued one-form defines a connection on the considered vector bundle? To answer such a question we obviously need first to define connections on generic vector bundles, which is what we do in the next section.

10.3 Connections on a vector bundle

Let us now consider a generic vector bundle $E \overset{\pi}{\Longrightarrow} \mathscr{M}$ of rank r.[1] Its standard fiber is an r-dimensional vector space V and the transition functions from one local trivialization

---

1 Let us remind the reader that in the mathematical literature the rank of a vector bundle is just the dimension of its standard fiber (see eq. (10.45) and the following lines).

to another one are maps:

$$\psi_{\alpha\beta} : U_\alpha \cap U_\beta \mapsto \mathrm{Hom}(V, V) \sim \mathrm{GL}(r, \mathbb{R}) \tag{10.40}$$

In other words, without further restrictions the structural group of a generic vector bundle is just $\mathrm{GL}(r, \mathbb{R})$.

The notion of a connection on generic vector bundles is formulated according to the following:

Definition 10.3.1. Let $E \overset{\pi}{\Longrightarrow} \mathcal{M}$ be a vector bundle, $T\mathcal{M} \overset{\pi}{\Longrightarrow} \mathcal{M}$ the tangent bundle to the base manifold $\mathcal{M}$ ($\dim \mathcal{M} = m$) and let $\Gamma(E, \mathcal{M})$ be the space of sections of E: a **connection** ∇ on E is a rule that, to each vector field $\mathbf{X} \in \Gamma(T\mathcal{M}, \mathcal{M})$ associates a map:

$$\nabla_{\mathbf{X}} : \Gamma(E, \mathcal{M}) \mapsto \Gamma(E, \mathcal{M}) \tag{10.41}$$

satisfying the following defining properties:

a) $\nabla_{\mathbf{X}}(a_1 s_1 + b_1 s_2) = a_1 \nabla_{\mathbf{X}} s_1 + a_2 \nabla_{\mathbf{X}} s_2$
b) $\nabla_{a_1 \mathbf{X}_1 + a_2 \mathbf{X}_2} s = a_1 \nabla_{\mathbf{X}_1} s + a_2 \nabla_{\mathbf{X}_2} s$
c) $\nabla_{\mathbf{X}}(f \cdot s) = \mathbf{X}(f) \cdot s + f \cdot \nabla_{\mathbf{X}} s$
d) $\nabla_{f \cdot \mathbf{X}} s = f \cdot \nabla_{\mathbf{X}} s$

where $a_{1,2} \in \mathbb{R}$ are real numbers, $s_{1,2} \in \Gamma(E, \mathcal{M})$ are sections of the vector bundle and $f \in C^\infty(\mathcal{M})$ is a smooth function.

The abstract description of a connection provided by Definition 10.3.1 becomes more explicit if we consider bases of sections for the two involved vector bundles. To this effect let $\{s_i(p)\}$ be a basis of sections of the vector bundle $E \overset{\pi}{\Longrightarrow} \mathcal{M}$ ($i = 1, 2, \dots, r = \operatorname{rank} E$) and let $\{\mathbf{e}_\mu(p)\}$, ($\mu = 1, 2, \dots, m = \dim \mathcal{M}$) be a basis of sections for the tangent bundle to the base manifold $T\mathcal{M} \overset{\pi}{\Longrightarrow} \mathcal{M}$. Then we can write:

$$\nabla_{\mathbf{e}_\mu} s_i \equiv \nabla_\mu s_i = \Theta_{\mu i}{}^j s_j \tag{10.42}$$

where the local functions $\Theta_{\mu i}{}^j(p)$, ($\forall p \in \mathcal{M}$) are named the **coefficients of the connection.** They are necessary and sufficient to specify the connection on an arbitrary section of the vector bundle. To this effect let $s \in \Gamma(E, \mathcal{M})$. By definition of basis of sections we can write:

$$s = c^i(p) s_i(p) \tag{10.43}$$

and correspondingly we obtain:

$$\begin{aligned} \nabla_\mu s &= (\nabla_\mu c^i) \cdot s_i \\ \nabla_\mu c^i &\equiv \partial_\mu c^i + \Theta_{\mu j}{}^i c^j \end{aligned} \tag{10.44}$$

having defined $\partial_\mu c^i \equiv e_\mu(c^i)$. Confronting this with the discussions at the end of Section 10.2.1.1, we see that in the language of physicists $\nabla_\mu c^i$ can be identified as the covariant derivatives of the vector components c^i. Let us now observe that for any vector bundle $E \overset{\pi}{\Longrightarrow} \mathscr{M}$ if:

$$r \equiv \text{rank}(E) \tag{10.45}$$

is the rank, namely, the dimension of the standard fiber and:

$$m \equiv \dim \mathscr{M}$$

is the dimension of the base manifold, the connection coefficients can be viewed as a set of m matrices, each of them $r \times r$:

$$\forall e_\mu(p) : \Theta_{\mu j}{}^{i} = (\Theta_{e_\mu})_j^{\ i} \tag{10.46}$$

Hence more abstractly we can say that the connection on a vector bundle associates to each vector field defined on the base manifold a matrix of dimension equal to the rank of the bundle:

$$\begin{aligned}&\forall X \in \Gamma(T\mathscr{M}, \mathscr{M}) :\\ &\nabla : X \mapsto \Theta_X = r \times r\text{-matrix depending on the base-point } p \in \mathscr{M}\end{aligned} \tag{10.47}$$

The relevant question is the following: Can we relate the matrix Θ_X to a one-form and make a bridge between the above definition of a connection and that provided by the Ehresmann approach? The answer is positive and it is encoded in the following equivalent:

Definition 10.3.2. Let $E \overset{\pi}{\Longrightarrow} \mathscr{M}$ be a vector bundle, $T^*\mathscr{M} \overset{\pi}{\Longrightarrow} \mathscr{M}$ the cotangent bundle to the base manifold $\mathscr{M}$ ($\dim \mathscr{M} = m$) and let $\Gamma(E, \mathscr{M})$ be the space of sections of E: a **connection** ∇ on E is a map:

$$\nabla : \Gamma(E, \mathscr{M}) \mapsto \Gamma(E, \mathscr{M}) \otimes T^*\mathscr{M} \tag{10.48}$$

which to each section $s \in \Gamma(E, \mathscr{M})$ associates a one-form ∇s with values in $\Gamma(E, \mathscr{M})$ such that the following defining properties are satisfied:

a) $\nabla(a_1 s_1 + b_1 s_2) = a_1 \nabla s_1 + a_2 \nabla s_2$

b) $\nabla(f \cdot s) = df \cdot s + f \cdot \nabla s$

where $a_{1,2} \in \mathbb{R}$ are real numbers, $s_{1,2} \in \Gamma(E, \mathscr{M})$ are sections of the vector bundle and $f \in C^\infty(\mathscr{M})$ is a smooth function.

The relation between the two definitions of a connection on a vector bundle is now easily obtained by stating that for each section $s \in \Gamma(E, \mathcal{M})$ and for each vector field $\mathbf{X} \in \Gamma(T\mathcal{M}, \mathcal{M})$ we have:

$$\nabla_{\mathbf{X}} s = \nabla s(\mathbf{X}) \tag{10.49}$$

The reader can easily convince himself that using (10.49) and relying on the properties of ∇s established by Definition 10.3.2 those of $\nabla_X s$ required by Definition 10.3.1 are all satisfied.

Consider now, just as before, a basis of sections of the vector bundle $\{s_i(p)\}$. According to its second Definition 10.3.2 a connection ∇ singles out an $r \times r$ matrix-valued one-form $\Theta \in \Gamma(T^*\mathcal{M}, \mathcal{M})$ through the equation:

$$\nabla s_i = \Theta_i^j s_j \tag{10.50}$$

Clearly the connection coefficients introduced in eq. (10.47) are nothing else but the values of Θ on each vector field $\mathbf{X}$:

$$\Theta_{\mathbf{X}} = \Theta(\mathbf{X}) \tag{10.51}$$

A natural question which arises at this point is the following. In view of our comments at the end of Section 10.2.1.1 could we identify Θ with the matrix representation of $\mathcal{A}$, the principal connection on the corresponding principal bundle? In other words, could we write:

$$\Theta = D(\mathcal{A})? \tag{10.52}$$

For a single generic vector bundle this question seems tautological. Indeed, the structural group is simply $\mathrm{GL}(r, \mathbb{R})$ and being the corresponding Lie algebra $\mathfrak{gl}(r, \mathbb{R})$ made by the space of generic matrices it looks like obvious that Θ could be viewed as Lie algebra valued. The real question however is another. The identification (10.52) is legitimate if and only if Θ transforms as the matrix representation of $\mathcal{A}$. This is precisely the case. Let us demonstrate it. Consider the intersection of two local trivializations. On $U_\alpha \cap U_\beta$ the relation between two bases of sections is given by:

$$s_i^{(\alpha)} = D(\mathfrak{t}_{(\alpha\beta)})_i^j s_j^{(\beta)} \tag{10.53}$$

where $\mathfrak{t}_{(\alpha\beta)}$ is the transition function seen as an abstract group element. Equation (10.53) implies:

$$\Theta^{(\alpha)} = D(d\mathfrak{t}_{(\alpha\beta)}) \cdot D(\mathfrak{t}_{(\alpha\beta)})^{-1} + D(\mathfrak{t}_{(\alpha\beta)}) \cdot \Theta^{(\beta)} \cdot D(\mathfrak{t}_{(\alpha\beta)})^{-1} \tag{10.54}$$

which is consistent with the gauge transformation rule (10.27) if we identify Θ as in eq. (10.52).

The outcome of the above discussion is that once we have defined a connection one-form $\mathscr{A}$ on a principal bundle $P(\mathscr{M}, G)$, a connection is induced on any associated vector bundle $E \overset{\pi}{\Longrightarrow} \mathscr{M}$. It is simply defined as follows:

$$\forall s \in \Gamma(E, \mathscr{M}) : \quad \nabla s = ds + D(\mathscr{A})s \tag{10.55}$$

where $D()$ denotes the linear representation of both G and $\mathbb{G}$ that define the associated bundle.

10.4 An illustrative example of fiber-bundle and connection

In this section we discuss a simple example that illustrates both the general definition of fiber-bundle and that of principal connection. The chosen example has an intrinsic physical interest since it corresponds to the mathematical description of a magnetic monopole in the standard electromagnetic theory. It is also geometrically relevant since it shows how a differentiable manifold can sometimes be reinterpreted as the total space of a fiber-bundle constructed on a smaller manifold.

10.4.1 The magnetic monopole and the Hopf fibration of S^3

We introduce our case study, defining a family of principal bundles that depend on an integer $n \in \mathbb{Z}$. Explicitly, let $P(\mathbb{S}^2, \mathrm{U}(1))$ be a principal bundle where the base manifold is a 2-sphere $\mathscr{M} = \mathbb{S}^2$, the structural group is a circle $\mathbb{S}^1 \sim \mathrm{U}(1)$ and, by definition of principal bundle, the standard fiber coincides with the structural group $F = \mathbb{S}^1$. An element $f \in F$ of this fiber is just a complex number of modulus one $f = \exp[i\theta]$. To describe this bundle we need an atlas of local trivializations and hence, to begin with, an atlas of open charts of the base manifold $\mathbb{S}^2$. We use the two charts provided by the stereographic projection. Defining $\mathbb{S}^2$ via the standard quadratic equation in $\mathbb{R}^3$:

$$\mathbb{S}^2 \subset \mathbb{R}^3 : \{x_1, x_2, x_3\} \in \mathbb{S}^2 \leftrightarrow x_1^2 + x_2^2 + x_3^2 = 1 \tag{10.56}$$

we define the two open neighborhoods of the North and the South Pole, respectively named H^+ and H^- that correspond, in $\mathbb{R}^3$ language, to the exclusion of either the point $\{0, 0, -1\}$ or the point $\{0, 0, 1\}$. These neighborhoods are shown in Figure 10.2. As we

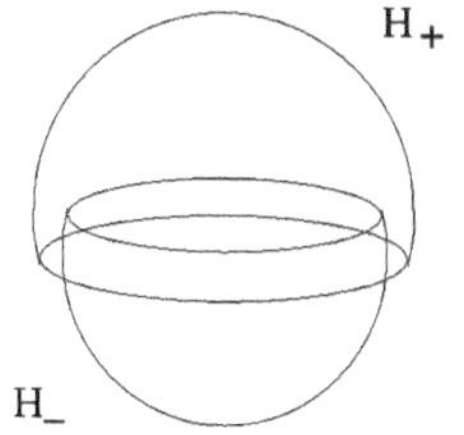

Figure 10.2: The two open charts H^+ and H^- covering the two-sphere.

already know from Chapter 5, the stereographic projections s_N and s_S mapping H^- and H^+ onto the complex plane $\mathbb{C}$ are as follows:

$$\begin{aligned} s_N \quad H^- \to \mathbb{C} \quad z_N \equiv s_N(x_1,x_2,x_3) = \frac{x_1 + ix_2}{1-x_3} \\ s_S \quad H^+ \to \mathbb{C} \quad z_S \equiv s_S(x_1,x_2,x_3) = \frac{x_1 - ix_2}{1+x_3} \end{aligned} \tag{10.57}$$

and on the intersection $H^- \cap H^+$ we have the transition function $z_N = 1/z_S$ (see eq. (5.19)).

We construct our fiber-bundle introducing two local trivializations, respectively associated with the two open charts H^- and H^+:

$$\begin{aligned} \phi_N^{-1} : \pi^{-1}(H^-) \to H^- \otimes U(1) : \quad \phi_N^{-1}(\{x_1,x_2,x_3\}) = (z_N, \exp[i\psi_N]) \\ \phi_S^{-1} : \pi^{-1}(H^+) \to H^+ \otimes U(1) : \quad \phi_S^{-1}(\{x_1,x_2,x_3\}) = (z_S, \exp[i\psi_S]) \end{aligned} \tag{10.58}$$

To complete the construction of the bundle we still need to define a transition function between the two local trivializations, namely, a map:

$$t_{SN} : H^- \cap H^+ \to \mathrm{U}(1) \tag{10.59}$$

such that

$$(z_S, \exp[i\psi_S]) = \left(\frac{1}{z_N}, t_{SN}(z_N)\exp[i\psi_N]\right) \tag{10.60}$$

In order to illustrate the geometrical interpretation of the transition function we are going to write, it is convenient to make use of polar coordinates on the two spheres. Following the standard conventions, we parameterize the points of $\mathbb{S}^2$ by means of the two angles $\theta \in [0,\pi]$ and $\phi \in [0,2\pi]$:

$$x_1 = \sin\theta\cos\phi; \quad x_2 = \sin\theta\sin\phi; \quad x_3 = \cos\theta \tag{10.61}$$

On the intersection $H^+ \cap H^-$ we obtain:

$$z_N = \frac{\sin\theta}{1-\cos\theta}\exp[i\phi]; \quad z_S = \frac{\sin\theta}{1+\cos\theta}\exp[-i\phi] \tag{10.62}$$

and we see that the azimuth angle ϕ parameterizes the points of the equator identified by the condition $\theta = \pi/2 \Rightarrow x_3 = 0$. Then we write the transition function as follows:

$$t_{SN}(z_N) = \exp[in\phi] = \left(\frac{z_N}{|z_N|}\right)^n = \left(\frac{|z_S|}{z_S}\right)^n; \quad n \in \mathbb{Z} \tag{10.63}$$

The catch in the above equation is that each choice of the integer number n identifies a different inequivalent principal bundle. The difference is that while going around the equator the transition function can cover the structural group $\mathbb{S}^1$ once, twice or an integer number of times. The transition function can also go clockwise or anti-clockwise and this accounts for the sign of n. We will see that the integer n characterizing the bundle can be interpreted as the magnetic charge of a magnetic monopole. Before going into that let us first consider the special properties of the case $n = 1$.

10.4.1.1 The case $n = 1$ and the Hopf fibration of $\mathbb{S}^3$

Let us name $P_{(n)}(\mathbb{S}^2, \mathrm{U}(1))$ the principal bundles we have constructed in the previous section. Here we want to show that the total space of the bundle $P_{(1)}(\mathbb{S}^2, \mathrm{U}(1))$ is actually the 3-sphere $\mathbb{S}^3$.

To this effect we define $\mathbb{S}^3$ as the locus in $\mathbb{R}^4$ of the standard quadric:

$$\mathbb{S}^3 \subset \mathbb{R}^4 : \{X_1, X_2, X_3, X_4\} \in \mathbb{S}^3 \leftrightarrow X_1^2 + X_2^2 + X_3^2 + X_4^2 = 1 \tag{10.64}$$

and we introduce the *Hopf map*

$$\pi : \mathbb{S}^3 \to \mathbb{S}^2 \tag{10.65}$$

which is explicitly given by:

$$\begin{aligned} x_1 &= 2(X_1X_3 + X_2X_4) \\ x_2 &= 2(X_2X_3 - X_1X_4) \\ x_3 &= X_1^2 + X_2^2 - X_3^2 - X_4^2 \end{aligned} \tag{10.66}$$

It is immediately verified that x_i $(i = 1, 2, 3)$ as given in eq. (10.66) satisfy (10.56) if X_i $(i = 1, 2, 3, 4)$ satisfy eq. (10.64). Hence this is indeed a projection map from $\mathbb{S}^3$ to $\mathbb{S}^2$ as claimed in (10.65). We want to show that π can be interpreted as the projection map in the principal fiber-bundle $P_{(1)}(\mathbb{S}^2, \mathrm{U}(1))$. To obtain this result we need first to show that for each $p \in \mathbb{S}^3$ the fiber over p is diffeomorphic to a circle, namely, $\pi^{-1}(p) \sim \mathrm{U}(1)$. To this effect we begin by complexifying the coordinates of $\mathbb{R}^4$:

$$Z_1 = X_1 + iX_2; \quad Z_2 = X_3 + iX_4 \tag{10.67}$$

so that the equation defining $\mathbb{S}^3$ (10.64) becomes the following equation in $\mathbb{C}^2$:

$$|Z_1|^2 + |Z_2|^2 = 1 \tag{10.68}$$

Then we name $U^- \subset \mathbb{S}^3$ and $U^+ \subset \mathbb{S}^3$ the two open neighborhoods where we respectively have $Z_2 \neq 0$ and $Z_1 \neq 0$. The Hopf map (10.66) can be combined with the stereographic projection of the North Pole in the neighborhood U^- and with the stereographic projection of the South Pole on the neighborhood U^+ obtaining:

$$\begin{aligned} \forall \{Z_1, Z_2\} \in U^- \subset \mathbb{S}^3; \quad s_N \circ \pi(Z_1, Z_2) &= \frac{Z_1}{Z_2} \\ \forall \{Z_1, Z_2\} \in U^+ \subset \mathbb{S}^3; \quad s_S \circ \pi(Z_1, Z_2) &= \frac{Z_2}{Z_1} \end{aligned} \tag{10.69}$$

From both local descriptions of the Hopf map we see that:

$$\forall \lambda \in \mathbb{C} \quad \pi(\lambda Z_1, \lambda Z_2) = \pi(Z_1, Z_2) \tag{10.70}$$

On the other hand, if $(Z_1, Z_2) \in \mathbb{S}^3$ we have that $(\lambda Z_1, \lambda Z_2) \in \mathbb{S}^3$ if and only if $\lambda \in \mathrm{U}(1)$, namely, if $|\lambda|^2 = 1$. This proves the desired result that $\pi^{-1}(p) \sim \mathrm{U}(1)$ for all $p \in \mathbb{S}^3$. The two equations (10.69) can also be interpreted as two local trivializations of a principal U(1) bundle with $\mathbb{S}^2$ base manifold whose total space is the 3-sphere $\mathbb{S}^3$. Indeed, we can rewrite them in the opposite directions as follows:

$$\begin{aligned} \phi_N^{-1} &: U^- \equiv \pi^{-1}(H^-) \to H^- \otimes U(1); \quad [Z_1, Z_2] \to \left[\frac{Z_1}{Z_2}, \frac{Z_2}{|Z_2|}\right] \\ \phi_S^{-1} &: U^+ \equiv \pi^{-1}(H^+) \to H^+ \otimes U(1); \quad [Z_1, Z_2] \to \left[\frac{Z_2}{Z_1}, \frac{Z_1}{|Z_1|}\right] \end{aligned} \tag{10.71}$$

This time the transition function does not have to be invented, rather it is decided a priori by the Hopf map. Indeed, we immediately read it from eqs (10.71):

$$t_{SN} = \frac{Z_1}{|Z_1|}\frac{|Z_2|}{Z_2} = \frac{Z_1/Z_2}{|Z_1/Z_2|} = \frac{z_N}{|z_N|} \tag{10.72}$$

which coincides with eq. (10.63) for $n = 1$. This is the result we wanted to prove: the 3-sphere is the total space for the principal fiber $P_{(1)}(\mathbb{S}^2, \mathrm{U}(1))$.

10.4.1.2 The U(1)-connection of the Dirac magnetic monopole

The Dirac monopole [36] of magnetic charge n corresponds to introducing a vector potential $\mathscr{A} = \mathscr{A}_i(x)\, dx^i$ that is a well-defined connection on the principal U(1)-bundle $P_{(n)}(\mathbb{S}^2, \mathrm{U}(1))$ we discussed above. Physically the discussion goes as follows. Let us work in ordinary three-dimensional space $\mathbb{R}^3$ and assume that in the origin of our coordinate system there is a magnetically charged particle.

Then we observe a magnetic field of the form:

$$\mathbf{H} = \frac{m}{r^3}\mathbf{r} \quad \Leftrightarrow \quad H_i = \frac{m}{r^3}x_i \tag{10.73}$$

From the relation $F_{ij} = \epsilon_{ijk}H_k$ we conclude that the electromagnetic field strength associated with such a magnetic monopole is the following 2-form on $\mathbb{R}^3 - 0$:

$$\mathfrak{F} = \frac{m}{r^3}\epsilon_{ijk}x^i\, dx^j \wedge dx^k \tag{10.74}$$

Indeed, eq. (10.74) makes sense everywhere in $\mathbb{R}^3$ except at the origin $\{0, 0, 0\}$. Using polar coordinates rather than Cartesian ones the 2-form $\mathfrak{F}$ assumes a very simple expression:

$$\mathfrak{F} = \frac{g}{2}\sin\theta\, d\theta \wedge d\phi \tag{10.75}$$

namely, it is proportional to the volume form on the 2-sphere $\mathrm{Vol}(\mathbb{S}^2) \equiv \sin\theta\, d\theta \wedge d\phi$. We leave it as an exercise for the reader to calculate the relation between the constant g

appearing in (10.75) and the constant m appearing in (10.74). Hence the 2-form F can be integrated on any 2-sphere of arbitrary radius and the result, according to Gauss law, is proportional to the charge of the magnetic monopole. Explicitly we have:

$$\int_{\mathbb{S}^2} \mathfrak{F} = \frac{g}{2} \int_0^{\pi} \sin\theta \, d\theta \int_0^{2\pi} d\phi = 2\pi g \tag{10.76}$$

Consider now the problem of writing a vector potential for this magnetic field. By definition, in each local trivialization (U_α, ϕ_α) of the underlying U(1) bundle we must have:

$$\mathfrak{F} = d\mathscr{A}_\alpha \tag{10.77}$$

where $\mathscr{A}_\alpha$ is a 1-form on U_α. Yet we have just observed that $\mathfrak{F}$ is proportional to the volume form on $\mathbb{S}^2$ which is a closed but not exact form. Indeed, if the volume form were exact, namely, if there existed a global section ω of $T^*\mathbb{S}^2$ such that $\mathrm{Vol} = d\omega$ than, using Stokes theorem we would arrive at the absurd conclusion that the volume of the sphere vanishes: indeed, $\int_{\mathbb{S}^2} \mathrm{Vol} = \int_{\partial\mathbb{S}^2} \omega = 0$ since the 2-sphere has no boundary. Hence there cannot be a vector potential globally defined on the 2-sphere. The best we can do is to have two different definitions of the vector potential, one in the North Pole patch and one in the South Pole patch. Recalling that the vector potential is the component of a connection on a U(1) bundle we multiply it by the Lie algebra generator of U(1) and we write the ansatz for the full connection in the two patches:

$$\begin{aligned} &\text{on } U^-; \quad \mathscr{A} = \mathscr{A}_N = i\frac{1}{2}g(1-\cos\theta)\,d\phi \\ &\text{on } U^+; \quad \mathscr{A} = \mathscr{A}_S = -i\frac{1}{2}g(1+\cos\theta)\,d\phi \end{aligned} \tag{10.78}$$

As one sees the first definition of the connection does not vanish as long as $\theta \neq 0$, namely, as long as we stay away from the North Pole, while the second definition does not vanish as long as $\theta \neq \pi$, namely, as long as we stay away from the South Pole. On the other hand, if we calculate the exterior derivative of either $\mathscr{A}_N$ or $\mathscr{A}_S$ we obtain the same result, namely i times the 2-form (10.75):

$$\mathfrak{F} = d\mathscr{A}_S = d\mathscr{A}_N = i \times \frac{g}{2} \sin\theta \, d\theta \wedge d\phi \tag{10.79}$$

The transition function between the two local trivializations can now be reconstructed from:

$$\mathscr{A}_S = \mathscr{A}_N + t_{SN}^{-1}\, dt_{NS} = -ig\, d\phi \tag{10.80}$$

which implies

$$t_{NS} = \exp[ig\phi] \tag{10.81}$$

and by comparison with eq. (10.63) we see that the magnetic charge g is the integer n classifying the principal U(1) bundle. The reason why it has to be an integer was already noted. On the equator of the 2-sphere the transition function realizes a map of the circle $\mathbb{S}^1$ into itself. Such a map *winds* a certain number of times but cannot wind a non-integer number of times: otherwise its image is no longer a closed circle.

In this way we have seen that the quantization of the magnetic charge is a topological condition.

10.5 Riemannian and pseudo-Riemannian metrics: the mathematical definition

We have discussed at length the notion of connections on fiber-bundles. From the historical review presented at the beginning of the chapter we know that connections appeared first in relation with the metric geometry introduced by Gauss and Riemann and developed by Ricci, Levi-Civita and Bianchi. It is now time to come to the rigorous modern mathematical definition of metrics.

The mathematical environment in which we will work is that provided by a differentiable manifold $\mathcal{M}$ of dimension $\dim \mathcal{M} = m$. As several times emphasized in Chapter 5, we can construct many different fiber-bundles on the same base manifold $\mathcal{M}$ but there are two which are intrinsically defined by the differentiable structure of $\mathcal{M}$, namely, the tangent bundle $T\mathcal{M} \stackrel{\pi}{\Longrightarrow} \mathcal{M}$ and its dual, the cotangent bundle $T^*\mathcal{M} \stackrel{\pi}{\Longrightarrow} \mathcal{M}$. Of these bundles we can take the tensor powers (symmetric or antisymmetric, or none) which are bundles whose sections transform with the corresponding tensor powers of the transition functions of the original bundle. The sections of these tensor-powers are the tensors introduced by Ricci and Levi-Civita in their *absolute differential calculus*, for whose differentiation Christoffel had invented his symbols (see eqs (10.37) and (10.36)).

For brevity let us use a special notation for the space of sections of the tangent bundle, namely, for the algebra of vector fields: $\mathfrak{X}(\mathcal{M}) \equiv \Gamma(T\mathcal{M}, \mathcal{M})$. Then we have:

Definition 10.5.1. A **pseudo-Riemannian** metric on a manifold $\mathcal{M}$ is a $C^\infty(\mathcal{M})$, symmetric, bilinear, non-degenerate form on $\mathfrak{X}(\mathcal{M})$, namely, it is a map:

$$g(,) : \mathfrak{X}(\mathcal{M}) \otimes \mathfrak{X}(\mathcal{M}) \longrightarrow C^\infty(\mathcal{M})$$

satisfying the following properties

(i) $\forall \mathbf{X}, \mathbf{Y} \in \mathfrak{X}(\mathcal{M}) : g(\mathbf{X}, \mathbf{Y}) = g(\mathbf{Y}, \mathbf{X}) \in C^\infty(\mathcal{M})$

(ii) $\forall \mathbf{X}, \mathbf{Y}, \mathbf{Z} \in \mathfrak{X}(\mathcal{M}),\ \forall f, g \in C^\infty(\mathcal{M}) : g(f\mathbf{X} + h\mathbf{Y}, \mathbf{Z}) = fg(\mathbf{X}, \mathbf{Z}) + hg(\mathbf{Y}, \mathbf{Z})$

(iii) $\{\forall X \in \mathfrak{X}(\mathcal{M}), g(\mathbf{X}, \mathbf{Y}) = 0\} \Rightarrow Y \equiv 0$

Definition 10.5.2. A **Riemannian** metric on a manifold $\mathcal{M}$ is a Riemannian metric which in addition satisfies the following two properties:

(iv) $\forall \mathbf{X} \in \mathfrak{X}(\mathcal{M}) : g(\mathbf{X}, \mathbf{X}) \geq 0$
(v) $g(\mathbf{X}, \mathbf{X}) = 0 \Rightarrow \mathbf{X} \equiv 0$

The two additional properties satisfied by a Riemannian metric state that it should be not only a symmetric non-degenerate form but also a positive definite one.

The definition of pseudo-Riemannian and Riemannian metrics can be summarized by saying that a metric is a section of the second symmetric tensor power of the cotangent bundle, namely, we have:

$$g \in \Gamma\Big(\bigotimes_{\text{symm}}{}^2 T^*\mathcal{M}, \mathcal{M}\Big) \tag{10.82}$$

In a coordinate patch the metric is described by a symmetric rank-2 tensor $g_{\mu\nu}(x)$. Indeed, we can write:

$$g = g_{\mu\nu}(x)\, dx^\mu \otimes dx^\nu \tag{10.83}$$

and we have:

$$\forall \mathbf{X}, \mathbf{Y} \in \mathfrak{X}(\mathcal{M}) : \quad g(\mathbf{X}, \mathbf{Y}) = g_{\mu\nu}(x) X^\mu(x) Y^\nu(x) \tag{10.84}$$

if

$$\mathbf{X} = X^\mu(x)\vec{\partial}_\mu; \quad \mathbf{Y} = Y^\nu(x)\vec{\partial}_\nu \tag{10.85}$$

are the local expression of the two vector fields in the considered coordinate patch. The essential point in the definition of the metric is the statement that $g(X, Y) \in C^\infty(\mathcal{M})$, namely, that $g(X, Y)$ should be a scalar. This implies that the coefficients $g_{\mu\nu}(x)$ should transform from a coordinate patch to the next one with the inverse of the transition functions with which transform the vector field components. This is what implies eq. (10.82). On the other hand, according to the viewpoint started by Gauss and developed by Riemann, Ricci and Levi-Civita, the metric is the mathematical structure which allows to define infinitesimal lengths and, by integration, the length of any curve. Indeed, given a curve $\mathcal{C} : [0, 1] \longrightarrow \mathcal{M}$ let $\mathbf{t}(s)$ be its tangent vector at any point of the curve $\mathcal{C}$, parameterized $s \in [0, 1]$: we define the length of the curve as:

$$\mathfrak{s}(\mathcal{C}) \equiv \int_0^1 g(\mathbf{t}(s), \mathbf{t}(s))\, ds \tag{10.86}$$

10.5.1 Signatures

At each point of the manifold $p \in \mathcal{M}$, the coefficients $g_{\mu\nu}(p)$ of a metric g constitute a symmetric non-degenerate real matrix. Such a matrix can always be diagonalized by means of an orthogonal transformation $\mathcal{O}(p) \in \mathrm{SO}(N)$, which obviously varies from point to point, namely, we can write:

$$g_{\mu\nu}(p) = \mathcal{O}^T(p) \begin{pmatrix} \lambda_1(p) & 0 & \dots & 0 & 0 \\ 0 & \lambda_2(p) & 0 & \dots & 0 \\ \vdots & \vdots & \vdots & \vdots & \vdots \\ 0 & \dots & \dots & \lambda_{N-1}(p) & 0 \\ 0 & \dots & \dots & \dots & \lambda_N(p) \end{pmatrix} \mathcal{O}(p) \tag{10.87}$$

where the real numbers $\lambda_i(p)$ are the eigenvalues. Each of them depends on the point $p \in \mathcal{M}$, but none of them can vanish, otherwise the determinant of the metric would be zero which contradicts one of the defining axioms. Consider next the diagonal matrix:

$$\mathcal{L}(p) = \begin{pmatrix} \frac{1}{\sqrt{|\lambda_1(p)|}} & 0 & \dots & 0 & 0 \\ 0 & \frac{1}{\sqrt{|\lambda_2(p)|}} & 0 & \dots & 0 \\ \vdots & \vdots & \vdots & \vdots & \vdots \\ 0 & \dots & \dots & \frac{1}{\sqrt{|\lambda_{N-1}(p)|}} & 0 \\ 0 & \dots & \dots & \dots & \frac{1}{\sqrt{|\lambda_{N-1}(p)|}} \end{pmatrix} \tag{10.88}$$

The matrix $M = \mathcal{O} \cdot \mathcal{L}$, where for brevity the dependence on the point p has been omitted, is such that:

$$g_{\mu\nu} = M^T \begin{pmatrix} \mathrm{sign}[\lambda_1] & 0 & \dots & 0 & 0 \\ 0 & \mathrm{sign}[\lambda_2] & 0 & \dots & 0 \\ \vdots & \vdots & \vdots & \vdots & \vdots \\ 0 & \dots & \dots & \mathrm{sign}[\lambda_{N-1}] & 0 \\ 0 & \dots & \dots & \dots & \mathrm{sign}[\lambda_N] \end{pmatrix} M \tag{10.89}$$

having denoted by $\mathrm{sign}[x]$ the function which takes the value 1 if $x > 0$ and the value -1 if $x < 0$. Hence we arrive at the following conclusion and definition.

Definition 10.5.3. Let $\mathcal{M}$ be a differentiable manifold of dimension m endowed with a metric g. At every point $p \in \mathcal{M}$ by means of a transformation $g \mapsto S^T \cdot g \cdot S$, the metric tensor $g_{\mu\nu}(p)$ can be reduced to a diagonal matrix with p entries equal to 1 and $m - p$ entries equal to -1. The pair of integers $(p, m - p)$ is named the **signature** of the metric g.

The rationale of the above definition is that the signature of a metric is an intrinsic property of g, independent both of the chosen coordinate patch and of the chosen point.

This issue was already discussed in Section 2.5.6 (see in particular eq. (2.39)) where it was recalled that it is the content of a theorem proved in 1852 by James Sylvester. According to Sylvester's theorem a symmetric non-degenerate matrix A can always be transformed into a diagonal one with ±1 entries by means of a substitution $A \mapsto B^T \cdot A \cdot B$. On the other hand, no such transformation can alter the signature $(p, m-p)$, which is intrinsic to the matrix A. This is what happens for a single matrix. Consider now a point-dependent matrix like the metric tensor g, whose entries are smooth functions. Defining $s = 2p - m$ the difference between the number of positive and negative eigenvalues of g it follows that also s is a smooth function. Yet s is an integer by definition. Hence it has to be a constant.

The metrics on a differentiable manifold are therefore intrinsically characterized by their signatures. Riemannian are the positive definite metrics with signature $(m, 0)$. Lorentzian are the metrics with signature $(1, m-1)$ just as the flat metric of Minkowski space. There are also metrics with more elaborate signatures which appear in many mathematical constructions related with supergravity.

10.6 The Levi-Civita connection

Having established the rigorous mathematical notion of both a metric and a connection we come back to the ideas of Riemannian curvature and Torsion which were heuristically touched upon in the course of our historical outline. In particular we are now in a position to derive from clear-cut mathematical principles of the Christoffel symbols anticipated in eq. (10.37), the Riemann tensor and the Torsion tensor. The starting point for the implementation of this plan is provided by a careful consideration of the special properties of affine connections.

10.6.1 Affine connections

In Definitions 10.3.1, 10.3.2 we fixed the notion of a connection on a generic vector bundle $E \stackrel{\pi}{\Longrightarrow} \mathscr{M}$. In particular we can consider the tangent bundle $T\mathscr{M} \stackrel{\pi}{\Longrightarrow} \mathscr{M}$. A connection on $T\mathscr{M}$ is named *affine*. It follows that we can give the following:

Definition 10.6.1. Let $\mathscr{M}$ be an m-dimensional differentiable manifold, an affine connection on $\mathscr{M}$ is a map

$$\nabla : \mathfrak{X}(\mathscr{M}) \times \mathfrak{X}(\mathscr{M}) \to \mathfrak{X}(\mathscr{M})$$

which satisfies the following properties:

(i) $\forall \mathbf{X}, \mathbf{Y}, \mathbf{Z} \in \mathfrak{X}(\mathscr{M}) : \nabla_{\mathbf{X}}(\mathbf{Y} + \mathbf{Z}) = \nabla_{\mathbf{X}}\mathbf{Y} + \nabla_{\mathbf{X}}\mathbf{Z}$

(ii) $\forall \mathbf{X}, \mathbf{Y}, \mathbf{Z} \in \mathfrak{X}(\mathscr{M}) : \nabla_{(\mathbf{X}+\mathbf{Y})}\mathbf{Z} = \nabla_{\mathbf{X}}\mathbf{Z} + \nabla_{\mathbf{Y}}\mathbf{Z}$

(iii) $\forall \mathbf{X}, \mathbf{Y} \in \mathfrak{X}(\mathscr{M}), \forall f \in C^\infty(\mathscr{M}) : \nabla_{f\mathbf{X}}\mathbf{Y} = f\nabla_{\mathbf{X}}\mathbf{Y}$

(iv) $\forall \mathbf{X}, \mathbf{Y} \in \mathfrak{X}(\mathscr{M}), \forall f \in C^\infty(\mathscr{M}) : \nabla_{\mathbf{X}}(f\mathbf{Y}) = \mathbf{X}[f]\,\mathbf{Y} + f\,\nabla_{\mathbf{X}}\mathbf{Y}$

Clearly also affine connections are encoded into corresponding connection one-forms, which are traditionally denoted by the symbol Γ. In the affine case, Γ is $\mathfrak{gl}(m,\mathbb{R})$-Lie algebra valued since the structural group of $T\mathcal{M}$ is $\mathrm{GL}(m,\mathbb{R})$. Let $\{\mathbf{e}_\mu\}$ be a basis of sections of the tangent bundle so that any vector field $\mathbf{X} \in \mathfrak{X}(\mathcal{M})$ can be written as follows:

$$\mathbf{X} = X^\mu(x)\mathbf{e}_\mu \tag{10.90}$$

The connection one-form is defined by calculating the *covariant differentials* of the basis elements:

$$\nabla\mathbf{e}_\nu = \Gamma_\nu{}^\rho\mathbf{e}_\rho \tag{10.91}$$

Let us introduce the dual basis of $T^*\mathcal{M}$, namely, the set of one-forms ω^μ such that:

$$\omega^\mu(\mathbf{e}_\nu) = \delta^\mu_\nu \tag{10.92}$$

The matrix-valued one-form Γ can be expanded along such a basis obtaining:

$$\Gamma = \omega^\mu\Gamma_\mu \quad \Rightarrow \quad \Gamma_\nu^\rho = \omega^\mu\Gamma^\rho_{\mu\nu} \tag{10.93}$$

The tri-index symbols $\Gamma^\rho_{\mu\nu}$ encode, patch by patch, the considered affine connection. According to Definition 10.6.1, these connection coefficients are equivalently defined by setting

$$\nabla_{\mathbf{e}_\mu}\mathbf{e}_\nu \equiv \nabla\mathbf{e}_\nu(\mathbf{e}_\mu) = \Gamma_{\mu\nu}{}^\rho\mathbf{e}_\rho \tag{10.94}$$

10.6.2 Curvature and torsion of an affine connection

To every connection one-form $\mathcal{A}$ on a principal bundle $P(\mathcal{M},G)$ we can associate a curvature 2-form:

$$\begin{aligned} \mathfrak{F} &\equiv d\mathcal{A} + \mathcal{A}\wedge\mathcal{A} \\ &\equiv \mathfrak{F}^I T_I = \left(d\mathcal{A}^I + \frac{1}{2}f_{JK}{}^I\mathcal{A}^J\wedge\mathcal{A}^K\right)T_I \end{aligned} \tag{10.95}$$

which is $\mathbb{G}$ Lie algebra valued. We note that, evaluated on any associated vector bundle, $E \overset{\pi}{\Longrightarrow} \mathcal{M}$, the connection $\mathcal{A}$ becomes a matrix and the same is true of the curvature $\mathcal{F}$. In that case the first line of eq. (10.95) is to be understood in the sense both of matrix multiplication and of wedge product, namely, the element (i,j) of $\mathcal{A}\wedge\mathcal{A}$ is calculated as $\mathcal{A}_i{}^k\wedge\mathcal{A}_k{}^j$, with summation over the dummy index k.

We can apply the general formula (10.95) to the case of an affine connection. In that case the curvature 2-form is traditionally denoted with the letter $\mathfrak{R}$ in honor of Riemann. We obtain:

$$\mathfrak{R} \equiv d\Gamma + \Gamma\wedge\Gamma \tag{10.96}$$

which, using the basis $\{\mathbf{e}_\mu\}$ for the tangent bundle and its dual $\{\omega^\nu\}$ for the cotangent bundle, becomes:

$$\begin{aligned}\mathfrak{R}_\mu{}^\nu &= d\Gamma_\mu{}^\nu + \Gamma_\mu{}^\rho \wedge \Gamma_\rho{}^\nu \\ &= \omega^\lambda \wedge \omega^\sigma \frac{1}{2} \mathscr{R}_{\lambda\sigma\mu}{}^\nu\end{aligned} \tag{10.97}$$

the four index symbols $\mathscr{R}_{\lambda\sigma\mu}{}^\nu$ being, by definition, twice the components of the 2-form along the basis $\{\omega^\nu\}$. In particular, in an open chart $U \subset \mathscr{M}$, whose coordinates we denote by x^μ, we can choose the holonomic basis of sections $\mathbf{e}_\mu = \partial_\mu \equiv \partial/\partial x^\mu$, whose dual is provided by the differentials $\omega^\nu = dx^\nu$ and we get:

$$\mathscr{R}_{\lambda\sigma\mu}{}^\nu = \partial_\lambda \Gamma_{\sigma\mu}^\nu - \partial_\sigma \Gamma_{\lambda\mu}^\nu + \Gamma_{\lambda\mu}^\rho \Gamma_{\sigma\rho}^\nu - \Gamma_{\sigma\mu}^\rho \Gamma_{\lambda\rho}^\nu \tag{10.98}$$

Comparing eq. (10.98) with the Riemann–Christoffel symbols of eq. (10.37), we see that the latter could be identified with the components of the curvature two-form of an affine connection Γ if the Christoffel symbols introduced in eq. (10.37) were the coefficients of such a connection. Which connection is the one described by the Christoffel symbols and how is it defined? The answer is: the *Levi-Civita connection*. Its definition follows in the next paragraph.

10.6.2.1 Torsion and torsionless connections

The notion of torsion applies to affine connections and distinguishes them from general connections on generic fibre bundles. Intuitively torsion has to do with the fact that when we parallel-transport vectors along a loop, the transported vector can differ from the original one not only through a rotation but also through a displacement. While the infinitesimal rotation angle is related to the curvature tensor, the infinitesimal displacement is related to the torsion tensor. Rigorously we have the following:

Definition 10.6.2. Let $\mathscr{M}$ be an m-dimensional manifold and ∇ denote an affine connection on its tangent bundle. The torsion T_∇ is a map:

$$\mathrm{T}_\nabla : \mathfrak{X}(\mathscr{M}) \times \mathfrak{X}(\mathscr{M}) \to \mathfrak{X}(\mathscr{M})$$

defined as follows:

$$\forall \mathbf{X}, \mathbf{Y} \in \mathfrak{X}(\mathscr{M}) : \quad \mathrm{T}_\nabla(\mathbf{X}, \mathbf{Y}) = -\mathrm{T}_\nabla(\mathbf{Y}, \mathbf{X}) \equiv \nabla_\mathbf{X} \mathbf{Y} - \nabla_\mathbf{Y} \mathbf{X} - [\mathbf{X}, \mathbf{Y}] \in \mathfrak{X}(\mathscr{M})$$

Given a basis of sections of the tangent bundle $\{\mathbf{e}_\mu\}$ we can calculate their commutators:

$$[\mathbf{e}_\mu, \mathbf{e}_\nu] = K_{\mu\nu}{}^\rho(p)\mathbf{e}_\rho \tag{10.99}$$

where the point-dependent coefficients $K_{\mu\nu}{}^\rho(p)$ are named the *contorsion* coefficients. They do not form a tensor, since they depend on the choice of basis. For instance, in the holonomic basis $\mathbf{e}_\mu = \partial_\mu$ the contorsion coefficients are zero, while they do not vanish in other bases. Notwithstanding their non-tensorial character they can be calculated in any basis and once this is done we obtain a true tensor, namely, the torsion from Definition 10.6.2. Explicitly we have:

$$\begin{aligned} T_\nabla(\mathbf{e}_\mu, \mathbf{e}_\nu) &= \mathcal{T}^\rho_{\mu\nu}\mathbf{e}_\rho \\ \mathcal{T}^\rho_{\mu\nu} &= \Gamma_{\mu\nu}{}^\rho - \Gamma_{\nu\mu}{}^\rho - K_{\mu\nu}{}^\rho \end{aligned} \tag{10.100}$$

Definition 10.6.3. An affine connection ∇ is named torsionless if its torsion tensor vanishes identically, namely, if $T_\nabla(\mathbf{X}, \mathbf{Y}) = 0$, $\forall \mathbf{X}, \mathbf{Y} \in \mathfrak{X}(\mathcal{M})$.

It follows from eq. (10.100) that the coefficients of a torsionless affine connection are symmetric in the lower indices in the holonomic basis. Indeed, if the contorsion vanishes, imposing zero torsion reduces to the condition:

$$\Gamma_{\mu\nu}{}^\rho = \Gamma_{\nu\mu}{}^\rho \tag{10.101}$$

10.6.2.2 The Levi-Civita metric connection

Consider now the case where the manifold $\mathcal{M}$ is endowed with a metric g. Independently from the signature of the latter (Riemannian or pseudo-Riemannian) we can define a unique affine connection which preserves the scalar products defined by g and is torsionless. That affine connection is the Levi-Civita connection. Explicitly we have the following:

Definition 10.6.4. Let $(\mathcal{M}, g)$ be a (pseudo-)Riemannian manifold. The associated Levi-Civita connection ∇^g is that unique affine connection which satisfies the following two conditions:

(i) ∇^g is torsionless, namely $T_{\nabla^g}(\,,\,) = 0$.

(ii) The metric is covariantly constant under the transport defined by ∇^g, that is: $\forall \mathbf{Z}, \mathbf{X}, \mathbf{Y} \in \mathfrak{X}(\mathcal{M}) : \mathbf{Z}g(\mathbf{X}, \mathbf{Y}) = g(\nabla^g_\mathbf{Z}\mathbf{X}, \mathbf{Y}) + g(\mathbf{X}, \nabla^g_\mathbf{Z}\mathbf{Y})$.

The idea behind such a definition is very simple and intuitive. Consider two vector fields $\mathbf{X}, \mathbf{Y}$. We can measure their scalar product and hence the angle they form by evaluating $g(\mathbf{X}, \mathbf{Y})$. Consider now a third vector field $\mathbf{Z}$ and let us parallel-transport the previously given vectors in the direction defined by $\mathbf{Z}$. For an infinitesimal displacement we have $\mathbf{X} \mapsto \mathbf{X} + \nabla_\mathbf{Z}\mathbf{X}$ and $\mathbf{Y} \mapsto \mathbf{Y} + \nabla_\mathbf{Z}\mathbf{Y}$. We can compare the scalar product of the parallel-transported vectors with that of the original ones. Imposing the second condition listed in Definition 10.6.4 corresponds to stating that the scalar product of the parallel-transported vectors should just be the increment along Z of the scalar product of the original ones. This is the very intuitive notion of parallelism.

It is now very easy to verify that the Christoffel symbols, defined in eq. (10.37) are just the coefficients of the Levi-Civita connection in the holonomic basis $\mathbf{e}_\mu = \partial_\mu \equiv$

$\partial/\partial x^\mu$. As we already remarked, in this case the contorsion vanishes and a torsionless connection has symmetric coefficients according to eq. (10.101). On the other hand, the second condition of Definition 10.6.4 translates into:

$$\partial_\lambda g_{\mu\nu} = \Gamma_{\lambda\mu}{}^{\sigma} g_{\sigma\nu} + \Gamma_{\lambda\nu}{}^{\sigma} g_{\mu\sigma} \tag{10.102}$$

which admits the Christoffel symbols (10.37) as unique solution. There is a standard trick to see this and solve eq. (10.102) for Γ. Just write three copies of the same equation with cyclically permuted indices:

$$\partial_\lambda g_{\mu\nu} = \Gamma_{\lambda\mu}{}^{\sigma} g_{\sigma\nu} + \Gamma_{\lambda\nu}{}^{\sigma} g_{\mu\sigma} \tag{10.103}$$

$$\partial_\mu g_{\nu\lambda} = \Gamma_{\mu\nu}{}^{\sigma} g_{\sigma\lambda} + \Gamma_{\mu\lambda}{}^{\sigma} g_{\nu\sigma} \tag{10.104}$$

$$\partial_\nu g_{\lambda\mu} = \Gamma_{\nu\lambda}{}^{\sigma} g_{\sigma\mu} + \Gamma_{\nu\mu}{}^{\sigma} g_{\lambda\sigma} \tag{10.105}$$

Next sum eq. (10.103) with eq. (10.104) and subtract eq. (10.105). In the result of this linear combination use the symmetry of $\Gamma_{\lambda\mu}{}^{\sigma}$ in its lower indices. With this procedure you obtain that $\Gamma_{\lambda\mu}{}^{\sigma}$ is equal to the Christoffel symbols.

Recalling eq. (10.36) which defines the covariant derivative of a generic tensor field according to the tensor calculus of Ricci and Levi-Civita, we discover the interpretation of eq. (10.102). It just states that the covariant derivative of the metric tensor should be zero:

$$\nabla_\lambda g_{\mu\nu} = 0 \tag{10.106}$$

Hence the Levi-Civita connection is that affine torsionless connection with respect to which the metric tensor is covariantly constant.

10.7 Geodesics

Once we have an affine connection we can answer the question that was at the root of the whole development of differential geometry, namely, *which lines are straight in a curved space?* To use a car driving analogy, the straight lines are obviously those that imply no turning of the steering wheel. In geometric terms steering the wheel corresponds to changing one's direction while proceeding along the curve and such a change is precisely measured by the parallel transport of the tangent vector to the curve along itself.

Let $\mathscr{C}(\lambda)$ be a curve $[0,1] \mapsto \mathscr{M}$ in a manifold of dimension m and let $\mathbf{t}$ be its tangent vector. In each coordinate patch the considered curve is represented as $x^\mu = x^\mu(\lambda)$ and the tangent vector has the following components:

$$t^\mu(\lambda) = \frac{d}{d\lambda} x^\mu(\lambda) \tag{10.107}$$

According to the above discussion we can rightly say that a curve is straight if we have:

$$\nabla_{\mathbf{t}} \mathbf{t} = 0 \tag{10.108}$$

The above condition immediately translates into a set of m differential equations of the second order for the functions $x^\mu(\lambda)$. Observing that:

$$\frac{d}{d\lambda} x^\rho(\lambda) \partial_\rho \left[\frac{d}{d\lambda} x^\mu(\lambda) \right] = \frac{d^2 x^\mu}{d\lambda^2} \tag{10.109}$$

we conclude that eq. (10.108) just coincides with:

$$\frac{d^2 x^\mu}{d\lambda^2} + \frac{dx^\rho}{d\lambda} \frac{dx^\sigma}{d\lambda} \Gamma_{\rho\sigma}{}^{\mu} = 0 \tag{10.110}$$

which is named the geodesic equation. The solutions of these differential equations are the straight lines of the considered manifold and are named the *geodesics*. A solution is completely determined by the initial conditions which, given the order of the differential system, are $2m$. These correspond to giving the values $x^\mu(0)$, namely, the initial point of the curve, and the values of $\frac{dx^\rho}{d\lambda}(0)$, namely, the initial tangent vector $\mathbf{t}(0)$. So we can conclude that at every point $p \in \mathcal{M}$ there is a geodesic departing along any chosen direction in the tangent space $T_p\mathcal{M}$.

We can define geodesics with respect to any affine connection Γ, yet nothing guarantees a priori that such straight lines should also be the shortest routes from one point to another of the considered manifold $\mathcal{M}$. In the case we have a metric structure, lengths are defined and we can consider the variational problem of calculating extremal curves for which any variation makes them longer. It suffices to implement the standard variational calculus to the length functional (see eq. (10.86)):[2]

$$\begin{aligned} \mathfrak{s} &= \int d\tau \equiv \int \sqrt{2\mathcal{L}}\, d\lambda \\ \mathcal{L} &\equiv \frac{1}{2} g_{\mu\nu}(x) \frac{dx^\mu}{d\lambda} \frac{dx^\nu}{d\lambda} \end{aligned} \tag{10.111}$$

Performing a variational calculation we get that the length is extremal if

$$\delta\mathfrak{s} = \int \frac{1}{\sqrt{2\mathcal{L}}} \delta\mathcal{L}\, d\lambda = 0 \tag{10.112}$$

We are free to use any parameter λ to parameterize the curves. Let us use the affine parameter $\lambda = \tau$ defined by the condition:

$$2\mathcal{L} = g_{\mu\nu}(x) \frac{dx^\mu}{d\tau} \frac{dx^\nu}{d\tau} = 1 \tag{10.113}$$

2 For a more detailed discussion of the use of the proper length parameter versus a generic affine parameter we refer the reader to Section 3.8 of the first volume of [51].

In this case eq. (10.112) reduces to $\delta\mathscr{L} = 0$ which is the standard variational equation for a Lagrangian $\mathscr{L}$ where the affine parameter τ is the time and x^μ are the Lagrangian coordinates q^μ. It is a straightforward exercise to verify that the Euler–Lagrange equations of this system:

$$\frac{d}{d\tau}\frac{\partial\mathscr{L}}{\partial\dot{x}^\mu} - \frac{\partial\mathscr{L}}{\partial x^\mu} = 0 \tag{10.114}$$

coincide with the geodesic equations (10.110) where for Γ we use the Christoffel symbols (10.37).

In this way we reach a very important conclusion. The Levi-Civita connection is that unique affine connection for which also in curved space the curves of extremal length (typically the shortest ones) are straight just as it happens in flat space. This being true, the geodesics can be directly obtained from the variational principle which is the easiest and fastest way.

10.8 Geodesics in Lorentzian and Riemannian manifolds: two simple examples

Let us now illustrate geodesics in some simple examples which will also be useful to emphasize the difference between Riemannian and Lorentzian manifolds. In a Riemannian manifold the metric is positive definite and there is only one type of geodesics. Indeed, the norm of the tangent vector is always positive and the auxiliary condition (10.113) defining the affine parameter is unique. In a Lorentzian case, on the other hand, we have three kinds of geodesics depending on the sign of the norm of the tangent vector. *Time-like geodesics* are those where $g(\mathbf{t},\mathbf{t}) > 0$ and the auxiliary condition is precisely stated as in eq. (10.113). However, we have also *space-like geodesics* where $g(\mathbf{t},\mathbf{t}) < 0$ and *null-like geodesics* where $g(\mathbf{t},\mathbf{t}) = 0$. In these cases the auxiliary condition defining the affine parameter is reformulated as $2\mathscr{L} = -1$ and $2\mathscr{L} = 0$, respectively.

In General Relativity, time-like geodesics are the world-lines traced in space-time by massive particles that move at a speed less than that of light. Null-like geodesics are the world-lines traced by massless particles moving at the speed of light, while space-like geodesics, corresponding to superluminal velocities violate causality and cannot be traveled by any physical particle.

10.8.1 The Lorentzian example of dS_2

An interesting toy example that can be used to illustrate in a pedagogical way many aspects of the so far developed theory is given by 2-dimensional de Sitter space. We can describe this pseudo-Riemannian manifold as an algebraic locus in $\mathbb{R}^3$, writing the following quadratic equation:

$$\mathbb{R}^3 \supset \mathrm{dS}_2 : -X^2 + Y^2 + Z^2 = 1 \tag{10.115}$$

A parametric solution of the defining locus equation (10.115) is easily obtained by the following position:

$$X = \sinh t; \quad Y = \cosh t \sin\theta; \quad Z = \cosh t \cos\theta \tag{10.116}$$

and an overall picture of the manifold is given in Figure 10.3.

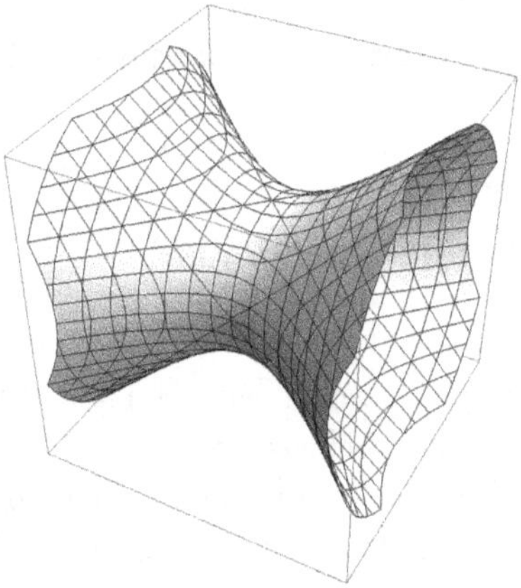

Figure 10.3: Two-dimensional de Sitter space is a hyperbolic rotational surface that can be visualized in three dimensions as in the picture above.

The parameters t and θ can be taken as coordinates on the dS_2 surface on which we can define a Lorentzian metric by means of the pullback of the standard SO(1, 2) metric on three-dimensional Minkowski space, namely:

$$\begin{aligned} ds^2_{\mathrm{dS}_2} &= -dX^2 + dY^2 + dZ^2 \\ &= -dt^2 + \cosh^2 t\, d\theta^2 \end{aligned} \tag{10.117}$$

The first thing to note about the above metric is that it describes an expanding two-dimensional universe where the spatial sections at constant time t = const are circles $\mathbb{S}^1$. Indeed, the angle θ can be regarded as the coordinate on $\mathbb{S}^1$ and $d\theta^2 = ds^2_{\mathbb{S}^1}$ is the corresponding metric, so that we can write:

$$ds^2_{\mathrm{dS}_2} = -dt^2 + a^2(t)\, ds^2_{\mathbb{S}^1} \tag{10.118}$$

where:

$$a(t) = \cosh t \tag{10.119}$$

Equation (10.118) provides a paradigm. This is precisely the structure one meets in the discussion of relativistic cosmology (see Chapter 5 of Volume Two of [51]). The second important thing to note about the metric (10.117) is that it has Lorentzian signature. Hence we are not supposed to find just one type of geodesics, rather we have to discuss three types of them:

1. The null geodesics for which the tangent vector is light-like.
2. The time-like geodesics for which the tangent vector is time-like.
3. The space-like geodesics for which the tangent vector is space-like.

According to our general discussion, the proper-length of any curve on dS_2 is given by the value of the following integral:

$$\mathfrak{s} = \int \sqrt{-\left(\frac{dt}{d\lambda}\right)^2 + \cosh^2 t \left(\frac{d\theta}{d\lambda}\right)^2}\, d\lambda \equiv \int \sqrt{2\mathcal{L}}\, d\lambda \tag{10.120}$$

where λ is any parameter labeling the points along the curve. Performing a variational calculation we get that the length is extremal if

$$\delta\mathfrak{s} = \int \frac{1}{\sqrt{2\mathcal{L}}}\delta\mathcal{L}\, d\lambda = 0 \quad \Rightarrow \quad \delta\mathcal{L} = 0 \tag{10.121}$$

Hence, as long as we use for λ an *affine parameter*, defined by the auxiliary condition

$$g_{\mu\nu}\frac{dx^\mu}{d\lambda}\frac{dx^\nu}{d\lambda} = -\dot{t}^2 + \cosh^2 t\dot{\theta}^2 = k = \begin{cases} 1; & \text{space–like} \\ 0; & \text{null–like} \\ -1; & \text{time–like} \end{cases} \tag{10.122}$$

we can just treat:

$$\mathcal{L} = -\dot{t}^2 + \cosh^2 t\dot{\theta}^2 \tag{10.123}$$

as the Lagrangian of an ordinary mechanical problem. The corresponding Euler–Lagrange equations of motion are:

$$\begin{aligned} 0 &= \partial_\lambda \frac{\partial\mathcal{L}}{\partial\dot{\theta}} - \frac{\partial\mathcal{L}}{\partial\theta} = \partial_\lambda(\cosh^2 t\dot{\theta}) \\ 0 &= \partial_\lambda \frac{\partial\mathcal{L}}{\partial\dot{t}} - \frac{\partial\mathcal{L}}{\partial t} = (\partial_\lambda \dot{t} - \cosh t \sinh t\dot{\theta}^2) \end{aligned} \tag{10.124}$$

The first of the above equations shows that θ is a cyclic variable and hence we have a first integral of the motion:

$$\text{const} = \ell \equiv \cosh^2 t\dot{\theta} \tag{10.125}$$

which deserves the name of *angular momentum*. Indeed, the existence of this first-integral follows from the SO(2) rotational symmetry of the metric (10.117). For the concept of symmetry of a metric, namely of isometry, we refer the reader to Chapter 11.

Thanks to ℓ and to the auxiliary condition (10.122), the geodesic equations are immediately reduced to quadratures. Let us discuss the resulting three types of geodesics separately.

10.8.1.1 Null geodesics

For null geodesics we have:

$$0 = -\dot{t}^2 + \cosh^2 t\dot{\theta}^2 \tag{10.126}$$

Combining this information with (10.125) we immediately get:

$$\begin{aligned} \dot{t} &= \pm\frac{\ell}{\cosh t} \\ \dot{\theta} &= \frac{\ell}{\cosh^2 t} \end{aligned} \tag{10.127}$$

The ratio of the above two equations yields the differential equation of the null-orbits:

$$\frac{d\theta}{dt} = \pm\frac{1}{\cosh t} \tag{10.128}$$

which is immediately integrated in the following form:

$$\tan\frac{\theta+\alpha}{2} = \pm\tanh\frac{t}{2} \tag{10.129}$$

where the arbitrary angle α is the integration constant that parameterizes the family of all possible null-like curves on dS_2. In order to visualize the structure of such curves in the ambient three-dimensional space, it is convenient to use the following elliptic and hyperbolic trigonometric identities:

$$\begin{aligned} \sinh t &= \frac{2\tanh\frac{t}{2}}{1-\tanh^2\frac{t}{2}}; \quad \cosh t = \frac{1+\tanh^2\frac{t}{2}}{1-\tanh^2\frac{t}{2}} \\ \sin\phi &= \frac{2\tan\frac{\phi}{2}}{1+\tan^2\frac{\phi}{2}}; \quad \cos t = \frac{1-\tan^2\frac{\phi}{2}}{1+\tan^2\frac{\phi}{2}} \end{aligned} \tag{10.130}$$

Setting $y = \tanh\frac{t}{2} = \tan\frac{\theta+\alpha}{2}$, utilizing the parametric solution of the locus equations (10.116) and also (10.130), we obtain the form of the null geodesics in $\mathbb{R}^3$:

$$\begin{aligned} &X = x; \quad Y = x\cos\alpha - \sin\alpha; \quad Z = \cos\alpha + x\sin\alpha \\ &x \equiv \frac{2y}{1-y^2} \end{aligned} \tag{10.131}$$

It is evident from (10.131) that null geodesics are straight lines, yet straight lines that lie on the hyperbolic dS_2 surface (see Figure 10.4).

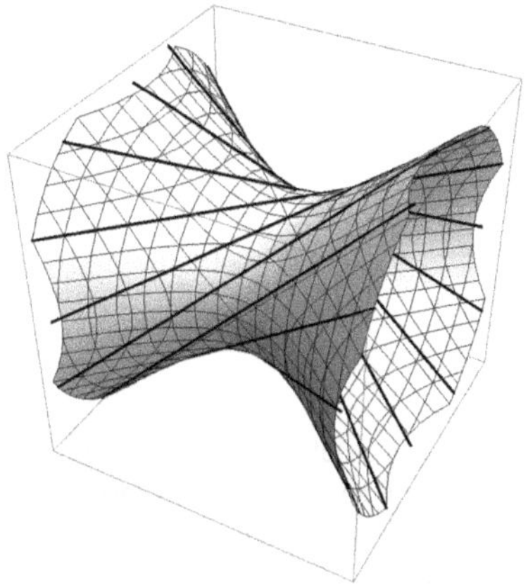

Figure 10.4: The null geodesics on dS_2 are straight lines lying on the hyperbolic surface. In this figure we show a family of these straight lines parameterized by the angle α.

10.8.1.2 Time-like geodesics

For time geodesics we have:

$$-1 = -\dot{t}^2 + \cosh^2 t\dot{\theta}^2 \tag{10.132}$$

and following the same steps as in the previous case we obtain the following differential equation for time-like orbits

$$\frac{dt}{d\theta} = \pm\frac{1}{\ell}\cosh t\sqrt{\ell^2 + \cosh^2 t} \tag{10.133}$$

which is immediately reduced to quadratures and integrated as follows:

$$\tan\frac{\theta + \alpha}{2} = \frac{\ell\sinh(t)}{\sqrt{4\ell^2 + 2\cosh(2t) + 2}} \tag{10.134}$$

Equation (10.134) provides the analytic form of all time-like geodesics in dS_2. The two integration constants are ℓ (the angular momentum) and the angle α.

It is instructive to visualize also the time geodesics as 3D curves that lie on the hyperbolic surface. To this effect we set once again $y = \tan\frac{\theta+\alpha}{2}$ and, using both the orbit equation (10.134) and the identities (10.130), we obtain:

$$\begin{aligned}
\sin(\theta + \alpha) &= \frac{4\sqrt{2}\ell\sqrt{2\ell^2 + \cosh(2t) + 1}\sinh(t)}{7\ell^2 + (\ell^2 + 4)\cosh(2t) + 4} \\
\cos(\theta + \alpha) &= \frac{4}{\frac{\ell^2\sinh^2(t)}{2\ell^2+\cosh(2t)+1} + 2} - 1
\end{aligned} \tag{10.135}$$

Changing variable:

$$t = \operatorname{arcsinh} x \tag{10.136}$$

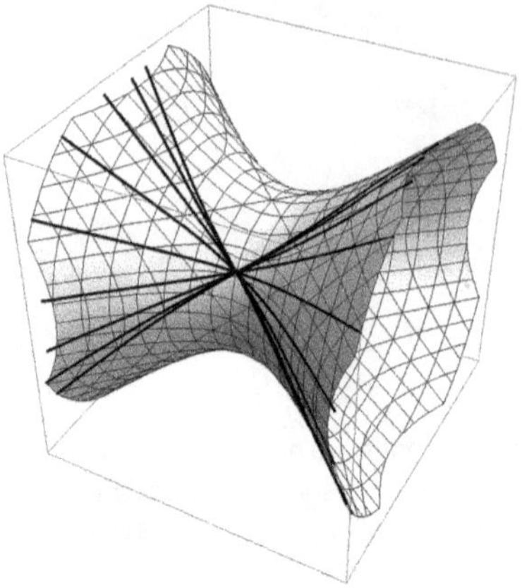

Figure 10.5: The time-like geodesics on dS_2. In this figure we show a family of geodesics parameterized by the value of the angular momentum ℓ and all passing through the same point.

Equations (10.135), combined with the parametric description of the surface (10.116), yield the parametric form of the time geodesics in $3D$ space:

$$\begin{aligned}
X &= x \\
Y &= \frac{4x\sqrt{x^2+1}\ell\sqrt{x^2+\ell^2+1}\cos(\alpha) - \sqrt{x^2+1}(4(\ell^2+1) - x^2(\ell^2-4))\sin(\alpha)}{(\ell^2+4)x^2+4(\ell^2+1)} \\
Z &= \frac{\sqrt{x^2+1}(4(\ell^2+1) - x^2(\ell^2-4))\cos(\alpha) + 4x\sqrt{x^2+1}\ell\sqrt{x^2+\ell^2+1}\sin(\alpha)}{(\ell^2+4)x^2+4(\ell^2+1)}
\end{aligned} \tag{10.137}$$

Some time-like geodesics of this metric are displayed in Figure 10.5.

10.8.1.3 Space-like geodesics

For space-like geodesics we have:

$$1 = -\dot{t}^2 + \cosh^2 t\dot{\theta}^2 \tag{10.138}$$

and we obtain the following differential equation for space-like orbits

$$\frac{dt}{d\theta} = \pm\frac{1}{\ell}\cosh t\sqrt{\ell^2 - \cosh^2 t} \tag{10.139}$$

which is integrated as follows:

$$\tan\frac{\theta+\alpha}{2} = \frac{\ell\sinh(t)}{\sqrt{4\ell^2 - 2\cosh(2t) - 2}} \tag{10.140}$$

As one sees, the difference with respect to the equation describing time-like orbits resides only in two signs. By means of algebraic substitutions completely analogous to those used in the previous case we obtain the parameterization of the space-like geodesics as 3D curves. We find

$$\begin{aligned}
X &= x \\
Y &= \frac{4x\sqrt{x^2+1}\ell\sqrt{-x^2+\ell^2-1}\cos(\alpha)+\sqrt{x^2+1}((\ell^2+4)x^2-4\ell^2+4)\sin(\alpha)}{(x^2+4)\ell^2-4(x^2+1)} \\
Z &= \frac{4x\sqrt{x^2+1}\ell\sqrt{-x^2+\ell^2-1}\sin(\alpha)-\sqrt{x^2+1}((\ell^2+4)x^2-4\ell^2+4)\cos(\alpha)}{(x^2+4)\ell^2-4(x^2+1)}
\end{aligned} \tag{10.141}$$

The sign-changes with respect to the time-like case have significant consequences. For a given value ℓ of the angular momentum the range of the X coordinate and hence of the x parameter is limited by:

$$-\sqrt{\ell^2-1} < x < \sqrt{\ell^2-1} \tag{10.142}$$

Out of this range of coordinates becomes imaginary as it is evident from eqs (10.141). In Figure 10.6 we display the shape of a family of space-like geodesics.

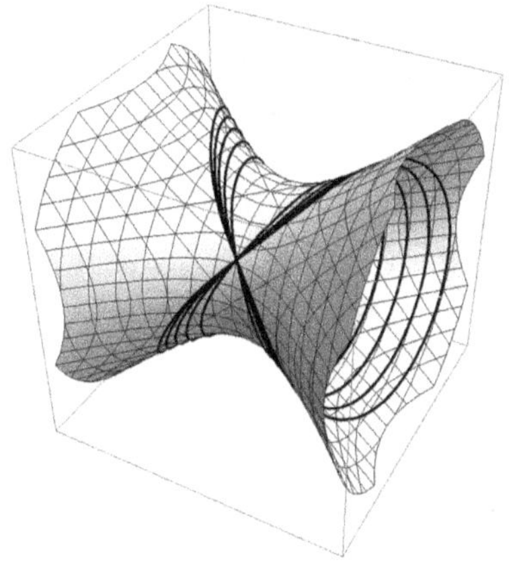

Figure 10.6: The space-like geodesics on dS_2. In this figure we show a family of space time geodesics all going through the same point.

10.8.2 The Riemannian example of the Lobachevsky–Poincaré plane

The second example we present of geodesic calculation is that relative to the hyperbolic upper plane model of Lobachevsky geometry found by Poincaré.

As many readers already know, the question of whether non-Euclidean geometries did or did not exist was a central issue of mathematical and philosophical thought for almost two-thousand years. The crucial question was whether the Vth postulate of Euclid about parallel lines was independent from the previous ones or not. Many mathematicians tried to demonstrate the Vth postulate and typically came to erroneous or tautological conclusions since, as now we know, the Vth postulate is indeed independent and distinguishes Euclidean from other equally self-consistent geometries. The first attempt of a proof dates back to *Posidonius of Rhodes* (*135–51 B.C.*) as early as the first century B.C. This encyclopedic scholar, acclaimed as one of the most erudite men of his epoch, tried to modify the definition of parallelism in order to prove the postulate, but came to inconclusive and contradictory statements. In the modern era the most interesting and deepest attempt at the proof of the postulate is that of the Italian Jesuit *Giovanni Girolamo Saccheri* (*1667–1733*). In his book *Euclides ab omni naevo vindicatus*, Saccheri tried to demonstrate the postulate with a *reductio ad absurdum*. So doing he actually proved a series of theorems in non-Euclidean geometry whose implications seemed so unnatural and remote from sensorial experience that Saccheri considered them absurd and flattered himself with the presumption of having proved the Vth postulate. The first to discover a consistent model of non-Euclidean geometry was probably Gauss around 1828. However, he refrained from publishing his result since he did not wish to hear the *screams of Beotians*. With this name he referred to the German philosophers of the time who, following Kant, considered Euclidean Geometry an *a priori truth* of human thought. Less influenced by post-Kantian philosophy in the remote town of Kazan in university of whose he was for many years the rector, the Russian mathematician *Nicolai Ivanovich Lobachevsky* (*1793–1856*) discovered and formulated a consistent axiomatic setup of non-Euclidean geometry where the Vth postulate did not hold true and where the sum of internal angles of a triangle was less than π. An explicit model of Lobachevsky geometry was first created by *Eugenio Beltrami* (*1836–1900*) by means of lines drawn on the hyperbolic surface known as the pseudo-sphere and then analytically realized by *Henri Poincaré* (*1854–1912*) some years later.

In 1882 Poincaré defined the following two-dimensional Riemannian manifold $(\mathcal{M}, g)$, where $\mathcal{M}$ is the upper plane:

$$\mathbb{R}^2 \supset \mathcal{M} : (x, y) \in \mathcal{M} \quad \Leftrightarrow \quad y > 0 \tag{10.143}$$

and the metric g is defined by the following infinitesimal line-element:

$$d\mathfrak{s}^2 = \frac{dx^2 + dy^2}{y^2}. \tag{10.144}$$

Lobachevsky geometry is realized by all polygons in the upper plane $\mathcal{M}$ whose sides are arcs of geodesics with respect to the Poincaré metric (10.144). Let us derive the general form of such geodesics.

This time the metric has Euclidean signature and there is just one type of geodesic curves. Following our variational method the effective Lagrangian is:

$$\mathcal{L} = \frac{1}{2} \frac{\dot{x}^2 + \dot{y}^2}{y^2} \tag{10.145}$$

where the dot denotes derivatives with respect to length parameter s. The Lagrangian variable x is cyclic (namely appears only under derivatives) and from this fact we immediately obtain a first order integral of motion:

$$\frac{\dot{x}}{y^2} = \frac{1}{R} = \text{const} \tag{10.146}$$

The name R given to this conserved quantity follows from its geometrical interpretation that we will next discover. Using the information (10.146) in the auxiliary condition:

$$2\mathcal{L} = 1 \tag{10.147}$$

which defines the affine length parameter we obtain:

$$\dot{y}^2 = \left(1 - \frac{1}{R^2}y^2\right)y^2 \tag{10.148}$$

and by eliminating ds between equations (10.146) and (10.148) we obtain:

$$\frac{1}{R}\,dx = \frac{y\,dy}{\sqrt{1-\frac{y^2}{R^2}}} \tag{10.149}$$

which upon integration yields:

$$\frac{1}{R}(x - x_0) = \sqrt{1 - \frac{y^2}{R^2}} \tag{10.150}$$

where x_0 is the integration constant. Squaring the above relation we get the following one:

$$(x - x_0)^2 + y^2 = R^2 \tag{10.151}$$

that has an immediate interpretation. A geodesic is just the arc lying in the upper plane of any circle of radius R having center in $(x_0, 0)$, namely, on some point lying on real axis.

With this result Lobachevsky geometry is easily visualized. Examples of planar figures with geodesical sides are presented in Figure 10.7.

10.8.3 Another Riemannian example: the catenoid

As a further Riemannian example we consider the catenoid revolution surface introduced in 1744 by Leonhard Euler [40].

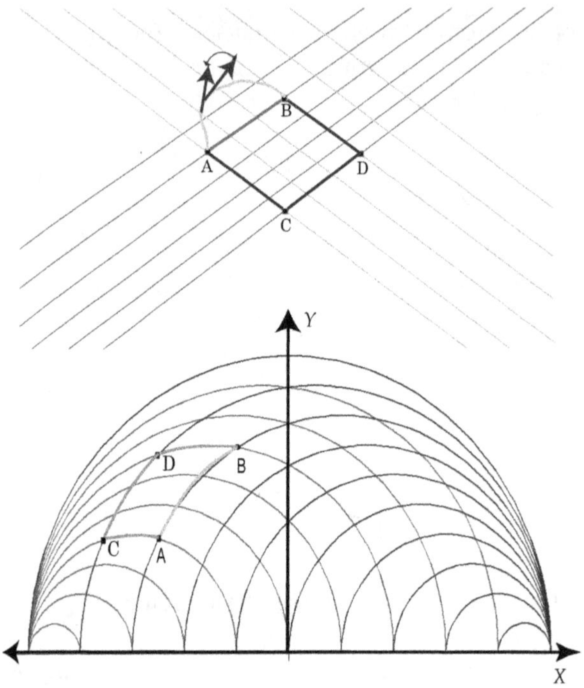

Figure 10.7: The geodesics of Poincaré metric in the upper plane compared to the geodesics of the Euclidean metric, namely, the straight lines.

The catenoid is parametrically described as follows:

$$\begin{aligned} X &= \frac{1}{\sqrt{2}} \cos(\theta) \cosh(C) \\ Y &= \frac{1}{\sqrt{2}} \sin(\theta) \cosh(C) \\ Z &= \frac{1}{\sqrt{2}} C \end{aligned} \tag{10.152}$$

where $\theta \in [0, 2\pi]$ is an angle and $C \in [-\infty, +\infty]$ is a real parameter. Its appearance in three-dimensional space is displayed in Figure 10.8.

The pullback of the ambient flat metric $\mathrm{d}s^2_{\mathbb{E}^3} = dX^2 + dY^2 + dZ^2$ on the locus (10.152) yields the following two-dimensional Riemannian metric:

$$\mathrm{d}s^2_{\text{catenoid}} = \frac{1}{2} \cosh^2 C (\mathrm{d}C^2 + \mathrm{d}\theta^2) \tag{10.153}$$

We derive the geodesics of the above metric.

As in the previous cases we realize that there is a cyclic variable which appears in the metric only through its own differential, namely, the angle θ. This tells us that every geodetic is characterized by a first integral of motion that we can name the *angular*

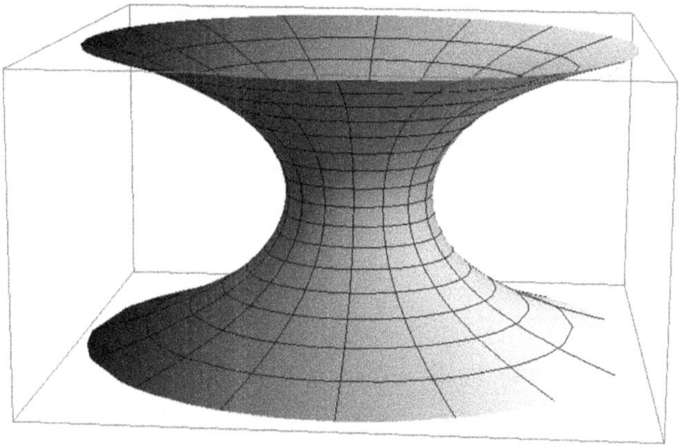

Figure 10.8: The representation of the catenoid as a surface immersed in three-dimensional space.

momentum of the *particle* moving on the catenoid surface:

$$J = \cosh^2 C \frac{d\theta}{d\tau} \tag{10.154}$$

Substituting back the condition (10.154) in the metric (10.153) we obtain:

$$\left(\frac{ds}{d\tau}\right)^2 = \frac{1}{2}\operatorname{sech}^2(C)\left(\left(\frac{dC}{d\tau}\right)^2 \cosh^4(C) + J^2\right) \tag{10.155}$$

Utilizing as parameter along the geodesic the arc length, namely, setting $\tau = s$, we obtain:

$$\frac{dC}{ds} = \pm\sqrt{\operatorname{sech}^4(C)(\cosh(2C) - J^2 + 1)} \tag{10.156}$$

and eliminating the variable s between eq. (10.156) and eq. (10.154) we get the orbit equation:

$$\frac{d\theta}{dC} = \frac{J}{\sqrt{\cosh(2C) - J^2 + 1}} \tag{10.157}$$

which, by integration, yields:

$$\begin{aligned} \theta &= \mathfrak{f}(C,J) + \theta_0 \\ \mathfrak{f}(C,J) &\equiv -\frac{iJ\sqrt{\frac{-\cosh(2C)+J^2-1}{J^2-2}}F(iC| - \frac{2}{J^2-2})}{\sqrt{\cosh(2C) - J^2 + 1}} \end{aligned} \tag{10.158}$$

$F(\phi|m) \equiv \int_0^\phi (1 - m\sin^2\theta)^{-1/2}\, d\theta$ denoting the elliptic integral of specified parameters.

In this way we have solved the geodesic problem completely. The integration constants are two, namely, J and the reference angle θ_0, and the three-dimensional image of the geodesic curves is parametrically provided by:

$$\begin{aligned} X &= \frac{1}{\sqrt{2}}\cos(\mathfrak{f}(C,J)+\theta_0)\cosh(C) \\ Y &= \frac{1}{\sqrt{2}}\sin(\mathfrak{f}(C,J)+\theta_0)\cosh(C) \\ Z &= \frac{1}{\sqrt{2}}C; \quad C \in [-\infty,+\infty] \end{aligned} \tag{10.159}$$

and a few of them are displayed in Figure (10.9).

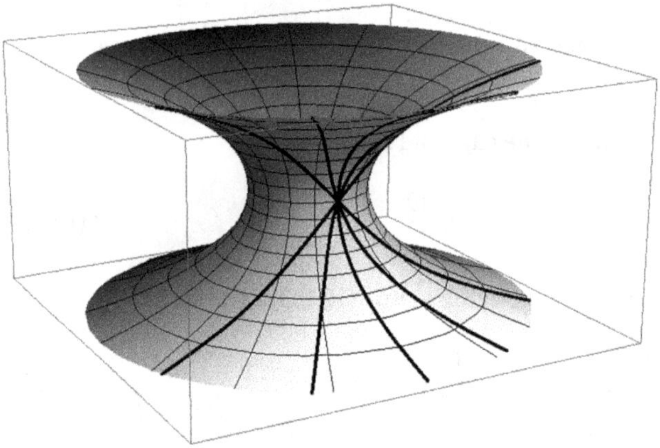

Figure 10.9: A family of geodesics on the catenoid surface all passing through the same point.

10.9 Bibliographical note

As for the present chapter bibliographical sources of the presented material are:
1. volume 1 of [51]
2. volume 1 of the three monographs on supergravity [22]
3. the textbook [84]
4. the textbook [116]
5. the textbook [117]

Furthermore, we stress that in Appendix B we describe two MATHEMATICA NoteBooks, respectively presented in Sections B.8 and B.9 that allow for the electronic evaluation of the spin connection, Levi-Civita connection, Riemann tensor and of its descendants for an arbitrary metric in an arbitrary dimension, with arbitrary signature.

11 Isometries and the geometry of coset manifolds

The art of doing mathematics consists in finding that special case which contains all the germs of generality.
David Hilbert

11.1 Conceptual and historical introduction

The word isometry comes from the Greek word ή ισομετρία which means the equality of measures.

The origin of the modern concept of isometry is rooted in that of *congruence* of geometrical figures that Euclid never introduced explicitly, yet implicitly assumed when he proceeded to identify those triangles that can be superimposed one onto the other.

It was indeed the question about what are the transformations that define such congruences what led Felix Klein to the famous Erlangen Programme.[1] Klein understood that Euclidean congruences are based on the transformations of the Euclidean Group and he came to the idea that other geometries are based on different groups of transformations with respect to which we consider congruences.

Such a concept, however, would have been essentially empty without the additional element which was the subject of Chapter 10, namely the *metric*. The area and the volume of geometrical figures, the length of sides and the relative angles have to be measured in order to compare them. These measurements can be performed if and only if we have a metric g, in other words if the *substratum* of the considered geometry is a *Riemannian* or a *pseudo-Riemannian* manifold $(\mathcal{M},g)$.

Therefore the group of transformations, which, according to the vision of the Erlangen Programme, defines a geometry, is the *group of isometries* $\mathrm{G}_{\mathrm{iso}}$ of a given Riemannian space $(\mathcal{M},g)$, the elements of this group being diffeomorphisms:

$$\phi : \mathcal{M} \to \mathcal{M} \tag{11.1}$$

such that their pullback on the metric form leaves it invariant:

$$\forall \phi \in \mathrm{G}_{\mathrm{iso}} : \quad \phi^\star \left[g_{\mu\nu}(x)\, dx^\mu\, dx^\nu \right] = g_{\mu\nu}(x)\, dx^\mu\, dx^\nu \tag{11.2}$$

Quite intuitively it becomes clear that the structure of $\mathrm{G}_{\mathrm{iso}}$ is determined by the manifold $\mathcal{M}$ and by its metric g, so that the concept of geometries is now identified with that of Riemannian spaces $(\mathcal{M},g)$.

1 For a historical account of the conceptual developments linked with the Erlangen Programme, see the forthcoming book [53].

https://doi.org/10.1515/9783110551204-011

A generic metric g has no isometries and hence there are no congruences to study. (Pseudo-)Riemannian manifolds with no isometry, or with few isometries, are relevant to several different problems pertaining to physics and also to other sciences, yet they are not in the vein of the Erlangen Programme, aiming at the classification of geometries in terms of groups. Hence we can legitimately ask ourselves the question of whether such a programme can be ultimately saved, notwithstanding our discovery that a geometry is necessarily based on a (pseudo-)Riemannian manifold $(\mathcal{M}, g)$. The answer is obviously positive if we can invert the relation between the metric g and its isometry group G_{iso}. Given a Lie group G, can we construct the Riemannian manifold $(\mathcal{M}, g)$ which admits G as its own isometry group G_{iso}? Indeed, we can; the answers are also exhaustive if we add an additional request, that of transitivity.

Definition 11.1.1. A group G acting on a manifold $\mathcal{M}$ by means of diffeomorphisms:

$$\forall \gamma \in \mathrm{G} \quad \gamma : \mathcal{M} \to \mathcal{M} \tag{11.3}$$

has a transitive action if and only if

$$\forall p, q \in \mathcal{M}, \quad \exists \gamma \in \mathrm{G}/\gamma(q) = p \tag{11.4}$$

If the Riemannian manifold $(\mathcal{M}, g)$ admits a transitive group of isometries, it means that any point of $\mathcal{M}$ can be mapped to any other by means of a transformation that is an isometry. In this case the very manifold $\mathcal{M}$ and its metric g are completely determined by group theory: $\mathcal{M}$ is necessarily a *coset manifold* G/H, namely, the space of equivalence classes of elements of G with respect to multiplication (either on the right or on the left) by elements of a subgroup $\mathrm{H} \subset \mathrm{G}$ (see Section 3.2.2). The metric g is induced on the equivalence classes by the Killing metric of the Lie algebra, defined on $\mathbb{G}$.

The present chapter, after a study of Killing vector fields, namely, of the infinitesimal generators that realize the Lie algebra $\mathbb{G}$ of the isometry group, will be devoted to the geometry of coset manifolds. Among them particular attention will be given to the so-named *symmetric spaces* characterized by an additional reflection symmetry whose nature will become clear to the reader in the following sections.

11.1.1 Symmetric spaces and Élie Cartan

The full-fledged classification of all symmetric spaces was the gigantic achievement of Élie Cartan. The classification of symmetric spaces is at the same time a classification of the real forms of the complex Lie algebras and it is the conclusive step in the path initiated with the classification of complex Lie algebras to which Chapter 7 was devoted. At the same time the geometries of non-compact symmetric spaces can be formulated in terms of other quite interesting algebraic structures, the *normed solvable Lie algebras*. The class of these latter is wider than that of symmetric spaces and this provides a generalization path leading to a wider class of geometries, all of them under firm algebraic control. This will be the topic of the last two sections of the present chapter which is propaedeutic to the most advanced topics in Lie Algebra Theory currently utilized in modern theoretical physics.[2]

11.1.2 Where and how do coset manifolds come into play?

By now it should be clear to the reader that, just as we have the whole spectrum of linear representations of a Lie algebra $\mathbb{G}$ and of its corresponding Lie group G, in the same way we have the set of *non-linear representations* of the same Lie algebra $\mathbb{G}$ and of the same Lie group G. These are encoded in all possible coset manifolds G/H with their associated G-invariant metrics.

Where and how do these geometries pop up?

The answer is that they appear at several different levels of analysis and in connection with different aspects of physical theories. Let us enumerate them and discover a conceptual hierarchy.

A) A first context of utilization of coset manifolds G/H is in the quest for solutions of Einstein Equations in $d = 4$ or in higher dimensions. One is typically interested in space-times with a prescribed *isometry* and one tries to fit into the equations G/H metrics whose parameters depend on some residual coordinate like the time t in cosmology or the radius r in black-hole physics. The field equations of the theory reduce to few-parameter differential equations in the residual space.

B) Another instance of utilization of coset manifolds is in the context of σ-models. In physical theories that include scalar fields $\phi^I(x)$ the kinetic term is necessarily of the following form:

$$\mathscr{L}_{\text{kin}} = \frac{1}{2}\gamma_{IJ}(\phi)\partial_\mu\phi^I(x)\partial_\nu\phi^J(x)g^{\mu\nu}(x) \tag{11.5}$$

where $g^{\mu\nu}(x)$ is the metric of space-time, while $\gamma_{IJ}(\phi)$ can be interpreted as the metric of some manifold $\mathscr{M}_{\text{target}}$ of which the fields ϕ^I are the coordinates and whose dimension is just equal to the number of scalar fields present in the theory. If we require the field theory to have some Lie Group symmetry G, either we have linear representations or non-linear ones. In the first case the metric γ_{IJ} is constant and invariant under the linear transformations of G acting on the $\phi^I(x)$. In the second case the manifold $\mathscr{M}_{\text{target}} = \text{G/H}$ is some coset of the considered group and $\gamma_{IJ}(\phi)$ is the corresponding G-invariant metric.

C) In mathematics and sometimes in physics you can consider structures that depend on a continuous set of parameters, for instance, the solutions of certain differen-

2 For an account of the advances in Lie Algebra and Geometry streaming from modern theoretical physics, see the forthcoming book [54].

tial equations, like the self-duality constraint for gauge-field strengths or the Ricci-flat metrics on certain manifolds, or the algebraic surfaces of a certain degree in some projective spaces. The parameters corresponding to all the possible deformations of the considered structure constitute themselves a manifold $\mathcal{M}$ which typically has some symmetries and in many cases is actually a coset manifold. A typical example is provided by the so-named Kummer surface K3 whose Ricci-flat metric no one has so far constructed, yet we know a priori that it depends on 3×19 parameters that span the homogeneous space $\frac{SO(3,19)}{SO(3)\times SO(19)}$.

D) In many instances of field theories that include scalar fields there is a scalar potential term $V(\phi)$ which has a certain group of symmetries G. The vacua of the theory, namely, the set of extrema of the potential, usually fill up a coset manifold G/H where $H \subset G$ is the residual symmetry of the vacuum configuration $\phi = \phi_0$.

E) In condensed matter theories quite often it happens that the *order parameter* takes values in a G/H symmetric space that encodes the symmetries of the physical system.

11.1.3 The deep insight of supersymmetry

In supersymmetric field theories, in particular in supergravities that are supersymmetric extensions of Einstein Gravity coupled to matter multiplets, all the listed above uses of coset manifolds do occur, but there is an additional ingredient whose consequences are very deep and far-reaching for geometry: supersymmetry itself. Consistency with supersymmetry introduces further restrictions on the geometry of target manifolds $\mathcal{M}_{\text{target}}$ that are required to fall in specialized categories like *Kähler manifolds, special Kähler manifolds, quaternionic Kähler manifolds* and so on. These geometries, that we collectively dub *Special Geometries*, require the existence of complex structures and encompass both manifolds that do not have transitive groups of isometries and homogeneous manifolds G/H. In the second case, which is one of the main focuses of interest in the essay [54], the combination of the special structures with the theory of Lie algebras produces new insights in homogeneous geometries that would have been inconceivable outside the framework of supergravity. This is what we call the deep geometrical insight of supersymmetry.

In the present book, which is more introductory, having as main target undergraduate students or beginners in Ph.D. studies, we just provide a primary on the notions of G/H geometries.

11.2 Isometries and Killing vector fields

The existence of continuous isometries is related to the existence of Killing vector fields. Here we explain the underlying mathematical theory which leads to the study of coset manifolds and symmetric spaces.

Suppose that the diffeomorphism considered in eq. (11.1) is infinitesimally close to the identity:

$$x^\mu \to \phi^\mu(x) \simeq x^\mu + k^\mu(x) \tag{11.6}$$

The condition for this diffeomorphism to be an isometry is a differential equation for the components of the vector field $\mathbf{k} = k^\mu \partial_\mu$ which immediately follows from (11.2):

$$\nabla_\mu k_\nu + \nabla_\nu k_\mu = 0 \tag{11.7}$$

Hence given a metric one can investigate the nature of its isometries by trying to solve the linear homogeneous equation (11.7) determining its general integral. The important point is that, if we have two Killing vectors $\mathbf{k}$ and $\mathbf{w}$, also their commutator $[\mathbf{k}, \mathbf{w}]$ is a Killing vector. This follows from the fact that the product of two finite isometries is also an isometry. Hence Killing vector fields form a finite dimensional Lie algebra $\mathbb{G}_{\text{iso}}$ and one can turn the question around. Rather than calculating the isometries of a given metric one can address the problem of constructing (pseudo-)Riemannian manifolds that have a prescribed isometry algebra. Due to the well established classification of semisimple Lie algebras this becomes a very fruitful point of view.

In particular, also in view of the Cosmological Principle, one is interested in homogeneous spaces, namely, in (pseudo-)Riemannian manifolds where each point of the manifold can be reached from a reference one by the action of an isometry.

Homogeneous spaces are identified with coset manifolds, whose differential geometry can be thoroughly described and calculated in pure Lie algebra terms.

11.3 Coset manifolds

Coset manifolds are a natural generalization of group manifolds and play a very important, ubiquitous, role both in Mathematics and in Physics.

In group-theory (irrespectively whether the group G is finite or infinite, continuous or discrete) we have the concept of *coset space* G/H which is just the set of equivalence classes of elements $g \in \mathrm{G}$, where the equivalence is defined by right multiplication with elements $h \in \mathrm{H} \subset \mathrm{G}$ of a subgroup:

$$\forall g, g' \in \mathrm{G}: \quad g \sim g' \quad \text{iff}\, \exists h \in \mathrm{H} \quad \backslash \quad gh = g' \tag{11.8}$$

Namely, two group elements are equivalent if and only if they can be mapped into each other by means of some element of the subgroup. The equivalence classes, which constitute the elements of G/H, are usually denoted $g\mathrm{H}$, where g is any representative of the class, namely, any one of the equivalent G-group elements of which the class is composed. The definition we have just provided by means of right multiplication can be obviously replaced by an analogous one based on left multiplication. In this case we construct the coset H\G composed of *right lateral classes* Hg while gH are named the *left lateral classes*. For non-abelian groups G and generic subgroups H the left G/H and right H\G coset spaces have generically different elements (see Exercise 3.2). Working with one or with the other definition is just a matter of convention. We choose to work with *left classes*.

Coset manifolds arise in the context of Lie group theory when G is a Lie group and H is a Lie subgroup thereof. In that case the set of lateral classes gH can be endowed with a manifold structure inherited from the manifold structure of the parent group G. Furthermore, on G/H we can construct *invariant metrics* such that all elements of the original group G are isometries of the constructed metric. As we show below, the curvature tensor of invariant metrics on coset manifolds can be constructed in purely algebraic terms starting from the structure constants of the Lie algebra $\mathbb{G}$, bypassing all analytic differential calculations.

For instance, the reason why coset manifolds are relevant to Cosmology is encoded in the concept of *homogeneity*, that is one of the two pillars of the Cosmological Principle. Indeed, coset manifolds are easily identified with *homogeneous spaces* which we presently define.

Definition 11.3.1. A Riemannian or pseudo-Riemannian manifold $\mathcal{M}_g$ is said to be homogeneous if it admits as an isometry the transitive action of a group G. A group acts transitively if any point of the manifold can be reached from any other by means of the group action.

A notable and very common example of such homogeneous manifolds is provided by the spheres $\mathbb{S}^n$ and by their non-compact generalizations, the pseudo-spheres $\mathbb{H}_{\pm}^{(n+1-m,m)}$. Let x^I denote the Cartesian coordinates in $\mathbb{R}^{n+1}$ and let:

$$\eta_{IJ} = \mathrm{diag}(\underbrace{+,+\dots,+}_{n+1-m},\underbrace{-,-,\dots,-}_{m}) \tag{11.9}$$

be the coefficient of a non-degenerate quadratic form with signature $(n+1-m,m)$:

$$\langle \mathbf{x},\mathbf{x}\rangle_\eta \equiv x^I x^J \eta_{IJ} \tag{11.10}$$

We obtain a pseudo-sphere $\mathbb{H}_{\pm}^{(n+1-m,m)}$ by defining the algebraic locus:

$$\mathbf{x} \in \mathbb{H}_{\pm}^{(n+1-m,m)} \quad \Leftrightarrow \quad \langle \mathbf{x},\mathbf{x}\rangle_\eta \equiv \pm 1 \tag{11.11}$$

which is a manifold of dimension n. The spheres $\mathbb{S}^n$ correspond to the particular case $\mathbb{H}_{+}^{n+1,0}$ where the quadratic form is positive definite and the sign on the right-hand side of eq. (11.11) is positive. Obviously with a positive definite quadratic form this is the only possibility.

All these algebraic loci are invariant under the transitive action of the group $\mathrm{SO}(n+1, n+1-m)$ realized by matrix multiplication on the vector $\mathbf{x}$ since:

$$\forall g \in \mathrm{G}: \quad \langle \mathbf{x}, \mathbf{x} \rangle_\eta = \pm 1 \quad \Leftrightarrow \quad \langle g\mathbf{x}, g\mathbf{x} \rangle_\eta = \pm 1 \tag{11.12}$$

namely, the group maps solutions of the constraint (11.11) to solutions of the same and, furthermore, all solutions can be generated starting from a standard reference vector:

$$\langle \mathbf{x}, \mathbf{x} \rangle_\eta = 0 \quad \Rightarrow \quad \exists g \in \mathrm{G} \quad \backslash \quad \mathbf{x} = g\mathbf{x}_0^{\pm} \tag{11.13}$$

where:

$$x_0^+ = \begin{pmatrix} 1 \\ 0 \\ \vdots \\ 0 \\ \hline 0 \\ 0 \\ \vdots \\ 0 \end{pmatrix} \; ; \quad x_0^- = \begin{pmatrix} 0 \\ 0 \\ \vdots \\ 0 \\ \hline 1 \\ 0 \\ \vdots \\ 0 \end{pmatrix} \tag{11.14}$$

the line separating the first $n+1-m$ entries from the last m. Equation (11.13) guarantees that the locus is invariant under the action of G, while eq. (11.14) states that G is transitive.

Definition 11.3.2. In a homogeneous space $\mathcal{M}_g$, the subgroup $\mathrm{H}_p \subset \mathrm{G}$ which leaves a point $p \in \mathcal{M}_g$ fixed ($\forall h \in \mathrm{H}_p$, $hp = p$) is named the **isotropy subgroup** of the point.[3] Because of the transitive action of G, any other point $p' = gp$ has an isotropy subgroup $\mathrm{H}_{p'} = g\mathrm{H}_p g^{-1}$ which is conjugate to H_p and therefore isomorphic to it.

It follows that, up to conjugation, the isotropy group of a homogeneous manifold $\mathcal{M}_g$ is unique and corresponds to an intrinsic property of such a space. It suffices to calculate the isotropy group H_0 of a conventional properly chosen reference point p_0: all other isotropy groups will immediately follow. For brevity H_0 will be just renamed H.

In our example of the spaces $\mathbb{H}_\pm^{(n+1-m,m)}$ the isotropy group is immediately derived by looking at the form of the vectors $\mathbf{x}_0^\pm$: all elements of G which rotate the vanishing entries of these vectors among themselves are clearly elements of the isotropy group. Hence we find:

$$\begin{aligned} \mathrm{H} &= \mathrm{SO}(n, m) && \text{for } \mathbb{H}_+^{(n+1-m,m)} \\ \mathrm{H} &= \mathrm{SO}(n+1, m-1) && \text{for } \mathbb{H}_-^{(n+1-m,m)} \end{aligned} \tag{11.15}$$

3 Let us remark that the notion of isotropy subgroup utilized while discussing coset manifolds, abstractly coincides with the notion of stability subgroup generically used when a group G acts as a transformation group on a set (see Section 3.2.14).

It is natural to label any point p of a homogeneous space by the parameters describing the G-group element which carries a conventional point p_0 into p. These parameters, however, are redundant: because of the H-isotropy there are infinitely many ways to reach p from p_0. Indeed, if g does that job, any other element of the lateral class $g\mathrm{H}$ does the same. It follows by this simple discussion that the homogeneous manifold $\mathcal{M}_g$ can be identified with the coset manifold G/H defined by the transitive group G divided by the isotropy group H.

Focusing once again on our example we find:

$$\mathbb{H}_+^{(n+1-m,m)} = \frac{\mathrm{SO}(n+1-m,m)}{\mathrm{SO}(n-m,m)}; \quad \mathbb{H}_-^{(n+1-m,m)} = \frac{\mathrm{SO}(n+1-m,m)}{\mathrm{SO}(n+1-m,m-1)} \tag{11.16}$$

In particular the spheres correspond to:

$$\mathbb{S}^n = \mathbb{H}_+^{(n+1,0)} = \frac{\mathrm{SO}(n+1)}{\mathrm{SO}(n)} \tag{11.17}$$

Other important examples relevant in cosmology and in many other physical theories are:

$$\mathbb{H}_+^{(n+1,1)} = \frac{\mathrm{SO}(n+1,1)}{\mathrm{SO}(n,1)}; \quad \mathbb{H}_-^{(n+1,1)} = \frac{\mathrm{SO}(n+1,1)}{\mathrm{SO}(n+1)} \tag{11.18}$$

The general classification of homogeneous (pseudo-)Riemannian spaces corresponds therefore to the classification of the coset manifolds G/H for all Lie groups G and for their closed Lie subgroups $\mathrm{H} \subset \mathrm{G}$.

The equivalence classes constituting the points of the coset manifold can be labeled by a set of d coordinates $y \equiv \{y^1, \dots, y^d\}$ where:

$$d = \dim \frac{\mathrm{G}}{\mathrm{H}} \equiv \dim \mathrm{G} - \dim \mathrm{H} \tag{11.19}$$

There are of course many different ways of choosing the y-parameters since, just as in any other manifold, there are many possible coordinate systems. What is specific of coset manifolds is that, given any coordinate system y by means of which we label the equivalence classes, within each equivalence class we can choose a representative group element $\mathbb{L}(y) \in \mathrm{G}$. The choice must be done in such a way that $\mathbb{L}(y)$ should be a smooth function of the parameters y. Furthermore, for different values y and y', the group elements $\mathbb{L}(y)$ and $\mathbb{L}(y')$ should never be equivalent, in other words no $h \in \mathrm{H}$ should exist such that $\mathbb{L}(y) = \mathbb{L}(y')h$. Under left multiplication by $g \in \mathrm{G}$, $\mathbb{L}(y)$ is in general carried into another equivalence class with coset representative $\mathbb{L}(y')$. Yet the g image of $\mathbb{L}(y)$ is not necessarily $\mathbb{L}(y')$: it is typically some other element of the same class, so that we can write:

$$\forall g \in G: \quad g\mathbb{L}(y) = \mathbb{L}(y')h(g,y); \quad h(g,y) \in \mathrm{H} \tag{11.20}$$

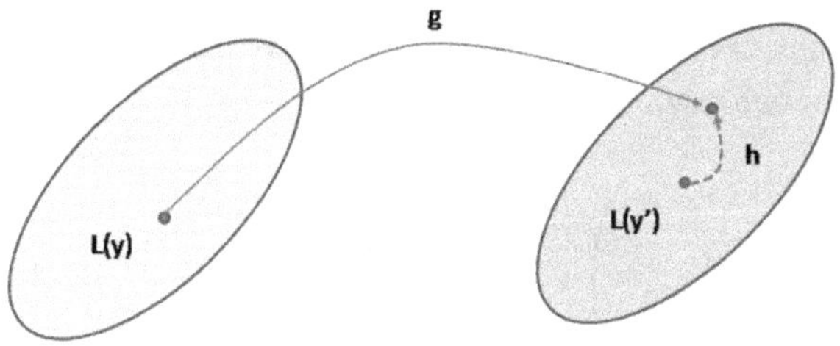

Figure 11.1: Pictorial description of the action of the group G on the coset representatives.

where we emphasized that the H-element necessary to map $\mathbb{L}(y')$ to the g-image of $\mathbb{L}(y)$ depends, in general, both on the point y and on the chosen transformation g. Equation (11.20) is pictorially described in Figure 11.1.

For the spheres a possible set of coordinates y can be obtained by means of the stereographic projection whose conception is recalled here. Considering the $\mathbb{S}^n$ sphere immersed in $\mathbb{R}^{n+1}$, from the North Pole $\{1, 0, 0, \dots, 0\}$ one draws the line that goes through the point $p \in \mathbb{S}^n$ and considers the point $\pi(p) \in \mathbb{R}^n$ where such a line intersects the $\mathbb{R}^n$ plane tangent to sphere in the South Pole and orthogonal to the line that joins the North and the South Pole. The n-coordinates $\{y^1, \dots, y^n\}$ of $\pi(p)$ can be taken as labels of an open chart in $\mathbb{S}^n$. (See Figure 5.3.)

As another explicit example, we consider the case of the Euclidean hyperbolic spaces $\mathbb{H}^{(n,1)}_{-}$ identified as coset manifolds in eq. (11.18). In this case, to introduce a coset parameterization means to write a family of $\mathrm{SO}(n,1)$ matrices $\mathbb{L}(\mathbf{y})$ depending smoothly on an n-component vector $\mathbf{y}$ in such a way that for different values of $\mathbf{y}$ such matrices cannot be mapped one to the other by means of right multiplication with any element h of the subgroup $\mathrm{SO}(n) \subset \mathrm{SO}(n,1)$:

$$\mathrm{SO}(n,1) \supset \mathrm{SO}(n) \ni h = \left(\begin{array}{c|c} \mathcal{O} & 0 \\ \hline 0 & 1 \end{array}\right); \quad \mathcal{O}^T \mathcal{O} = \mathbf{1}_{n\times n} \tag{11.21}$$

An explicit parameterization of this type can be written as follows:

$$\mathbb{L}(\mathbf{y}) = \left(\begin{array}{c|c} \mathbf{1}_{n\times n} + 2\frac{\mathbf{y}\mathbf{y}^T}{1-\mathbf{y}^2} & -2\frac{\mathbf{y}}{1-\mathbf{y}^2} \\ \hline -2\frac{\mathbf{y}^T}{1-\mathbf{y}^2} & \frac{1+\mathbf{y}^2}{1-\mathbf{y}^2} \end{array}\right) \tag{11.22}$$

where $\mathbf{y}^2 \equiv \mathbf{y} \cdot \mathbf{y}$ denotes the standard $\mathrm{SO}(n)$ invariant scalar product in $\mathbb{R}^n$. Why the matrices $\mathbb{L}(\mathbf{y})$ form a good parameterization of the coset? The reason is simple, first of all observe that:

$$\mathbb{L}(\mathbf{y})^T \eta \mathbb{L}(\mathbf{y}) = \eta \tag{11.23}$$

where

$$\eta = \mathrm{diag}(+, +, \dots, +, -) \tag{11.24}$$

This guarantees that $\mathbb{L}(\mathbf{y})$ are elements of $\mathrm{SO}(n,1)$. Secondly, observe that the image $\mathbf{x}(\mathbf{y})$ of the standard vector $\mathbf{x}_0$ through $\mathbb{L}(\mathbf{y})$,

$$\mathbf{x}(\mathbf{y}) \equiv \mathbb{L}(\mathbf{y})\mathbf{x}_0 = \mathbb{L}(\mathbf{y}) \begin{pmatrix} 0 \\ \vdots \\ 0 \\ \hline\hline 1 \end{pmatrix} = \frac{1}{1-\mathbf{y}^2} \begin{pmatrix} 2y^1 \\ \vdots \\ 2y^n \\ \hline\hline \frac{1+\mathbf{y}^2}{1-\mathbf{y}^2} \end{pmatrix} \tag{11.25}$$

lies, as it should, in the algebraic locus $\mathbb{H}_-^{(n,1)}$,

$$\mathbf{x}(\mathbf{y})^T \eta \mathbf{x}(\mathbf{y}) = -1 \tag{11.26}$$

and has n linearly independent entries (the first n) parameterized by $\mathbf{y}$. Hence the lateral classes can be labeled by y and this concludes our argument to show that (11.22) is a good coset parameterization. $\mathbb{L}(0) = \mathbf{1}_{(n+1)\times(n+1)}$ corresponds to the identity class which is usually named the *origin* of the coset.

11.3.1 The geometry of coset manifolds

In order to study the geometry of a coset manifold G/H, the first important step is provided by the orthogonal decomposition of the corresponding Lie algebra, namely by

$$\mathbb{G} = \mathbb{H} \oplus \mathbb{K} \tag{11.27}$$

where $\mathbb{G}$ is the Lie algebra of G and the subalgebra $\mathbb{H} \subset \mathbb{G}$ is the Lie algebra of the subgroup H and where $\mathbb{K}$ denotes a vector space orthogonal to $\mathbb{H}$ with respect to the Cartan Killing metric of $\mathbb{G}$. By definition of subalgebra we always have:

$$[\mathbb{H}, \mathbb{H}] \subset \mathbb{H} \tag{11.28}$$

while in general one has:

$$[\mathbb{H}, \mathbb{K}] \subset \mathbb{H} \oplus \mathbb{K} \tag{11.29}$$

Definition 11.3.3. Let G/H be a Lie coset manifold and let the orthogonal decomposition of the corresponding Lie algebra be as in eq. (11.27). If the condition:

$$[\mathbb{H}, \mathbb{K}] \subset \mathbb{K} \tag{11.30}$$

applies, the coset G/H is named **reductive**.

Equation (11.30) has an obvious and immediate interpretation. The complementary space $\mathbb{K}$ forms a linear representation of the subalgebra $\mathbb{H}$ under its adjoint action within the ambient algebra $\mathbb{G}$.

Almost all of the "reasonable" coset manifolds which occur in various provinces of Mathematical Physics are reductive. Violation of *reductivity* is a sort of pathology whose study we can disregard in the scope of this book. We will consider only reductive coset manifolds.

Definition 11.3.4. Let G/H be a reductive coset manifold. If in addition to (11.30) also the following condition:

$$[\mathbb{K}, \mathbb{K}] \subset \mathbb{H} \tag{11.31}$$

applies, then the coset manifold G/H is named a **symmetric space**.

Let T_A $(A = 1, \dots, n)$ denote a complete basis of generators for the Lie algebra $\mathbb{G}$:

$$[T_A, T_B] = C^C{}_{AB} T_C \tag{11.32}$$

and T_i $(i = 1, \dots, m)$ denote a complete basis for the subalgebra $\mathbb{H} \subset \mathbb{G}$. We also introduce the denotation T_a $(a = 1, \dots, n - m)$ for a set of generators that provide a basis of the complementary subspace $\mathbb{K}$ in the orthogonal decomposition (11.27). We name T_a the *coset generators*. Using such denotations, eq. (11.32) splits into the following three ones:

$$[T_j, T_k] = C^i{}_{jk} T_i \tag{11.33}$$
$$[T_i, T_b] = C^a{}_{ib} T_a \tag{11.34}$$
$$[T_b, T_c] = C^i{}_{bc} T_i + C^a{}_{bc} T_a \tag{11.35}$$

Equation (11.33) encodes the property of $\mathbb{H}$ of being a subalgebra. Equation (11.34) encodes the property of the considered coset of being reductive. Finally if in eq. (11.35) we have $C^a{}_{bc} = 0$, the coset is not only reductive but also symmetric.

We are able to provide explicit formulae for the Riemann tensor of reductive coset manifolds equipped with G-invariant metrics in terms of such structure constants. Prior to that we consider the infinitesimal transformation and the very definition of the Killing vectors with respect to which the metric has to be invariant.

11.3.1.1 Infinitesimal transformations and Killing vectors

Let us consider the transformation law (11.20) of the coset representative. For a group element g infinitesimally close to the identity, we have:

$$g \simeq 1 + \epsilon^A T_A \tag{11.36}$$

$$h(y,g) \simeq 1 - \epsilon_A W^i_A(y) T_i \tag{11.37}$$
$$y'^\alpha \simeq y^\alpha + \epsilon^A k^\alpha_A \tag{11.38}$$

The induced h transformation in eq. (11.20) depends in general on the infinitesimal G-parameters ϵ^A and on the point in the coset manifold y, as shown in eq. (11.37). The y-dependent rectangular matrix $W^i_A(y)$ is usually named the $\mathbb{H}$-compensator. The shift in the coordinates y^α is also proportional to ϵ^A and the vector fields:

$$\mathbf{k}_A = k^\alpha_A(y) \frac{\partial}{\partial y^\alpha} \tag{11.39}$$

are named the *Killing vectors of the coset*. The reason for such a name will be justified when we show that on G/H we can construct a (pseudo-)Riemannian metric which admits the vector fields (11.39) as generators of infinitesimal isometries. For the time being those in (11.39) are just a set of vector fields that, as we prove few lines below, close the Lie algebra of the group G.

Inserting eqs (11.36)–(11.38) into the transformation law (11.20) we obtain:

$$T_A \mathbb{L}(y) = \mathbf{k}_A \mathbb{L}(y) - W^i_A(y) \mathbb{L}(y) T_i \tag{11.40}$$

Consider now the commutator $g_2^{-1} g_1^{-1} g_2 g_1$ acting on $\mathbb{L}(y)$. If both group elements $g_{1,2}$ are infinitesimally close to the identity in the sense of eq. (11.36), then we obtain:

$$g_2^{-1} g_1^{-1} g_2 g_1 \mathbb{L}(y) \simeq (1 - \epsilon_1^A \epsilon_2^B [T_A, T_B]) \mathbb{L}(y) \tag{11.41}$$

By explicit calculation we find:

$$\begin{aligned}[T_A, T_B]\mathbb{L}(y) &= T_A T_B \mathbb{L}(y) - T_B T_A \mathbb{L}(y) \\ &= [\mathbf{k}_A, \mathbf{k}_B]\mathbb{L}(y) - (\mathbf{k}_A W^i_B - \mathbf{k}_B W^i_A + 2C^i{}_{jk} W^j_A W^k_B)\mathbb{L}(y) T_i \end{aligned} \tag{11.42}$$

On the other hand, using the Lie algebra commutation relations, we obtain:

$$[T_A, T_B]\mathbb{L}(y) = C^C{}_{AB} T_C \mathbb{L}(y) = C^C{}_{AB}(\mathbf{k}_C \mathbb{L}(y) - W^i_C \mathbb{L}(y) T_i) \tag{11.43}$$

By equating the right-hand sides of eq. (11.42) and eq. (11.43) we conclude that:

$$[\mathbf{k}_A, \mathbf{k}_B] = C^C{}_{AB} \mathbf{k}_C \tag{11.44}$$
$$\mathbf{k}_A W^i_B - \mathbf{k}_B W^i_A + 2C^i{}_{jk} W^j_A W^k_B = C^C{}_{AB} W^i_C \tag{11.45}$$

where we separately compared the terms with and without Ws, since the decomposition of a group element into $\mathbb{L}(y)h$ is unique.

Equation (11.44) shows that the Killing vector fields defined above close the commutation relations of the $\mathbb{G}$-algebra.

Instead, eq. (11.45) will be used to construct a consistent $\mathbb{H}$-covariant Lie derivative.

In the case of the spaces $\mathbb{H}_{-}^{(n,1)}$, which we choose as illustrative example, the Killing vectors can be easily calculated by following the above described procedure step by step. For pedagogical purposes we find it convenient to present such a calculation in a slightly more general setup by introducing the following coset representative that depends on a discrete parameter $\kappa = \pm 1$:

$$\mathbb{L}_\kappa(\mathbf{y}) = \left(\begin{array}{c|c} \mathbf{1}_{n\times n} + 2\mathbf{y}\mathbf{y}^T \frac{\kappa}{1+\kappa\mathbf{y}^2} & -2\frac{\mathbf{y}}{1+\kappa\mathbf{y}^2} \\ \hline 2\kappa\frac{\mathbf{y}^T}{1+\kappa\mathbf{y}^2} & \frac{1-\kappa\mathbf{y}^2}{1+\kappa\mathbf{y}^2} \end{array}\right) \tag{11.46}$$

An explicit calculation shows that:

$$\mathbb{L}_\kappa(\mathbf{y})^T \underbrace{\left(\begin{array}{ccccc|c} 1 & 0 & \dots & 0 & 0 \\ 0 & 1 & \dots & 0 & 0 \\ \vdots & \vdots & \vdots & \vdots & \vdots \\ 0 & \dots & 0 & 1 & 0 \\ \hline 0 & \dots & 0 & 0 & \kappa \end{array}\right)}_{\eta_\kappa} \mathbb{L}_\kappa(\mathbf{y}) = \underbrace{\left(\begin{array}{cccc|c} 1 & 0 & \dots & 0 & 0 \\ 0 & 1 & \dots & 0 & 0 \\ \vdots & \vdots & \vdots & \vdots & \vdots \\ 0 & \dots & 0 & 1 & 0 \\ \hline 0 & \dots & 0 & 0 & \kappa \end{array}\right)}_{\eta_\kappa} \tag{11.47}$$

Namely, $\mathbb{L}_{-1}(\mathbf{y})$ is an $\mathrm{SO}(n,1)$ matrix, while $\mathbb{L}_1(\mathbf{y})$ is an $\mathrm{SO}(n+1)$ group element. Furthermore defining, as in eq. (11.25):

$$\mathbf{x}_\kappa(\mathbf{y}) \equiv \mathbb{L}_\kappa(\mathbf{y}) \begin{pmatrix} 0 \\ \vdots \\ 0 \\ \hline 1 \end{pmatrix} \tag{11.48}$$

we find that:

$$\mathbf{x}_\kappa(\mathbf{y})^T \eta_\kappa \mathbf{x}_\kappa(\mathbf{y}) = \kappa \tag{11.49}$$

Hence by means of $\mathbb{L}_1(\mathbf{y})$ we parameterize the points of the n-sphere $\mathbb{S}^n$, while by means of $\mathbb{L}_{-1}(\mathbf{y})$ we parameterize the points of $\mathbb{H}_{-}^{(n,1)}$ named also the n-pseudo-sphere or the n-hyperboloid. In both cases the stability subalgebra is $\mathfrak{so}(n)$ for which a basis of generators is provided by the following matrices:

$$J_{ij} = \mathscr{I}_{ij} - \mathscr{I}_{ji}; \quad i,j = 1,\dots,n \tag{11.50}$$

having named:

$$\mathscr{I}_{ij} = \left(\begin{array}{cccc|c} 0 & \dots & \dots & 0 & 0 \\ 0 & \dots & 1 & 0 & 0 \; \}\, i\text{-th row} \\ 0 & \dots & \dots & 0 & 0 \\ \hline 0 & \dots & \underbrace{\dots}_{j\text{-th column}} & 0 & 0 \end{array}\right). \tag{11.51}$$

the $(n+1)\times(n+1)$ matrices whose only non-vanishing entry is the ij-th one, equal to 1.

The commutation relations of the $\mathfrak{so}(n)$ generators are very simple. We have:

$$[J_{ij}, J_{k\ell}] = -\delta_{ik}J_{j\ell} + \delta_{jk}J_{i\ell} - \delta_{j\ell}J_{ik} + \delta_{i\ell}J_{jk} \tag{11.52}$$

The coset generators can instead be chosen as the following matrices:

$$P_i = \left(\begin{array}{cccc|cl} 0 & \dots & \dots & 0 & 0 & \\ 0 & \dots & 0 & 0 & 1 & \}\, i\text{-th row} \\ 0 & \dots & \dots & 0 & 0 & \\ \hline 0 & \dots & \underbrace{-\kappa}_{i\text{-th column}} & 0 & 0 & \end{array}\right) \tag{11.53}$$

and satisfy the following commutation relations:

$$[J_{ij}, P_k] = -\delta_{ik}P_j + \delta_{jk}P_i \tag{11.54}$$

$$[P_i, P_j] = -\kappa J_{ij} \tag{11.55}$$

Equation (11.54) states that the generators P_i transform as an n-vector under $\mathfrak{so}(n)$ rotations (reductivity) while eq. (11.55) shows that for both signs $\kappa = \pm 1$ the considered coset manifold is a symmetric space. Correspondingly, we name $\mathbf{k}_{ij} = k^\ell_{ij}(y)\frac{\partial}{\partial y^\ell}$ the Killing vector fields associated with the action of the generators J_{ij}:

$$J_{ij}\mathbb{L}_\kappa(\mathbf{y}) = \mathbf{k}_{ij}\mathbb{L}_\kappa(\mathbf{y}) + \mathbb{L}_\kappa(\mathbf{y})J_{pq}W^{pq}_{ij}(y) \tag{11.56}$$

while we name $\mathbf{k}_i = k^\ell_i(y)\frac{\partial}{\partial y^\ell}$ the Killing vector fields associated with the action of the generators P_i:

$$P_i\mathbb{L}_\kappa(\mathbf{y}) = \mathbf{k}_i\mathbb{L}_\kappa(\mathbf{y}) + \mathbb{L}_\kappa(\mathbf{y})J_{pq}W^{pq}_i(y) \tag{11.57}$$

Resolving conditions (11.56) and (11.57) we obtain:

$$\mathbf{k}_{ij} = y_i\partial_j - y_j\partial_i \tag{11.58}$$

$$\mathbf{k}_i = \frac{1}{2}(1 - \kappa\mathbf{y}^2)\partial_i + \kappa y_i \mathbf{y}\cdot\vec{\partial} \tag{11.59}$$

The $\mathbb{H}$-compensators W^{pq}_i and W^{pq}_{ij} can also be extracted from the same calculation but since their explicit form is not essential, in this context we skip them.

11.3.1.2 Vielbeins, connections and metrics on G/H

Consider next the following 1-form defined over the reductive coset manifold G/H:

$$\Sigma(y) = \mathbb{L}^{-1}(y)\,\mathrm{d}\mathbb{L}(y) \tag{11.60}$$

which generalizes the Maurer–Cartan form defined over the group manifold G. As a consequence of its own definition the 1-form Σ satisfies the equation:

$$0 = d\Sigma + \Sigma \wedge \Sigma \tag{11.61}$$

which provides the clue to the entire (pseudo-)Riemannian geometry of the coset manifold. To work out this latter we start by decomposing Σ along a complete set of generators of the Lie algebra $\mathbb{G}$. According to the denotations introduced in the previous subsection we put:

$$\Sigma = V^a T_a + \omega^i T_i \tag{11.62}$$

The set of $(n-m)$ 1-forms $V^a = V^a_\alpha(y)\, dy^\alpha$ provides a covariant frame for the cotangent bundle $T^\star(G/H)$, namely, a complete basis of sections of this vector bundle that transform in a proper way under the action of the group G. On the other hand, $\omega = \omega^i T_i = \omega^i_\alpha(y)\, dy^\alpha\, T_i$ is called the $\mathbb{H}$-connection. Indeed, according to the theory exposed in Section 10.2, ω turns out to be the 1-form of a bona fide principal connection on the principal fiber-bundle:

$$\mathscr{P}\left(\frac{G}{H}, H\right) : G \overset{\pi}{\rightarrow} \frac{G}{H} \tag{11.63}$$

which has the Lie group G as total space, the coset manifold G/H as base space and the closed Lie subgroup $H \subset G$ as structural group. The bundle $\mathscr{P}(\frac{G}{H}, H)$ is uniquely defined by the projection that associates to each group element $g \in G$ the equivalence class gH it belongs to.

Introducing the decomposition (11.62) into the Maurer–Cartan equation (11.61), this latter can be rewritten as the following pair of equations:

$$dV^a + C^a{}_{ib}\omega^i \wedge V^b = -\frac{1}{2}C^a{}_{bc}V^b \wedge V^c \tag{11.64}$$

$$d\omega^i + \frac{1}{2}C^i{}_{jk}\omega^j \wedge \omega^k = -\frac{1}{2}C^i{}_{bc}V^b \wedge V^c \tag{11.65}$$

where we have used the Lie algebra structure constants organized as in eqs (11.33)–(11.35).

Let us now consider the transformations of the 1-forms we have introduced.

Under left multiplication by a constant group element $g \in G$ the 1-form $\Sigma(y)$ transforms as follows:

$$\begin{aligned}\Sigma(y') &= h(y,g)\mathbb{L}^{-1}(y)g^{-1}\, d(g\mathbb{L}(y)h^{-1}) \\ &= h(y,g)^{-1}\Sigma(y)h(y,g) + h(y,g)^{-1}\, dh(y,g)\end{aligned} \tag{11.66}$$

where $y' = g.y$ is the new point in the manifold G/H whereto y is moved by the action of g. Projecting the above equation on the coset generators T_a we obtain:

$$V^a(y') = V^b(y)\mathscr{D}_b{}^a(h(y,g)) \tag{11.67}$$

where $\mathscr{D} = \exp[\mathfrak{D}_{\mathbb{H}}]$, having denoted by $\mathfrak{D}_{\mathbb{H}}$ the $(n-m)$-dimensional representation of the subalgebra $\mathbb{H}$ which occurs in the decomposition of the adjoint representation of $\mathbb{G}$:

$$\operatorname{adj}\mathbb{G} = \underbrace{\operatorname{adj}\mathbb{H}}_{=\mathfrak{A}_{\mathbb{H}}} \oplus \mathfrak{D}_{\mathbb{H}} \tag{11.68}$$

Projecting on the other hand on the $\mathbb{H}$-subalgebra generators T_i we get:

$$\omega(y') = \mathscr{A}[h(y,g)]\omega(y)\mathscr{A}^{-1}[h(y,g)] + \mathscr{A}[h(y,g)]\,\mathrm{d}\mathscr{A}^{-1}[h(y,g)] \tag{11.69}$$

where we have set:

$$\mathscr{A} = \exp[\mathfrak{A}_{\mathbb{H}}] \tag{11.70}$$

Considering a complete basis T_A of generators for the full Lie algebra $\mathbb{G}$, the adjoint representation is defined as follows:

$$\forall g \in \mathrm{G}: \quad g^{-1}T_A g \equiv \operatorname{adj}(g)_A{}^B T_B \tag{11.71}$$

In the explicit basis of T_A generators the decomposition (11.68) means that, once restricted to the elements of the subgroup $\mathrm{H} \subset \mathrm{G}$, the adjoint representation becomes block-diagonal:

$$\forall h \in \mathrm{H}: \quad \operatorname{adj}(h) = \left(\begin{array}{c|c} \mathscr{D}(h) & 0 \\ \hline 0 & \mathscr{A}(h) \end{array}\right) \tag{11.72}$$

Note that for such decomposition to hold true the coset manifold has to be reductive according to definition (11.30).

The infinitesimal form of eq. (11.67) is the following one:

$$V^a(y+\delta y) - V^a(y) = -\epsilon^A W^i_A(y) C^a{}_{ib} V^b(y) \tag{11.73}$$

$$\delta y^a = \epsilon^A k^a_A(y) \tag{11.74}$$

for a group element $g \in \mathrm{G}$ very close to the identity as in eq. (11.36).

Similarly the infinitesimal form of eq. (11.69) is:

$$\omega^i(y+\delta y) - \omega^i(y) = -\epsilon^A (C^i{}_{kj} W^k_A \omega^j + \mathrm{d}W^i_A) \tag{11.75}$$

11.3.1.3 Lie derivatives

The Lie derivative of a tensor $T_{\alpha_1\dots\alpha_p}$ along a vector field v^μ provides the change in shape of that tensor under an infinitesimal diffeomorphism:

$$y^\mu \mapsto y^\mu + v^\mu(y) \tag{11.76}$$

Explicitly one sets:

$$\ell_{\mathbf{v}} T_{\alpha_1\dots\alpha_p}(y) = v^\mu \partial_\mu T_{\alpha_1\dots\alpha_p} + (\partial_{\alpha_1} v^\gamma) T_{\gamma\alpha_2\dots\alpha_p} + \cdots \\ + (\partial_{\alpha_p} v^\gamma) T_{\alpha_1\alpha_2\dots\gamma} \tag{11.77}$$

In the case of p-forms, namely, of antisymmetric tensors, the definition (11.77) of Lie derivative can be recast into a more intrinsic form using both the exterior differential d and the *contraction operator.*

Definition 11.3.5. Let $\mathcal{M}$ be a differentiable manifold and let $\Lambda_k(\mathcal{M})$ be the vector bundles of differential k-forms on $\mathcal{M}$. Let $\mathbf{v} \in \Gamma(T\mathcal{M}, \mathcal{M})$ be a vector field. The contraction $\mathfrak{i}_{\mathbf{v}}$ is a linear map:

$$\mathfrak{i}_{\mathbf{v}} : \Lambda_k(\mathcal{M}) \to \Lambda_{k-1}(\mathcal{M}) \tag{11.78}$$

such that for any $\omega^{(k)} \in \Lambda_k(\mathcal{M})$ and for any set of $k-1$ vector fields $\mathbf{w}_1, \dots, \mathbf{w}_{k-1}$, we have:

$$\mathfrak{i}_{\mathbf{v}} \omega^{(k)}(\mathbf{w}_1, \dots, \mathbf{w}_{k-1}) \equiv k\omega^{(k)}(\mathbf{v}, \mathbf{w}_1, \dots, \mathbf{w}_{k-1}) \tag{11.79}$$

Then by going to components we can verify that the tensor definition (11.77) is equivalent to the following one:

Definition 11.3.6. Let $\mathcal{M}$ be a differentiable manifold and let $\Lambda_k(\mathcal{M})$ be the vector bundles of differential k-forms on $\mathcal{M}$. Let $\mathbf{v} \in \Gamma(T\mathcal{M}, \mathcal{M})$ be a vector field. The Lie derivative $\ell_{\mathbf{v}}$ is a linear map:

$$\ell_{\mathbf{v}} : \Lambda_k(\mathcal{M}) \to \Lambda_k(\mathcal{M}) \tag{11.80}$$

such that for any $\omega^{(k)} \in \Lambda_k(\mathcal{M})$ we have:

$$\ell_{\mathbf{v}} \omega^{(k)} \equiv \mathfrak{i}_{\mathbf{v}}\, d\omega^{(k)} + d\mathfrak{i}_{\mathbf{v}}\, \omega^{(k)} \tag{11.81}$$

On the other hand, for vector fields the tensor definition (11.77) is equivalent to the following one.

Definition 11.3.7. Let $\mathcal{M}$ be a differentiable manifold and let $T\mathcal{M} \to \mathcal{M}$ be the tangent bundle, whose sections are the vector fields. Let $\mathbf{v} \in \Gamma(T\mathcal{M}, \mathcal{M})$ be a vector field. The Lie derivative $\ell_{\mathbf{v}}$ is a linear map:

$$\ell_{\mathbf{v}} : \Gamma(T\mathcal{M}, \mathcal{M}) \to \Gamma(T\mathcal{M}, \mathcal{M}) \tag{11.82}$$

such that for any $\mathbf{w} \in \Gamma(T\mathcal{M}, \mathcal{M})$ we have:

$$\ell_{\mathbf{v}} \mathbf{w} \equiv [\mathbf{v}, \mathbf{w}] \tag{11.83}$$

The most important properties of the Lie derivative, which immediately follow from its definition, are the following ones:

$$\begin{aligned} [\ell_{\mathbf{v}}, \mathrm{d}] &= 0 \\ [\ell_{\mathbf{v}}, \ell_{\mathbf{w}}] &= \ell_{[\mathbf{v},\mathbf{w}]} \end{aligned} \tag{11.84}$$

The first of the above equations states that the Lie derivative commutes with exterior derivative. This is just a consequence of the invariance of the exterior algebra of k-forms with respect to diffeomorphisms. On the other hand, the second equation states that the Lie derivative provides an explicit representation of the Lie algebra of vector fields on tensors.

The Lie derivatives along the Killing vectors of the frames V^a and of the $\mathbb{H}$-connection ω^i introduced in the previous subsection are:

$$\ell_{\mathbf{v}_A} V^a = W^i_A C^a{}_{ib} V^b \tag{11.85}$$

$$\ell_{\mathbf{v}_A} \omega^i = -(\mathrm{d}W^i_A + C^i{}_{kj} W^k_A \omega^j) \tag{11.86}$$

This result can be interpreted by saying that, associated with every Killing vector $\mathbf{k}_A$, there is a an infinitesimal $\mathbb{H}$-gauge transformation:

$$\mathbf{W}_A = W^i_A(y) T_i \tag{11.87}$$

and that the Lie derivative of both V^a and ω^i along the Killing vectors is just such local gauge transformation pertaining to their respective geometrical type. The frame V^a is a section of an H-vector bundle and transforms as such, while ω^i is a connection and it transforms as a connection should do.

11.3.1.4 Invariant metrics on coset manifolds

The result (11.85), (11.86) has a very important consequence which constitutes the fundamental motivation to consider coset manifolds. Indeed, this result instructs us to construct G-invariant metrics on G/H, namely, metrics that admit all the above discussed Killing vectors as generators of true isometries.

The argument is quite simple. We saw that the 1-forms V^a transform in a linear representation $\mathfrak{D}_{\mathbb{H}}$ of the isotropy subalgebra $\mathbb{H}$ (and group H). Hence if τ_{ab} is a symmetric H-invariant constant two-tensor, by setting:

$$ds^2 = \tau_{ab} V^a \otimes V^b = \underbrace{\tau_{ab} V^a_\alpha(y) V^b_\beta(y)}_{g_{\alpha\beta}(y)}\, dy^\alpha \otimes dy^\beta \tag{11.88}$$

we obtain a metric for which all the above constructed Killing vectors are indeed Killing vectors, namely:

$$\ell_{\mathbf{k}_A} ds^2 = \tau_{ab}(\ell_{\mathbf{k}_A} V^a \otimes V^b + V^a \otimes \ell_{\mathbf{k}_A} V^b) \tag{11.89}$$

$$= \underbrace{\tau_{ab}([\mathfrak{D}_{\mathbb{H}}(W_A)]^a{}_c \delta^b_d + [\mathfrak{D}_{\mathbb{H}}(W_A)]^b{}_c \delta^a_d)}_{=0 \text{ by invariance}} V^c \otimes V^d$$
$$= 0 \tag{11.90}$$

The key point, in order to utilize the above construction, is the decomposition of the representation $\mathfrak{D}_{\mathbb{H}}$ into irreducible representations. Typically, for most common cosets, $\mathfrak{D}_{\mathbb{H}}$ is already irreducible. In this case there is just one invariant H-tensor τ and the only free parameter in the definition of the metric (11.88) is an overall scale constant. Indeed, if τ_{ab} is an invariant tensor, any multiple thereof $\tau'_{ab} = \lambda \tau_{ab}$ is also invariant. In the case $\mathfrak{D}_{\mathbb{H}}$ splits into $\mathfrak{r}$ irreducible representations:

$$\mathfrak{D}_{\mathbb{H}} = \left(\begin{array}{c|c|c|c|c} \mathfrak{D}_1 & 0 & \dots & 0 & 0 \\ \hline 0 & \mathfrak{D}_2 & 0 & \dots & 0 \\ \hline \vdots & \vdots & \vdots & \vdots & \vdots \\ \hline 0 & \dots & 0 & \mathfrak{D}_{\mathfrak{r}-1} & 0 \\ \hline 0 & 0 & \dots & 0 & \mathfrak{D}_{\mathfrak{r}} \end{array}\right) \tag{11.91}$$

we have $\mathfrak{r}$ irreducible invariant tensors $\tau^{(i)}_{a_i b_i}$ in correspondence of such irreducible blocks and we can introduce $\mathfrak{r}$ independent scale factors:

$$\tau = \left(\begin{array}{c|c|c|c|c} \lambda_1 \tau^{(1)} & 0 & \dots & 0 & 0 \\ \hline 0 & \lambda_2 \tau^{(2)} & 0 & \dots & 0 \\ \hline \vdots & \vdots & \vdots & \vdots & \vdots \\ \hline 0 & \dots & 0 & \lambda_{p-1} \tau^{(p-1)} & 0 \\ \hline 0 & 0 & \dots & 0 & \lambda_p \tau^{(p)} \end{array}\right) \tag{11.92}$$

Correspondingly we arrive at a continuous family of G-invariant metrics on G/H depending on $\mathfrak{r}$-parameters or, as it is customary to say in this context, of $\mathfrak{r}$ *moduli*. The number $\mathfrak{r}$ defined by eq. (11.91) is named the *rank of the coset manifold* G/H.

In this section we confine ourself to the most common case of rank-1 cosets ($\mathfrak{r} = 1$), assuming, furthermore, that the algebras $\mathbb{G}$ and $\mathbb{H}$ are both semisimple. By an appropriate choice of basis for the coset generators T^a, the invariant tensor τ_{ab} can always be reduced to the form:

$$\tau_{ab} = \eta_{ab} = \operatorname{diag}(\underbrace{+,+,\dots,+}_{n_+},\underbrace{-,-,\dots,-}_{n_-}) \tag{11.93}$$

where the two numbers n_+ and n_- sum up to the dimension of the coset:

$$n_+ + n_- = \dim \frac{\mathrm{G}}{\mathrm{H}} = \dim \mathbb{K} \tag{11.94}$$

and provide the dimensions of the two eigenspaces, $\mathbb{K}_\pm \subset \mathbb{K}$, respectively corresponding to real and pure imaginary eigenvalues of the matrix $\mathfrak{D}_{\mathbb{H}}(W)$ which represents a generic element W of the isotropy subalgebra $\mathbb{H}$.

Focusing on our example (11.46), that encompasses both the spheres and the pseudo-spheres, depending on the sign of κ, we find that:

$$n_+ = 0; \quad n_- = n \tag{11.95}$$

so that in both cases ($\kappa = \pm 1$) the invariant tensor is proportional to a Kronecker delta:

$$\eta_{ab} = \delta_{ab} \tag{11.96}$$

The reason is that the subalgebra $\mathbb{H}$ is the compact $\mathfrak{so}(n)$, hence the matrix $\mathfrak{D}_{\mathfrak{H}}(W)$ is antisymmetric and all of its eigenvalues are purely imaginary.

If we consider cosets with non-compact isotropy groups, then the invariant tensor τ_{ab} develops a non-trivial Lorentzian signature η_{ab}. In any case, if we restrict ourselves to rank-1 cosets, the general form of the metric is:

$$ds^2 = \lambda^2 \eta_{ab} V^a \otimes V^b \tag{11.97}$$

where λ is a scale factor. This allows us to introduce the *vielbein*

$$E^a = \lambda V^a \tag{11.98}$$

and calculate the *spin connection* from the vanishing torsion equation:

$$0 = \mathrm{d}E^a - \omega^{ab} \wedge E^c \eta_{bc} \tag{11.99}$$

Using the Maurer–Cartan equations (11.64)–(11.65), eq. (11.99) can be immediately solved by:

$$\omega^{ab}\eta_{bc} \equiv \omega^a{}_c = \frac{1}{2\lambda} C^a{}_{cd} E^d + C^a{}_{ci}\omega^i \tag{11.100}$$

Inserting this in the definition of the curvature 2-form

$$\mathfrak{R}^a{}_b = \mathrm{d}\omega^a{}_b - \omega^a{}_c \wedge \omega^c{}_b \tag{11.101}$$

allows to calculate the Riemann tensor defined by:

$$\mathfrak{R}^a{}_b = \mathscr{R}^a{}_{bcd} E^c \wedge E^d \tag{11.102}$$

Using once again the Maurer–Cartan equations (11.64)–(11.65), we obtain:

$$\mathscr{R}^a{}_{bcd} = \frac{1}{\lambda^2}\left(-\frac{1}{4} C^a{}_{be} C^e{}_{cd} - \frac{1}{8} C^a{}_{ec} C^e{}_{bd} + \frac{1}{8} C^a{}_{ed} C^e{}_{bc} - \frac{1}{2} C^a{}_{bi} C^i{}_{cd}\right) \tag{11.103}$$

which, as previously announced, provides the expression of the Riemann tensor in terms of structure constants.

In the case of symmetric spaces $C^a{}_{be} = 0$ formula (11.103) simplifies to:

$$\mathscr{R}^a{}_{bcd} = -\frac{1}{2\lambda^2} C^a{}_{bi} C^i{}_{cd} \tag{11.104}$$

11.3.1.5 For spheres and pseudo-spheres

In order to illustrate the structures presented in the previous section, we consider the explicit example of the spheres and pseudo-spheres. Applying the outlined procedure to this case we immediately get:

$$\begin{aligned} E^a &= -\frac{2}{\lambda}\frac{\mathrm{d}y^a}{1+\kappa \mathbf{y}^2} \\ \omega^{ab} &= 2\frac{\kappa}{\lambda^2}E^a \wedge E^b \end{aligned} \tag{11.105}$$

This means that for spheres and pseudo-spheres the Riemann tensor is proportional to an antisymmetrized Kronecker delta:

$$R^{ab}{}_{cd} = \frac{\kappa}{\lambda^2}\delta^{[a}_{[c}\delta^{b]}_{d]} \tag{11.106}$$

11.4 The real sections of a complex Lie algebra and symmetric spaces

In the context of coset manifolds a very interesting class that finds important applications in supergravity and superstring theories is the following one:

$$\mathcal{M}_{\mathrm{G}_R} = \frac{\mathrm{G}_R}{\mathrm{H}_c} \tag{11.107}$$

where G_R is some semisimple Lie group and $\mathrm{H}_c \subset \mathrm{G}_R$ is its maximal compact subgroup. The Lie algebra $\mathbb{H}_c$ of the denominator H_c is the maximal compact subalgebra $\mathbb{H} \subset \mathbb{G}_R$ which has typically rank $r_{\text{compact}} > r$. Denoting, as usual, by $\mathbb{K}$ the orthogonal complement of $\mathbb{H}_c$ in $\mathbb{G}_R$:

$$\mathbb{G}_R = \mathbb{H}_c \oplus \mathbb{K} \tag{11.108}$$

and defining as non-compact rank or rank of the coset $\mathrm{G}_\mathrm{R}/\mathrm{H}$ the dimension of the non-compact Cartan subalgebra:

$$r_{nc} = \mathrm{rank}(\mathrm{G}_\mathrm{R}/\mathrm{H}) \equiv \dim \mathcal{H}^{n.c.}; \quad \mathcal{H}^{n.c.} \equiv \mathrm{CSA}_{\mathbb{G}(\mathbb{C})} \cap \mathbb{K} \tag{11.109}$$

we obtain that $r_{nc} < r$.

By definition the Lie algebra $\mathbb{G}_R$ is a real section of a complex semisimple Lie algebra. In Section 8.1.1.1 we met the first two universal instances of real sections of a simple complex Lie algebra $\mathbb{G}(\mathbb{C})$, namely, the *maximally split* and the *maximally compact real sections*.

All other possible real sections are obtained by studying the available Cartan involutions of the complex Lie algebra. So consider:

Definition 11.4.1. Let:

$$\theta : \mathfrak{g} \to \mathfrak{g} \tag{11.110}$$

be a linear automorphism of the compact Lie algebra $\mathfrak{g} = \mathbb{G}_c$, where $\mathbb{G}_c$ is the maximal compact real section of a complex semisimple Lie algebra $\mathbb{G}(\mathbb{C})$. By definition we have:

$$\forall \alpha, \beta \in \mathbb{R}, \ \forall \mathbf{X}, \mathbf{Y} \in \mathfrak{g} : \quad \begin{cases} \theta(\alpha\mathbf{X} + \beta\mathbf{Y}) = \alpha\theta(\mathbf{X}) + \beta\theta(\mathbf{Y}) \\ \theta([\mathbf{X}, \mathbf{Y}]) = [\theta(\mathbf{X}), \theta(\mathbf{Y})] \end{cases} \tag{11.111}$$

If $\theta^2 = \mathbf{Id}$ then θ is named a Cartan involution of the Lie algebra $\mathfrak{g}$.

For any Cartan involution θ the possible eigenvalues are ± 1. This allows us to split the entire Lie algebra $\mathfrak{g}$ into two subspaces corresponding to the eigenvalues 1 and -1, respectively:

$$\mathfrak{g} = \mathfrak{H}_\theta \oplus \mathfrak{p}_\theta \tag{11.112}$$

One immediately realizes that:

$$\begin{aligned} [\mathfrak{H}_\theta, \mathfrak{H}_\theta] &\subset \mathfrak{H}_\theta \\ [\mathfrak{H}_\theta, \mathfrak{p}_\theta] &\subset \mathfrak{p}_\theta \\ [\mathfrak{p}_\theta, \mathfrak{p}_\theta] &\subset \mathfrak{H}_\theta \end{aligned} \tag{11.113}$$

Hence for any Cartan involution, $\mathfrak{H}_\theta$ is a subalgebra and θ singles out a symmetric homogeneous compact coset manifold:

$$\mathcal{M}_\theta = \frac{\mathrm{G_c}}{\mathrm{H}_\theta} \quad \text{where } \mathrm{H}_\theta \equiv \exp[\mathfrak{H}_\theta]; \quad \mathrm{G_c} \equiv \exp[\mathfrak{g}] \tag{11.114}$$

The structure (11.113) has also another important consequence. If we define the vector space:

$$\mathfrak{g}^\star_\theta = \mathfrak{H}_\theta \oplus \mathfrak{p}^\star_\theta; \quad \mathfrak{p}^\star_\theta \equiv i\mathfrak{p}_\theta \tag{11.115}$$

we see that $\mathfrak{g}^\star_\theta$ is closed under the Lie bracket and hence it is a Lie algebra. It is some real section of the complex Lie algebra $\mathbb{G}(\mathbb{C})$ and we can consider a new, generally non-compact coset manifold:

$$\mathcal{M}^\star_\theta = \frac{\mathrm{G}^\star_\theta}{\mathrm{H}_\theta}; \quad \mathrm{H}_\theta \equiv \exp[\mathfrak{H}_\theta]; \quad \mathrm{G}^\star_\theta \equiv \exp[\mathfrak{g}^\star_\theta] \tag{11.116}$$

An important theorem for which we refer the reader to classical textbooks [72, 84, 101][4] states that all real forms of a Lie algebra, up to isomorphism, are obtained in this way.

4 The proof is also summarized in appendix B of [85].

Furthermore, as part of the same theorem one has that θ can always be chosen in such a way that it maps the compact Cartan subalgebra onto itself:

$$\theta : \mathscr{H}_c \to \mathscr{H}_c \tag{11.117}$$

This short discussion reveals that the classification of real forms of a complex Lie Algebra $\mathbb{G}(\mathbb{C})$ is in one-to-one correspondence with the classification of symmetric spaces, the complexification of whose Lie algebra of isometries is $\mathbb{G}(\mathbb{C})$. For this reason we have postponed the discussion of the real forms to the present chapter devoted to homogeneous coset manifolds.

Let us now consider the action of the Cartan involution on the Cartan subalgebra: $\mathscr{H}_c = \text{span}\{iH_i\}$ of the maximal compact section $\mathbb{G}_c$. Choosing a basis of $\mathscr{H}_c$ aligned with the simple roots:

$$\mathscr{H}_c = \text{span}\{iH_{\alpha_i}\} \tag{11.118}$$

we see that the action of the Cartan involution θ is by duality transferred to the simple roots α_i and hence to the entire root lattice. As a consequence we can introduce the notion of real and imaginary roots. One argues as follows.

We split the Cartan subalgebra into its compact and non-compact subalgebras:

$$\begin{array}{rcl} \text{CSA}_{\mathbb{G}_R} & = & i\mathscr{H}^{\text{comp}} \oplus \mathscr{H}^{n.c.} \\ & & \Updownarrow \qquad\quad \Updownarrow \\ \text{CSA}_{\mathbb{G}_{\max}} & = & \mathscr{H}^{\text{comp}} \oplus \mathscr{H}^{n.c.} \end{array} \tag{11.119}$$

defining:

$$\begin{array}{rcl} h \in \mathscr{H}^{\text{comp}} & \Leftrightarrow & \theta(h) = h \\ h \in \mathscr{H}^{n.c} & \Leftrightarrow & \theta(h) \neq h \end{array} \tag{11.120}$$

Then every vector in the dual of the full Cartan subalgebra, in particular every root α, can be decomposed into its parallel and its transverse part to $\mathscr{H}^{n.c.}$:

$$\alpha = \alpha_{\parallel} \oplus \alpha_{\perp} \tag{11.121}$$

A root α is named *imaginary* if $\alpha_{\parallel} = 0$. On the contrary, a root α is called real if $\alpha_{\perp} = 0$. Generically a root is complex.

Given the original Dynkin diagram of a complex Lie algebra we can characterize a real section by mentioning which of the simple roots are imaginary. We do this by painting black the imaginary roots. The result is a Tits–Satake diagram like that in Figure 11.2 which corresponds to the real Lie Algebra $\mathfrak{so}(p, 2\ell - p + 1)$ for $p > 2$, $\ell > 2$.

$\mathfrak{b}_\ell$

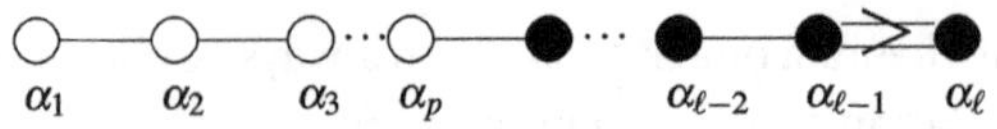

Figure 11.2: The Tits–Satake diagram representing the real form $\mathfrak{so}(p, 2\ell - p + 1)$ of the complex $\mathfrak{so}(2\ell + 1)$ Lie algebra.

11.5 The solvable group representation of non-compact coset manifolds

Definition 11.5.1. A Riemannian space $(\mathcal{M}, g)$ is named **normal** if it admits a completely solvable[5] Lie group $\exp[\text{Solv}(\mathcal{M})]$ of isometries that acts on the manifold in a simply transitive manner (i.e., for every two points in the manifold there is one and only one group element connecting them). The group $\exp[\text{Solv}(\mathcal{M})]$ is generated by a so-called **normal metric Lie algebra**, $\text{Solv}(\mathcal{M})$, that is a completely solvable Lie algebra endowed with a Euclidean metric.

The main tool to classify and study homogeneous spaces of the type (11.116) is provided by a theorem [84] that states that if a Riemannian manifold $(\mathcal{M}, g)$ is normal, according to Definition 11.5.1, then it is metrically equivalent to the solvable group manifold

$$\begin{aligned} \mathcal{M} &\simeq \exp[\text{Solv}(\mathcal{M})] \\ g\,|_{e\in\mathcal{M}} &= \langle , \rangle \end{aligned} \tag{11.122}$$

where $\langle , \rangle$ is a Euclidean metric defined on the normal solvable Lie algebra $\text{Solv}(\mathcal{M})$. The key point is that non-compact coset manifolds of the form (11.116) are all normal. This is so because there is always, *for all real forms except the maximally compact one*, a **solvable subalgebra** with the following features:

$$\begin{aligned} &\forall \text{Solv}\left(\frac{\mathrm{G}_R}{\mathrm{H}_c}\right) \subset \mathrm{G}_R \\ &\dim\left[\text{Solv}\left(\frac{\mathrm{G}_R}{\mathrm{H}_c}\right)\right] = \dim\left(\frac{\mathrm{G}_R}{\mathrm{H}_c}\right) \\ &\exp\left[\text{Solv}\left(\frac{\mathrm{G}_R}{\mathrm{H}_c}\right)\right] = \text{transitive on } \frac{\mathrm{G}_R}{\mathrm{H}_c} \end{aligned} \tag{11.123}$$

It is very easy to single out the appropriate solvable algebra in the case of the maximally split real form $\mathbb{G}_{\mathbf{max}}$. In that case, as we know, the maximal compact subalgebra has the following form:

$$\mathbb{H}_c = \text{span}\{(E^\alpha - E^{-\alpha})\}; \quad \forall \alpha \in \Phi_+ \tag{11.124}$$

5 A solvable Lie algebra s is completely solvable if the adjoint operation ad_X for all generators $X \in s$ has only real eigenvalues. The nomenclature of the Lie algebra is carried over to the corresponding Lie group in general in this chapter.

The solvable algebra that does the required job is the Borellean subalgebra:

$$\mathrm{Bor}(\mathbb{G}_{\mathbf{max}}) \equiv \mathscr{H} \oplus \mathrm{span}(E^{\alpha}); \quad \forall \alpha \in \mathbf{\Phi}_{+} \tag{11.125}$$

where $\mathscr{H}$ is the complete Cartan subalgebra and E^{α} are the step operators associated with the positive roots. That $\mathrm{Bor}(\mathbb{G}_{\mathbf{max}})$ is a solvable Lie algebra follows from the canonical structure of Lie algebras (7.37) presented in Section 7.2.1. If you exclude the negative roots, you immediately see that the Cartan generators are not in the first derivative of the algebra. The second derivative excludes all the simple roots: the third derivative excludes the roots of height 2 and so on until you end up in a derivative that makes zero. Hence the Lie algebra is solvable. Furthermore, it is obvious that any equivalence class of $\frac{\mathrm{G}_R}{\mathrm{H}_c}$ has a representative that is an element of the solvable Lie group $\exp[\mathrm{Bor}(\mathbb{G}_{\mathbf{max}})]$. This is intuitive at the infinitesimal level from the fact that each element of the complementary space:

$$\mathbb{K} = \mathscr{H} \oplus \mathrm{span}[(E^{\alpha} + E^{-\alpha})] \tag{11.126}$$

which generates the coset, can be uniquely rewritten as an element of $\mathrm{Bor}(\mathbb{G}_{\mathbf{max}})$ plus an element of the subalgebra $\mathbb{H}_c$. At the finite level an exact formula which connects the solvable representative $\exp[\mathbf{s}]$ (with $\mathbf{s} \in \mathrm{Bor}$) to the orthogonal representative $\exp[\mathbf{k}]$ (with $\mathbf{k} \in \mathbb{K}$) of the same equivalence class was derived by P. Frè and A. Sorin and it is presented in [54]. Here it suffices to us to understand that the action of the Borel group is transitive on the coset manifold, so that the coset manifold $\mathrm{G_R}/\mathrm{H_c}$ is indeed normal and its metric can be obtained from the non-degenerate Euclidean metric $\langle , \rangle$ defined over $\mathrm{Bor}(\mathrm{G}_{\mathbf{max}}) = \mathrm{Solv}(\frac{\mathrm{G}_{\mathrm{max}}}{\mathrm{H}_c})$.

The example of the maximally split case clearly suggests what is the required solvable algebra for other normal forms. We have:

$$\mathrm{Solv}\left(\frac{\mathrm{G}_R}{\mathrm{H}_c}\right) = \mathscr{H}^{n.c.} \oplus \mathrm{span}(\mathscr{E}^{\alpha}); \quad \forall \alpha \in \mathbf{\Phi}_{+}/\alpha_{\parallel} \neq 0 \tag{11.127}$$

where $\mathscr{H}^{n.c.}$ is the non-compact part of the Cartan subalgebra and $\mathscr{E}^{\alpha}$ denotes the combination of step operators pertaining to the positive roots α that appear in the real form $\mathbb{G}_R$ and the sum is extended only to those roots that are not purely imaginary. Indeed, the step operators pertaining to imaginary roots are included into the maximal compact subalgebra that now is larger than the number of positive roots.

For any solvable group manifold with a non-degenerate invariant metric[6] the differential geometry of the manifold is completely rephrased in algebraic language through the relation of the Levi-Civita connection and the *Nomizu operator* acting on

6 See [60, 4, 3, 2, 50] for reviews on the solvable Lie algebra approach to supergravity scalar manifolds and the use of the Nomizu operator.

the solvable Lie algebra. The latter is defined as

$$\mathbb{L} : \mathrm{Solv}(\mathcal{M}) \otimes \mathrm{Solv}(\mathcal{M}) \to \mathrm{Solv}(\mathcal{M}) \tag{11.128}$$

$$\forall X, Y, Z \in \mathrm{Solv}(\mathcal{M}) : \quad 2\langle \mathbb{L}_X Y, Z\rangle = \langle [X, Y], Z\rangle - \langle X, [Y, Z]\rangle - \langle Y, [X, Z]\rangle \tag{11.129}$$

The *Riemann curvature operator* on this group manifold can be expressed as

$$\mathrm{Riem}(X, Y) = [\mathbb{L}_X, \mathbb{L}_Y] - \mathbb{L}_{[X,Y]} \tag{11.130}$$

This implies that the covariant derivative explicitly reads:

$$\mathbb{L}_X Y = \Gamma^Z_{XY} Z \tag{11.131}$$

where

$$\Gamma^Z_{XY} = \frac{1}{2}(\langle Z, [X, Y]\rangle - \langle X, [Y, Z]\rangle - \langle Y, [X, Z]\rangle)\frac{1}{\langle Z, Z\rangle} \quad \forall X, Y, Z \in \mathrm{Solv} \tag{11.132}$$

Equation (11.132) is true for any solvable Lie algebra, but in the case of **maximally non-compact, split algebras** we can write a general form for Γ^Z_{XY}, namely:

$$\begin{aligned}
\Gamma^i_{jk} &= 0 \\
\Gamma^i_{\alpha\beta} &= \frac{1}{2}(-\langle E_\alpha, [E_\beta, H^i]\rangle - \langle E_\beta, [E_\alpha, H^i]\rangle) = \frac{1}{2}\alpha^i \delta_{\alpha\beta} \\
\Gamma^\alpha_{ij} &= \Gamma^\alpha_{i\beta} = \Gamma^i_{j\alpha} = 0 \\
\Gamma^\alpha_{\beta i} &= \frac{1}{2}(\langle E^\alpha, [E_\beta, H_i]\rangle - \langle E_\beta, [H_i, E^\alpha]\rangle) = -\alpha_i \delta^\alpha_\beta \\
\Gamma^{\alpha+\beta}_{\alpha\beta} &= -\Gamma^{\alpha+\beta}_{\beta\alpha} = \frac{1}{2}N_{\alpha\beta} \\
\Gamma^\alpha_{\alpha+\beta\beta} &= \Gamma^\alpha_{\beta\alpha+\beta} = \frac{1}{2}N_{\alpha\beta}
\end{aligned} \tag{11.133}$$

where $N^{\alpha\beta}$ is defined by the commutator

$$[E_\alpha, E_\beta] = N_{\alpha\beta} E_{\alpha+\beta} \tag{11.134}$$

The explicit form (11.133) follows from the following choice of the non-degenerate metric:

$$\begin{aligned}
\langle \mathcal{H}_i, \mathcal{H}_j\rangle &= 2\delta_{ij} \\
\langle \mathcal{H}_i, E_\alpha\rangle &= 0 \\
\langle E_\alpha, E_\beta\rangle &= \delta_{\alpha,\beta}
\end{aligned} \tag{11.135}$$

$\mathcal{H}_i \in$ CSA and E_α are the step operators associated with positive roots $\alpha \in \Phi_+$. For any other **non-split case**, the Nomizu connection exists nonetheless although it does not take the form (11.133). It follows from eq. (11.132) upon the choice of an invariant positive metric on Solv and the use of the structure constants of Solv.

11.5.1 The Tits–Satake projection: just a flash

Let us now come back to eq. (11.121). Setting all $\alpha_\perp = 0$ corresponds to a projection:

$$\Pi_{TS} : \Phi_{\mathbb{G}} \mapsto \overline{\Phi} \tag{11.136}$$

of the original root system $\Phi_{\mathbb{G}}$ onto a new system of vectors living in an Euclidean space of dimension equal to the non-compact rank r_{nc}. A priori this is not obvious, but it is nonetheless true that $\overline{\Phi}$, with only one exception, is by itself the root system of a simple Lie algebra $\mathbb{G}_{TS}$, the Tits–Satake subalgebra of $\mathbb{G}_R$:

$$\overline{\Phi} = \text{root system of } \mathbb{G}_{TS} \subset \mathbb{G}_R \tag{11.137}$$

The Tits–Satake subalgebra $\mathbb{G}_{TS} \subset \mathbb{G}_R$ is always the maximally non-compact real section of its own complexification. For this reason, considering its maximal compact subalgebra $\mathbb{H}_{\mathrm{TS}} \subset \mathbb{G}_{\mathrm{TS}}$, we have a new smaller coset $\frac{\mathrm{G}_{\mathrm{TS}}}{\mathrm{H}_{\mathrm{TS}}}$, which is maximally split. What is the relation between the two solvable Lie algebras $\mathrm{Solv}(\frac{\mathrm{G}}{\mathrm{H}})$ and $\mathrm{Solv}(\frac{\mathrm{G}_{\mathrm{TS}}}{\mathrm{H}_{\mathrm{TS}}})$ is the natural question which arises. The explicit answer to this question and the systematic illustration of the geometrical relevance of the Tits–Satake projection is a topic extensively discussed in [54]. Indeed, this projection plays a very significant role in supergravity theories and in the search for black-hole solutions, for cosmological potentials and for gaugings of supergravity models. The Tits–Satake projection finds its deepest and most useful interpretation in the context of Special Geometries and of the c-map, these items being the most conspicuous contributions of Supergravity to Modern Geometry. We refer the reader to the forthcoming book [54] for the development of such topics.

11.6 Bibliographical note

For the topics discussed in the present chapter, main references are [51, 22, 84] and the forthcoming [54].

12 An anthology of group theory physical applications

12.1 Introduction

We conclude this primary on Group Theory and Manifolds for Physicists with a restricted anthology of applications of the mathematical conceptions developed in previous chapters to various areas of theoretical physics.

We stress that, from the perspective of contemporary theoretical visions of the world, Group Theory is not just a tool for physics, but rather it is an essential constituent of its inner structure.

Differential Calculus was developed by Newton in order to define the fundamental concepts of mechanics and arrive at the laws of gravitation. In the same way one cannot formulate the standard model of fundamental interactions without Lie groups, fiber-bundles and representation theory. Neither can one understand the atomic spectra, or digest the concept of quantum numbers that label physical states, without knowing group representations. Similarly, finite group theory is equally essential to understand and classify crystals, molecular dynamics and most topics in condensed matter physics.

In mathematics and especially in geometry the theory of groups, both finite and continuous, is equally central. It is fundamental in algebraic topology, in differential geometry and in algebraic geometry.

In advanced modern quantum field theory, and in particular in supergravity and in superstring theory, groups and coset manifolds enter in an essential full-fledged way. Exceptional Lie algebras found in this context a multifaceted primary role that was unsuspected before.

These scant remarks, which might be largely extended, already demonstrate that Group Theory must be a cornerstone in the education of a theoretical physics student. To give such a student just a taste of the ramified applications of Group Theory that permeate the entire fabrics of Physics at all levels, in this chapter we provide three examples that illustrate the issue from different viewpoints:

A) The use of the tetrahedral extended group and of its irreducible representations in the treatment of molecular vibrational normal modes for molecules with such a symmetry like the methane CH_4. This example illustrates the role of finite group theory in chemical-physics.

B) The hidden extended symmetry of the Hydrogen atom quantum system which allows to understand its spectrum group theoretically. This is a primary example of the role of group theory in relation to integrable systems.

C) The group theoretical basis of the AdS_4/CFT_3 correspondence. This is just a flash in very hot topics of current advanced research. It provides the opportunity to illustrate three important issues:

https://doi.org/10.1515/9783110551204-012

1. The very much physical concept of *particle* coincides at the quantum level with the mathematical concept of *Unitary Irreducible Representation* of the isometry group of space-time or at least of its asymptotic geometry.
2. The additional ingredient of supersymmetry that promotes Lie algebras to super Lie algebras
3. The physical role of different real sections of the same complex Lie algebras coexisting as different subalgebras of the same ambient Lie algebra. The rationale of the AdS/CFT correspondence that has produced some of the most existing advances in contemporary research is just rooted in such rotation of real sections!

12.2 The full tetrahedral group and the vibrations of XY_4 molecules

A classical application of group theory in chemical-physics is the classification of eigen-modes for the vibrations of molecules with prescribed discrete symmetry. One paradigmatic case is that of molecules of type XY_4 with four atoms of an element Y located at the vertices of a tetrahedron and one atom of the element X located at the center of the tetrahedron. One such molecule is *methane* whose chemical formula is CH_4 (see Figure 12.1). The general problem of molecular vibrations can be formulated in the following way. We have five atoms in the considered molecule; let us name $\boldsymbol{q}_I = \{x_I, y_I, z_I\}$ their positions. The Hamiltonian of the corresponding dynamical system is of the form:

$$H = \frac{1}{2}\sum_{I=0}^{4} m_I \left(\frac{d\boldsymbol{q}_I}{dt}\right)^2 + V(\boldsymbol{q}_I) \tag{12.1}$$

where $V(\boldsymbol{q}_I)$ is the potential describing interatomic forces. By hypothesis, in the case of XY_4 molecules since the molecule exists, the positions $\boldsymbol{q}_I = \boldsymbol{p}_I$, where $\boldsymbol{p}_i$ $(i = 1, \dots, 4)$

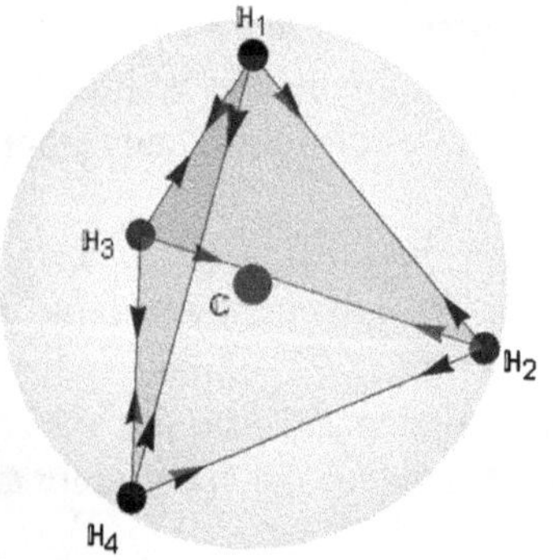

Figure 12.1: The molecule of methane is made by four hydrogen atoms located at the vertices of a tetrahedron and one carbon atom located at the center of the tetrahedron. Each of the four hydrogens can move into three directions with respect to their rest position.

are the vertices of the tetrahedron for the Y atoms, and $\boldsymbol{q}_0 = \boldsymbol{p}_0$, where $\boldsymbol{p}_0$ is the centroid of the same for the X atom, correspond to a stable minimum of the potential:

$$\partial_{\boldsymbol{q}_I} V(\boldsymbol{q}_I)|_{\boldsymbol{q}_I=\boldsymbol{p}_I} = 0 \quad I = 0, 1, \dots, 4 \tag{12.2}$$

Hence we consider small fluctuations around this equilibrium position:

$$\boldsymbol{q}_I = \boldsymbol{p}_I + \boldsymbol{\xi}_I \tag{12.3}$$

At second order in ξ_I we have the following Hamiltonian:

$$H = \frac{1}{2}\sum_{I=0}^{4} m_I \left(\frac{d\boldsymbol{\xi}_I}{dt}\right)^2 + \sum_{I,J=0}^{4}\sum_{\mu,\nu=1}^{3} A_{I,\mu,J,\nu}\xi_I^\mu \xi_J^\nu \tag{12.4}$$

where

$$A_{I,\mu,J,\nu} \equiv \partial^2_{\boldsymbol{q}_I^\mu,\boldsymbol{q}_J^\nu} V|_{\boldsymbol{q}_I=\boldsymbol{p}_I} \tag{12.5}$$

The problem is that of finding the eigenvalues and the eigenvectors of the 15×15 matrix $A_{I,\mu,J,\nu}$. Since the original Hamiltonian is invariant under the extended tetrahedral symmetry, this matrix must commute with elements of the group $\mathrm{T_d}$ and therefore must be diagonal on each irreducible subspace in force of Schur's lemma. The eigenvalues are the vibrational frequencies that are physically observable. Mathematically we just have to decompose the 15-dimensional representation of the extended tetrahedral group into irreducible representations. In this section we just sketch such a procedure.

12.2.1 Group theory of the full tetrahedral group

We begin with the proper tetrahedral group T_{12} that is a subgroup of SO(3) and it is a symmetry of the tetrahedron. Consider the standard tetrahedron (see Figure 12.2) the coordinates of whose vertices are given by the following four three-vectors:

$$\begin{aligned} \boldsymbol{V}_1 &= \{0, 0, 1\} \\ \boldsymbol{V}_2 &= \left\{\sqrt{\frac{2}{3}}, \frac{\sqrt{2}}{3}, -\frac{1}{3}\right\} \\ \boldsymbol{V}_3 &= \left\{-\sqrt{\frac{2}{3}}, \frac{\sqrt{2}}{3}, -\frac{1}{3}\right\} \\ \boldsymbol{V}_4 &= \left\{0, -\frac{2\sqrt{2}}{3}, -\frac{1}{3}\right\} \end{aligned} \tag{12.6}$$

Next we introduce the four rotations of 120 degrees around the four axes defined as the line that joins the center $\{0, 0, 0\}$ with each of the four vertices of the tetrahedron.

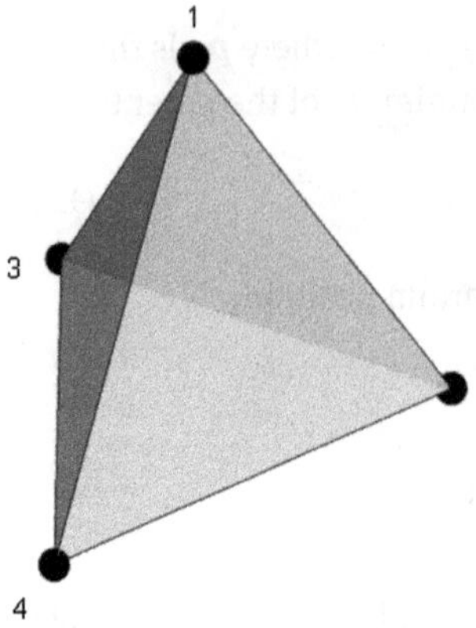

Figure 12.2: The standard tetrahedron the coordinates of whose vertices are displayed in eq. (12.6).

These finite elements of SO(3) are easily calculated recalling that the one parameter group of rotations around any given axis $\boldsymbol{v} = \{v_1, v_2, v_3\}$ is given by the matrices:

$$R_{\boldsymbol{v}}(\theta) \equiv \exp[\theta \mathbf{A}_{\boldsymbol{v}}] \quad \text{where } \mathbf{A}_{\boldsymbol{v}} = \left(\begin{array}{c|c|c} 0 & -v_3 & v_2 \\ \hline v_3 & 0 & -v_1 \\ \hline -v_2 & v_1 & 0 \end{array}\right) \tag{12.7}$$

Setting:

$$R_i = R_{\boldsymbol{V}_i}\left(\frac{2\pi}{3}\right) \tag{12.8}$$

where $\boldsymbol{V}_i$ are the above defined vertices of the standard tetrahedron, we obtain the following four matrices:

$$\begin{aligned} R_1 &= \begin{pmatrix} -\frac{1}{2} & -\frac{\sqrt{3}}{2} & 0 \\ \frac{\sqrt{3}}{2} & -\frac{1}{2} & 0 \\ 0 & 0 & 1 \end{pmatrix}; \quad & R_2 &= \begin{pmatrix} \frac{1}{2} & \frac{\sqrt{3}}{2} & 0 \\ \frac{1}{2\sqrt{3}} & -\frac{1}{6} & -\frac{2\sqrt{2}}{3} \\ -\sqrt{\frac{2}{3}} & \frac{\sqrt{2}}{3} & -\frac{1}{3} \end{pmatrix} \\ R_3 &= \begin{pmatrix} \frac{1}{2} & -\frac{1}{2\sqrt{3}} & \sqrt{\frac{2}{3}} \\ -\frac{\sqrt{3}}{2} & -\frac{1}{6} & \frac{\sqrt{2}}{3} \\ 0 & -\frac{2\sqrt{2}}{3} & -\frac{1}{3} \end{pmatrix}; \quad & R_4 &= \begin{pmatrix} -\frac{1}{2} & \frac{1}{2\sqrt{3}} & -\sqrt{\frac{2}{3}} \\ -\frac{1}{2\sqrt{3}} & \frac{5}{6} & \frac{\sqrt{2}}{3} \\ \sqrt{\frac{2}{3}} & \frac{\sqrt{2}}{3} & -\frac{1}{3} \end{pmatrix} \end{aligned} \tag{12.9}$$

If we multiply these four matrices in all possible ways and we multiply their products among themselves, we generate a finite group with 12 elements that is named the proper tetrahedral group T_{12}. Abstractly we can show that it is isomorphic to a 12-element subgroup of the octahedral group extensively discussed in Section 4.3.5.

Let us now introduce also reflections, namely, elements of O(3) with determinant different from 1. In particular let us add the following reflection:

$$S = \begin{pmatrix} -1 & 0 & 0 \\ 0 & 1 & 0 \\ 0 & 0 & 1 \end{pmatrix} \tag{12.10}$$

which changes the sign of the first component of every vector. Looking at the explicit form of the vertices (12.6) we easily see that all of the five operations R_1, R_2, R_3, R_4, S are symmetry of the tetrahedron since they transform its vertices one into the other and so they do with its faces and edges. Hence also any product of these symmetry operations is equally a symmetry. The group generated by R_1, R_2, R_3, R_4, S has 24 elements. Organizing it into conjugacy classes we realize that it has the same structure as the octahedral group and actually it is isomorphic to $O_{24} \sim S_4$, whose structure was extensively discussed in Section 3.4.3. The explicit form of the 24 group elements identified by their action on the three Euclidean coordinates $\{x, y, z\}$ and organized into conjugation classes was displayed in eq. (3.108); the multiplication table was provided in eq. (3.110). Here we have a new representation of the same abstract group, where the same group elements, following the same multiplication table and therefore falling in the same conjugation classes, are identified by a different action on the three Euclidean coordinates displayed in Table 12.1.

12.2.1.1 Generators of the extended tetrahedral group

Being isomorphic to the proper octahedral group, the extended tetrahedral group admits a presentation in terms of two generators that satisfy the same relations as given in eq. (3.107), namely,

$$\mathscr{A}, \mathscr{B} : \mathscr{A}^3 = \mathbf{e}; \quad \mathscr{B}^2 = \mathbf{e}; \quad (\mathscr{B}\mathscr{A})^4 = \mathbf{e} \tag{12.11}$$

The explicit form of these generators is the following one

$$\begin{aligned} \mathscr{A} = 2_3 &= \begin{pmatrix} -\frac{1}{2} & -\frac{\sqrt{3}}{2} & 0 \\ \frac{\sqrt{3}}{2} & -\frac{1}{2} & 0 \\ 0 & 0 & 1 \end{pmatrix} \\ \mathscr{B} = 5_3 &= \begin{pmatrix} \frac{1}{2} & \frac{1}{2\sqrt{3}} & -\sqrt{\frac{2}{3}} \\ \frac{1}{2\sqrt{3}} & \frac{5}{6} & \frac{\sqrt{2}}{3} \\ -\sqrt{\frac{2}{3}} & \frac{\sqrt{2}}{3} & -\frac{1}{3} \end{pmatrix} \end{aligned} \tag{12.12}$$

Relying on the established isomorphism we can utilize the irreducible representations of O_{24} discussed in Section 4.3.6 and the character Table 4.2.

12.2.1.2 Creation of a basis of vectors for the displacements of the atoms

We begin by introducing a basis of vectors for the displacements of the four hydrogen atoms occupying the vertices of the tetrahedron. Instead of an orthonormal system, in each vertex of the tetrahedron we introduce three linear independent vectors aligned with the three hedges of the tetrahedron departing from that vertex and pointing to

Table 12.1: *Action of the extended tetrahedral group* T_d *on* $\mathbb{R}^3$.

Conjugacy class = 1
1_1 ; $\{x,y,z\}$
Conjugacy class = 2
$2_1; \{\frac{1}{6}(-3x-\sqrt{3}y+2\sqrt{6}z), \frac{1}{6}(\sqrt{3}x+5y+2\sqrt{2}z), \frac{1}{3}(-\sqrt{6}x+\sqrt{2}y-z)\}$
$2_2; \{\frac{1}{6}(-3x+\sqrt{3}y-2\sqrt{6}z), \frac{1}{6}(-\sqrt{3}x+5y+2\sqrt{2}z), \frac{1}{3}(\sqrt{6}x+\sqrt{2}y-z)\}$
$2_3; \{\frac{1}{2}(-x-\sqrt{3}y), \frac{1}{2}(\sqrt{3}x-y), z\}$
$2_4; \{\frac{1}{2}(-x+\sqrt{3}y), \frac{1}{2}(-\sqrt{3}x-y), z\}$
$2_5; \{\frac{1}{6}(3x-\sqrt{3}y+2\sqrt{6}z), \frac{1}{6}(-3\sqrt{3}x-y+2\sqrt{2}z), \frac{1}{3}(-2\sqrt{2}y-z)\}$
$2_6; \{\frac{1}{6}(3x+\sqrt{3}y-2\sqrt{6}z), \frac{1}{6}(3\sqrt{3}x-y+2\sqrt{2}z), \frac{1}{3}(-2\sqrt{2}y-z)\}$
$2_7; \{\frac{1}{2}(x-\sqrt{3}y), \frac{1}{6}(-\sqrt{3}x-y-4\sqrt{2}z), \frac{1}{3}(\sqrt{6}x+\sqrt{2}y-z)\}$
$2_8; \{\frac{1}{2}(x+\sqrt{3}y), \frac{1}{6}(\sqrt{3}x-y-4\sqrt{2}z), \frac{1}{3}(-\sqrt{6}x+\sqrt{2}y-z)\}$
Conjugacy class = 3
$3_1; \{-x, \frac{1}{3}(y-2\sqrt{2}z), \frac{1}{3}(-2\sqrt{2}y-z)\}$
$3_2; \{-\frac{y+\sqrt{2}z}{\sqrt{3}}, \frac{1}{3}(-\sqrt{3}x-2y+\sqrt{2}z), \frac{1}{3}(-\sqrt{6}x+\sqrt{2}y-z)\}$
$3_3; \{\frac{y+\sqrt{2}z}{\sqrt{3}}, \frac{1}{3}(\sqrt{3}x-2y+\sqrt{2}z), \frac{1}{3}(\sqrt{6}x+\sqrt{2}y-z)\}$
Conjugacy class = 4
$4_1; \{\frac{1}{6}(-3x-\sqrt{3}y+2\sqrt{6}z), \frac{1}{6}(3\sqrt{3}x-y+2\sqrt{2}z), \frac{1}{3}(-2\sqrt{2}y-z)\}$
$4_2; \{\frac{1}{6}(-3x+\sqrt{3}y-2\sqrt{6}z), \frac{1}{6}(-3\sqrt{3}x-y+2\sqrt{2}z), \frac{1}{3}(-2\sqrt{2}y-z)\}$
$4_3; \{\frac{1}{2}(-x-\sqrt{3}y), \frac{1}{6}(\sqrt{3}x-y-4\sqrt{2}z), \frac{1}{3}(-\sqrt{6}x+\sqrt{2}y-z)\}$
$4_4; \{\frac{1}{2}(-x+\sqrt{3}y), \frac{1}{6}(-\sqrt{3}x-y-4\sqrt{2}z), \frac{1}{3}(\sqrt{6}x+\sqrt{2}y-z)\}$
$4_5; \{-\frac{y+\sqrt{2}z}{\sqrt{3}}, \frac{1}{3}(\sqrt{3}x-2y+\sqrt{2}z), \frac{1}{3}(\sqrt{6}x+\sqrt{2}y-z)\}$
$4_6; \{\frac{y+\sqrt{2}z}{\sqrt{3}}, \frac{1}{3}(-\sqrt{3}x-2y+\sqrt{2}z), \frac{1}{3}(-\sqrt{6}x+\sqrt{2}y-z)\}$
Conjugacy class = 5
$5_1; \{-x,y,z\}$
$5_2; \{\frac{1}{6}(3x-\sqrt{3}y+2\sqrt{6}z), \frac{1}{6}(-\sqrt{3}x+5y+2\sqrt{2}z), \frac{1}{3}(\sqrt{6}x+\sqrt{2}y-z)\}$
$5_3; \{\frac{1}{6}(3x+\sqrt{3}y-2\sqrt{6}z), \frac{1}{6}(\sqrt{3}x+5y+2\sqrt{2}z), \frac{1}{3}(-\sqrt{6}x+\sqrt{2}y-z)\}$
$5_4; \{\frac{1}{2}(x-\sqrt{3}y), \frac{1}{2}(-\sqrt{3}x-y), z\}$
$5_5; \{\frac{1}{2}(x+\sqrt{3}y), \frac{1}{2}(\sqrt{3}x-y), z\}$
$5_6; \{x, \frac{1}{3}(y-2\sqrt{2}z), \frac{1}{3}(-2\sqrt{2}y-z)\}$

the other three vertices. These 12 vectors are defined as follows:

$$\begin{array}{lll} \boldsymbol{\xi}_{12} = \boldsymbol{V}_1 - \boldsymbol{V}_2; & \boldsymbol{\xi}_{13} = \boldsymbol{V}_1 - \boldsymbol{V}_3; & \boldsymbol{\xi}_{14} = \boldsymbol{V}_1 - \boldsymbol{V}_4 \\ \boldsymbol{\xi}_{21} = \boldsymbol{V}_2 - \boldsymbol{V}_1; & \boldsymbol{\xi}_{23} = \boldsymbol{V}_2 - \boldsymbol{V}_3; & \boldsymbol{\xi}_{24} = \boldsymbol{V}_2 - \boldsymbol{V}_4 \\ \boldsymbol{\xi}_{31} = \boldsymbol{V}_3 - \boldsymbol{V}_1; & \boldsymbol{\xi}_{32} = \boldsymbol{V}_3 - \boldsymbol{V}_2; & \boldsymbol{\xi}_{34} = \boldsymbol{V}_3 - \boldsymbol{V}_4 \end{array} \tag{12.13}$$

Let us organize them into a 12-array as follows:

$$\mathbb{V} = \{\boldsymbol{\xi}_{12}, \boldsymbol{\xi}_{13}, \boldsymbol{\xi}_{14}, \boldsymbol{\xi}_{23}, \boldsymbol{\xi}_{24}, \boldsymbol{\xi}_{21}, \boldsymbol{\xi}_{31}, \boldsymbol{\xi}_{32}, \boldsymbol{\xi}_{34}, \boldsymbol{\xi}_{41}, \boldsymbol{\xi}_{42}, \boldsymbol{\xi}_{43}\} \tag{12.14}$$

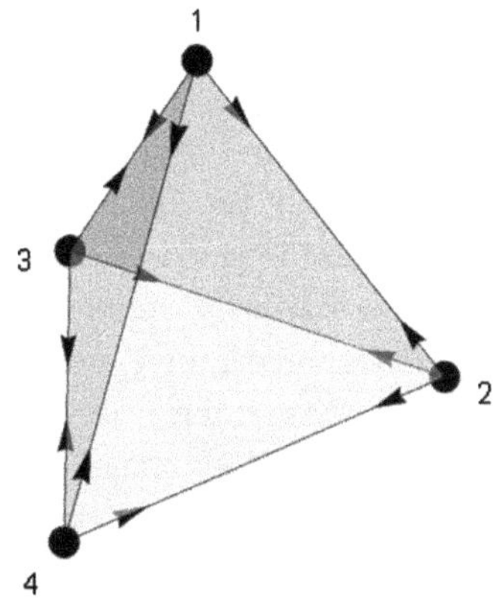

Figure 12.3: The 12 vectors given in eq. (12.13) provide a complete basis for the displacements of the hydrogen atoms and are depicted here.

Acting on these vectors the transformations of the extended tetrahedral group permute them in a way that can be easily reconstructed for all 24 elements of the group and in particular for the two generators $\mathscr{A}$ and $\mathscr{B}$ whose effect is displayed below (see also Figure 12.3).

$$\mathscr{A}\begin{pmatrix}\xi_{12}\\ \xi_{13}\\ \xi_{14}\\ \xi_{23}\\ \xi_{24}\\ \xi_{21}\\ \xi_{31}\\ \xi_{32}\\ \xi_{34}\\ \xi_{41}\\ \xi_{42}\\ \xi_{43}\end{pmatrix}=\begin{pmatrix}\xi_{13}\\ \xi_{14}\\ \xi_{12}\\ \xi_{34}\\ \xi_{32}\\ \xi_{31}\\ \xi_{41}\\ \xi_{43}\\ \xi_{42}\\ \xi_{21}\\ \xi_{23}\\ \xi_{24}\end{pmatrix};\quad \mathscr{B}\begin{pmatrix}\xi_{12}\\ \xi_{13}\\ \xi_{14}\\ \xi_{23}\\ \xi_{24}\\ \xi_{21}\\ \xi_{31}\\ \xi_{32}\\ \xi_{34}\\ \xi_{41}\\ \xi_{42}\\ \xi_{43}\end{pmatrix}=\begin{pmatrix}\xi_{32}\\ \xi_{31}\\ \xi_{34}\\ \xi_{21}\\ \xi_{24}\\ \xi_{23}\\ \xi_{13}\\ \xi_{12}\\ \xi_{14}\\ \xi_{43}\\ \xi_{42}\\ \xi_{41}\end{pmatrix} \tag{12.15}$$

The above result is encoded in two 12×12 integer-valued matrices and altogether what we have done is constructing a 12-dimensional representation of the tetrahedral group that we name $\boldsymbol{D}_{12}$.

Using the standard formula (3.88) for multiplicities in the decomposition of reducible group representations into irreducible ones and the character Table 4.2, we immediately obtain:

$$\boldsymbol{D}_{12}=\boldsymbol{D}_1\oplus\boldsymbol{D}_2\oplus 2\boldsymbol{D}_4\oplus\boldsymbol{D}_5 \tag{12.16}$$

where we remind the reader that $\boldsymbol{D}_1$ is the singlet trivial representation, $\boldsymbol{D}_2$ is the unique 2-dimensional representation, while $\boldsymbol{D}_4$ and $\boldsymbol{D}_5$ are the two inequivalent 3-dimensional representations.

12.2.1.3 Physical interpretation

The physical implications of this decomposition are of the utmost relevance. It means that the 12×12 matrix (12.5) can depend at most on five parameters corresponding to the five representations mentioned in eq. (12.16). So far we have not yet considered the X-atom in the centroid of the tetrahedron. We have discussed the motions of the Y-atoms in the coordinate system centered at X. One of the two irreducible representations $\boldsymbol{D}_4$ clearly must correspond with the translation modes where all the Y atoms move coherently in the same direction. Interpreted in the reverse order these are the vibrations of the X-atom with respect to the Y-ones and this accounts for one of the five parameters. The other four parameters can be easily given a physical interpretation looking at the structure of the projection onto the irreducible representations. We consider a pair of them as an illustration.

12.2.1.3.1 The breathing mode

The identity representation $\boldsymbol{D}_1$ has a unique physical interpretation as breathing mode, namely, the overall contraction or enlargement of the molecule (dilatation or expansion). To see this it suffices to consider the projection operator on the singlet representation. We obtain that the invariant space is provided by a 12-vector whose 12 entries are all equal. Physically interpreted this means that in this representation each of the four Y-atoms (the hydrogens in the case of the methane) are displaced of the same length along the direction from the vertex where they sit when at equilibrium towards the centroid of the tetrahedron (the location of the carbon atom in the methane). Explicitly this means that the four displacement vectors are:

$$
\begin{aligned}
\delta \boldsymbol{V}_1 &= \boldsymbol{\xi}_{12} + \boldsymbol{\xi}_{13} + \boldsymbol{\xi}_{14} \\
\delta \boldsymbol{V}_2 &= \boldsymbol{\xi}_{23} + \boldsymbol{\xi}_{24} + \boldsymbol{\xi}_{21} \\
\delta \boldsymbol{V}_3 &= \boldsymbol{\xi}_{32} + \boldsymbol{\xi}_{34} + \boldsymbol{\xi}_{31} \\
\delta \boldsymbol{V}_4 &= \boldsymbol{\xi}_{41} + \boldsymbol{\xi}_{42} + \boldsymbol{\xi}_{42}
\end{aligned}
\tag{12.17}
$$

The effect on the molecule is represented in Figure (12.4).

12.2.1.3.2 The rotational modes

Next we provide the physical interpretation of the irreducible representation $\boldsymbol{D}_5$. Let us define the following four vectors (inside the 12-dimensional space spanned by linear combinations of the basis vectors (12.13)):

$$
\begin{aligned}
\boldsymbol{p}_1 &= \boldsymbol{\xi}_{23} - \boldsymbol{\xi}_{32} + \boldsymbol{\xi}_{34} - \boldsymbol{\xi}_{43} + \boldsymbol{\xi}_{42} - \boldsymbol{\xi}_{24} \\
\boldsymbol{p}_2 &= \boldsymbol{\xi}_{31} - \boldsymbol{\xi}_{13} + \boldsymbol{\xi}_{14} - \boldsymbol{\xi}_{41} + \boldsymbol{\xi}_{43} - \boldsymbol{\xi}_{34} \\
\boldsymbol{p}_3 &= \boldsymbol{\xi}_{12} - \boldsymbol{\xi}_{21} + \boldsymbol{\xi}_{41} - \boldsymbol{\xi}_{14} + \boldsymbol{\xi}_{24} - \boldsymbol{\xi}_{42} \\
\boldsymbol{p}_4 &= \boldsymbol{\xi}_{21} - \boldsymbol{\xi}_{12} + \boldsymbol{\xi}_{13} - \boldsymbol{\xi}_{31} + \boldsymbol{\xi}_{32} - \boldsymbol{\xi}_{23}
\end{aligned}
\tag{12.18}
$$

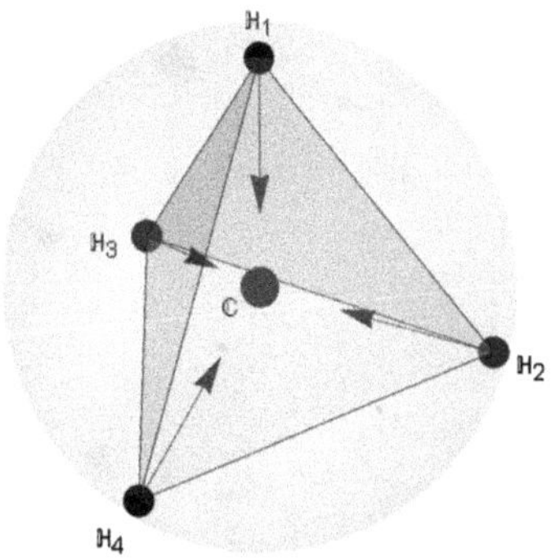

Figure 12.4: The 1-dimensional singlet representation $\boldsymbol{D}_1$ corresponds to the breathing mode where the molecule just expands or contracts in a conformal fashion preserving all the angles. Geometrically one can inscribe the tetrahedron into a sphere $\mathbb{S}^2$, whose surface intersects all of the four vertices $\boldsymbol{\xi}_i$. Increasing or decreasing the radius of such a sphere is what the breathing mode does.

By using eq. (12.15) we easily check that they span an irreducible subspace and that they sum up to zero:

$$\sum_{i=1}^{4} \boldsymbol{p}_i = 0 \tag{12.19}$$

Calculation of the character of this 3-dimensional representation shows that it corresponds to the irreducible $\boldsymbol{D}_5$. The physical interpretation of the four vectors (12.18) is visualized in Figure 12.5.

The displacements of the Y encoded in the vector $\boldsymbol{p}_i$ correspond to a rotation of the molecule around the axis passing through the centroid and the vertex i. As a result of such a rotation the face of the tetrahedron opposite to the vertex i rotates in the plane orthogonal to the mentioned axis.

12.2.1.3.3 The other modes

In a similar way one can study the geometry of the deformations associated with the remaining two irreducible representations, $\boldsymbol{D}_2$ and $\boldsymbol{D}_4$. They correspond to asymmetrical compressions and dilatations of the molecule that change its shape, preserving however some of its features. We do not enter further into details being confident that what we have so far illustrated is already quite a convincing illustration of the foremost relevance of finite group theory in solid-state and molecular physics. Observable data like vibrational normal modes are in one-to-one correspondence with the irreducible representations of the basic symmetry either of the crystal or of the molecule.

Molecular vibrations are further addressed in the solution to Exercise 3.42 where the decomposition into irreps is utilized also with other symmetry groups relevant for different types of molecules.

12.3 The hidden symmetry of the hydrogen atom spectrum

In classical mechanics, Bertrand's theorem singles out the Newtonian potential $U(r) = -\alpha/r$ and that of the isotropic harmonic oscillator as the only central potentials whose

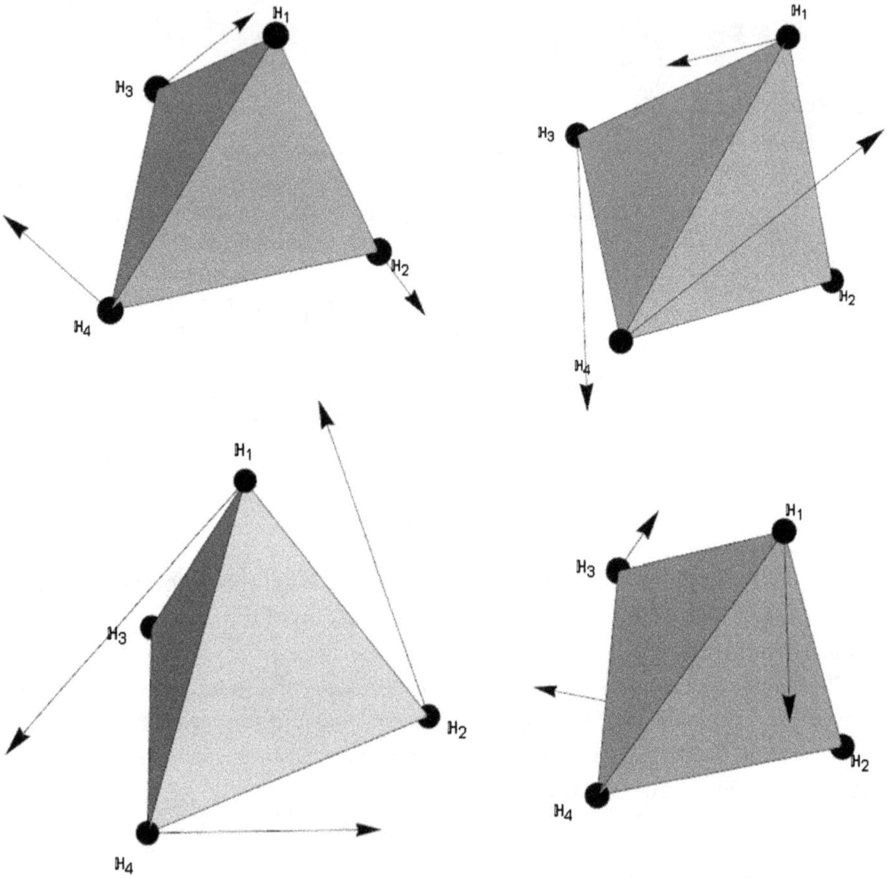

Figure 12.5: The 3-dimensional representation $\boldsymbol{D}_5$ corresponds to the rotation modes. Each vector $\mathbf{p}_i$ ($i = 1, 2, 3, 4$) corresponds to the rotation of the entire molecule around the axis that joins the center with the ith vertex as displayed in the figure. The four vectors are not independent, since their sum equals zero and this makes the representation 3-dimensional.

bounded orbits are all closed [128, 75]. In both cases this degeneracy was recognized to be a direct consequence of a *hidden* symmetry. Yet while for the isotropic oscillator revealing such a symmetry is quite straightforward, this is much harder for the Kepler problem. The additional ('accidental') integral of motion

$$\mathbf{A} = \mathbf{v} \times \mathbf{L} - \frac{\alpha}{r}\mathbf{r}$$

which provides integrability of the system was independently discovered by Jakob Hermann, Johann Bernoulli and Pierre-Simon de Laplace. Later it was further studied by William Hamilton and Josiah Gibbs. Now it is known as the Laplace–Runge–Lenz vector [73, 74]. Such a vector points from the center of force towards the periapsis, i.e., the point of closest approach, and its magnitude fixes the eccentricity of the orbit.

Making use of both the angular momentum and the Laplace–Runge–Lenz vectors, the solution to the classical Kepler problem can be given without working out the solution of the differential equations of motion.

In the quantum theory of the hydrogen atom, the same degeneracy is manifested by the independence of the energy levels $E_n = -m\alpha^2/2\hbar^2 n^2$ $(n = 1, 2, \ldots)$ from the azimuthal quantum number l. Just as in the classical Kepler problem, W. Pauli worked out a derivation of the hydrogen energy levels which was based on the conservation of the Laplace–Runge–Lenz vector rather than on the direct solution of Schrödinger equation [123]. The *hidden* symmetry was recognized to be SO(4) for bound states and SO(3, 1) for scattering states at about the same time by O. Klein, who identified the Lie algebra of all the conserved observables [87], and by V. A. Fock [49], who showed how to reveal the symmetry by using its explicit representation in momentum space. The equivalence of the two approaches was later established by V. Bargmann [7]. Since then, different aspects of this hidden symmetry have been reconsidered by many authors, e.g., the Fock transformation was later identified as a canonical transformation and popularized among the mathematicians by J. Moser [114]. It was further realized in the sixties that the *hidden* symmetry SO(4) or SO(3, 1) can be actually extended to a larger dS/AdS group SO(4, 1) or even to the conformal group SO(4, 2) [119, 9, 111, 68]. Note that unlike SO(4) or SO(3, 1), these larger groups act transitively on the entire space of states, not only on the subspaces of states having the same energy level, hence they are called *dynamical* (or *spectrum generating*) groups.

12.3.1 Classical Kepler problem

First we would like to explain why Kepler problem (classical motion of a nonrelativistic particle moving in a central attractive potential $U(r) = -\alpha/r$) is singled out by Bertrand's theorem. To this effect, let us start from Maupertuis' principle stating that the equations of motion of classical mechanics can be derived by requiring that the abbreviated action $S_0 = \int_1^2 \mathbf{p}\, d\mathbf{r} = \int_1^2 \sqrt{2m[E - U(r)]}|d\mathbf{r}|$ should attain an extremum. Obviously this formulation corresponds to a geodesic flow (motion along geodesic lines) on a configuration space equipped with the fictitious Riemannian metric $ds^2 = 2m[E - U(r)]\, d\mathbf{r}^2$. For Kepler problem, such a formulation, as it stands, does not provide genuine new insights, yet we can develop it further by taking a closer look at its rephrasing in momentum space [132, 133].

Let us recall the conventional Lagrangian and the least action principle

$$L(\mathbf{r}, \dot{\mathbf{r}}) = \frac{1}{2} m\dot{\mathbf{r}}^2 - U(r), \quad S = \int_{t_1}^{t_2} L(\mathbf{r}, \dot{\mathbf{r}})\, dt = \min \tag{12.20}$$

that provide the equations of motion in the well-known Euler form

$$\frac{d}{dt}\left(\frac{\partial L}{\partial \dot{\mathbf{r}}}\right) - \frac{\partial L}{\partial \mathbf{r}} = m\ddot{\mathbf{r}} + \frac{\alpha}{r^3}\mathbf{r} = 0 \tag{12.21}$$

By means of a Legendre transformation $H(\mathbf{r},\mathbf{p}) = \dot{\mathbf{r}}\mathbf{p} - L(\mathbf{r},\dot{\mathbf{r}}) = \mathbf{p}^2/2m + U(r)$, with the momentum defined by $\mathbf{p} = \partial L/\partial\dot{\mathbf{r}} = m\dot{\mathbf{r}}$, one goes to an equivalent description in phase-space (coordinate-momenta) or, put in a geometrical way, in the cotangent bundle, where the equations of motion take the form of Hamiltonian equations

$$\frac{d\mathbf{r}}{dt} = \frac{\partial H}{\partial \mathbf{p}} = \frac{\mathbf{p}}{m}, \quad \frac{d\mathbf{p}}{dt} = -\frac{\partial H}{\partial \mathbf{r}} = -\frac{\alpha}{r^3}\mathbf{r} \tag{12.22}$$

Mnemonically, in this case the Legendre transformation provides the replacement $\dot{\mathbf{r}} \leftrightarrow \mathbf{p}$, $L \leftrightarrow H$. In this way we construct the Hamiltonian function which represents energy and, because of that, is conserved. Next, let us go a step further and let us apply a second Legendre transformation, this time in order to replace $\mathbf{r}$ with $\dot{\mathbf{p}}$:

$$\begin{aligned} \dot{\mathbf{p}} = -\frac{\partial H}{\partial \mathbf{r}} = -\nabla U = -\frac{\alpha}{r^3}\mathbf{r} \quad &\Longrightarrow \quad \mathbf{r} = -\frac{\sqrt{\alpha}}{|\dot{\mathbf{p}}|^{3/2}}\dot{\mathbf{p}} \\ \tilde{L}(\dot{\mathbf{p}},\mathbf{p}) = \dot{\mathbf{p}}\mathbf{r} + H(\mathbf{r},\mathbf{p}) &= \frac{\mathbf{p}^2}{2m} - 2\sqrt{\alpha|\dot{\mathbf{p}}|} \end{aligned} \tag{12.23}$$

Since L and $\tilde{L}$ differ only by sign and a total derivative (of the virial), the initial least action principle takes the new form $\int_{t_1}^{t_2} \tilde{L}(\dot{\mathbf{p}},\mathbf{p})\,dt = \max$, and hence the equations of motion are cast into the form

$$\frac{d}{dt}\left(\frac{\partial \tilde{L}}{\partial \dot{\mathbf{p}}}\right) - \frac{\partial \tilde{L}}{\partial \mathbf{p}} = \frac{d}{dt}\left(-\frac{\sqrt{\alpha}}{|\dot{\mathbf{p}}|^{3/2}}\dot{\mathbf{p}}\right) - \frac{\mathbf{p}}{m} = 0 \tag{12.24}$$

which, of course, is fully equivalent to eq. (12.21). In full analogy with the conventional derivation of Maupertuis' principle, we can now restate the least action principle by restricting it to those virtual paths that obey energy conservation and allowing for variation of the final time t_2 instead. This leads to extremality of the analog of the abbreviated action in momentum space,

$$\tilde{S}_0 = \int_1^2 \mathbf{r}\,d\mathbf{p} = -\int_{t_1}^{t_2} \sqrt{\alpha|\dot{\mathbf{p}}|}\,dt = \max \tag{12.25}$$

According to the spirit of Maupertuis' principle, we can eliminate time by using energy conservation,

$$H = \frac{p^2}{2m} - \sqrt{\alpha|\dot{\mathbf{p}}|} = E, \quad dt = \frac{4m^2\alpha|d\mathbf{p}|}{(p^2 - 2mE)^2} \tag{12.26}$$

and so doing we finally arrive at the following form of the least action principle:

$$\tilde{S}_0 = -\int_{t_1}^{t_2} \sqrt{\alpha|d\mathbf{p}|\,dt} = -2m\alpha\int_1^2 \frac{|d\mathbf{p}|}{p^2 - 2mE} = -\frac{m\alpha}{a^2}\Delta s = \max, \quad a^2 = -2mE \tag{12.27}$$

Equation (12.27) reveals the most relevant point that Kepler problem is actually equivalent to a geodesic flow $\Delta s = \min$ on a cotangent (momentum) bundle of the configuration manifold, equipped with the following metric [114, 120, 113]:

$$ds^2 = f^2(p)\,d\mathbf{p}^2 = f^2(p)(dp^2 + p^2\,d\Omega_{\mathbf{p}}^2), \quad f^2(p) = \frac{4}{(1 + p^2/a^2)^2} \tag{12.28}$$

After these preliminaries, let us show how this unexpected formulation, despite an apparent complexity when compared with eq. (12.21), provides the aforementioned new insight. The metric (12.28) is well known as a conformal representative of a space of constant curvature $K = 1/a^2$, i.e., is the sphere of radius a for $E < 0$, the flat space for $E = 0$, or the Lobachevsky space for $E > 0$. Whereas the above approach, when applied to whatsoever mechanical problem, always leads to its rephrasing as a geodesic flow in momentum space, *constant curvature* of the effective metric is the *hidden* specific intrinsic property of Kepler problem.

Let us now consider in more detail the case $E < 0$ (or $a^2 > 0$) of a bounded motion. In order to fully employ the symmetry of the problem, it is natural to switch to a more convenient parameterization. To this effect the stereographic projection

$$\mathbf{P} = f(p)\mathbf{p} = \frac{2\mathbf{p}}{p^2/a^2+1}, \quad P_0 = \int \sqrt{ds^2 - d\mathbf{P}^2} = a\frac{p^2/a^2-1}{p^2/a^2+1} \tag{12.29}$$

maps momentum space isometrically onto[1] the stereographically embedded sphere $\mathbb{S}^3 = \{(P_0, \mathbf{P}) : P_0^2 + \mathbf{P}^2 = a^2\} \subset \mathbb{R}^4$ with length element $ds^2 = dP_0^2 + d\mathbf{P}^2$. While the initial SO(3) symmetry of the problem corresponds here to rotations around the P_0 axis, the sphere has the larger symmetry group SO(4) named the *hidden symmetry* of Kepler problem. In this respect the notion of *hidden symmetry* has much in common with the more recent conception of *spontaneous symmetry breaking* (here the choice of a vacuum state is replaced by the choice of coordinates on the momentum space $\mathbb{S}^3$).

Since motion on $\mathbb{S}^3$ momentum space is just a geodesic flow, we can use the known properties of free motion and translate them into momentum space. For example, SO(4) symmetry implies conservation of the 4D analog of the angular momentum $\mathbf{M} = m\mathbf{r} \times \dot{\mathbf{r}}$, namely, $M^{(4D)}_{\alpha\beta} = P_\alpha(s)P'_\beta(s) - P'_\alpha(s)P_\beta(s)$ $(\alpha, \beta = 0, 1, 2, 3)$. By explicit evaluation of these quantities in terms of $\mathbf{p}$ and $\dot{\mathbf{p}}$ we obtain:

$$\mathbf{P} \times \mathbf{P}' = \frac{a^2}{m\alpha}\mathbf{M}, \quad P_0\mathbf{P}' - P_0'\mathbf{P} = -\frac{a}{\alpha}\mathbf{A} \tag{12.30}$$

Thus we see that the 4D rotations of SO(4) are respectively generated by $\mathbf{M}$ and $m\mathbf{A}/a$. Obviously, the squared 4D momentum

$$\frac{1}{2}M^{(4D)}_{\alpha\beta}M^{(4D)}_{\alpha\beta} = \left(\frac{a^2}{m\alpha}\right)^2 M^2 + \left(\frac{a}{\alpha}\right)^2 A^2 = a^2 \tag{12.31}$$

is a constant. This equation relates energy to the other conserved quantities M and A.

Since we know that a geodesic on a sphere is just a maximal circle, we can immediately read off the solution of Kepler problem from the above construction. For

1 With the exclusion of the North Pole, which corresponds to infinite momentum (collision). Compactification by addition of this point is known as the regularization of the Kepler problem.

example, starting with the geodesics

$$P_0(s) = ae\cos\left(\frac{s}{a}\right), \quad \mathbf{P}(s) = \left\{-a\sin\left(\frac{s}{a}\right), a\sqrt{1-e^2}\cos\left(\frac{s}{a}\right), 0\right\} \tag{12.32}$$

for which

$$\mathbf{M} = \frac{m\alpha}{a}\sqrt{1-e^2}\hat{\mathbf{z}}, \quad \mathbf{A} = e\alpha\hat{\mathbf{x}}$$

we arrive at the solution:

$$\begin{aligned} x = \frac{m\alpha}{a^2}\left(\cos\left(\frac{s}{a}\right) - e\right), \quad y = \frac{m\alpha}{a^2}\sqrt{1-e^2}\sin\left(\frac{s}{a}\right) \\ z = 0, \quad t = \frac{m\alpha^2}{a^3}\left(\frac{s}{a} - e\sin\left(\frac{s}{a}\right)\right) \end{aligned} \tag{12.33}$$

which, up to notation, coincides with the canonical one as given in Landau and Lifshitz the textbook. Note that our parameter a is solely determined by energy E and does not change under SO(4) rotations. However, for given energy these rotations are capable of changing both the location of the orbit and its eccentricity e.

In the case $E > 0$ ($a^2 < 0$) of unbounded motion, all the formulas remain valid provided the sphere $\mathbb{S}^3$ is replaced with a sheet $\mathbb{H}^3$ of the hyperboloid $\mathbf{P}^2 - P_0^2 = a^2$ in Minkowski space. In this case the *hidden symmetry* is described by the Lorentz group SO(3, 1). For $E = 0$ (i.e., in the limit $a \to 0$) the transformation (12.29) is reduced to inversion which maps momentum space onto the hyperplane $P_0 = 0$. In this case the *hidden symmetry* is given by the Galilean group ISO(3).

12.3.2 Hydrogen atom

The above discussion of the classical problem provides all the hints on how to deal with the quantum problem of a hydrogen atom. Consider the corresponding stationary Schrödinger equation

$$\hat{H}\Psi(\mathbf{r}) = \left(\frac{\hat{p}^2}{2m} - \frac{\alpha}{r}\right)\Psi(\mathbf{r}) = E\Psi(\mathbf{r}) \tag{12.34}$$

Since the expected *hidden symmetry* is explicitly revealed in momentum space, first we have to shift to the momentum representation. Fortunately, in quantum mechanics this machinery is even more familiar than in classical mechanics, yet doing this directly results in an integral equation whose further handling would require extensive usage of potential theory. Yet, one can avoid this turning directly to a differential equation which is obtained by squaring eq. (12.34) in a clever way:

$$\begin{aligned} (\hat{p}^2 + a^2)\Psi = \frac{2m\alpha}{r}\Psi, \quad r(\hat{p}^2 + a^2)\Psi = 2m\alpha\Psi \\ r(\hat{p}^2 + a^2)r(\hat{p}^2 + a^2)\Psi = 4m^2\alpha^2\Psi \end{aligned} \tag{12.35}$$

Starting from the above we just have to use operator algebra and bring the two factors r together thus forming $r^2 = \mathbf{r}^2$, which admits an easy representation in momentum space. These steps can be performed in several different ways. In particular let us derive the following identity:

$$r(\hat{p}^2 + a^2)r = (\hat{p}^2 + a^2)\mathbf{r}\frac{1}{\hat{p}^2 + a^2}\mathbf{r}(\hat{p}^2 + a^2) + \frac{4\hbar^2 a^2}{\hat{p}^2 + a^2}$$

Substituting the above into eq. (12.35) and multiplying the result by the additional factor $(\hat{p}^2 + a^2)^2$, we obtain:

$$(\hat{p}^2 + a^2)^3 \mathbf{r}\frac{1}{\hat{p}^2 + a^2}\mathbf{r}\Phi = 4(m^2\alpha^2 - \hbar^2 a^2)\Phi \qquad (12.36)$$

for the auxiliary function

$$\Phi = \frac{(\hat{p}^2 + a^2)^2}{4a^4}\Psi \qquad (12.37)$$

With this representation, the transition to momentum space is easily implemented by means of the substitution $\mathbf{r} \to i\hbar\partial/\partial\mathbf{p}$, $\hat{\mathbf{p}} \to p$, $\Phi(\mathbf{r}) \to \tilde{\Phi}(\mathbf{p})$. One readily notices that the resulting equation matches the form

$$-\Delta_{\mathbb{S}^3,p}\tilde{\Phi}(\mathbf{p}) = \left(\frac{m^2\alpha^2}{\hbar^2 a^2} - 1\right)\tilde{\Phi}(\mathbf{p}) \qquad (12.38)$$

where

$$\begin{aligned} \Delta_{\mathbb{S}^3,p} &= \frac{1}{\sqrt{g}}\frac{\partial}{\partial p^i}\left(\sqrt{g}g^{ik}\frac{\partial}{\partial p^k}\right), \quad g_{ik}(p) = f^2(p)\delta_{ik} \\ g^{ik}(p) &= \frac{1}{f^2(p)}\delta_{ik}, \quad g = \det(g_{ik}) = f^6(p) \end{aligned} \qquad (12.39)$$

is the Laplacian on the 3-sphere in the momentum space described in eq. (12.28). Hence Schrödinger equation for the hydrogen atom is equivalent to that associated with free motion on a 3-dimensional sphere (rigid rotator in four dimensions). Of course, this might be expected from our previous results on the corresponding classical case.

Consider next the normalization integral for the auxiliary function (12.37) in momentum representation

$$\int_{\mathbb{S}^3}|\Phi|^2 = \int d^3p\sqrt{g}|\tilde{\Phi}|^2 = \frac{1}{2}\int d^3p\left(1 + \frac{p^2}{a^2}\right)|\tilde{\Psi}|^2 \qquad (12.40)$$

The average of the commutator of any operator with the Hamiltonian is identically zero in all stationary states. When this consideration is applied to

$$\frac{i}{\hbar}[\hat{H}, \hat{\mathbf{p}}\mathbf{r}] = \frac{\hat{p}^2}{m} - \frac{\alpha}{r}$$

we obtain the quantum analog $\int d^3pp^2|\tilde{\Psi}|^2 = a^2 \int d^3p|\tilde{\Psi}|^2 = a^2$ of the virial theorem. Taking this into account, the function $\tilde{\Phi}$ is normalized to unity on $\mathbb{S}^3$.

The eigenvalues for eq. (12.38) can be identified by separating out radial variable in the 4-dimensional Laplace equation

$$\left(\frac{\partial^2}{\partial P_0^2} + \frac{\partial^2}{\partial \mathbf{P}^2}\right)\tilde{\Phi} = \frac{1}{a^3}\frac{\partial}{\partial a}\left(a^3\frac{\partial\tilde{\Phi}}{\partial a}\right) + \frac{1}{a^2}\Delta_{\mathbb{S}^3,p}\tilde{\Phi} = 0$$

The resulting eigenfunctions are the 4-dimensional spherical harmonics Y_{Llm}, which can be obtained by restricting homogeneous harmonic polynomials of degree L to $\mathbb{S}^3$. Obviously, the corresponding eigenvalues are $\Delta_{\mathbb{S}^3,p}Y_{Llm} = -L(L+2)Y_{Llm}$. Thus, by means of comparison with eq. (12.38) we conclude that the energy levels are given by

$$E_L = -\frac{a^2}{2m} = -\frac{m\alpha^2}{2\hbar^2(L+1)^2}, \quad L = 0, 1, 2\ldots \tag{12.41}$$

The degeneracy g_L of each level is the number of linearly independent harmonic polynomials in four variables of degree L. A homogeneous polynomial of degree L is determined by $(L+3)!/L!3!$ coefficients at each monomial and the 4-dimensional Laplace equation is equivalent to nullification of a polynomial of degree $L-2$ which imposes $(L+1)!/(L-2)!3!$ restrictions. Hence we finally get:

$$g_L = \frac{(L+3)!}{L!3!} - \frac{(L+1)!}{(L-2)!3!} = (L+1)^2 \tag{12.42}$$

As an illustration let us also show the derivation of the explicit form of the familiar lowest lying eigenstate with $L = 0$. In this case $\tilde{\Phi}(\mathbf{p})$ is of degree zero, i.e. a constant fixed by the normalization condition, $\tilde{\Phi}(\mathbf{p}) = (2\pi^2a^3)^{-1/2}$, hence according to eq. (12.37),

$$\tilde{\Psi}(\mathbf{p}) = \frac{2\sqrt{2}}{\pi}\frac{a^{5/2}}{(p^2+a^2)^2}, \quad \Psi(\mathbf{r}) = \int \frac{d^3p}{(2\pi\hbar)^{3/2}}\tilde{\Psi}(\mathbf{p})e^{i\mathbf{pr}/\hbar} = \sqrt{\frac{a^3}{\pi\hbar^3}}e^{-ar/\hbar} \tag{12.43}$$

Obviously, $\hbar/a = \hbar^2/m\alpha$ is identified with the Bohr radius.

12.4 The AdS/CFT correspondence and (super) Lie algebra theory

One of the most exciting developments in the recent history of theoretical physics has been the discovery of the holographic AdS/CFT correspondence:

$$\text{SCFT on } \partial(\text{AdS}_{p+2}) \leftrightarrow \text{SUGRA on AdS}_{p+2} \tag{12.44}$$

between a $d = p + 1$ *quantum (super)conformal field theory* on the boundary of anti de Sitter space and *classical (super)gravity* emerging from compactification of either superstrings or M-theory on the product space

$$\mathrm{AdS}_{p+2} \times X^{D-p-2}; \quad (D = 11 \text{ or } D = 10) \tag{12.45}$$

where X^{D-p-2} is a $(D-p-2)$-dimensional compact Einstein manifold.

The enormous relevance of this correspondence relies on the possibility it provides of reducing the calculation of exact correlation functions of a quantum field theory, in particular of certain gauge theories, to classical geometrical calculations within the framework of matter coupled versions of General Relativity in higher space-time dimensions. Furthermore, in the context of the formulation and of the applications of this correspondence we encounter one of the most outstanding and furthest reaching examples of what is a general truth, more and more pervading modern theoretical physics: *under many respects Group Theory, in all of its manifestations, is not a tool for the theory of fundamental interactions, but rather it encodes the essential items of its very structure*.

For this reason a quick look at the most relevant group theoretical features of the AdS/CFT correspondence is probably one of the best illustrations of the role of Group Theory in modern Theoretical Physics.

The vision of the AdS/CFT correspondence has its starting point in November 1997 with the publication on the ArXive of a paper by Juan Maldacena [110] on the large N limit of gauge theories.

From the viewpoint of the superstring scientific community this was seen as the first explicit example of the long sought *duality* between gauge theories and superstrings. Yet the scope of this correspondence was destined to be enlarged in many directions and to become, more generically, the *gauge/gravity correspondence* based on various declinations of the basic idea referred to as the *holographic principle*. According to this latter, fundamental information on the quantum behavior of fields leaving on some boundary of a larger space-time can be obtained from the classical gravitational dynamics of fields leaving in the bulk of that space-time. Such wider approach to the AdS/CFT correspondence, which can be regarded as a modern mathematical reformulation of Plato's myth of the shadows on the walls of the cavern (the myth of the *antrum platonicum*) diminishes the emphasis on strings and brings to higher relevance both supergravity theories and their perturbative and non-perturbative symmetries. In such a framework geometrical issues become the central focus of attention.

Maldacena's paper was followed from december 1997 to the late spring of 1998 by a series of fundamental papers by Ferrara, Fronsdal, Zaffaroni, Kallosh and Van Proeyen [46, 48, 96, 47], where the algebraic and field theoretical basis of the correspondence was clarified independently from microscopic string considerations. Indeed, as everything important and profound, the AdS/CFT correspondence has a relatively simple origin which, however, is extremely rich in ramified and powerful consequences. The key point is the double interpretation of any anti de Sitter group $\mathrm{SO}(2,p+1)$ as the isometry group of AdS_{p+2} space or as the conformal group on the $(p+1)$-dimensional boundary $\partial\mathrm{AdS}_{p+2}$. Such a double interpretation is inherited by the supersymmetric extensions of $\mathrm{SO}(2, p+1)$. This is what leads to consider *superconformal field theories* on the boundary. Two cases are of particular relevance because of concurrent reasons which are peculiar to them: from the algebraic side the essential use of one of the low-rank sporadic isomorphisms of orthogonal Lie algebras, from the supergravity side the existence of a spontaneous compactification of the Freund–Rubin type [66]. The two cases are:

A) The case $p = 3$ which leads to AdS_5 and to its 4-dimensional boundary. Here the sporadic isomorphism is $\mathrm{SO}(2,4) \sim \mathrm{SU}(2,2)$ which implies that the list of superconformal algebras is given by the superalgebras $\mathfrak{su}(2,2|\mathcal{N})$ for $1 \le \mathcal{N} \le 4$. On the other hand, in Type IIB Supergravity, there is a self-dual five-form field strength. Giving a v.e.v. to this latter ($F_{a_1a_2a_3a_4a_5} \ltimes \epsilon_{a_1a_2a_3a_4a_5}$), one splits the ambient 10-dimensional space into $5 \oplus 5$ where the first "5" stands for the AdS_5 space, while the second "5" stands for any compact 5-dimensional Einstein manifold $\mathcal{M}_5$. The holonomy of $\mathcal{M}_5$ decides the number of supersymmetries and on the 4-dimensional boundary $\partial\mathrm{AdS}_5$ we have a superconformal Yang–Mills gauge theory.

B) The case $p = 2$ which leads to AdS_4 and to its 3-dimensional boundary. Here the sporadic isomorphism is $\mathrm{SO}(2,3) \sim \mathrm{Sp}(4,\mathbb{R})$ which implies that the list of superconformal algebras is given by the superalgebras $\mathrm{Osp}(\mathcal{N}|4)$ for $\mathcal{N} = 1,2,3,6,8$. On the other hand, in $D = 11$ Supergravity, there is a four-form field strength. Giving a v.e.v. to this latter ($F_{a_1a_2a_3a_4} \ltimes \epsilon_{a_1a_2a_3a_4}$), one splits the ambient 10-dimensional space into $4 \oplus 7$ where "4" stands for the AdS_4 space, while "7" stands for any compact 7-dimensional Einstein manifold $\mathcal{M}_7$. The holonomy of $\mathcal{M}_7$ decides the number of supersymmetries and on the 3-dimensional boundary $\partial\mathrm{AdS}_4$ we should have a superconformal gauge theory.

The first case was that mostly explored at the beginning of the AdS/CFT tale in 1998 and in successive years. Yet the existence of the second case was immediately evident to anyone who had experience in supergravity and particularly to those who had worked in Kaluza–Klein supergravity in the years 1982–1985. Thus in a series of papers [41, 26, 43, 58, 42, 11, 10], mostly produced by the Torino Group but in some instances in collaboration with the SISSA Group, the $\mathrm{AdS}_4/\mathrm{CFT}_3$ correspondence was proposed and intensively developed in the spring and in the summer of the year 1999.

One leading idea, motivating this outburst of activity, was that the entire corpus of results on Kaluza–Klein mass-spectra [66, 38, 33, 5, 12, 78, 142, 19, 34, 25, 121, 32, 122, 35, 20, 65, 21], which had been derived in the years 1982–1986, could now be recycled in the new superconformal interpretation. Actually it was immediately clear that the Kaluza–Klein towers of states, in particularly those corresponding to short representations of the superalgebra $\mathrm{Osp}(\mathcal{N}|4)$, provided an excellent testing ground for the $\mathrm{AdS}_4/\mathrm{CFT}_3$ correspondence. One had to conceive candidate superconformal field theories living on the boundary, that were able to reproduce all the infinite towers of Kaluza–Klein multiplets as corresponding towers of composite operators with the same quantum numbers.

The subject of the $AdS_3 \times CFT_3$ correspondence was resurrected ten years later in 2007–2009 by the work presented in papers [69, 1, 6] which stirred a great interest in the scientific community and obtained a very large number of citations. The ABJM-construction of [1] is very clear and the attentive reader, making the required changes of notation and names of the objects, can verify that the $\mathcal{N} = 3$ Lagrangian presented there is just the same as that constructed in papers [41, 43] by letting the Yang–Mills coupling constant go to infinity. What is new and extremely important in ABJM is the relative quantization of the Chern–Simons levels $k_{1,2}$ of the two gauge groups and its link to a quotient of the seven-sphere by means of a cyclic group $\mathbb{Z}_k$. The superconformal 3-dimensional field theories obtained in the ABJM construction and in its multifaceted generalizations are presently the target of great interest and have fruitful applications not only in string-theory but also in condensed matter physics.

From the point of view of the present textbook in Group and Lie Algebra Theory for Physicists, the most interesting aspects of the tale sketched above consist in the following. Some of the most fundamental concepts of physics, like that of a relativistic particle, have their proper mathematical definition in terms of Lie Algebra Theory. Indeed, a particle of mass m and spin s is nothing else but a unitary irreducible representation (UIR) of the isometry group G of space-time or of its asymptotic limit at large distances. These UIRs are induced representations from the representations of the subgroup $\mathrm{H} \subset \mathrm{G}$ that stabilizes a generic point of space-time $\mathcal{M}_{p+2}$, this latter (or its asymptotic limit) being identified with the coset manifold $\frac{\mathrm{G}}{\mathrm{H}}$. The double interpretation of G as the isometry group of $\mathcal{M}_{p+2}$ or as the conformal group on its $(p+1)$-dimensional boundary $\partial\mathcal{M}_{p+2}$ is reflected into well-defined structural properties of the Lie algebra $\mathbb{G}$ and leads to the double interpretation of UIRs either as particles in $\mathcal{M}_{p+2}$ or as quantum conformal fields on the boundary $\partial\mathcal{M}_{p+2}$. Hence the miracles of the gauge/gravity correspondence are algebraically tamed at a large extent by Group Theory.

The next section we begin by illustrating the concept of particles in AdS_4 as induced representations of $SO(2,3)$.

12.4.1 Particles in anti de Sitter space and induced representations

Anti de Sitter space is the following non-compact coset manifold:

$$\mathrm{AdS}_4 = \frac{\mathrm{SO}(2,3)}{\mathrm{SO}(1,3)} \tag{12.46}$$

and, in the normalizations we utilize, the intrinsic components of its Riemann tensor are the following ones:

$$\mathcal{R}^{ab}{}_{cd} = -4\nu^2(\delta^a_c\delta^b_d - \delta^a_d\delta^b_d) \tag{12.47}$$

The parameter ν^{-1} which has the dimensions of a length is named the anti de Sitter radius and measures the amount of curvature of the manifold. In the limit $\nu \to 0$ one retrieves flat Minkowski space.

The physical concept of a *particle* in standard Minkowski space (Mink_4) corresponds to the mathematical concept of a *Unitary Irreducible Representation* (UIR) of the Poincaré Group. Since the Poincaré Group is non-compact, all UIRs are necessarily infinite dimensional, namely, they span a functional space, which is the space of field configurations in field theory. Every UIR of the Poincaré group is characterized by a mass m and a spin s. The elements in the UIR carrying space are the solutions of a *wave equation*, appropriate to the chosen mass m and spin s which describes propagation of the corresponding free field in Mink_4.

In complete analogy, a UIR of $\mathrm{SO}(2,3)$ is what one calls a particle in anti de Sitter space.

The new features are related to the concept of mass. Indeed, the square momentum operator $P_a P^a$ is not in this case a group invariant, as it is in the Poincaré case. Hence in AdS_4 a particle is not characterized by the eigenvalue of $P_a P^a$, but rather by the eigenvalue of the true second Casimir operator of $\mathrm{SO}(2,3)$ which, in our normalization, has the following expression:

$$\mathfrak{c}_2 = -2J_{ab}J^{ab} - \frac{1}{4\nu^2}P_a P^a \tag{12.48}$$

J_{ab} being the generators of the Lorentz group $\mathrm{SO}(1,3)$ and P^a the momentum operator.

This result is easily retrieved by noticing that, if we introduce indices $\Lambda, \Sigma = 0,1,2,3,4$, $a,b = 0,1,2,3$ and we set:

$$M_{\Lambda\Sigma} = -M_{\Sigma\Lambda}; \quad J_{ab} = M_{ab}; \quad M_{4a} = \frac{1}{4\nu}P_a \tag{12.49}$$

then the ten generators satisfy the $\mathrm{SO}(2,3)$ Lie algebra in its standard form:

$$\begin{aligned} [M_{\Lambda\Sigma}, M_{\Gamma\Delta}] &= \frac{1}{2}(\eta_{\Sigma\Gamma}M_{\Lambda\Delta} + \eta_{\Lambda\Delta}M_{\Sigma\Gamma} - \eta_{\Sigma\Delta}M_{\Lambda\Gamma} - \eta_{\Lambda\Gamma}M_{\Sigma\Delta}) \\ \eta_{\Lambda\Sigma} &= \mathrm{diag}(+,-,-,-,+) \end{aligned} \tag{12.50}$$

and $\mathfrak{c}_2$, defined by eq. (12.48), coincides with the standard quadratic invariant of (pseudo)-orthogonal Lie algebras:

$$C_2 = -2M_{\Lambda\Sigma}M^{\Lambda\Sigma} \tag{12.51}$$

The problem is how to relate the eigenvalues of C_2 to something which we can call the *mass* and the *spin* of a particle. The answer to such a question is mainly a matter of comparison.

On one hand, as we presently show, we can construct an irreducible unitary representation of $\mathrm{SO}(2,3)$ via the Wigner induced representation method, starting from an irrep of the maximal compact subgroup $SO(2) \otimes SO(3) \subset SO(2,3)$. The SO(2) quantum number E_0 is the eigenvalue of the Hamiltonian operator and, as such, is worth the name of energy (the minimal energy level in the representation); the number J labeling the SO(3) irrep is instead what we call the spin.

On the other hand, an irreducible unitary representation must also be identified with the Hilbert space spanned by finite norm solutions of a *free field equation*, suitable to the spin-J particle that we consider. We know how to write field equations for arbitrary fields of arbitrary spin on an arbitrary space-time $\mathcal{M}_4$. In particular we choose:

$$\mathcal{M}_4 = \mathrm{AdS}_4 \tag{12.52}$$

and we have the result we look for, namely an equation of the form:

$$\boxtimes_{(s)}\psi_{(s)} = m^2_{(s)}\psi_{(s)} \tag{12.53}$$

where $\boxtimes_{(s)}$ is a second-order invariant differential operator (it commutes with the Lie derivatives along the SO(2,3) Killing vectors) which acts on the space of spin-s wavefunctions and whose eigenvalues we can name the mass-squared.

Since there is just one quadratic Casimir operator, we must have:

$$\boxtimes_{(s)} = \alpha C_2 + \beta_s = m_s^2 \tag{12.54}$$

where α, β_s are constants. Moreover, since C_2 is a function of E_0 and J, that is, of the labels of the vacuum state (lowest weight state) in the induced representation procedure, then eq. (12.54) provides a relation between these labels and m_s^2. Needless to say, we must choose $J = s$ and (12.54) becomes a relation between E_0 and m_s^2, relation which is different for different spins.

The delicate point in this game is the choice of the origin of the $m^2_{(s)}$ scale, or, in other words, the definition of massless particles. Indeed, we know that $m^2 = 0$ is a singular value for Poincaré representations, corresponding to a reduction of the number of states (multiplet shortening) and the same must be true of anti de Sitter representations. The best way to understand this shortening is from a symmetry viewpoint. At $m_s^2 = 0$ the wave equation (12.53) must acquire a larger symmetry than the Poincaré or anti de Sitter symmetry. This larger symmetry is conformal symmetry for $s = 0$ and $s = \frac{1}{2}$ while it is a gauge symmetry for $s \geq 1$; in any case, it is responsible for a reduction of the dynamical degrees of freedom and the associated particle is worth the name of massless.

Let us briefly review the wave equations for $s = 0, \frac{1}{2}, 1, \frac{3}{2}, 2$.

12.4.1.1 The scalar $s = 0$ particle

Let

$$\square_{\mathrm{cov}} \equiv \nabla_\mu \nabla^\mu \tag{12.55}$$

be the covariant Laplacian on the considered space-time manifold and let $\mathscr{R} = \mathscr{R}^{ab}{}_{ab}$ be the curvature scalar in our adopted normalization. Then one can check that the following generalized Klein–Gordon equation:

$$\left(\Box_{\text{cov}} + \frac{\mathscr{R}}{3}\right)\varphi(x) = 0 \tag{12.56}$$

in addition to invariance under whatever isometries the metric $g_{\mu\nu}$ might possess, has further invariance under scale transformations $x^\mu \to \lambda x^\mu$ that are instead broken by the equation:

$$\left(\Box_{\text{cov}} + \frac{\mathscr{R}}{3}\right)\varphi(x) = -m_0^2\varphi(x) \tag{12.57}$$

which therefore is the correct wave-equation of a spin $s = 0$ particle of mass m_0^2. If we choose anti de Sitter space as space-time, from eq. (12.47) we obtain $\mathscr{R} = -24v^2$ and eq. (12.57) becomes:

$$(\Box_{\text{cov}} + m_0^2 - 8v^2)\varphi(x) = 0 \tag{12.58}$$

12.4.1.2 The spinor $s = \frac{1}{2}$ particle

In the spin-$\frac{1}{2}$ case, we recall the form of the covariant derivative of the Lorentz group SO(1, 3) acting on an arbitrary spinor λ:

$$\mathscr{D}_\mu\lambda \equiv \partial_\mu\lambda - \frac{1}{4}\omega_\mu^{ab}\gamma_{ab}\lambda \tag{12.59}$$

where ω_μ^{ab} is the spin connection (see for instance [51] Vol. 1, page 193) and $\gamma_{ab} = \frac{1}{2}[\gamma_a, \gamma_b]$ is the commutator of two gamma-matrices. The commutator of two such derivatives produces the Riemann tensor, according to:

$$[\mathscr{D}_\mu, \mathscr{D}_\nu]\lambda = -\frac{1}{4}\mathscr{R}^{ab}{}_{\mu\nu}\gamma_{ab}\lambda \tag{12.60}$$

setting:

$$\not{\mathscr{D}} \equiv g^{\mu\nu}\gamma_\mu\mathscr{D}_\nu; \quad \{\gamma_\mu, \gamma_\nu\} = 2g_{\mu\nu} \times \mathbf{1} \tag{12.61}$$

The Dirac equation, written in the following form (where γ_μ denote the standard gamma-matrices):

$$i\not{\mathscr{D}}\lambda = 0 \tag{12.62}$$

is scale invariant in addition to being invariant under any possible isometry. Hence the correct spin-$\frac{1}{2}$ wave equation is:

$$i\not{\mathscr{D}}\lambda = -m_{1/2}\lambda \tag{12.63}$$

Squaring the operator $\not{\mathscr{D}}$ we obtain:

$$\not{\mathscr{D}}^2 = \Box_{\text{cov}} - \frac{1}{4}\gamma^{ab}\gamma_{cd}\mathscr{R}^{ab}{}_{cd} \tag{12.64}$$

and inserting the explicit form of the anti de Sitter Riemann-tensor (see eq. (12.47)), we get:

$$(\Box_{\text{cov}} + m_{1/2}^2 - 12\nu^2)\lambda(x) = 0 \tag{12.65}$$

12.4.1.3 The spin $s = 1$ particle

The wave equation of a spin $s = 1$ particle, which is described by a vector field W_μ, is the following one:

$$\mathscr{D}^\mu(\mathscr{D}_\mu W_\nu - \mathscr{D}_\nu W_\mu) = -m_1^2 W_\nu \tag{12.66}$$

since when $m_1 = 0$ the above equation becomes invariant under the following gauge transformation:

$$W_\mu \mapsto W_\mu + \partial_\mu \phi \tag{12.67}$$

where $\phi(x)$ is any scalar function. In the massive case from eq. (12.67) we derive the transversality constraint:

$$\mathscr{D}^\mu W_\mu = 0 \tag{12.68}$$

and on the transverse field eq. (12.67) reduces to:

$$\Box_{\text{cov}} W_\mu + m_1^2 W_\mu + 2\mathscr{R}^{\rho\sigma}{}_{\rho\mu} W_\sigma = 0 \tag{12.69}$$

Finally, by substituting the explicit form of the AdS_4 Riemann tensor (12.47) into eq. (12.69) we obtain the wave equation of a spin-1 field in anti de Sitter space-time:

$$(\Box_{\text{cov}} + m_1^2 - 12\nu^2)W_\mu = 0 \tag{12.70}$$

12.4.1.4 The spin $s = \frac{3}{2}$ particle

A spin-$\frac{3}{2}$ particle is described by a spinor-vector field χ_μ. In order to write its wave equation, named the Rarita–Schwinger equation, it is convenient to turn to the formalism of flat Latin indices appropriate to the formulation of General Relativity in the Vielbein approach (see [51] Volume I, Chapter 5). Setting $\chi_a \equiv V_a^\mu \chi_\mu$ where V_a^μ is the inverse vierbein, we write the Rarita–Schwinger equation as follows:

$$\eta_{ab}\epsilon^{bcde}\gamma_5\gamma_c\nabla_d\chi_e = m_{3/2}\chi_a \tag{12.71}$$

In the above writing, the anti de Sitter derivative is defined as follows:

$$\begin{aligned} \nabla_a \chi_b &\equiv \mathscr{D}_a \chi_b + i\nu\gamma_a \chi_b \\ \mathscr{D}_a \chi_b &= \partial_a \chi_b - \frac{1}{4}\omega_a{}^{pq}\gamma_{pq}\chi_b - \eta_{br}\omega_a{}^{rs}\chi_s \end{aligned} \tag{12.72}$$

$\omega_a{}^{pq} \equiv V^\mu_a \omega_\mu{}^{pq}$ being the flat components of the spin connection and $\partial_a \chi_b \equiv V^\mu_a \partial_\mu \chi_b$ the flat components of the partial derivatives. When $m_{3/2} \neq 0$ from eq. (12.72) one deduces both irreducibility-transversality constraints:

$$\mathscr{D}^a \chi_a = 0; \quad \gamma^a \chi_a = 0 \tag{12.73}$$

and the Dirac equation in the form:

$$i\not{\mathscr{D}}\chi_a = -(m_{3/2} - 2\nu)\chi_a \tag{12.74}$$

On the other hand, since in anti de Sitter space the anti de Sitter derivatives are commutative:

$$[\nabla_c, \nabla_d] = -\frac{1}{2}\gamma_{ab}\mathscr{R}^{ab}{}_{cd} - 2\nu^2\gamma_{cd} = (2\nu^2 - 2\nu^2)\gamma_{cd} = 0 \tag{12.75}$$

at $m_{3/2} = 0$, the Rarita–Schwinger equation (12.71) acquires the following gauge invariance:

$$\chi_a \mapsto \chi_a + \nabla_a \lambda \tag{12.76}$$

where λ is an arbitrary spinor function. This gauge invariance reduces the number of on-shell physical degrees of freedom of the spinor vector. A priori a spinor vector contains 4×4 functions, the first "4" being the span of the spinor index and the second "4" being the span of the vector index. In the massive case we have:

$$\text{massive} \quad \#\,\text{d.o.f} = \underbrace{\frac{1}{2}}_{\text{Dirac eq.}} \Big(4 \times 4 - \underbrace{4}_{\text{irr. cond.}} - \underbrace{4}_{\text{transv. cond.}}\Big) = 4 \tag{12.77}$$

The Dirac equation is a projection that halves the number of functions. Both the irreducibility constraint $\gamma^a \chi_a = 0$ and the transversality condition $\mathscr{D}^a \chi_a = 0$ suppress four functions, so we get the result (12.77) and we recognize that $4 = 2 \times \frac{3}{2} + 1$ is just the dimension of an irreducible spin $J = \frac{3}{2}$ representation of the little group SO(3) (the stability subgroup of a time-like momentum vector p_μ). In the massless case we have instead:

$$\text{massless} \quad \#\,\text{d.o.f} = \underbrace{\frac{1}{2}}_{\text{Dirac eq.}} \Big(4 \times 2 - \underbrace{4}_{\text{gauge invariance}}\Big) = 2 \tag{12.78}$$

because the vector index must lie in the space transverse to a null-like momentum vector. Here we recognize that "2" is the dimension of any irreducible representation of the massless little group SO(2).

Applying $i\not{\mathscr{D}}$ to both sides of eq. (12.74) we obtain:

$$\left(\Box_{\text{cov}} - \frac{1}{4}\gamma^{pq}\mathscr{R}^{rs}{}_{pq}\gamma_{rs}\right)\chi_a - \eta_{ab}\gamma^{pq}\mathscr{R}^{bm}{}_{pq}\chi_m = -(m_{3/2} - 2\nu)^2\chi_a \tag{12.79}$$

and substituting the explicit form of the anti de Sitter Riemann-tensor (12.47), we arrive at:

$$(\Box_{\text{cov}} - 16\nu^2 + (m_{3/2} - 2\nu)^2)\chi_a = 0 \tag{12.80}$$

12.4.1.5 The spin $s = 2$ particle

In any given space-time geometry a spin-2 field is a symmetric tensor $h_{ab} = h_{ba}$ and its wave-equation is the linearized Einstein equation for the metric small fluctuations. Using as before the flat Latin index formalism, we have:

$$\begin{aligned} &\frac{1}{2}\Box_{\text{cov}}\, h_{ab} - \mathscr{D}_{\{a}\mathscr{D}^m h_{b\}m} + \frac{1}{2}\mathscr{D}_a\mathscr{D}_b h_{mn}\eta^{mn} \\ &- 2\eta_{an}\mathscr{R}^{nm}{}_{bs}h_{mr}\eta^{rs} = -\frac{1}{2}m_2^2 h_{ab} \end{aligned} \tag{12.81}$$

Let us insert into (12.81) the explicit form of the AdS Riemann tensor (12.47). When $m_2 \neq 0$, from eq. (12.81) one deduces both irreducibility-transversality constraints:

$$\mathscr{D}^a h_{ab} = 0; \quad \eta^{rs}h_{rs} = 0 \tag{12.82}$$

and the wave equation:

$$(\Box_{\text{cov}} - 8\nu^2)h_{ab} = -m_2^2 h_{ab} \tag{12.83}$$

On the other hand, at $m_2 = 0$, eq. (12.81) acquires the following gauge invariance:

$$h_{mn} \mapsto h_{mn} + \mathscr{D}_{\{m}t_{n\}} \tag{12.84}$$

and eqs (12.82) can be imposed as gauge fixings. The same is true of the spin-$\frac{3}{2}$ and spin-1 equations. The irreducibility-transversality constraints become, in the massless case, gauge fixing choices.

12.4.1.6 Summary of the wave equations

Our results can be summarized by saying that the second order wave equation of a spin-s particle in AdS_4 has the following general form:

$$(-\Box_{\text{cov}} + 4\alpha_s\nu^2)\Psi_s = m_s^2\Psi_s \tag{12.85}$$

where the numbers α_s are the following ones:

$$\begin{aligned} &\alpha_0 = 2; \quad \alpha_{1/2} = 3; \quad \alpha_1 = 3 \\ &\alpha_{3/2} = 3; \quad \alpha_2 = 2 \end{aligned} \tag{12.86}$$

This result, combined with the results to be derived in the next section allows to express the C_2 Casimir operator of the Lie algebra SO(2, 3) in terms of the covariant Laplacian $\Box_{\text{cov}}$.

12.4.2 Unitary irreducible representations of SO(2, 3)

We address the problem of constructing the unitary irreducible representations of the anti de Sitter group from a purely algebraic point of view. Our starting point is a convenient decomposition of the SO(2, 3) Lie algebra (12.50) with respect to its maximal compact subgroup:

$$\mathrm{G}_0 = \mathrm{SO}(2) \otimes \mathrm{SO}(3) \subset \mathrm{SO}(2,3) \tag{12.87}$$

Since $M_{ab}^\dagger = -M_{ab}$ and $P_a = -P_a^\dagger$ are anti-Hermitian, we define:

$$\begin{aligned} &H = -\frac{i}{2v} P_0; \qquad\qquad J_3 = -2iM_{12} \\ &J_+ = -\sqrt{2}(iM_{23} + M_{13}); \quad J_- = -\sqrt{2}(iM_{23} - M_{13}) \end{aligned} \tag{12.88}$$

The generator H is Hermitian and compact. It generates the SO(2) subgroup. It can be identified with the Hamiltonian of the system and its eigenvalues are worth the name of energy:

$$H|\psi\rangle = E|\psi\rangle; \quad (E^\star = E) \tag{12.89}$$

The generators $J_\pm$ and J_3 commute with H and generate the spin subgroup SO(3) ≃ SU(2):

$$\begin{aligned} &[H, J_\pm] = [H, J_3] = 0 \\ &[J_3, J_\pm] = \pm J_\pm; \quad J_3^\dagger = J_3; \quad J_+^\dagger = J_- \\ &[J_+, J_-] = J_3 \end{aligned} \tag{12.90}$$

The remaining six generators spanning the tangent space to the coset $\frac{\mathrm{SO}(2,3)}{\mathrm{SO}(2)\otimes\mathrm{SO}(3)}$ can be arranged into the following combinations:

$$\begin{aligned} &\left.\begin{aligned} K_i^+ &= -2M_{0i} + 2iM_{4i} \\ K_i^- &= -2M_{0i} - 2iM_{4i} \end{aligned}\right\} \quad i = 1, 2, 3 \\ &K_i^- = -(K_i^+)^\dagger \end{aligned} \tag{12.91}$$

which have the following commutation relations with H (as it can be checked from eq. (12.50)):

$$[H, K_i^{\pm}] = \pm K_i^{\pm} \tag{12.92}$$

Hence the $K_i^{\pm}$ act as raising and lowering operators for the energy eigenvalues E. Furthermore, we can rearrange the $K_i^{\pm}$, which under SO(3) transform as three-vectors, in the following way:

$$\begin{aligned} K_{\pm}^{+} &= \frac{1}{\sqrt{2}}(K_1^{+} \pm iK_2^{+}) \\ K_{\mp}^{-} &= \frac{1}{\sqrt{2}}(K_1^{-} \mp iK_2^{-}) = (K_{\pm}^{+})^{\dagger} \\ K_3^{+} &= K_3^{+} \end{aligned} \tag{12.93}$$

The commutation relations of these operators with J_3 are as follows:

$$[J_3, K_{\pm}^{+}] = \pm K_{\pm}^{+}; \quad [J_3, K_3^{+}] = 0 \tag{12.94}$$

This means that K_{+}^{+} raises both the energy and the third component of the spin, while K_{-}^{+} raises E and lowers J_3. Finally K_3^{+} raises E but leaves J_3 unchanged.

In this basis the Casimir invariant (12.51) can be rewritten as follows:

$$C_2 = H^2 + \mathbf{J}^2 + \frac{1}{2}\sum_{i=1}^{3}\{K_i^{+}, K_i^{-}\} \tag{12.95}$$

Let $\mathscr{H}$ be the Hilbert space carrying the typical unitary irreducible representation we are looking for. It is convenient to label the states $|\psi\rangle \in \mathscr{H}$ by means of the eigenvalues of E, $\mathbf{J}^2$ and J_3:

$$\begin{aligned} H|(\dots)E, j, m\rangle &= E|(\dots)E, j, m\rangle \\ \mathbf{J}^2|(\dots)E, j, m\rangle &= j(j+1)|(\dots)E, j, m\rangle \\ J_3|(\dots)E, j, m\rangle &= m|(\dots)E, j, m\rangle \end{aligned} \tag{12.96}$$

where $(\dots)$ denotes a representation label so far unspecified.

The representations we are interested in must have an energy spectrum bounded from below. Hence we introduce a multiplet of vacuum states:

$$|(E_0, s)E_0, s, m\rangle \tag{12.97}$$

which form an SU(2) irreducible representation of spin s:

$$\begin{aligned} &\mathbf{J}^2 = s(s+1); \quad s = \begin{cases} \text{integer} \\ \text{half integer} \end{cases} \\ &-s \le m \le s \end{aligned} \tag{12.98}$$

and are eigenstates of the Hamiltonian H with eigenvalue $E_0 > 0$:

$$H|(E_0,s)E_0,s,m\rangle = E_0|(E_0,s)E_0,s,m\rangle \tag{12.99}$$

Furthermore, by definition the vacuum is annihilated by all energy lowering operators:

$$K_i^-|(E_0,s)E_0,s,m\rangle = 0 \tag{12.100}$$

In the spirit of all the other irreducible representation constructions, E_0 and s label the irreducible unitary representation generated by applying to the vacuum $|(E_0,s)E_0,s,m\rangle$ the rising operators K_i^+ as many times as we like and regarding the Hilbert space spanned by such ket vectors as the carrier space. For this reason E_0 and s have been inserted in the slot we had prepared for the representation labels.

Evaluating C_2 on the vacuum $|(E_0,s)E_0,s,m\rangle$ we get:

$$C_2 = E_0(E_0-3) + s(s+1) \tag{12.101}$$

This result follows from the previously specified commutation relations and from the additional one:

$$[K_i^+, K_j^-] = 2\delta_{ij}H + 2i\epsilon_{ijk}J_k \tag{12.102}$$

all following from eq. (12.50).

The explicit structure of the Hilbert space $\mathcal{H}$ is then given by:

$$\mathcal{H} = \bigoplus_{n=0}^{\infty} \mathcal{H}_n \tag{12.103}$$

where $\mathcal{H}_n$ is the span of all the vectors of the form:

$$\sum_{n_1+n_2+n_3=n} c_{n_1n_2n_3}(K_1^+)^{n_1}(K_2^+)^{n_2}(K_3^+)^{n_3}|(E_0,s)E_0,s,m\rangle \tag{12.104}$$

with $c_{n_1n_2n_3} \in \mathbb{C}$. $\mathcal{H}_n$ is the finite dimensional subspace ($\dim\mathcal{H}_n < \infty$) of those states whose energy E is $E_0 + n$:

$$|\psi_n\rangle \in H_n \quad \Leftrightarrow \quad H|\psi_n\rangle = (E_0+n)|\psi_n\rangle \tag{12.105}$$

The crucial point is that, in order for $\mathcal{H}$ to be a true Hilbert space, its states must have a positive norm:

$$|\psi\rangle = \bigoplus_{n=0}^{\infty} |\psi_n\rangle \tag{12.106}$$

$$\|\psi\|^2 = \sum_{n=0}^{\infty} \|\psi_n\|^2 > 0 \tag{12.107}$$

$$\|\psi_n\|^2 = \langle\psi_n|\psi_n\rangle \tag{12.108}$$

This is guaranteed if the scalar products in all the $\mathscr{H}_n$ subspaces are positive definite, namely if:

$$\forall\psi_n \in \mathscr{H}_n; \quad \langle\psi_n|\psi_n\rangle > 0 \tag{12.109}$$

In this case the Hilbert space H is composed of those series (12.106) whose norm is convergent:

$$\sum_n^{\infty} \|\psi_n\|^2 < \infty \tag{12.110}$$

We may also tolerate the presence of zero-norm states. If these exist, we define a Hilbert space $\mathscr{H}_{\text{phys}}$ composed of the equivalence classes of all states modulo the addition of zero-norm states:

$$\begin{aligned} &\mathscr{H}_{\text{phys}} = \mathscr{H}/\mathscr{H}_0 \\ &|\psi\rangle \sim |\psi'\rangle \quad \Leftrightarrow \quad |\psi'\rangle = |\psi\rangle + |\psi_0\rangle; \quad |\psi_0\rangle \in \mathscr{H}_0 \\ &\mathscr{H}_0 = \{|\psi\rangle \in \mathscr{H}, \|\psi\|^2 = 0\} \end{aligned} \tag{12.111}$$

Such a situation is typical of all massless theories and in particular of gauge theories. The zero-norm states which are removed by the standard procedure (12.111) are gauge degrees of freedom and their subtraction leads to a *shortening of the representation*.

What we can never accept is the presence of negative norm states (*ghosts*).

Hence before declaring that we have found the unitary irreducible representations of SO(2, 3) we must ascertain under which conditions the space $\mathscr{H}$ does not contain negative norm states and is therefore a Hilbert space. These conditions are simply expressed as lower bounds on the energy label E_0, relative to the spin s.

Let us state these bounds that are obtained calculating the norms of the states (12.104) by means of repeated use of the Lie algebra commutation relations. This is a straightforward but very tedious derivation that we skip: the interested reader can find all the details in [118].

a) For $s \geq 1$ there are no ghosts if and only if

$$E_0 \geq s + 1 \tag{12.112}$$

When $E_0 > s + 1$, there are no zero-norm states and no representation shortening occurs. The representation is massive. For $E_0 = s + 1$, we have zero-norm states which can be decoupled. The corresponding representation is massless and it is described by the appropriate massless wave-equation. The zero-norm states are associated with a corresponding gauge invariance: for $s = 1$ it is the gauge invariance of standard gauge fields, for $s = 2$ it is the general coordinate invariance of General Relativity, for $s = \frac{3}{2}$ it is local supersymmetry of supergravity.

b) For $s = \frac{1}{2}$ there are no ghosts if and only if

$$E_0 \geq 1 \tag{12.113}$$

Decoupling of zero-norm states takes place for $E_0 = \frac{3}{2}$ and $E_0 = 1$. The first value corresponds to a massless representation described by the massless wave equation that acquires conformal invariance, while the limiting value $E_0 = 1$ is the so-called *Dirac singleton* that has no counterpart in Poincaré theory. It is not a particle in the AdS_4 bulk manifold, but rather it is a particle leaving only on the 3-dimensional boundary ∂AdS_4. It is relevant for the AdS_4/CFT_3 correspondence.

c) For $s = 0$ there are no ghosts if and only if:

$$E_0 \geq \frac{1}{2} \tag{12.114}$$

The zero-norm states are found for the special values $E_0 = 2$, $E_0 = 1$ and $E_0 = \frac{1}{2}$. Both values $E_0 = 1$ and $E_0 = 2$ yield the standard massless representation described by the conformal invariant wave equation (12.57), while the lowest value $E_0 = \frac{1}{2}$ is again a *Dirac singleton* representation with no counterpart in the Poincaré case and no field theory interpretation in the bulk of AdS_4-space.

The results of this section can be summarized by mentioning the relations between the square masses of the various particles and the corresponding energy labels E. Setting $\nu = \frac{2}{\ell}$ where ℓ is the scale length of the curvature radius of anti de Sitter space, they are obtained from the above discussion in the following form:

$$\begin{aligned} \ell^2 m_{(0)}^2 &= 16(E_{(0)} - 2)(E_{(0)} - 1) \\ \ell|m_{(\frac{1}{2})}| &= 4E_{(\frac{1}{2})} - 6 \\ \ell^2 m_{(1)}^2 &= 16(E_{(1)} - 2)(E_{(1)} - 1) \\ \ell|m_{(\frac{3}{2})} + 4| &= 4E_{(\frac{3}{2})} - 6 \\ \ell^2 m_{(2)}^2 &= 16E_{(2)}(E_{(2)} - 3) \end{aligned} \tag{12.115}$$

12.4.3 The Osp($\mathcal{N}$|4) superalgebra: definition, properties and notation

Super Lie algebras, that are the basis of supersymmetry, correspond to an extension of the concept of Lie algebra based on a $\mathbb{Z}_2$-grading of the elements. Each element of the super Lie algebra $\mathbb{G}$ is either even or odd, denoted by a letter e or o. The super Lie bracket, now denoted $[,\}$, is either antisymmetric or symmetric depending on the grading:

$$[X_e, Y_e\} \equiv [X_e, Y_e] = -[Y_e, X_e]$$

$$[X_e, Y_o\} \equiv [X_e, Y_o] = -[Y_o, X_e] \qquad (12.116)$$
$$[X_o, Y_o\} \equiv \{X_o, Y_o\} = \{Y_o, X_o\}$$

The even elements of a super Lie algebra form a standard Lie algebra, named the even algebra.

The superalgebras relevant to physics are those that involve as subalgebra of the even algebra $\mathbb{G}_e$ either the Poincaré or the anti de Sitter algebra in dimension d and where the odd-elements Q_I^α, named supercharges, transform as spinors with respect to the Poincaré or anti de Sitter subalgebra. These supercharges generate the new very strong symmetry that exchanges the role of bosons and fermions in the *supersymmetric field theory* to which they pertain.

In the case of anti de Sitter in $d = 4$ which is that of interest to us here, the relevant superalgebra is of the type $\mathrm{Osp}(\mathcal{N}|4)$.

The non-compact superalgebra $\mathrm{Osp}(\mathcal{N}|4)$ relevant to the $\mathrm{AdS}_4/\mathrm{CFT}_3$ correspondence is a real section of the complex orthosymplectic superalgebra $\mathrm{Osp}(\mathcal{N}|4,\mathbb{C})$ that admits the Lie algebra

$$\mathrm{G}_{\text{even}} = \mathrm{Sp}(4,\mathbb{R}) \times \mathrm{SO}(\mathcal{N},\mathbb{R}) \qquad (12.117)$$

as even subalgebra. Alternatively, due to the isomorphism $\mathrm{Sp}(4,\mathbb{R}) \equiv \mathrm{Usp}(2,2)$ we can rewrite the same real section of $\mathrm{Osp}(\mathcal{N}|4,\mathbb{C})$ in a basis where the even subalgebra is identified as follows:

$$\mathrm{G}'_{\text{even}} = \mathrm{Usp}(2,2) \times \mathrm{SO}(\mathcal{N},\mathbb{R}) \qquad (12.118)$$

Here we mostly rely on the second formulation (12.118) which is more convenient to discuss unitary irreducible representations. The two formulations are related by a unitary transformation that, in spinor language, corresponds to a different choice of the gamma matrix representation. Formulation (12.117) is obtained in a Majorana representation where all the gamma matrices are real (or purely imaginary), while formulation (12.118) is related to a Dirac representation.

Our choice for the gamma matrices in a Dirac representation is the following one:[2]

$$\Gamma^0 = \begin{pmatrix} \mathbf{1} & 0 \\ 0 & -\mathbf{1} \end{pmatrix}, \quad \Gamma^{1,2,3} = \begin{pmatrix} 0 & \tau^{1,2,3} \\ -\tau^{1,2,3} & 0 \end{pmatrix}, \quad C_{[4]} = i\Gamma^0\Gamma^3 \qquad (12.119)$$

having denoted by $C_{[4]}$ the charge conjugation matrix in four dimensions $C_{[4]}\Gamma^\mu C_{[4]}^{-1} = -(\Gamma^\mu)^T$.

2 As explicit representation of the $\mathrm{SO}(3)\tau$ matrices we adopt a permutation of the canonical Pauli matrices σ^a: $\tau^1 = \sigma^3$, $\tau^2 = \sigma^1$ and $\tau^3 = \sigma^2$.

Then the $\mathrm{Osp}(\mathcal{N}|4)$ superalgebra is defined as the set of graded $(4+\mathcal{N})\times(4+\mathcal{N})$ matrices μ that satisfy the following two conditions:

$$\begin{aligned}\mu^T\begin{pmatrix}C_{[4]} & 0\\ 0 & \mathbf{1}_{\mathcal{N}\times\mathcal{N}}\end{pmatrix}+\begin{pmatrix}C_{[4]} & 0\\ 0 & \mathbf{1}_{\mathcal{N}\times\mathcal{N}}\end{pmatrix}\mu=0\\ \mu^\dagger\begin{pmatrix}\Gamma^0 & 0\\ 0 & -\mathbf{1}_{\mathcal{N}\times\mathcal{N}}\end{pmatrix}+\begin{pmatrix}\Gamma^0 & 0\\ 0 & -\mathbf{1}_{\mathcal{N}\times\mathcal{N}}\end{pmatrix}\mu=0\end{aligned} \tag{12.120}$$

the first condition defining the complex orthosymplectic algebra, the second one the real section with even subalgebra as in eq. (12.118). Equations (12.120) are solved by setting:

$$\mu=\begin{pmatrix}\varepsilon^{AB}\frac{1}{4}[\Gamma_A,\Gamma_B] & \epsilon^i\\ \overline{\epsilon}^i & \mathrm{i}\varepsilon_{ij}t^{ij}\end{pmatrix} \tag{12.121}$$

In eq. (12.121), $\varepsilon_{ij}=-\varepsilon_{ji}$ is an arbitrary real antisymmetric $\mathcal{N}\times\mathcal{N}$ tensor, $t^{ij}=-t^{ji}$ is the antisymmetric $\mathcal{N}\times\mathcal{N}$ matrix:

$$(t^{ij})_{\ell m}=\mathrm{i}(\delta^i_\ell\delta^j_m-\delta^i_m\delta^j_\ell) \tag{12.122}$$

namely, a standard generator of the $\mathrm{SO}(\mathcal{N})$ Lie algebra,

$$\Gamma_A=\begin{cases}\mathrm{i}\Gamma_5\Gamma_\mu & A=\mu=0,1,2,3\\ \Gamma_5\equiv\mathrm{i}\Gamma^0\Gamma^1\Gamma^2\Gamma^3 & A=4\end{cases} \tag{12.123}$$

denotes a realization of the $\mathrm{SO}(2,3)$ Clifford algebra:

$$\begin{aligned}\{\Gamma_A,\Gamma_B\}&=2\eta_{AB}\\ \eta_{AB}&=\mathrm{diag}(+,-,-,-,+)\end{aligned} \tag{12.124}$$

and

$$\epsilon^i=C_{[4]}(\overline{\epsilon}^i)^T\quad(i=1,\dots,\mathcal{N}) \tag{12.125}$$

are $\mathcal{N}$ anticommuting Majorana spinors.

The index conventions we have so far introduced can be summarized as follows. Capital indices $A,B=0,1,\dots,4$ denote $\mathrm{SO}(2,3)$ vectors. The Latin indices of type $i,j,k=1,\dots,\mathcal{N}$ are $\mathrm{SO}(\mathcal{N})$ vector indices. The indices $a,b,c,\dots=1,2,3$ are used to denote spatial directions of AdS_4: $\eta_{ab}=\mathrm{diag}(-,-,-)$, while the indices of type $m,n,p,\dots=0,1,2$ are space-time indices for the Minkowskian boundary $\partial(\mathrm{AdS}_4)$: $\eta_{mn}=\mathrm{diag}(+,-,-)$. To write the $\mathrm{Osp}(\mathcal{N}|4)$ algebra in abstract form it suffices to read the graded matrix (12.121) as a linear combination of generators:

$$\mu\equiv-\mathrm{i}\varepsilon^{AB}M_{AB}+\mathrm{i}\varepsilon_{ij}T^{ij}+\overline{\epsilon}_iQ^i \tag{12.126}$$

where $Q^i = C_{[4]}(\overline{Q}^i)^T$ are also Majorana spinor operators. Then the superalgebra reads as follows:

$$\begin{aligned}
[M_{AB}, M_{CD}] &= \mathrm{i}(\eta_{AD}M_{BC} + \eta_{BC}M_{AD} - \eta_{AC}M_{BD} - \eta_{BD}M_{AC}) \\
[T^{ij}, T^{kl}] &= -\mathrm{i}(\delta^{jk}T^{il} - \delta^{ik}T^{jl} - \delta^{jl}T^{ik} + \delta^{il}T^{jk}) \\
[M_{AB}, Q^i] &= -\mathrm{i}\frac{1}{4}[\mathbb{\Gamma}_A, \mathbb{\Gamma}_B]Q^i \\
[T^{ij}, Q^k] &= -\mathrm{i}(\delta^{jk}Q^i - \delta^{ik}Q^j) \\
\{Q^{\alpha i}, \overline{Q}^j_\beta\} &= \mathrm{i}\delta^{ij}\frac{1}{4}[\mathbb{\Gamma}^A, \mathbb{\Gamma}^B]^\alpha_\beta M_{AB} + \mathrm{i}\delta^\alpha_\beta T^{ij}
\end{aligned} \tag{12.127}$$

In the gamma matrix basis (12.119) the Majorana supersymmetry charges have the following form:

$$Q^i = \begin{pmatrix} a^i_\alpha \\ \varepsilon_{\alpha\beta}\overline{a}^{\beta i} \end{pmatrix}, \quad \overline{a}^{\alpha i} \equiv (a^i_\alpha)^\dagger \tag{12.128}$$

where a^i_α are 2-component $\mathrm{SL}(2,\mathbb{C})$ spinors: $\alpha, \beta, \ldots = 1, 2$. We do not use dotted and undotted indices to denote conjugate $\mathrm{SL}(2,\mathbb{C})$ representations; we rather use higher and lower indices. Raising and lowering is performed by means of the ε-symbol:

$$\psi_\alpha = \varepsilon_{\alpha\beta}\psi^\beta, \quad \psi^\alpha = \varepsilon^{\alpha\beta}\psi_\beta \tag{12.129}$$

where $\varepsilon_{12} = \varepsilon^{21} = 1$, so that $\varepsilon_{\alpha\gamma}\varepsilon^{\gamma\beta} = \delta^\beta_\alpha$. Unwritten indices are contracted according to the rule "*from eight to two.*"

12.4.3.1 Compact and non-compact five gradings of the Osp($\mathcal{N}$|4) superalgebra

As it is extensively explained in [80], a non-compact group G admits unitary irreducible representations of the lowest weight type if it has a maximal compact subgroup G^0 of the form $\mathrm{G}^0 = \mathrm{H} \times \mathrm{U}(1)$ with respect to whose Lie algebra $\mathfrak{g}^0$ there exists a *three-grading* of the Lie algebra $\mathfrak{g}$ of G. This was precisely the key token utilized in Section 12.4.2 to derive the UIRs of the anti de Sitter group SO(2, 3).

In the case of a non-compact superalgebra the lowest weight UIRs can be constructed if the three-grading is generalized to a *five-grading* where the even (odd) elements are integer (half-integer) graded:

$$\mathfrak{g} = \mathfrak{g}^{-1} \oplus \mathfrak{g}^{-\frac{1}{2}} \oplus \mathfrak{g}^0 \oplus \mathfrak{g}^{+\frac{1}{2}} \oplus \mathfrak{g}^{+1} \tag{12.130}$$

$$[\mathfrak{g}^k, \mathfrak{g}^l] \subset \mathfrak{g}^{k+l} \quad \mathfrak{g}^{k+l} = 0 \text{ for } |k+l| > 1 \tag{12.131}$$

For the supergroup Osp($\mathcal{N}$|4) this grading can be made in two ways, choosing as grade zero subalgebra either the maximal compact subalgebra

$$\mathfrak{g}^0 \equiv \mathfrak{so}(3) \times \mathfrak{so}(2) \times \mathfrak{so}(\mathcal{N}) \subset \mathrm{Osp}(\mathcal{N}|4) \tag{12.132}$$

or the non-compact subalgebra

$$\tilde{\mathfrak{g}}^0 \equiv \mathfrak{so}(1,2) \times \mathfrak{so}(1,1) \times \mathfrak{so}(\mathcal{N}) \subset \mathrm{Osp}(\mathcal{N}|4) \tag{12.133}$$

which also exists, has the same complex extension and is also maximal.

The existence of the double five-grading is the algebraic core of the $\mathrm{AdS}_4/\mathrm{CFT}_3$ correspondence. Decomposing a UIR of $\mathrm{Osp}(\mathcal{N}|4)$ into representations of $\mathfrak{g}^0$ shows its interpretation as a supermultiplet of *particles states* in the bulk of AdS_4, while decomposing it into representations of $\tilde{\mathfrak{g}}^0$ shows its interpretation as a supermultiplet of *conformal primary fields* on the boundary $\partial(\mathrm{AdS}_4)$.

In both cases the grading is determined by the generator X of the abelian factor SO(2) or SO(1,1):

$$[X, \mathfrak{g}^k] = k\mathfrak{g}^k \tag{12.134}$$

In the compact case the SO(2) generator X is given by M_{04}. As we have seen in Section 12.4.2, it is interpreted as the energy generator of the 4-dimensional AdS theory. It was used in [25] and [41] for the construction of the Osp(2|4) representations, yielding the long multiplets of [25] and the short and ultra-short multiplets constructed there for the first time. We repeat such decompositions here.

We call E the energy generator of SO(2), L_a the rotations of SO(3):

$$\begin{aligned} E &= M_{04} \\ L_a &= \frac{1}{2}\varepsilon_{abc}M_{bc} \end{aligned} \tag{12.135}$$

and $M_a^{\pm}$ the boosts:

$$\begin{aligned} M_a^+ &= -M_{a4} + iM_{0a} \\ M_a^- &= M_{a4} + iM_{0a} \end{aligned} \tag{12.136}$$

The supersymmetry generators are a_α^i and $\overline{a}^{\alpha i}$. Rewriting the $\mathrm{Osp}(\mathcal{N}|4)$ superalgebra (12.127) in this basis we obtain:

$$\begin{aligned}
[E, M_a^+] &= M_a^+ \\
[E, M_a^-] &= -M_a^- \\
[L_a, L_b] &= i\varepsilon_{abc}L_c \\
[M_a^+, M_b^-] &= 2\delta_{ab}E + 2i\varepsilon_{abc}L_c \\
[L_a, M_b^+] &= i\varepsilon_{abc}M_c^+ \\
[L_a, M_b^-] &= i\varepsilon_{abc}M_c^- \\
[T^{ij}, T^{kl}] &= -i(\delta^{jk}T^{il} - \delta^{ik}T^{jl} - \delta^{jl}T^{ik} + \delta^{il}T^{jk}) \\
[T^{ij}, \overline{a}^{\alpha k}] &= -i(\delta^{jk}\overline{a}^{\alpha i} - \delta^{ik}\overline{a}^{\alpha j})
\end{aligned}$$

$$\begin{aligned}
[T^{ij}, a^k_\alpha] &= -i(\delta^{jk} a^i_\alpha - \delta^{ik} a^i_\alpha) \\
[E, a^i_\alpha] &= -\frac{1}{2} a^i_\alpha \\
[E, \overline{a}^{\alpha i}] &= \frac{1}{2} \overline{a}^{\alpha i} \\
[M^+_a, a^i_\alpha] &= (\tau_a)_{\alpha\beta} \overline{a}^{\beta i} \\
[M^-_a, \overline{a}^{\alpha i}] &= -(\tau_a)^{\alpha\beta} a^i_\beta \\
[L_a, a^i_\alpha] &= \frac{1}{2} (\tau_a)_\alpha{}^\beta a^i_\beta \\
[L_a, \overline{a}^{\alpha i}] &= -\frac{1}{2} (\tau_a)^\alpha{}_\beta \overline{a}^{\beta i} \\
\{a^i_\alpha, a^j_\beta\} &= \delta^{ij} (\tau^k)_{\alpha\beta} M^-_k \\
\{\overline{a}^{\alpha i}, \overline{a}^{\beta j}\} &= \delta^{ij} (\tau^k)^{\alpha\beta} M^+_k \\
\{a^i_\alpha, \overline{a}^{\beta j}\} &= \delta^{ij} \delta_\alpha{}^\beta E + \delta^{ij} (\tau^k)_\alpha{}^\beta L_k + i \delta_\alpha{}^\beta T^{ij}
\end{aligned} \tag{12.137}$$

The five-grading structure of the algebra (12.137) is shown in Figure 12.6. In the superconformal field theory context we are interested in the action of the $\mathrm{Osp}(\mathcal{N}|4)$ generators on superfields living on the Minkowskian boundary $\partial(\mathrm{AdS}_4)$. To be precise, the boundary is a compactification of $d = 3$ Minkowski space and admits a conformal family of metrics $g_{mn} = \phi(z)\eta_{mn}$ conformally equivalent to the flat Minkowski metric

$$\eta_{mn} = (+,-,-), \quad m,n,p,q = 0,1,2 \tag{12.138}$$

Precisely, since we are interested in conformal field theories, the choice of the representative metric inside the conformal family is immaterial and the flat one (12.138) is certainly the most convenient. The requested action of the superalgebra generators is obtained upon starting from the non-compact grading with respect to (12.133). To this effect we define the *dilatation* $\mathrm{SO}(1,1)$ generator D and the *Lorentz* $\mathrm{SO}(1,2)$ generators J^m as follows:

$$D \equiv \mathrm{i} M_{34}, \quad J^m = \frac{\mathrm{i}}{2} \varepsilon^{mpq} M_{pq} \tag{12.139}$$

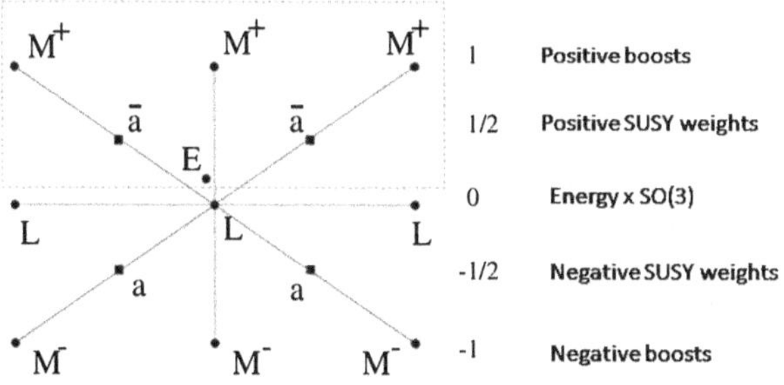

Figure 12.6: Schematic representation of the root diagram of $\mathrm{Osp}(\mathcal{N}|4)$ in the $\mathrm{SO}(2) \times \mathrm{SO}(3)$ basis. The grading w.r.t. the energy E is given on the right.

In addition we define the $d = 3$ *translation generators* P_m and *special conformal boosts* K_m as follows:

$$\begin{aligned} P_m &= M_{m4} - M_{3m} \\ K_m &= M_{m4} + M_{3m} \end{aligned} \tag{12.140}$$

Finally we define the generators of $d = 3$ *ordinary* and *special conformal supersymmetries*, respectively given by:

$$\begin{aligned} q^{\alpha i} &= \frac{1}{\sqrt{2}}(a^i_\alpha + \overline{a}^{\alpha i}) \\ s^i_\alpha &= \frac{1}{\sqrt{2}}(-a^i_\alpha + \overline{a}^{\alpha i}) \end{aligned} \tag{12.141}$$

The SO($\mathcal{N}$) generators are left unmodified as above. In this new basis the Osp($\mathcal{N}|4$)-algebra (12.127) reads as follows:

$$\begin{aligned}
[D, P_m] &= -P_m \\
[D, K_m] &= K_m \\
[J_m, J_n] &= \varepsilon_{mnp} J^p \\
[K_m, P_n] &= 2\eta_{mn} D - 2\varepsilon_{mnp} J^p \\
[J_m, P_n] &= \varepsilon_{mnp} P^p \\
[J_m, K_n] &= \varepsilon_{mnp} K^p \\
[T^{ij}, T^{kl}] &= -i(\delta^{jk} T^{il} - \delta^{ik} T^{jl} - \delta^{jl} T^{ik} + \delta^{il} T^{jk}) \\
[T^{ij}, q^{\alpha k}] &= -i(\delta^{jk} q^{\alpha i} - \delta^{ik} q^{\alpha j}) \\
[T^{ij}, s^k_\alpha] &= -i(\delta^{jk} s^i_\alpha - \delta^{ik} s^i_\alpha) \\
[D, q^{\alpha i}] &= -\frac{1}{2} q^{\alpha i} \\
[D, s^i_\alpha] &= \frac{1}{2} s^i_\alpha \\
[K^m, q^{\alpha i}] &= -i(\gamma^m)^{\alpha\beta} s^i_\beta \\
[P^m, s^i_\alpha] &= -i(\gamma^m)_{\alpha\beta} q^{\beta i} \\
[J^m, q^{\alpha i}] &= -\frac{i}{2}(\gamma^m)^\alpha{}_\beta q^{\beta i} \\
[J^m, s^i_\alpha] &= \frac{i}{2}(\gamma^m)_\alpha{}^\beta s^i_\beta \\
\{q^{\alpha i}, q^{\beta j}\} &= -i\delta^{ij}(\gamma^m)^{\alpha\beta} P_m \\
\{s^i_\alpha, s^j_\beta\} &= i\delta^{ij}(\gamma^m)_{\alpha\beta} K_m \\
\{q^{\alpha i}, s^j_\beta\} &= \delta^{ij}\delta^\alpha{}_\beta D - i\delta^{ij}(\gamma^m)^\alpha{}_\beta J_m + i\delta^\alpha{}_\beta T^{ij}
\end{aligned} \tag{12.142}$$

and the five-grading structure of eqs (12.142) is displayed in Figure 12.7. In both cases of Figure 12.6 and Figure 12.7, if one takes the subset of generators of positive grading plus the abelian grading generator $X = \{^E_D$, one obtains a *solvable superalgebra* of

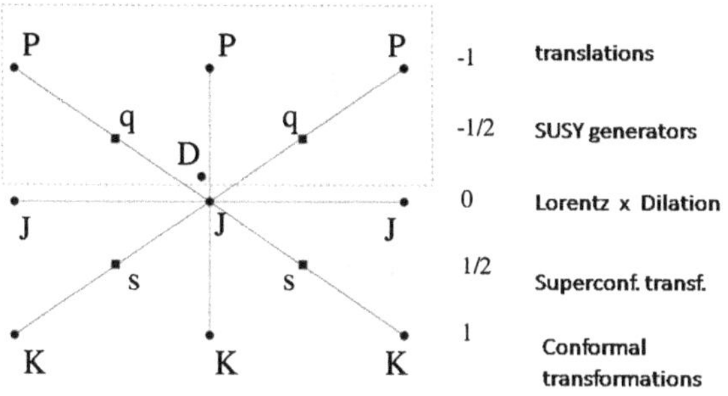

Figure 12.7: Schematic representation of the root diagram of $\mathrm{Osp}(\mathcal{N}|4)$ in the $\mathrm{SO}(1,1)\times\mathrm{SO}(1,2)$ basis. The grading w.r.t. the dilatation D is given on the right.

dimension $4+2\mathcal{N}$. It is however only in the non-compact case of Figure 12.7 that the bosonic subalgebra of the solvable superalgebra generates anti de Sitter space AdS_4 as a solvable group manifold. We have the following:

$$\mathrm{AdS}^{(\mathrm{Solv})}_{4|2\mathcal{N}} \equiv \exp[\mathrm{SSolv}_{\mathrm{AdS}}] \tag{12.143}$$

The supermanifold (12.143) is a supercoset of the same supergroup $\mathrm{Osp}(\mathcal{N}|4)$ but with respect to a different subgroup:

$$\mathrm{AdS}^{(\mathrm{Solv})}_{4|2\mathcal{N}} = \frac{\mathrm{Osp}(4|\mathcal{N})}{\mathrm{CSO}(1,2|\mathcal{N})} \tag{12.144}$$

where $\mathrm{CSO}(1,2|\mathcal{N})\subset\mathrm{Osp}(\mathcal{N}|4)$ is an algebra containing $3+3+\frac{\mathcal{N}(\mathcal{N}-1)}{2}$ bosonic generators and $2\times\mathcal{N}$ fermionic ones. This algebra is the semidirect product:

$$\mathrm{CSO}(1,2|\mathcal{N}) = \underbrace{\mathrm{ISO}(1,2|\mathcal{N})\times\mathrm{SO}(\mathcal{N})}_{\text{semidirect}} \tag{12.145}$$

of $\mathcal{N}$-extended *superPoincaré* algebra in $d=3$ ($\mathrm{ISO}(1,2|\mathcal{N})$) with the orthogonal group $\mathrm{SO}(\mathcal{N})$.

Therefore the solvable superalgebra $\mathrm{Ssolv}_{\mathrm{AdS}}$ mentioned in eq. (12.143) is the vector span of the following generators:

$$\mathrm{Ssolv}_{\mathrm{adS}} \equiv \mathrm{span}\{P_m, D, q^{\alpha i}\} \tag{12.146}$$

12.4.3.2 The lowest weight UIRs as seen from the compact and non-compact five-grading viewpoint

In the following we mostly concentrate on the simplest case $\mathcal{N}=2$. The structure of all the $\mathrm{Osp}(2|4)$ supermultiplets relevant to the $\mathrm{AdS}_4/\mathrm{CFT}_3$ correspondence are known. Their spin content is upper bounded by $s=2$ and they fall into three classes: *long*, *short* and *ultrashort*. Such a result was obtained in the 1999 paper [41] by explicit harmonic analysis with a particular choice of the internal manifold[3] $X^7 = M^{111}$, namely, through the analysis of a specific example of $\mathcal{N}$=2 compactification on $AdS_4 \times X^7$. Up to a certain stage, however, we find it more convenient to discuss Osp($\mathcal{N}$|4) for generic $\mathcal{N}$.

We start by briefly recalling the procedure of [65, 82] to construct UIRs of Osp($\mathcal{N}$|4) in the compact grading (12.132). Then, in a parallel way to what was done in [79] for the case of the SU(2, 2|4) superalgebra we show that also for Osp($\mathcal{N}$|4) in each UIR carrier space there exists a unitary rotation that maps eigenstates of E, L^2, L_3 to eigenstates of D, J^2, J_2. By means of such a rotation the decomposition of the UIR into SO(2) × SO(3) representations is mapped to an analogous decomposition into SO(1, 1) × SO(1, 2) representations. While SO(2) × SO(3) representations describe the *on-shell* degrees of freedom of a *bulk particle* with an energy E_0 and a spin-s, irreducible representations of SO(1, 1) × SO(1, 2) describe the *off-shell* degrees of freedom of a *boundary field* with scaling weight D and Lorentz character J. Relying on this we show how to construct the on-shell 4-dimensional superfield multiplets that generate the states of these representations and the off-shell 3-dimensional superfield multiplets that build the conformal field theory on the boundary.

In full analogy with the construction UIR representation of the purely bosonic group SO(2, 3) ~ Sp(4, ℝ) that we performed in Section 12.4.2, lowest weight representations of Osp($\mathcal{N}$|4) are constructed starting from the basis (12.137) and choosing a *vacuum state* such that

$$\begin{aligned} M_i^- |(E_0, s, \Lambda)\rangle &= 0 \\ a_\alpha^i |(E_0, s, \Lambda)\rangle &= 0 \end{aligned} \tag{12.147}$$

where E_0 denotes the eigenvalue of the energy operator M_{04} while s and Λ are the labels of an irreducible SO(3) and SO($\mathcal{N}$) representations, respectively. In particular we have:

$$\begin{aligned} M_{04}|(E_0, s, \Lambda)\rangle &= E_0|(E_0, s, \Lambda)\rangle \\ L^a L^a|(E_0, s, \Lambda)\rangle &= s(s+1)|(E_0, s, \Lambda)\rangle \\ L^3|(E_0, s, \Lambda)\rangle &= s|(E_0, s, \Lambda)\rangle \end{aligned} \tag{12.148}$$

The states filling up the UIR are then built by applying the operators M^+ and the antisymmetrized products of the operators $\overline{a}_\alpha^i$:

$$(M_1^+)^{n_1}(M_2^+)^{n_2}(M_3^+)^{n_3}[\overline{a}_{\alpha_1}^{i_1} \dots \overline{a}_{\alpha_p}^{i_p}]|(E_0, s, \Lambda)\rangle \tag{12.149}$$

3 The manifold $M^{111} \sim \frac{SU(3)\times SU(2)\times U(1)}{SU(2)\times U(1)\times U(1)}$ is a particular Sasakian coset manifold discovered by the authors of [19] within a class originally proposed by Witten in [142]. M^{111} used as compactification seven manifold in M-theory is the only one which has the isometry SU(3) × SU(2) × U(1) of the standard model and preserves some supersymmetry, to be precise $\mathcal{N} = 2$.

Lowest weight representations are similarly constructed with respect to the five–grading (12.142). One starts from a vacuum state that is annihilated by the conformal boosts and by the special conformal supersymmetries

$$\begin{aligned} K_m|(D_0, j, \Lambda)\rangle &= 0 \\ s^i_\alpha|(D_0, j, \Lambda)\rangle &= 0 \end{aligned} \tag{12.150}$$

and that is an eigenstate of the dilatation operator D and an irreducible SO(1, 2) representation of spin j:

$$\begin{aligned} D|(D_0, j, \Lambda)\rangle &= D_0|(D_0, j, \Lambda)\rangle \\ J^m J^n \eta_{mn}|(D_0, j, \Lambda)\rangle &= j(j+1)|(D_0, j, \Lambda)\rangle \\ J_2|(D_0, j, \Lambda)\rangle &= j|(D_0, j, \Lambda)\rangle \end{aligned} \tag{12.151}$$

As for the SO($\mathcal{N}$) representation, the new vacuum is the same as before. The states filling the UIR are now constructed by applying to the vacuum the operators P_m and the anti-symmetrized products of $q^{\alpha i}$,

$$(P_0)^{p_0}(P_1)^{p_1}(P_2)^{p_2}[q^{\alpha_1 i_1} \dots q^{\alpha_q i_q}]|(D_0, j, \Lambda)\rangle \tag{12.152}$$

In the language of conformal field theories the vacuum state satisfying eq. (12.150) is named a *primary state* (corresponding to the value at $z^m = 0$ of a primary conformal field). The states (12.152) are called the *descendants*.

The rotation between the SO(3) × SO(2) basis and the SO(1, 2) × SO(1, 1) basis is performed by the operator:

$$U \equiv \exp\left[\frac{i}{\sqrt{2}}\pi(E - D)\right] \tag{12.153}$$

which has the following properties,

$$\begin{aligned} DU &= -UE \\ J_0 U &= iUL_3 \\ J_1 U &= UL_1 \\ J_2 U &= UL_2 \end{aligned} \tag{12.154}$$

with respect to the grade 0 generators. Furthermore, with respect to the non-vanishing grade generators we have:

$$\begin{aligned} K_0 U &= -iUM_3^- \\ K_1 U &= -UM_1^- \\ K_2 U &= -UM_2^- \\ P_0 U &= iUM_3^+ \end{aligned}$$

$$\begin{aligned} P_1 U &= U M_1^+ \\ P_2 U &= U M_2^+ \\ q^{\alpha i} U &= -iU\overline{a}^{\alpha i} \\ s_\alpha^i U &= iU a_\alpha^i \end{aligned} \tag{12.155}$$

As one immediately sees from (12.155), U interchanges the compact five-grading structure of the superalgebra with its non-compact one. In particular the SO(3) × SO(2)-vacuum with energy E_0 is mapped onto an SO(1,2) × SO(1,1) primary state and one obtains all the descendants (12.152) by acting with U on the particle states (12.149). Furthermore, from (12.154) we read the conformal weight and the Lorentz group representation of the primary state $U|(E_0,s,J)\rangle$. Indeed, its eigenvalue with respect to the dilatation generator D is:

$$D_0 = -E_0 \tag{12.156}$$

and we find the following relation between the Casimir operators of SO(1,2) and SO(3):

$$J^2 U = U L^2, \quad J^2 \equiv -J_0^2 + J_1^2 + J_2^2 \tag{12.157}$$

which implies that

$$j = s \tag{12.158}$$

Hence under the action of U a particle state of energy E_0 and spin-s of the bulk is mapped onto a *primary conformal field* of conformal weight $-E_0$ and Lorentz spin-s on the boundary. This discussion is visualized in Figure 12.8.

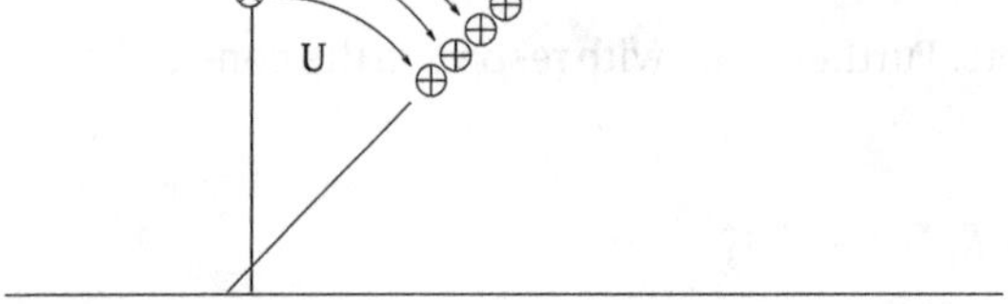

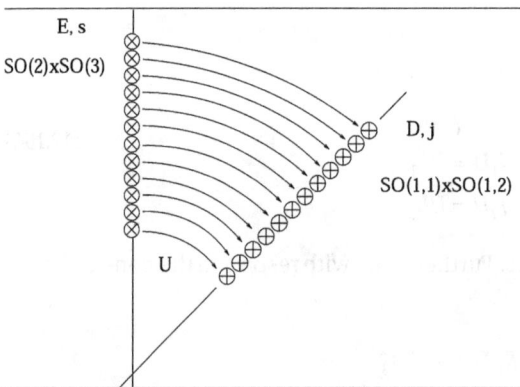

Figure 12.8: The operator $U = \exp\{i\pi/\sqrt{2}(E - D)\}$ rotates the Hilbert space of the physical states. It takes states labeled by the Casimirs (E,s) of the SO(2) × SO(3) ⊂ Osp($\mathcal{N}$|4) into states labeled by the Casimirs (D,j) of SO(1,1) × SO(1,2).

12.4.4 Structure of the Osp(2|4) multiplets

Utilizing the approach outlined in the previous section, the relevant $\mathcal{N} = 2$ supermultiplets in AdS_4 superspace were derived in [41] and their structure is summarized in Tables 12.2–12.10 whose content we discuss in the present section. To describe the organization of such tables we recall the basic fact that in this case the even subalgebra of the superalgebra is the following one:

$$G_{even} = Sp(4,\mathbb{R}) \oplus SO(2) \subset Osp(2|4) \tag{12.159}$$

where $Sp(4,\mathbb{R}) \sim SO(2,3)$ is the isometry algebra of AdS_4 while the compact subalgebra SO(2) generates R-symmetry. The maximally compact subalgebra of G_{even} is

$$G_{compact} = SO(2)_E \oplus SO(3)_S \oplus SO(2)_R \subset G_{even} \tag{12.160}$$

As we explained in the previous subsection, the generator of $SO(2)_E$ is interpreted as the Hamiltonian of the system when Osp(2|4) acts as the isometry group of anti de Sitter superspace. This was discussed extensively in Section 12.4.2 where we identified the UIR of $SO(2,3)$ with the particle-states propagating in anti de Sitter space. Consequently also in the supersymmetric setup the eigenvalues E are the energy levels of possible states for the system. The group $SO(3)_S$ is the ordinary rotation group and similarly its representation labels s describe the possible spin states of the system as we illustrated above. Finally the eigenvalue y of the generator of $SO(2)_R$ is the hypercharge of a state.

A supermultiplet, namely a UIR of the superalgebra Osp(2|4), is composed of a *finite* number of UIRs of the even subalgebra G_{even} (12.159), each of them being, as we just recalled, what in physical language we call a particle state, characterized by a spin "s," a mass "m" and a hypercharge "y." As we explained above at length, from a mathematical viewpoint each UIR representation of the non-compact even subalgebra G_{even} is an *infinite* tower of finite dimensional UIR representations of the compact subalgebra $G_{compact}$ (12.160). The lowest lying representation of such a tower $|E,s,y\rangle$ is the Clifford vacuum. The mass, spin and the hypercharge of the corresponding particle are read from the labels of the Clifford vacuum by use of the relations between mass and energy that we have recalled in eq. (12.115).

In the same way there is a Clifford vacuum $|E_0,s_0,y_0\rangle$ for the entire supermultiplet out of which we not only construct the corresponding particle state but also, through the action of the SUSY charges, we construct the Clifford vacua $|E_0+\cdots,s_0+\cdots,y_0+\dots\rangle$ of the other members of the same supermultiplet. Hence the structure of a supermultiplet is conveniently described by listing the energy E, the spin s and the hypercharge y of all the Clifford vacua of the multiplet.

In Tables 12.2–12.10 we provide such information.

Given these preliminaries let us discuss our result for the structure of the Osp(2|4) supermultiplets with $s \leq 2$.

In $\mathcal{N}$-extended AdS_4 superspace there are three kinds of supermultiplets:
- the long multiplets
- the short multiplets
- the massless multiplets

For $\mathcal{N} = 2$ the long multiplets satisfy the following unitarity relation, without saturation:

$$E_0 > |y_0| + s_0 + 1 \tag{12.161}$$

Furthermore, we distinguish three kinds of long multiplets depending on the highest spin state they contain: $s_{max} = 2, \frac{3}{2}, 1$. These multiplets are respectively named *long graviton, long gravitino* and *long vector multiplets.*

The *long graviton multiplet*, satisfying $E_0 > |y_0| + 2$, has the structure displayed in Table 12.2.

The *long gravitino multiplet*, satisfying $E_0 > |y_0| + \frac{3}{2}$, has the structure displayed in Table 12.3.

The *long vector multiplet*, satisfying $E_0 > |y_0| + 1$, has the structure displayed in Table 12.4.

The short multiplets are of two kinds: the *short graviton, gravitino* and *vector multiplet*, that saturate the bound

$$E_0 = |y_0| + s_0 + 1 \tag{12.162}$$

Table 12.2: $\mathcal{N} = 2$ long graviton multiplet.

Spin	Energy	Hypercharge	Mass²
2	$E_0 + 1$	y_0	$16(E_0 + 1)(E_0 - 2)$
$\frac{3}{2}$	$E_0 + \frac{3}{2}$	$y_0 - 1$	$-4E_0 - 4$
$\frac{3}{2}$	$E_0 + \frac{3}{2}$	$y_0 + 1$	$-4E_0 - 4$
$\frac{3}{2}$	$E_0 + \frac{1}{2}$	$y_0 - 1$	$4E_0 - 8$
$\frac{3}{2}$	$E_0 + \frac{1}{2}$	$y_0 + 1$	$4E_0 - 8$
1	$E_0 + 2$	y_0	$16E_0(E_0 + 1)$
1	$E_0 + 1$	$y_0 - 2$	$16E_0(E_0 - 1)$
1	$E_0 + 1$	$y_0 + 2$	$16E_0(E_0 - 1)$
1	$E_0 + 1$	y_0	$16E_0(E_0 - 1)$
1	$E_0 + 1$	y_0	$16E_0(E_0 - 1)$
1	E_0	y_0	$16(E_0 - 1)(E_0 - 2)$
$\frac{1}{2}$	$E_0 + \frac{3}{2}$	$y_0 - 1$	$4E_0$
$\frac{1}{2}$	$E_0 + \frac{3}{2}$	$y_0 + 1$	$4E_0$
$\frac{1}{2}$	$E_0 + \frac{1}{2}$	$y_0 - 1$	$-4E_0 + 4$
$\frac{1}{2}$	$E_0 + \frac{1}{2}$	$y_0 + 1$	$-4E_0 + 4$
0	$E_0 + 1$	y_0	$16E_0(E_0 - 1)$

Table 12.3: $\mathcal{N} = 2$ long gravitino multiplet.

Spin	Energy	Hypercharge	Mass²
$\frac{3}{2}$	$E_0 + 1$	y_0	$4E_0 - 6$
1	$E_0 + \frac{3}{2}$	$y_0 - 1$	$16(E_0 - \frac{1}{2})(E_0 + \frac{1}{2})$
1	$E_0 + \frac{3}{2}$	$y_0 + 1$	$16(E_0 - \frac{1}{2})(E_0 + \frac{1}{2})$
1	$E_0 + \frac{1}{2}$	$y_0 - 1$	$16(E_0 - \frac{3}{2})(E_0 - \frac{1}{2})$
1	$E_0 + \frac{1}{2}$	$y_0 + 1$	$16(E_0 - \frac{3}{2})(E_0 - \frac{1}{2})$
$\frac{1}{2}$	$E_0 + 2$	y_0	$4E_0 + 2$
$\frac{1}{2}$	$E_0 + 1$	$y_0 - 2$	$-4E_0 + 2$
$\frac{1}{2}$	$E_0 + 1$	y_0	$-4E_0 + 2$
$\frac{1}{2}$	$E_0 + 1$	$y_0 + 2$	$-4E_0 + 2$
$\frac{1}{2}$	$E_0 + 1$	y_0	$-4E_0 + 2$
$\frac{1}{2}$	E_0	y_0	$4E_0 - 6$
0	$E_0 + \frac{3}{2}$	$y_0 - 1$	$16(E_0 - \frac{1}{2})(E_0 + \frac{1}{2})$
0	$E_0 + \frac{3}{2}$	$y_0 + 1$	$16(E_0 - \frac{1}{2})(E_0 + \frac{1}{2})$
0	$E_0 + \frac{1}{2}$	$y_0 - 1$	$16(E_0 - \frac{3}{2})(E_0 - \frac{1}{2})$
0	$E_0 + \frac{1}{2}$	$y_0 + 1$	$16(E_0 - \frac{3}{2})(E_0 - \frac{1}{2})$

Table 12.4: $\mathcal{N} = 2$ long vector multiplet.

Spin	Energy	Hypercharge	Mass²
1	$E_0 + 1$	y_0	$16E_0(E_0 - 1)$
$\frac{1}{2}$	$E_0 + \frac{3}{2}$	$y_0 - 1$	$-4E_0$
$\frac{1}{2}$	$E_0 + \frac{3}{2}$	$y_0 + 1$	$-4E_0$
$\frac{1}{2}$	$E_0 + \frac{1}{2}$	$y_0 - 1$	$4E_0 - 4$
$\frac{1}{2}$	$E_0 + \frac{1}{2}$	$y_0 + 1$	$4E_0 - 4$
0	$E_0 + 2$	y_0	$16E_0(E_0 + 1)$
0	$E_0 + 1$	$y_0 - 2$	$16E_0(E_0 - 1)$
0	$E_0 + 1$	$y_0 + 2$	$16E_0(E_0 - 1)$
0	$E_0 + 1$	y_0	$16E_0(E_0 - 1)$
0	E_0	y_0	$16(E_0 - 2)(E_0 - 1)$

and the *hypermultiplets* (spin-$\frac{1}{2}$ multiplets), that saturate the other bound

$$E_0 = |y_0| \text{ with } |y_0| \geq \frac{1}{2} \tag{12.163}$$

The *short graviton multiplet*, satisfying $E_0 = |y_0| + 2$, has the structure displayed in Table 12.5.

The *short gravitino multiplet*, satisfying $E_0 = |y_0| + \frac{3}{2}$, has the structure displayed in Table 12.6.

Table 12.5: $\mathcal{N} = 2$ short graviton multiplet with positive hypercharge $y_0 > 0$.

Spin	Energy	Hypercharge	Mass²
2	$y_0 + 3$	y_0	$16y_0(y_0 + 3)$
$\frac{3}{2}$	$y_0 + \frac{7}{2}$	$y_0 - 1$	$-4y_0 - 12$
$\frac{3}{2}$	$y_0 + \frac{5}{2}$	$y_0 + 1$	$4y_0$
$\frac{3}{2}$	$y_0 + \frac{5}{2}$	$y_0 - 1$	$4y_0$
1	$y_0 + 3$	$y_0 - 2$	$16(y_0 + 2)(y_0 + 1)$
1	$y_0 + 3$	y_0	$16(y_0 + 2)(y_0 + 1)$
1	$y_0 + 2$	y_0	$16y_0(y_0 + 1)$
$\frac{1}{2}$	$y_0 + \frac{5}{2}$	$y_0 - 1$	$-4y_0 - 4$

Table 12.6: $\mathcal{N} = 2$ short gravitino multiplet with positive hypercharge $y_0 > 0$.

Spin	Energy	Hypercharge	Mass²
$\frac{3}{2}$	$y_0 + \frac{5}{2}$	y_0	$4y_0$
1	$y_0 + 3$	$y_0 - 1$	$16(y_0 + 1)(y_0 + 2)$
1	$y_0 + 2$	$y_0 + 1$	$16y_0(y_0 + 1)$
1	$y_0 + 2$	$y_0 - 1$	$16y_0(y_0 + 1)$
$\frac{1}{2}$	$y_0 + \frac{5}{2}$	y_0	$-4y_0 - 4$
$\frac{1}{2}$	$y_0 + \frac{5}{2}$	$y_0 - 2$	$-4y_0 - 4$
$\frac{1}{2}$	$y_0 + \frac{3}{2}$	y_0	$4y_0$
0	$y_0 + 3$	$y_0 \pm 1$	$16(y_0 + 1)(y_0 + 2)$

Table 12.7: $\mathcal{N} = 2$ short vector multiplet with positive hypercharge $y_0 > 0$.

Spin	Energy	Hypercharge	Mass²
1	$y_0 + 2$	y_0	$16y_0(y_0 + 1)$
$\frac{1}{2}$	$y_0 + \frac{5}{2}$	$y_0 \pm 1$	$-4y_0 - 4$
$\frac{1}{2}$	$y_0 + \frac{3}{2}$	$y_0 + 1$	$4y_0$
$\frac{1}{2}$	$y_0 + \frac{3}{2}$	$y_0 - 1$	$4y_0$
0	$y_0 + 2$	$y_0 - 2$	$16y_0(y_0 + 1)$
0	$y_0 + 2$	y_0	$16y_0(y_0 + 1)$
0	$y_0 + 1$	y_0	$16y_0(y_0 - 1)$

The *short vector multiplet*, satisfying $E_0 = |y_0| + 1$, has the structure displayed in Table 12.7.

We must stress that the multiplets displayed in Tables 12.5, 12.6, 12.7 are only half of the story, since they can be viewed as the BPS states where $E_0 = y_0 + s_0 + 1$ and $y_0 > 0$. In addition one has also the anti BPS states. These are the short multiplets

Table 12.8: $\mathcal{N} = 2$ hypermultiplet, $y_0 > 0$.

Spin	Energy	Hypercharge	Mass²
$\frac{1}{2}$	$y_0 + \frac{1}{2}$	$y_0 - 1$	$4y_0 - 4$
0	$y_0 + 1$	$y_0 - 2$	$16y_0(y_0 - 1)$
0	y_0	y_0	$16(y_0 - 2)(y_0 - 1)$
$\frac{1}{2}$	$y_0 + \frac{1}{2}$	$-y_0 + 1$	$4y_0 - 4$
0	$y_0 + 1$	$-y_0 + 2$	$16y_0(y_0 - 1)$
0	y_0	$-y_0$	$16(y_0 - 1)(y_0 - 2)$

Table 12.9: $\mathcal{N} = 2$ massless graviton multiplet.

Spin	Energy	Hypercharge	Mass²
2	3	0	0
$\frac{3}{2}$	$\frac{5}{2}$	−1	0
$\frac{3}{2}$	$\frac{5}{2}$	+1	0
1	2	0	0

Table 12.10: $\mathcal{N} = 2$ massless vector multiplet.

Spin	Energy	Hypercharge	Mass²
1	2	0	0
$\frac{1}{2}$	$\frac{3}{2}$	−1	0
$\frac{1}{2}$	$\frac{3}{2}$	+1	0
0	2	0	0
0	1	0	

where $E_0 = -y_0 + s_0 + 1$ with $y_0 < 0$. The structure of these anti short multiplets can be easily read off from Tables 12.5, 12.6, 12.7 by reversing the signs of all hypercharges.

The *hypermultiplet*, satisfying $E_0 = |y_0| \geq \frac{1}{2}$, has the structure displayed in Table 12.8. This structure is different from the others because this multiplet is complex. This means that for each field there is another field with same energy and spin but opposite hypercharge. So it is built with two $\mathcal{N} = 1$ Wess–Zumino multiplets. The four real scalar fields can be arranged into a quaternionic complex form.

The massless multiplets are either short graviton or short vector multiplets satisfying the further condition

$$E_0 = s_0 + 1 \quad \text{equivalent to } y_0 = 0 \tag{12.164}$$

The *massless graviton multiplet*, satisfying $E_0 = 2y_0 = 0$, has the structure displayed in Table 12.9.

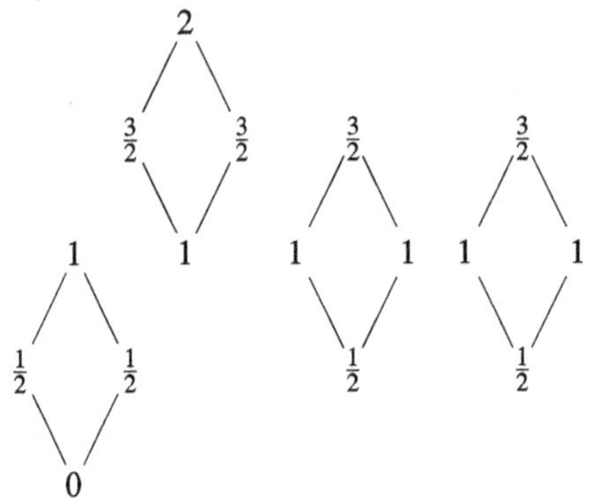

Figure 12.9: $\mathcal{N} = 2 \to \mathcal{N} = 1$ decomposition of the long graviton multiplet.

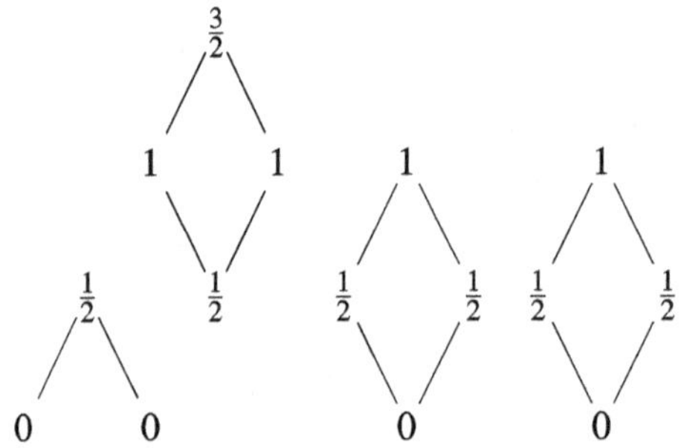

Figure 12.10: $\mathcal{N} = 2 \to \mathcal{N} = 1$ decomposition of the long gravitino multiplet.

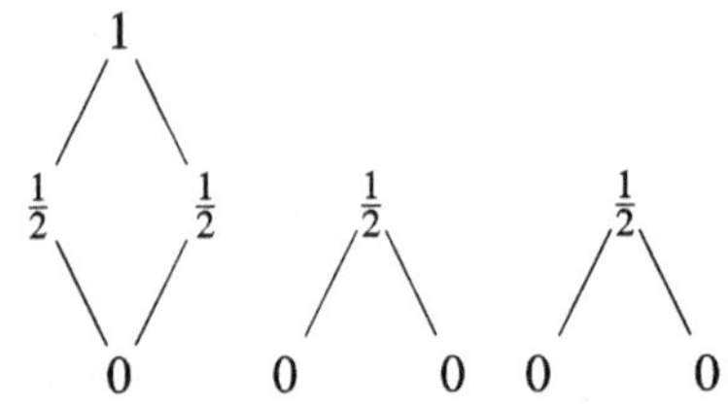

Figure 12.11: $\mathcal{N} = 2 \to \mathcal{N} = 1$ decomposition of the long vector multiplet.

The *massless vector multiplet*, satisfying $E_0 = 1y_0 = 0$, has the structure displayed in Table 12.10. These are the $\mathcal{N} = 2$ supermultiplets in anti de Sitter space that can occur in Kaluza–Klein supergravity.

The structure of the long multiplets was derived in the eighties (see [25]), whereas the structure of the short and massless multiplets we have given here was derived in [41]. In establishing this result the authors of [41] used as a tool the necessary decomposition of the $\mathcal{N} = 2$ multiplets into $\mathcal{N} = 1$ multiplets (see Figures 12.9–12.15) for long and short multiplets. Finally we note that the structure of massless multiplets is identical in the anti de Sitter and in the Poincaré case, as is well known.

The reinterpretation of these UIRs as primary conformal fields together with their descendants was described in Section 12.4.3.2 and we do not return on that point.

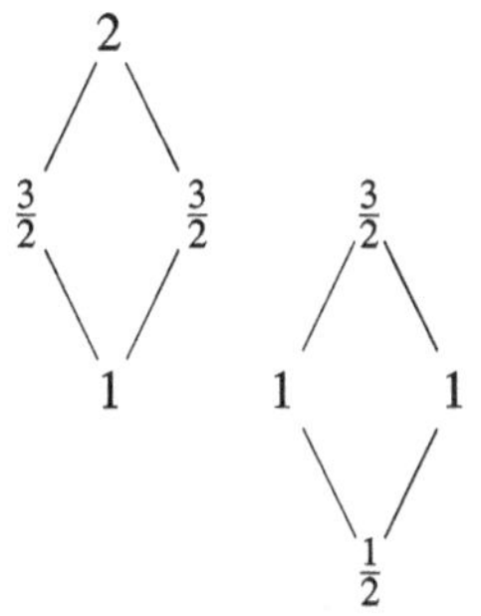

Figure 12.12: $\mathcal{N} = 2 \to \mathcal{N} = 1$ decomposition of the short graviton multiplet.

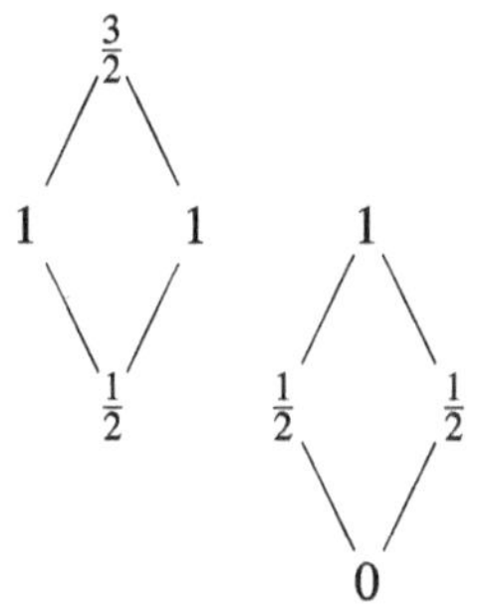

Figure 12.13: $\mathcal{N} = 2 \to \mathcal{N} = 1$ decomposition of the short gravitino multiplet.

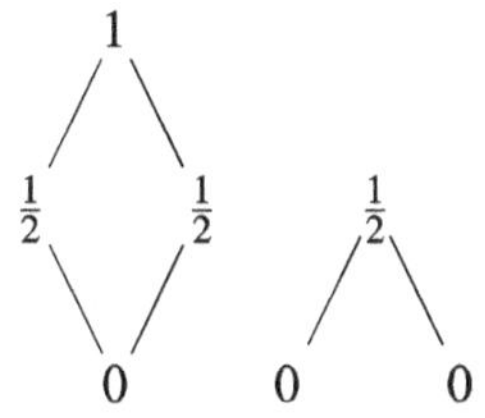

Figure 12.14: $\mathcal{N} = 2 \to \mathcal{N} = 1$ decomposition of the short vector multiplet.

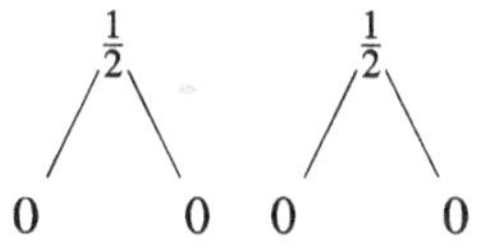

Figure 12.15: $\mathcal{N} = 2 \to \mathcal{N} = 1$ decomposition of the hypermultiplet.

12.5 Bibliographical note

Several references have already been quoted in the course of the various sections of the present chapter. We just stress that for Section 12.2 the main source is [93], while for the first subsections of Section 12.4 the main source is [22].

A Exercises

Learning group theory requires from the part of the students to perform many explicit constructions and work out several explicit examples. Furthermore, a broad understanding is promoted when the theory is seen, so to say, in action applied to the solution of interesting problems of relevance to theoretical physics, chemistry and pure mathematics. Exercises and examples are therefore an integral part of the tuition process and correspondingly of a textbook like the present one.

As for the two main separate, yet strongly connected, topics pursued in this book, namely,

A) Finite Group Theory

B) Lie Group Theory and Manifolds

we have followed slightly different practices.

A) Relative to discrete Group Theory we present in the current section a collection of interesting and inspiring problems whose solutions are separately provided. The exercises listed below are taken from various sources and have been worked out in full in the class of master and Ph.D. students who took this course in the academic year 2016–2017. The exercises are mostly not elementary (even for a student educated in mathematics, not to say in physics). They were selected in order to give a taste and a feeling of some of the most important properties and of the typical applications of finite groups. Let us mention especially the original lecture notes [147] and the web forum [148], both of which were extensively used to prepare the solution part.

B) As for Lie algebras, Lie groups, manifolds, metrics and connections we decided to rely more on the working out of explicit examples that are dispersed through the main text. In particular let us mention:

1. The explicit construction of the root systems of classical Lie algebras in Section 7.5.
2. The explicitly worked out example of selected representations of the A_2 Lie algebra in Section 8.2.2.
3. The explicitly worked out example of selected representations of the $\mathfrak{b}_2 \sim \mathfrak{c}_2$ Lie algebra in Section 8.2.3.
4. The entire Chapter 9 devoted to the explicit construction of four out of the five exceptional Lie algebras, namely, $\mathfrak{g}_2, \mathfrak{f}_4, \mathfrak{e}_7, \mathfrak{e}_8$.
5. The explicit construction of the Hopf fibration and the discussion of the magnetic monopole in Section 10.4.1 which provides an inspiring exemplification of the notions of fiber-bundles and of connections.
6. The three examples of explicit construction of geodesics for both Lorentzian and Euclidean manifolds presented in Sections 10.8.1 (2-dimensional anti de Sitter space dS_2), 10.8.2 (Lobachevsky–Poincaré plane), 10.8.3 (catenoid).

https://doi.org/10.1515/9783110551204-013

7. The construction of Unitary Irreducible Representations of the anti de Sitter group SO(2, 3) contained in Section 12.4.2.

In addition, as we emphasized in the Introduction, the practical computational skills of the students can be developed and self-tested by the use of the special dedicated MATHEMATICA NoteBooks that are distributed in association with this volume and are described in detail in Appendix B.

A.1 Exercises for Chapter 1

Exercise 1.1. Show that in Definition 1.2.1 the axioms 2–3 can be formally weaken to the requirement of existence of *only right* (or rather *only left*) identity and inverse:
2′ $\exists e_R \in G$ such that $a \cdot e_R = a \; \forall a \in G$;
3′ $\forall a \in G \; \exists b \in G$ (called a_R^{-1}), such that $a \cdot b = e$.

Exercise 1.2. Prove that in any group G: a) the identity element e is unique; b) the inverse g^{-1} of any element is unique; c) for any element $g \in G$, $g^2 = g$ entails $g = e$.

Exercise 1.3. Show that a group G of even order contains an odd number of elements of order 2 (in particular, at least one).

Exercise 1.4. For cyclic groups $C_n = \langle g | g^n = e \rangle$ of the orders $n = 2, 3, 4, 6$ write down the multiplication tables. Looking at the tables, list all the proper subgroups.

Exercise 1.5. Give an estimation on the upper bound of the number $N_{\text{groups}}(n)$ of finite groups of order n (hint: just enumerate a variety of multiplication tables disregarding all the restrictions imposed by the other group axioms).

Exercise 1.6. Write down the multiplication tables and list the subgroups for the groups $S_3 = \langle s, t | s^2 = t^2 = e, (st)^3 = e \rangle$ and $S_3' = \langle r, i | r^3 = i^2 = e, iri^{-1} = r^{-1} \rangle$. Show that they are in fact *isomorphic* and relate both sets of generators explicitly.

Exercise 1.7. Write down the multiplication tables and list the subgroups for the groups $S_3 = \langle r, i | r^3 = i^2 = e, ir = r^{-1} i \rangle$ and $C_3 \otimes C_2 = \langle r, i | r^3 = i^2 = e, ir = ri \rangle$. Show that they are *non-isomorphic*. Observe and summarize similarities and differences.

Exercise 1.8. Does a set $\mathbb{R}[x]$ of real valued polynomials in one variable x, with the usual definition of sum $(P + Q)(x) = P(x) + Q(x)$ and multiplication $(PQ)(x) = P(x)Q(x)$, constitute a ring? What if multiplication in the ring is taken to be a composition $(P \circ Q)(x) = P(Q(x))$?

Exercise 1.9. Prove that the number q of elements of any finite field $\mathbb{F}_q$ can be only a power $q = p^n$ of a prime.

Exercise 1.10. Construct explicitly the finite fields (i.e., derive the addition, subtraction, multiplication and division tables): a) $\mathbb{F}_2$; b) $\mathbb{F}_3$; c) $\mathbb{F}_4$.

A.2 Exercises for Chapter 2

Exercise 2.1. Calculate the orders of the finite linear groups: a) $\mathrm{GL}_n(\mathbb{F}_q)$; b) $\mathrm{SL}_n(\mathbb{F}_q)$; c) $\mathrm{PSL}_n(\mathbb{F}_q) \equiv \mathrm{SL}_n(\mathbb{F}_q)/Z(\mathrm{SL}_n(\mathbb{F}_q))$ over a finite field $\mathbb{F}_{q=p^k}$.

Exercise 2.2. For

$$\sigma = \begin{pmatrix} 1 & 2 & 3 & 4 & 5 & 6 \\ 6 & 4 & 3 & 2 & 1 & 5 \end{pmatrix} \quad \text{and} \quad \tau = \begin{pmatrix} 1 & 2 & 3 & 4 & 5 & 6 \\ 3 & 2 & 1 & 4 & 5 & 6 \end{pmatrix}$$

compute: a) σ^{-1}; b) τ^{-1}; c) σ^2; d) τ^2; e) $\sigma \circ \tau$; f) $\tau \circ \sigma$.

Exercise 2.3. Expand the permutations into products of independent cycles (draw them graphically if needed):

a) $\sigma = \begin{pmatrix} 1 & 2 & 3 & 4 & 5 \\ 4 & 3 & 2 & 5 & 1 \end{pmatrix}$;

b) $\tau = \begin{pmatrix} 1 & 2 & 3 & 4 & 5 & 6 \\ 5 & 6 & 1 & 4 & 3 & 2 \end{pmatrix}$;

c) $\rho = \begin{pmatrix} 1 & 2 & 3 & 4 & 5 & 6 & 7 & 8 & 9 \\ 7 & 9 & 3 & 1 & 5 & 8 & 4 & 2 & 6 \end{pmatrix}$.

Exercise 2.4. Convince yourself with the rule $g \circ (i_1 i_2 \dots i_r) \circ g^{-1} = (g(i_1)g(i_2) \dots g(i_r))$ and using it try to quickly compute orally without writing: a) $(12) \circ (123) \circ (12)$; b) $(123) \circ (12) \circ (132)$; c) $(1234) \circ (12)(34) \circ (4321)$.

Exercise 2.5. Show that the group S_n for $n \geq 3$ is generated by: a) $n-1$ transpositions $(1i)$ $(2 \leq i \leq n)$; b) $n-1$ transpositions $(i\,i+1)$ $(1 \leq i \leq n-1)$; c) two cycles $\tau = (12)$ and $\sigma = (12\dots n)$.

Exercise 2.6. Show that the group A_n for $n \geq 4$ is generated by: a) $n(n-1)(n-2)/3$ 3-cycles (ijk); b) $n-2$ consecutive 3-cycles $(i\,i+1\,i+2)$ $(1 \leq i \leq n-2)$; c) two cycles: $\tau = (123)$ and $\sigma = (12\dots n)$ for odd n, or $\tau = (123)$ and $\sigma' = (23\dots n)$ for even n.

Exercise 2.7. Find the orders of the elements

$$A = \begin{pmatrix} 0 & -1 \\ 1 & 0 \end{pmatrix}, \quad B = \begin{pmatrix} 0 & 1 \\ -1 & -1 \end{pmatrix}$$

of $\mathrm{SL}_2(\mathbb{Z})$. What is the order of the element $AB \in \mathrm{SL}_2(\mathbb{Z})$?

Exercise 2.8. Discuss how the general properties of an equivalence relation: $F' \sim F$ iff $F' = g(F)$ for some transformation $g \in G$ (reflexivity: $F \sim F$, symmetry: $F_1 \sim F_2 \Leftrightarrow F_2 \sim F_1$, and transitivity: $F_1 \sim F_2$ and $F_2 \sim F_3 \Longrightarrow F_1 \sim F_3$) are in correspondence with the group axioms 1–3 for G.

Exercise 2.9. Show that for any invertible transformation g:
a) if r_l is a rotation around the axis l, then $r_{g(l)} = g r_l g^{-1}$ is a rotation around the axis $g(l)$;
b) if i_Π is a reflection in the plane Π, then $i_{g(\Pi)} = g i_\Pi g^{-1}$ is a reflection in the plane $g(\Pi)$.

Exercise 2.10. Consider the finite cyclic group $C_n = \langle g | g^n = e \rangle$: a) show that within S_n cycle permutations of length n form a subgroup isomorphic to C_n; b) show that the group of rotations around a fixed axis by angles $\vartheta_k = 2\pi k/n$ ($k = 1, 2, \ldots n-1$) is isomorphic to C_n. Give examples of figures invariant under the action of the groups C_n; c) determine the group of symmetries of a rectangular. Fill in the multiplication table. Show that it is *not isomorphic* to any C_n; d) describe symmetry group of the polynomial $P(x_1, x_2, x_3, x_4) = x_1 x_2 + x_3 x_4$.

Exercise 2.11. Derive the symmetry group Dih_3 of an equilateral triangle: represent each element by 2×2 matrix and show that $\mathrm{Dih}_3 \approx S_3$.

Exercise 2.12. By considering and generalizing the particular cases $n = 3$ (equi lateral *triangle*), $n = 4$ (*square*), $n = 5$ (equilateral *pentagon* or star), $n = 6$ (equilateral *hexagon*), derive a presentation of a general *dihedral group* Dih_n (symmetry group of a regular n-sided polygon).

Exercise 2.13. Discuss the symmetry group $\mathrm{T_d}$ of a *tetrahedron* (describe its action on vertices, edges and faces and show that $\mathrm{T_d} \approx S_4$).

Exercise 2.14. Discuss the symmetry group O_d of a *cube* (describe its action on vertices, edges and faces and show that it is the same as the symmetry group of an *octahedron*). Show that $\mathrm{O}_d \approx S_4 \otimes C_2$, so that $|\mathrm{O}_d| = 48$.

A.3 Exercises for Chapter 3

Exercise 3.1. One way to visualize multiplication table of a finite group is the Cayley graph. It encodes each of the group elements by a vertex whereas its left shift by each of the group generators is represented by an arrow connecting the vertices before and after the shift (the arrows corresponding to different generators are colored differently). Draw and compare the Cayley graphs for the groups presented in Exs 1.6 and 1.7.

Exercise 3.2. List right and left cosets in a group $G = S_3$ of a) $H = A_3 \subset G$; b) $H = \langle(12)\rangle \subset G$.

Exercise 3.3. List right and left cosets in a group $G = C_6 = \langle\zeta|\zeta^6 = 1\rangle$ of
a) $H = \langle\zeta^3\rangle \approx C_2 \subset G$;
b) $H = \langle\zeta^2\rangle \approx C_3 \subset G$.

Exercise 3.4. Prove that a subgroup $H \subset G$, such that $|G : H| = 2$, is normal and that in such a case for each $g \in G$ one has $g^2 \in H$. Provide examples.

Exercise 3.5. Prove that for abelian group each element represents a separate conjugacy class and that each subgroup is normal. For $G = C_6$ describe the quotient groups C_6/C_2 and C_6/C_3.

Exercise 3.6. While slicing an arbitrary group into conjugacy classes, prove that: a) identity e constitutes a separate class; b) all elements of the same class necessarily have the same order; c) their inverses either belong to the same class or form a separate class.

Exercise 3.7. List the conjugacy classes of the symmetric group: a) S_3; b) S_4; c) S_5.

Exercise 3.8. Derive (with exponential accuracy only) an asymptotic formula for the number of conjugacy classes of S_n for $n \to \infty$ and compare it with the size of S_n.

Exercise 3.9. Show that a conjugacy class of S_n consisting of even permutations is either a conjugacy class of A_n (iff there exists an odd permutation commuting with a representative of that class), or else it splits, within A_n, into exactly two classes of equal size. Show that splitting occurs if the cycle decomposition of the elements of the class comprises cycles of distinct odd lengths, and does not occur if the cycle decomposition of the elements of the class contains either a cycle of even length, or two cycles of equal odd lengths. Relying on that, list the number of permutations in each conjugacy class of a) A_4; b) A_5.

Exercise 3.10. List all the conjugacy classes of the groups Dih_n (hints: for convenience start first with $n = 3$ and $n = 4$; next consider the cases $n =$ even and $n =$ odd separately).

Exercise 3.11. Identify all normal subgroups $H \lhd G$ in the groups G: a) Dih_3; b) Dih_4; c) A_4; d) S_4, and for each the corresponding quotient group G/H.

Exercise 3.12. Show that the group A_5 is *simple*, i.e., contains no normal subgroups (hint: combine the results of Ex. 3.9 with Lagrange theorem). Think of proving the same result for general A_n, $n \geq 5$.

Exercise 3.13. Classify all homomorphisms of the group S_3. In each case identify the kernel and the image subgroups.

Exercise 3.14. By choosing an appropriate homomorphism and using the homomorphism theorem $G/\ker f \approx \operatorname{im} f$, find the quotient groups: a) $\mathbb{Z}/n\mathbb{Z}$; b) $\mathbb{Z}_4/\mathbb{Z}_2$; c) $\mathbb{Z}_6/\mathbb{Z}_2$; d) $\mathbb{Z}_6/\mathbb{Z}_3$; e) A_4/N_4; f) S_n/A_n; g) Dih_n/C_n.

Exercise 3.15. How many homomorphisms $C_n \to C_m$ are there? Show that $C_{pq}/C_p \approx C_q$.

Exercise 3.16. Show that if $\gcd(p,q) = 1$ then $C_{pq} \approx C_p \otimes C_q$.

Exercise 3.17. Show that the three groups C_8, $C_4 \otimes C_2$, $C_2 \otimes C_2 \otimes C_2$ of the same order $2^3 = 8$ are mutually non-isomorphic.

Exercise 3.18. Prove that a finite abelian group is a direct product of cyclic groups.

Exercise 3.19. Show that a) $A_4 \approx \mathrm{N}_4 \rtimes C_3$; b) $\mathrm{Dih}_n \approx C_n \rtimes C_2$ $(n \geq 2)$; c) $S_n \approx A_n \rtimes C_2$ $(n \geq 3)$.

Exercise 3.20. Classify all semidirect products $C_8 \rtimes C_2$.

Exercise 3.21. For the quaternion group $Q_8 = \langle -1, i, j, k | (-1)^2 = 1, i^2 = j^2 = k^2 = ijk = -1 \rangle$ find all the subgroups, center $Z(Q_8)$ and conjugacy classes. Show that all the subgroups are normal (despite that Q_8 is non-abelian). Find the quotient $Q_8/Z(Q_8)$ and show that it is not isomorphic to any subgroup of Q_8. Show that Q_8 is not a semidirect product.

Exercise 3.22. Find the successive commutator subgroups $(k = 1, 2, \ldots)$: a) $S_3^{(k)}$; b) $S_4^{(k)}$; c) $\mathrm{Dih}_n^{(k)}$. In each case write the subnormal series.

Exercise 3.23. Show that the number of finite groups of order n is bounded by

$$N_{\text{groups}}(n) \leq n^{n \log_2(n)}$$

(compare to Ex. 1.5, hint: relying on the above provide an estimate of the number of group generators, then apply the Cayley's theorem).

Exercise 3.24. When considering the action of a finite group G as a transformation group on a set X one often uses the following notation: $Gx = \{gx \mid g \in G\}$ is called an *orbit* of $x \in X$; a subgroup $G_x = \{g \in G \mid gx = x\} \subset G$ is called *stability subgroup* of $x \in X$ (see Section 3.2.14). Prove:

a) X is foliated by orbits (meaning that each two orbits either coincide or never intersect);

b) the 'orbit-stabilizer theorem' $|Gx| = |G : G_x|$ (points of an orbit are enumerated by left cosets $G : G_x$);
c) the number of elements of X is expressed by

$$|X| = \sum_{\text{orbits}} |G : G_x| = \sum_{\text{orbits}} \frac{|G|}{|G_x|}$$

(in particular, derive that if the order of G is a power of a prime p, $|G| = p^n$, then $|X| \equiv |\text{fixed points}| \bmod p$);
d) the Burnside's lemma: the number $|X/G|$ of orbits $\{gx|g \in G\}$ can be counted as

$$|X/G| = \frac{1}{|G|} \sum_{g \in G} |X_g|$$

where $X_g = \{x \in X : gx = x\} \subset X$ is the set of *fixed points* for $g \in G$.

Exercise 3.25. Let $|G| = p^n$ for a prime number n. Show that: a) the center $Z(G)$ is non-trivial; b) G is solvable; c) if $n = 2$ then G is abelian.

Exercise 3.26. Show that in general the converse of Lagrange theorem is false (hint: consider A_4). Prove instead a very partial converse, known as Cauchy theorem: for any group G and for each prime factor p of $|G|$ there always exists an element $g \in G$ of order p (and hence, a cyclic subgroup of that order). To this effect:
a) consider the set $X = \{(g_1, g_2, \dots, g_p) : g_i \in G,\ g_1 \dots g_p = e\}$ and show that $|X| = |G|^{p-1}$;
b) observe that the group C_p acts on X by cyclic permutations, thus $|G|^{p-1} = |\text{fixed points}| \pmod p$;
c) identify the fixed points with either the identity or the elements $g \in G$ of order p and conclude that since p divides their total number, the number of such nontrivial elements g is nonzero.

Exercise 3.27. Using the method suggested in Ex. 3.26 and specifying $G = S_p$ (with p prime), prove the Wilson theorem $(p-1)! = -1 \pmod p$.

Exercise 3.28. Using the method suggested in Ex. 3.26 and specifying $G = C_m$ (this time assume that the prime p does not divide $|G| = m$), prove the *Fermat's little theorem* $m^{p-1} \equiv 1 \pmod p$.

Exercise 3.29. Classify all groups of order $|G| \leq 12$.

Exercise 3.30. Show that a finite group G is generated by a collection of representatives from all conjugacy classes.

Exercise 3.31. Consider the rotational symmetry group G of a regular polyhedron. Characterize its action on flags $\{V \in E \subset F\}$ and by attempting to assemble G from stabilizers derive both the Euler formula $|V| - |E| + |F| = 2$ and a full classification of platonic solids and their symmetry groups (hint: observe that each rotation except the identity leaves invariant *a pair* of elements among vertices, edges, and faces).

Exercise 3.32. List all finite groups with a) $n = 2$; b) $n = 3$; c) $n = 4$ conjugacy classes (hint: use the numerology provided by the class equation, then exclude irrelevant candidates).

Exercise 3.33. Find the number of ways to paint the vertices of a square using m colors if the squares differing by both rotations and reflections are considered equivalent.

Exercise 3.34. In how many ways can one paint the a) faces of a cube in three colors (red, blue and green); b) vertices of a cube in black and white colors, if in both cases the colorings that differ by rotations are not distinguished?

Exercise 3.35. (Di-, tri-,...)nitronaphthalene has a structure of a double benzene ring with $n = 1, 2, \dots$ hydrogen atoms replaced with a nitrite (NO_2) group, e.g.:

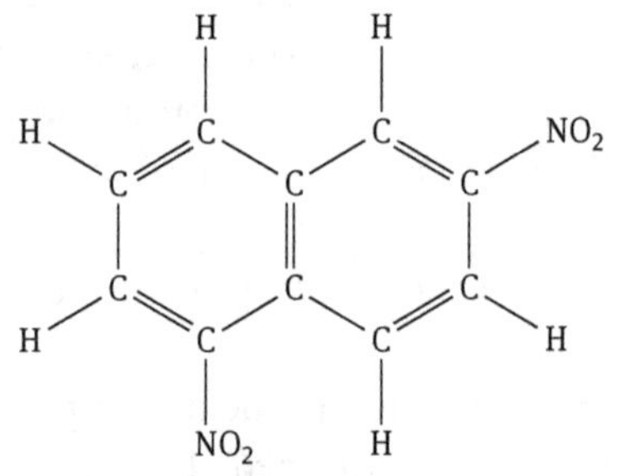

Stereoisomers differ by various locations of replacements. Enumerate all possible stereoisomers of n-nitronaphthalenes ($n = 1, \dots 7$).

Exercise 3.36. Show that a) $\mathrm{GL}_n(\mathbb{F}_2) \approx \mathrm{SL}_n(\mathbb{F}_2) \approx \mathrm{PSL}_n(\mathbb{F}_2)$; b) there exists a homomorphism $\mathrm{PSL}_2(\mathbb{F}_q) \to S_{q+1}$ (hint: consider the action on a projective line $\mathbb{F}_q\mathrm{P}^1$ and count its elements); c) $\mathrm{GL}_2(\mathbb{F}_2) \approx \mathrm{PSL}_2(\mathbb{F}_2) \approx S_3$; d) $\mathrm{PSL}_2(\mathbb{F}_3) \approx A_4$; e) $\mathrm{PSL}_2(\mathbb{F}_4) \approx A_5$.

Exercise 3.37. By considering a natural action on Fano plane $\mathbb{F}_2\mathrm{P}^2 \equiv \mathbb{F}_2^3 \backslash \{0\}$ of the group $\mathrm{PSL}_3(\mathbb{F}_2)$, find the numbers of elements in all its conjugacy classes and show that this group is simple. Calculate the number of inequivalent colorings of the Fano plane by use of three colors.

Exercise 3.38. Compute the character tables of the abelian group a) C_2; b) C_3; c) C_4; d) N_4.

Exercise 3.39. Compute the character table of the group S_3, reconstruct (up to equivalence) the matrices of the irreducible representations and decompose the regular representation $D^{(\mathrm{reg})}$ into irreps.

Exercise 3.40. Compute the character tables of the groups a) Dih_4; b) Q_8; c) A_4; d) S_4.

Exercise 3.41. Show that characters of rotational symmetries of a crystal lattice (constituting a *Point Group*) can take on only particular integer values and list those values. Based on that, prove that the order of a Point Group G divides 24 (hint: show that for any polynomial $P(x)$, $\sum_{g\in G} P(\chi(g))$ is a multiple of $|G|$ and select $P(x)$ suitable for counting the sum).

Exercise 3.42. Without writing the equations of motion, determine the number of different frequencies of normal vibrations of the molecule a) H_2O; b) $CHCl_3$; c) CH_4. You can use notation and character tables for the demanded Point Groups as provided by the standard literature (see, e.g., [137]. See also Section 12.2).

Exercise 3.43. Prove that the characters for n-th symmetric $\vee^n V$ and antisymmetric $\wedge^n V$ power of a representation V of group G are given by

$$\chi_{\vee^n V}(g) = \sum_{\{r_i,l_i\}} \prod_i \frac{\chi_V^{r_i}(g^{l_i})}{r_i! l_i^{r_i}}, \quad \chi_{\wedge^n V}(g) = (-1)^n \sum_{\{r_i,l_i\}} \prod_i (-1)^{l_i} \frac{\chi_V^{r_i}(g^{l_i})}{r_i! l_i^{r_i}}$$

where $\chi_V(g)$ is the character of V, the sums extend to all the classes of S_n enumerated by the number r_i of cycles of length l_i, constrained by $n = \sum_i r_i l_i$ (hint: consider a natural tensor representation of the group $S_n \otimes G$ in $\otimes^n V$ and extract the representation of G with the desired symmetry property). For each case write down explicit formulas for the lowest powers $n = 2, 3$.

Exercise 3.44. Tensor $C_{ik\ldots}$ definition and symmetry Point Group G of a medium are given in the variant table:

Point Group G	$\Phi = C_{ik}P_iP_k$	$\Phi = C_{ikl}A_iA_kA_l$	$\Phi = C_{(ik)(lm)}S_{ik}S_{lm}$
C_{2v}	a)	b)	c)
C_{3v}	d)	e)	f)
T_d	g)	h)	i)

where Φ is scalar, P_i – polar vector, A_i – axial vector, and $S_{ik} = S_{ki}$ – polar symmetric tensor. Taking aside time reversal symmetry, how many independent components can the tensor $C_{ik\ldots}$ have at most? You can use character table for the demanded Point Group from the literature.

A.4 Solutions and answers to selected problems

1.1. Using only 1, 2′ and 3′ we obtain $a_R^{-1} = a_R^{-1}e = a_R^{-1}(aa_R^{-1}) = (a_R^{-1}a)a_R^{-1}$, hence

$$e_R = a_R^{-1}(a_R^{-1})_R^{-1} = ((a_R^{-1}a)a_R^{-1})(a_R^{-1})_R^{-1} = (a_R^{-1}a)(a_R^{-1}(a_R^{-1})_R^{-1})$$
$$= (a_R^{-1}a)e_R = a_R^{-1}a, \quad \text{hence also} \quad e_R a = (aa_R^{-1})a = a(a_R^{-1}a) = ae_R = a$$

1.2. a) Assuming there are two identities e_1 and e_2, we obviously have $e_2 = e_2e_1 = e_1$.
b) Similarly, $a_2^{-1} = a_2^{-1}e = a_2^{-1}(aa_1^{-1}) = (a_2^{-1}a)a_1^{-1} = ea_1^{-1} = a_1^{-1}$.
c) Just multiply $g^2 = g$ by g^{-1}, then use the group axioms 2 and 3.

1.3. Hint: consider splitting G into pairs $\{g, g^{-1}\}$, $g \in G$.

1.5. To fix a group, one should fill in the $n \times n$ multiplication table with n elements. Assuming there were no other restrictions there would be n^{n^2} different ways to do the job. But since restrictions are obviously there and since some of the resulting tables may define isomorphic groups, we have $N_{\text{groups}}(n) < n^{n^2}$. For large n this naive estimate gives extremely large numbers, but this estimate is obviously extremely weak and actually it can be dramatically improved. Just as an elementary illustration, the row and the column in the table containing unity are both fixed, hence $N_{\text{groups}}(n) < n^{(n-1)^2}$ (unfortunately, this sole argument alone does not reduce the number essentially, it turns out that most of the restrictions come from combination of all the group axioms working together). As a further attempt, see Exercise 3.23.

1.6. The suggested two sets of generators are related, e.g., as $s = i$, $t = rir^{-1}$, or backwards $i = s$, $r = st$.

1.8. For the first case, yes. For the second case, an issue comes for the distribution law: $P(Q_1(x) + Q_2(x)) = P(Q_1(x)) + P(Q_2(x))$ only for $P(x) = ax$. By restriction to such polynomials one obtains the ring of endomorphisms of the field $\mathbb{R}$.

1.9. Let p be the least (hence obviously prime) integer, such that $\underbrace{1 + 1 + \cdots + 1}_{p \text{ times}} = 0$ (since the field is finite, such p, called the characteristic of the field, should necessarily exist). Then one can define multiplication $k \cdot x$ of any $k \in \mathbb{F}_p$ and any element $x \in \mathbb{F}_q$ by choosing integer representation for k and using repeated addition. One can easily check that by this $\mathbb{F}_q$ turns into a linear vector space over the finite field $\mathbb{F}_p$. Suppose its dimension is n, then $\mathbb{F}_q$ consists of exactly p^n elements.

1.10. a) For $\mathbb{F}_2$, addition and subtraction are essentially the same,

±	**0**	**1**
0	0	1
1	1	0

×	**0**	**1**
0	0	0
1	0	1

c) The process is similar to the extension from real to complex numbers: according to Ex. 1.9 the field $\mathbb{F}_4$ is a 2-dimensional vector field over $\mathbb{F}_2$, $\mathbb{F}_4 \approx \mathbb{F}_2 \oplus \mathbb{F}_2$. By identifying the first summand with $\mathbb{F}_2$ itself we identify the first basis vector with unity 1 of the field. Denote the second one by a (this would be an analog of the imaginary unit i), then all the elements of the field $\mathbb{F}_4$ are listed as $\{0, 1, a, 1 + a\}$. The addition/subtraction table is determined uniquely by the vector space structure. The multiplication table is fully determined by the field axioms once the product $a \cdot a$ is fixed. Since $a \neq 1$, a^2 cannot be a. It also cannot be 1, since otherwise $(1 + a)^2 = (1 + a^2) \bmod 2 = 0$. Hence the only option is $a^2 = 1 + a$. This is enough to determine the multiplication (hence also division) table completely:

±	0	1	a	$1+a$
0	0	1	a	$1+a$
1	1	0	$1+a$	a
a	a	$1+a$	0	1
$1+a$	$1+a$	a	1	0

×	0	1	a	$1+a$
0	0	0	0	0
1	0	1	a	$1+a$
a	0	a	$1+a$	1
$1+a$	0	$1+a$	1	a

Note that the equality $a^2 = a + 1$ is an analog of the well-known identity $i^2 + 1 = 0$. This observation is in fact the origin of a far reaching universal construction of field extensions.

2.1. a) The group $\mathrm{GL}_n(\mathbb{F}_q)$ consists of all the non-degenerate $n \times n$ matrices with elements from $\mathbb{F}_q$. Their first row has arbitrary nonzero entries, which gives $q^n - 1$ possibilities. For any of them, the second row includes arbitrary entries with exception of q multiples of the first row, giving $q^n - q$ possibilities, the third one cannot be among q^2 linear combinations of the first two, giving $q^n - q^2$ possibilities, etc. Hence $|\mathrm{GL}_n(\mathbb{F}_q)| = \prod_{i=0}^{n-1}(q^n - q^i)$. b) Since the determinant of a non-degenerate matrix can take anyone of the available $q - 1$ nonzero values, $|\mathrm{SL}_n(\mathbb{F}_q)| = |\mathrm{GL}_n(\mathbb{F}_q)|/(q - 1)$. c) The center $Z(\mathrm{SL}_n(\mathbb{F}_q))$ is a group of scalar matrices with unit determinant, hence it is enumerated by the n-th roots of unity. Since the multiplicative group $\mathbb{F}_q^* = \mathbb{F}_q \setminus \{0\}$ is cyclic of order $q - 1$, there are $\gcd(n, q - 1)$ such different roots. Hence $|\mathrm{PSL}_n(\mathbb{F}_q)| = |\mathrm{SL}_n(\mathbb{F}_q)|/\gcd(n, q - 1)$.

2.2. a) $\sigma^{-1} = \begin{pmatrix} 1 & 2 & 3 & 4 & 5 & 6 \\ 5 & 4 & 3 & 2 & 6 & 1 \end{pmatrix}$;

b) $\tau^{-1} = \begin{pmatrix} 1 & 2 & 3 & 4 & 5 & 6 \\ 3 & 2 & 1 & 4 & 5 & 6 \end{pmatrix}$;

c) $\sigma^2 = \begin{pmatrix} 1 & 2 & 3 & 4 & 5 & 6 \\ 5 & 2 & 3 & 4 & 6 & 1 \end{pmatrix}$;

d) $\tau^2 = \begin{pmatrix} 1 & 2 & 3 & 4 & 5 & 6 \\ 1 & 2 & 3 & 4 & 5 & 6 \end{pmatrix}$;

e) $\sigma \circ \tau = \begin{pmatrix} 1 & 2 & 3 & 4 & 5 & 6 \\ 3 & 4 & 6 & 2 & 1 & 5 \end{pmatrix}$;

f) $\tau \circ \sigma = \begin{pmatrix} 1 & 2 & 3 & 4 & 5 & 6 \\ 6 & 4 & 1 & 2 & 3 & 5 \end{pmatrix}$.

2.3. a) $\sigma = (145)(23)$; b) $\tau = (153)(26)$; c) $\rho = (174)(2968)$.

2.4. a) $(2,1,3) \equiv (1,3,2)$; b) (23); c) $(14)(23)$.

2.5. a) Assuming as known (see any textbook on linear algebra) that any permutation is decomposed in a product of transpositions, it is enough to observe that $(ij) = (1i)(1j)(1i)$.

b) Observe that $(ij) = (i\,i+1)(i+1\,j)(i\,i+1)$ and apply induction in distance between the transposed elements.

c) Observe that $\sigma^k \tau \sigma^{-k} = (\sigma^k(1)\,\sigma^k(2)) = (k+1\,k+2)$ and use the result from the previous item.

2.6. a) Assuming as known (see any textbook on linear algebra) that any even permutation is decomposed in a product of even number of transpositions, it is enough to observe that the product of any pair of adjacent transpositions is expressed in 3-cycles: $(ij)(ik) = (ikj)$, $(ij)(kl) = (ij)(ik)(ik)(kl) = (ikj)(ikl)$.

b) By taking a conjugate of a 3-cycle by consecutive ones we can successively reduce it to a consecutive one. A typical step of this procedure is given by $(i\,i+1\,i+2)(ijk)(i\,i+1\,i+2)^{-1} = (i+1\,j\,k)$ (under the assumption that both $i+1$ and $i+2$ do not coincide with either k or l). Similar conjugations shifting i by -1 or shifting j and k by ± 1 can be readily adjusted. Producing finally a consecutive 3-cycle and by iteration, one expresses (ijk) in terms of consecutive 3-cycles and their inverses.

c) If n is odd then the long cycle $\sigma \in A_n$ and any consecutive 3-cycle can be produced as $(i\,i+1\,i+2) = (\sigma^i(1)\,\sigma^i(2)\,\sigma^i(3)) = \sigma^i \tau \sigma^{-i}$. If n is even, then $\sigma \notin A_n$ but $\sigma' \in A_n$. In this case $\sigma'^i \tau \sigma'^{-i} = (\sigma'^i(1)\,\sigma'^i(2)\,\sigma'^i(3)) = (1\,i+2\,i+3)$ and an arbitrary consecutive 3-cycle can be assembled as $(i\,i+1\,i+2) = (1\,i\,i+1)(1\,i+1\,i+2)$.

2.7. By simple evaluation, $A^4 = B^3 = 1$ and

$$AB = \begin{pmatrix} 1 & 1 \\ 0 & 1 \end{pmatrix}, \quad (AB)^n = \begin{pmatrix} 1 & n \\ 0 & 1 \end{pmatrix}$$

hence the element $AB \in \mathrm{SL}_2(\mathbb{Z})$ is of infinite order.

2.10. c) The symmetry group is denoted by Dih_2 (sometimes called the Klein group and denoted by N_4) and consists of the identity, of two reflections with respect to symmetry axes and of their product (rotation by π). If the vertices are enumerated by $1, \ldots, 4$ then the group is listed by permutations $\mathrm{Dih}_2 = \{e, (12)(34), (13)(24), (14)(23)\}$.

Since all non-trivial elements are of order 2, the group is not cyclic. d) The symmetry group is the Klein group Dih_2, see the previous item.

2.11. By enumerating the vertices by 1, 2, 3 the group consists of the identity e, of the rotations (123) and (132) by angles $2\pi/3$ and $4\pi/3$ around the center, and of the reflections (23), (13) and (12) along the symmetry axes passing through each of the vertices 1, 2, 3 and the centers of opposite sides. Representation by permutations clearly demonstrates that the group coincides with S_3. By picking up the vertices 1, 2, 3 at positions $(0, \frac{1}{\sqrt{3}})$, $(-\frac{1}{2}, -\frac{1}{2\sqrt{3}})$ and $(\frac{1}{2}, -\frac{1}{2\sqrt{3}})$, respectively, the discussed transformations are given by the matrices

$$\begin{pmatrix} 1 & 0 \\ 0 & 1 \end{pmatrix}, \begin{pmatrix} -\frac{1}{2} & -\frac{\sqrt{3}}{2} \\ \frac{\sqrt{3}}{2} & -\frac{1}{2} \end{pmatrix}, \begin{pmatrix} -\frac{1}{2} & \frac{\sqrt{3}}{2} \\ -\frac{\sqrt{3}}{2} & -\frac{1}{2} \end{pmatrix}, \begin{pmatrix} -1 & 0 \\ 0 & 1 \end{pmatrix},$$
$$\begin{pmatrix} \frac{\sqrt{3}}{2} & -\frac{1}{2} \\ -\frac{1}{2} & -\frac{\sqrt{3}}{2} \end{pmatrix}, \begin{pmatrix} -\frac{\sqrt{3}}{2} & -\frac{1}{2} \\ -\frac{1}{2} & \frac{\sqrt{3}}{2} \end{pmatrix}$$

2.12. The general group Dih_n contains identity e, $n-1$ rotations r^k by angles $2\pi k/n$ around the center $(1 \le k \le n-1)$, and n reflections along each of the n symmetry axes. Note a distinction between the cases of n odd and n even. Indeed, if n is odd, the symmetry axes pass through the vertices and the centers of the sides opposite to them, while if n is even, they pass either through opposite vertices or through opposite sides. In the first case, each reflection can be expressed in terms of one of them $(i \equiv i_0)$ as $i_k = r^k i r^{-k}$ $(k = 0, 1 \dots, n-1)$. In the second case, reflections of each type are expressed as a product of a rotation and a reflection of another type. Hence in both cases the group can be equivalently presented in terms of a single reflection and a primary rotation as $\mathrm{Dih}_n = \{e, r, \dots, r^{n-1}, i, ri, \dots, r^{n-1}i\}$. Thus, it is generated by r and i. Since $i^2 = r^n = e$ and $iri = r^{-1}$, the group presentation is $\mathrm{Dih}_n = \langle r, i | r^n = i^2 = e,\ iri = r^{-1}\rangle$ (independently, whether n is even or odd).

2.13. A tetrahedron has four vertices (which we enumerate as 1, 2, 3, 4), six edges and four faces. The subgroup of rotational symmetries consists of the identity e, of $2 \cdot 4 = 8$ rotations by angles $2\pi/3$ and $4\pi/3$ around the four symmetry axes passing through a vertex and the center of the opposite face (corresponding to 3-cycle permutations (ijk)), and three rotations around the $6/2 = 3$ lines passing through the centers of opposite edges (corresponding to permutations $(ij)(kl)$). This subgroup, named the proper tetrahedral group $\mathcal{T}_p$, is evidently isomorphic to A_4. In addition to rotations, the full tetrahedral group $\mathrm{T_d}$ contains also six reflections in symmetry planes passing through an edge and the midpoint of the opposite edge (corresponding to transpositions (ij) etc.) and six roto-reflections described by the 4-cycles $(ijkl)$, e.g., $(134)(12) = (1234)$. The whole group is thus obviously isomorphic to S_4 (see Section 12.2 where we emphasize the difference between the octahedral proper group O_{24} and the tetrahedron extended group $\mathrm{T_d}$ described here. They are isomorphic as abstract groups yet only the proper tetrahedral subgroup of order 12, $\mathcal{T}_p \subset \mathcal{T}_p$, is a subgroup of the octahedral group of order 24 in the 3-dimensional defining representation of O_{24}). While its action on faces is essentially similar to the action on vertices, one could realize $\mathcal{T}_p$ alternatively as a proper subgroup of S_6 by enumerating the edges and looking at the effect of symmetry transformations on them (we leave the details of such a dual description to the reader).

2.14. A cube has eight vertices, twelve edges and four faces. To write symmetries as permutations, let us enumerate its $8/2 = 4$ long diagonals connecting opposite vertices as $1, 2, 3, 4$ and, in addition, assign to each of them some orientation. Then each symmetry transformation is characterized by a permutation of diagonals *and* a list of four signs indicating whether their orientation is altered or not. The group of rotational symmetries preserves their orientation and includes: identity e; $3 \cdot 3 = 9$ rotations around the axes passing through the centers of three different pairs of opposite faces by angles $\pi/2$, π and $3\pi/2$ (corresponding to either 4-cycles $(ijkl)$ or $(ij)(kl)$); $4 \cdot 2 = 8$ rotations around the axes passing through four different pairs of opposite vertices by angles $2\pi/3$ and $4\pi/3$ (corresponding to 3-cycles (ijk)); six rotations around the axes passing through the midpoints of six different pairs of opposite edges by an angle π (corresponding to transpositions (ij)). Such a group, named O_{24}, is the proper octahedral group and it is isomorphic to S_4. Instead, each reflection changes the orientation of all diagonals. They include: three reflections (described by permutations $(ij)(kl)$) in planes perpendicular to each of the 4-fold axes; $12/2 = 6$ reflections (described by transpositions (ij)) in planes passing through the pairs of opposite edges; eight roto-reflections (described by 3-cycles) by $\pi/3$; 6 roto-reflections by an angle $\pi/2$ (described by 4-cycles). Since under successive transformations both diagonal permutations and orientation signs are multiplied, the whole group O_d is isomorphic to the product $S_4 \otimes C_2$. The case of octahedron differs only by replacing vertices with faces and vice versa, and by considering instead of diagonals the lines passing through the centers of opposite faces.

3.1. The resulting graphs are:

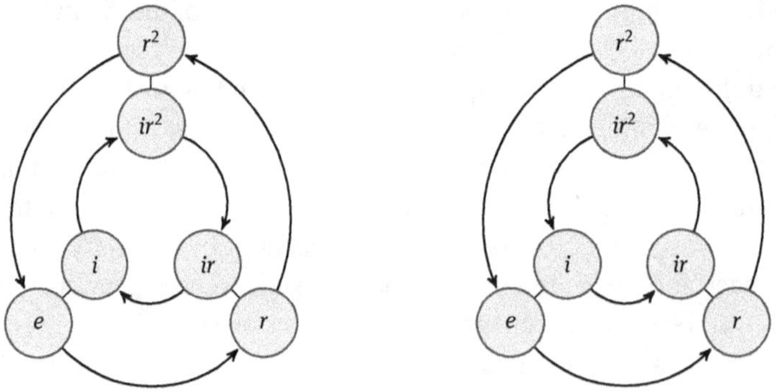

(we omit arrows for edges corresponding to action of a generator of order 2, since it coincides with its inverse; note especially different direction of arrows in the inner circles of both graphs!). Further applications of Cayley graphs in visualizing various results of finite group theory can be found in Ref. [18].

3.2. We have: $G = \{e, (123), (132), (12), (13), (23)\}$. Hence,

a) $G = H \cup (12)H = H \cup H(12)$, where

$$H = \{e, (123), (132)\}, \quad (12)H = \{(12), (23), (13)\}$$
$$H(12) = \{(12), (13), (23)\}$$

One can observe that the right and left classes coincide, $(12)H = H(12)$. This is because $H \lhd G$ is a normal subgroup.

b) $G = H \cup (13)H \cup (23)H = H \cup H(13) \cup H(23)$, where

$$H = \{e, (12)\}, \quad (13)H = \{(13), (123)\}, \quad (23)H = \{(23), (132)\}$$
$$H(13) = \{(13), (132)\}, \quad H(23) = \{(23), (123)\}$$

In this case splittings into right and left classes are completely different.

3.3. Since G is abelian, the right and left classes coincide.

We have $G = \{1, \zeta, \zeta^2, \zeta^3, \zeta^4, \zeta^5\}$, hence a) $G = H \cup \zeta H \cup \zeta^2 H$, where $H = \{1, \zeta^3\}$, $\zeta H = \{\zeta, \zeta^4\}$ and $\zeta^2 H = \{\zeta^2, \zeta^5\}$; b) $G = H \cup \zeta H$, where $H = \{1, \zeta^2, \zeta^4\}$ and $\zeta H = \{\zeta, \zeta^3, \zeta^5\}$.

3.4. By definition, the number of left and right cosets is 2. If $g \in H$ then so does g^2. If $g \notin H$ then the only left cosets of H can be H and gH, and the only right cosets can be H and Hg. Hence $gH = Hg$ so that H has to be normal. The class g^2H should coincide with either gH or H, but since $g \notin H$ it cannot be gH, hence it is H and thus $g^2 \in H$. A typical example is provided by $G = S_n$ and $H = A_n$.

3.5. The conclusions trivially follow from commutativity; $C_6/C_2 \approx C_3$ and $C_6/C_3 \approx C_2$.

3.6. Hints: b) follows from $(h^{-1}gh)^n = h^{-1}g^n h$; c) follows from $(h^{-1}gh)^{-1} = h^{-1}g^{-1}h$.

3.7. As explained in the main text, conjugacy classes of the symmetric group S_n are enumerated by the Young tableaux and join all permutations of the same cycle structure. Calculation of classes is illustrated in the tables (for brevity, the last column listing the elements of a class for S_5 is excluded):

	Young tableau	$\{r_l\}$			# elements, $\lvert\{r_l\}\rvert$	Cycle structure	Representatives
		r_1	r_2	r_3			
a)	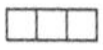	3	0	0	$\frac{3!}{3!} = 1$	e	e
		1	1	0	$\frac{3!}{1!2^1 1!} = 3$	(ij)	(12), (13), (23)
		0	0	1	$\frac{3!}{3^1 1!} = 2$	(ijk)	(123), (132)

b)

Young tableau	$\{r_l\}$				# elements, $\lvert\{r_l\}\rvert$	Cycle structure	Representatives
	r_1	r_2	r_3	r_4			
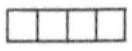	4	0	0	0	$\frac{4!}{4!}=1$	e	e
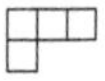	2	1	0	0	$\frac{4!}{2!2^1 1!}=6$	(ij)	(12), (13), (14), (23), (24), (34)
	0	2	0	0	$\frac{4!}{2^2 2!}=3$	$(ij)(kl)$	(12)(34), (13)(24), (14)(23)
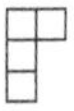	1	0	1	0	$\frac{4!}{1!3^1 1!}=8$	(ijk)	(123), (132), (124), (142), (134), (143), (234), (243)
	0	0	0	1	$\frac{4!}{4^1 1!}=6$	$(ijkl)$	(1234), (1342), (1423), (1324), (1243), (1432)

c)

Young tableau	$\{r_l\}$					# elements, $\lvert\{r_l\}\rvert$	Cycle structure
	r_1	r_2	r_3	r_4	r_5		
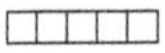	5	0	0	0	0	$\frac{5!}{5!}=1$	e
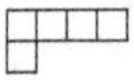	3	1	0	0	0	$\frac{5!}{3!2^1 1!}=10$	(ij)
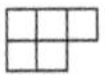	1	2	0	0	0	$\frac{5!}{1!2^2 2!}=15$	$(ij)(kl)$
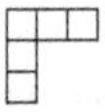	2	0	1	0	0	$\frac{5!}{2!3^1 1!}=20$	(ijk)
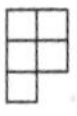	0	1	1	0	0	$\frac{5!}{2^1 1! 3^1 1!}=20$	$(ij)(klm)$
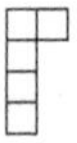	1	0	0	1	0	$\frac{5!}{1!4^1 1!}=30$	$(ijkl)$
	0	0	0	0	5	$\frac{5!}{5^1 1!}=24$	$(ijklm)$

3.8. By means of self-explanatory algebra and calculus, the number $|\mathscr{C}(S_n)|$ of conjugacy classes of S_n can be represented as follows:

$$|\mathscr{C}(S_n)| = \sum_{\{r_l\}} \delta_{n,\sum_l l r_l} = \oint_{|z|=\rho<1} \frac{dz}{2\pi i} \sum_{\{r_l\}} z^{\sum_l l r_l - n-1} = \frac{1}{2\pi i} \oint_{|z|=\rho<1} \frac{dz}{z^{n+1}} \prod_{l=1}^{\infty} \sum_{r_l=0}^{\infty} z^{l r_l}$$

$$= \frac{1}{2\pi i} \oint_{|z|=\rho<1} \frac{dz}{z^{n+1}} \prod_{l=1}^{\infty} \frac{1}{1-z^l} = \int_0^{2\pi} \frac{d\varphi}{2\pi} \exp\left\{-in\varphi - n\ln\rho - \sum_{l=1}^{\infty} \ln(1-\rho e^{il\varphi})\right\}$$

where the last representation is obtained by the change of variable $z=\rho e^{i\varphi}$. For $n \gg 1$ this integral grows exponentially. If we are interested only in the exponential factor, it is enough to take a limit $\rho \to 1$ (note that determination of the exponential pre-factor is much more sophisticated [126]). For such n only small φ are important, hence the sum can be replaced by an integral

$$\sum_{l=1}^{\infty} \ln(1 - pe^{il\varphi}) \approx \frac{i}{\varphi}\int_0^{\infty} \ln(1 - e^{-\xi})\, d\xi = -i\frac{\pi^2}{6\varphi}$$

$$|C(S_n)| \approx \int_{-\infty}^{+\infty} \frac{d\varphi}{2\pi} e^{-if(\varphi)}, \quad f(\varphi) = n\varphi - \frac{\pi^2}{6\varphi}$$

By looking at the stationary points:

$$f'(\varphi_*) = n + \frac{\pi^2}{6\varphi_*^2} = 0, \quad \varphi_* = \pm i\frac{\pi}{\sqrt{6n}}; \quad f(\varphi_*) = \pm i\pi\sqrt{\frac{2n}{3}}, \quad f''(\varphi_*) = \mp i\frac{(6n)^{3/2}}{3\pi}$$

and taking the largest contribution (i.e., with the upper sign), we finally obtain

$$|\mathscr{C}(S_n)| \simeq e^{-if(\varphi_*)} = e^{\pi\sqrt{2n/3}}$$

This number should be compared with the order $|S_n| = n! \simeq \sqrt{2\pi n}e^{n\ln n - n}$. The number of classes grows exponentially, though much slower than the order.

3.9. Permutations of different cycle structures are non-conjugate in S_n, not to say in A_n. Two permutations of the same cycle structure are conjugate in S_n but can be still non-conjugate in A_n. This alternative leads to the possible splitting of a class into two classes when we restrict the group to its alternating subgroup $A_n \subset S_n$. Let $\sigma, \sigma' \in A_n$ have the same cycle structure, then $\sigma' = \tau^{-1}\sigma\tau$ for some $\tau \in S_n$. If $\tau \in A_n$ then they are also conjugate in A_n. Now suppose that $\tau \in S_n \backslash A_n$ but there also exists some $\tau' \in A_n$ such that $\sigma' = \tau'^{-1}\sigma\tau'$. Then we have $\tau^{-1}\sigma\tau = \tau'^{-1}\sigma\tau'$ and hence σ necessarily commutes with the odd permutation $\tau'\tau^{-1} \in S_n \backslash A_n$. In particular,

(i) if $\sigma = c_{l_1}c_{l_2}\dots c_{l_k}$ where c_{l_i} are all cycles with different odd lengths l_i, then any $\tau \in S_n$ commuting with σ must fix each of the c_{l_i}, hence it has to be of the form $\tau = c_{l_1}^{q_1}c_{l_2}^{q_2}\dots c_{l_k}^{q_k}$ ($q_i \in \mathbb{Z}$, $0 \le q_i < l_i$), and since c_{l_i} are all even, we necessarily have $\tau \in A_n$;

(ii) if σ has a cycle c_l of even length l then σ commutes with $c_l \in S_n \backslash A_n$;

(iii) if σ has two cycles $(i_1\, i_2 \dots i_l)$ and $(j_1\, j_2 \dots j_l)$ of the same odd lengths, then it commutes with the odd permutation $(i_1 j_1)(i_2 j_2)\dots(i_l j_l) \in S_n \backslash A_n$, which just swaps them under conjugation.

By looking at the tables from solution of Ex. 3.7, selecting the classes of even permutations and applying the above results, we obtain:

a) four classes: $[e]$ (1), $[(ij)(kl)]$ (3), $[(ijk)]$ (4), $[(ijk)]'$ (4) – the latter since (ijk) contains two cycles of distinct odd lengths 1 and 3;

b) five classes: $[e]$ (1), $[(ij)(kl)]$ (15), $[(ijk)]$ (20), $[(ijklm)]$ (12), $[(ijklm)]'$ (12) – the latter since $(ijklm)$ is a single cycle of odd length;

where the number of elements of each class is given in brackets.

3.10. The group Dih_n consists of the elements $\mathrm{Dih}_n = \{e, r, \ldots, r^{n-1}, i, ri, \ldots, r^{n-1}i\}$ (see Ex. 2.12). Conjugations $\tau^{-1}\sigma\tau$ of different types (rotations and reflections) of elements are given in the table:

τ	σ	
	r^k	$r^k i$
r^j	r^k	$r^{k-2j} i$
$r^j i$	r^{-k}	$r^{2j-k} i$

Let us first apply it to a couple of suggestive cases:
(i) In $\mathrm{Dih}_3 = \{e, r, r^2, i, ri, r^2 i\} \approx S_3$ we have: $r^2 \sim r$, $r^2 i \sim i$ and $ri \sim r^{-1}i \equiv r^2 i$, hence the class equation reads: $6 = 1 + 2 + 3$.
(ii) In $\mathrm{Dih}_4 = \{e, r, r^2, r^3, i, ri, r^2 i, r^3 i\}$ we have: $r^3 \sim r$, $r^2 i \sim i$ and $r^3 i \sim ri$, hence the class equation reads: $8 = 1 + 2 + 1 + 2 + 2$.

Likewise, in general the cases of odd and even n should be also considered separately. If n is odd, then all the non-trivial rotations are grouped in $(n-1)/2$ 2-element classes $\{r^k, r^{n-k}\}$ $(1 \le k \le (n-1)/2)$, while all the reflections are within the same class:

$$|\mathrm{Dih}_n| = 2n = 1 + \underbrace{2 + 2 + \cdots + 2}_{(n-1)/2} + n$$

If n is even, then there are two 1-element classes $[e]$ and $[r^{n/2}]$, $(n-2)/2$ 2-element classes $\{r^k, r^{n-k}\}$ $(1 \le k \le (n-2)/2)$, and two classes $[i] = \{i, r^2 i, r^4 i, \ldots, r^{n-2} i\}$ and $[ri] = \{ri, r^3 i, r^5 i, \ldots, r^{n-1} i\}$ of $(n-2)/2$ reflections. In this case the class equation reads:

$$|\mathrm{Dih}_n| = 2n = 1 + 1 + \underbrace{2 + 2 + \cdots + 2}_{(n-2)/2} + n/2 + n/2$$

3.11. The key argument is that a normal subgroup should be built of entire conjugacy classes and at that necessarily it should contain the 1-element class of the neutral element (the identity).

a) Since $\mathrm{Dih}_3 \approx S_3$ (see Ex. 2.11), we can apply the results either of Ex. 3.7a), or of Ex. 3.10: $|\mathrm{Dih}_3| = 1 + 2 + 3$. Hence a proper normal subgroup, if exists, can be either of order $1 + 2 = 3$ or $1 + 3 = 4$. However, since 4 does not divide $|\mathrm{Dih}_3| = 6$, this possibility is excluded by Lagrange theorem. Since 3 is prime, the desired normal subgroup should be cyclic. Indeed, the only 3-element cyclic subgroup $\langle r \rangle$ is normal.
b) In this case the class equation is $|\mathrm{Dih}_4| = 8 = 1 + 2 + 1 + 2 + 2$ which is less restrictive than Lagrange theorem, stating that there might be proper subgroups only of order either 2 or 4. Obviously, one of these elements should be identity, so that the remaining ones should have order(s) either 2 or 4. However, such remaining elements cannot be reflections, since the latter are grouped in 2-element classes. Hence they should be rotations, and indeed, there are rotations r and r^2 of orders 4 and 2, generating normal subgroups isomorphic to C_4 and C_2, respectively.

c) The class equation $|A_4| = 12 = 1 + 3 + 4 + 4$ (see Ex. 3.9a)) suggests that any normal subgroup might contain only $1+3=4$, $1+4=5$, $1+3+4=8$ or $1+4+4=9$ elements. However, among these numbers, only 4 is a divisor the order of the group. Indeed, the permutations $(ij)(kl)$ generate a 4-element normal (Klein) subgroup in A_4.

d) The class equation $|S_4| = 24 = 1 + 3 + 6 + 6 + 8$ (see Ex. 3.7b)) suggests that the normal subgroup might contain only $1 + 3 = 4$, $1 + 6 = 7$, $1 + 8 = 9$, $1 + 3 + 6 = 10$, $1 + 3 + 8 = 12$, $1 + 6 + 6 = 13$, $1 + 6 + 8 = 15$, $1 + 3 + 6 + 6 = 16$, $1 + 3 + 6 + 8 = 18$ or $1+6+6+8 = 21$ elements. Among these numbers, only 4 and 12 are divisors of the order of the group. Looking at the corresponding classes, we easily identify the subgroups $\mathrm{N}_4 = \langle (ij)(kl) \rangle$ and A_4 and we easily check that both of them are indeed normal.

3.12. A normal subgroup of A_5, if it existed, should be built of entire conjugacy classes and necessarily contain the class $[e]$. Hence, according to the class equation $60 = 1 + 12 + 12 + 15 + 20$ (see Ex. 3.9), any such proper subgroup should have one of the following orders: 13, 16, 21, 25, 28, 33, 36, 40, 45 or 48. Yet none of these numbers is a divisor of $|A_5| = 40$, thus contradicting Lagrange theorem. Since we have not explicitly derived the class equation for A_n, the general case requires a more refined treatment. Let $N \lhd A_n$ is a non-trivial normal subgroup and $\sigma \in N$ is non-identity. Then for all $\tau \in A_n$ $\tau^{-1}\sigma\tau \in N$ and $\sigma^{-1} \in N$, hence $[\tau^{-1}, \sigma] = \tau^{-1}\sigma\tau\sigma^{-1} \in N$. If we pick up τ to be a 3-cycle then both τ^{-1} and $\sigma\tau\sigma^{-1}$ are also 3-cycles, so that $[\tau^{-1}, \sigma] \in N$ is in fact a product of two 3-cycles, thus fixing $n - 6$ elements. But since σ is non-identity, we can specify τ so that the supports of τ^{-1} and $\sigma\tau\sigma^{-1}$ overlap, in such a case $[\tau^{-1}, \sigma] \in N$ is in fact either a 5-cycle or a product of two 2-cycles. In both cases it fixes more than $n - 6$ elements and for a while we restrict our attention to a 5-element set that is permuted. Since we already know that A_5 is simple, N should also contain all the 3-cycles with supports in that set, hence all the 3-cycles from A_n which are conjugated to them in A_n (see Ex. 3.9). But we know that a set of all 3-cycles generates the whole A_n (see Ex. 2.6), thus N should necessarily coincide with it.

3.13. S_3 is generated, e.g., by two transpositions (see Ex. 1.6). Since they are conjugate, a homomorphism f can send both of them either to identity (then $\ker f = S_3$ and $\operatorname{im} f = e$), or to element(s) of order 2. A product of transpositions is a 3-cycle, and in the second case it can be sent either to identity (then $\ker f = A_3 \approx C_3$ and $\operatorname{im} f = C_2$) or to some element of order 3 (then $\ker f = e$ and $\operatorname{im} f = S_3$). If the target group is abelian, then only the first two options can take place, since the generators are non-commuting.

3.14. a) $f(x) = x \bmod n$; $\ker f = n\mathbb{Z}$; $\operatorname{im} f = n\mathbb{Z}$; $\mathbb{Z}/n\mathbb{Z} \approx \mathbb{Z}_n$.

b) $f([0]_4) = f([2]_4) = [0]_2$, $f([1]_4) = f([3]_4) = [1]_2$; $\ker f = \{[0]_4, [2]_4\} \approx \mathbb{Z}_2$; $\operatorname{im} f = \{[0]_2, [1]_2\} \approx \mathbb{Z}_2$; $\mathbb{Z}_4/\mathbb{Z}_2 \approx \mathbb{Z}_2$.

c) $f([0]_6) = f([3]_6) = [0]_3$, $f([1]_6) = f([3]_6) = [1]_3$; $f([2]_6) = f([5]_6) = [2]_3$; $\ker f = \{[0]_6, [3]_6\} \approx \mathbb{Z}_2$; $\operatorname{im} f = \{[0]_3, [1]_3, [2]_3\} \approx \mathbb{Z}_3$; $\mathbb{Z}_6/\mathbb{Z}_2 \approx \mathbb{Z}_3$.
d) $f([0]_6) = f([2]_6) = f([4]_6) = [0]_2, f([1]_6) = f([3]_6) = f([5]_6) = [1]_2$; $\ker f = \{[0]_6, [2]_6, [4]_6\} \approx \mathbb{Z}_3$; $\operatorname{im} f = \{[0]_2, [1]_2\} \approx \mathbb{Z}_2$; $\mathbb{Z}_6/\mathbb{Z}_3 \approx \mathbb{Z}_2$.
e) It is enough to define a homomorphism on generators and A_4 is generated by $\sigma = (123)$ and $\sigma' = (234)$ (see Ex. 2.6). Its complete presentation can be written as $A_4 = \langle \sigma, \sigma' | \sigma^3 = \sigma'^3 = (\sigma\sigma')^2 = e \rangle$. Hence we can set $f(\sigma) = \sigma$ and $f(\sigma') = \sigma^2$, then $\operatorname{im} f = \langle \sigma \rangle = A_3 \approx C_3$, $\ker f = \mathrm{N}_4$ and $A_4/\mathrm{N}_4 \approx C_3$.
f) For all $\sigma \in A_n$ set $f(\sigma) = \operatorname{sgn}(\sigma)$. Obviously, this is a homomorphism onto $C_2 = \{1, -1\}$ with $\ker f = A_n$. Hence $S_n/A_n \approx C_2$.
g) It is enough to define a homomorphism on generators and Dih_n is generated by a rotation r and inversion i (see Ex. 2.12). It is easy to see that by setting $f(r) = e$, $f(i) = i$, we obtain a correct homomorphism onto $\langle i \rangle$ with $\ker f = \langle r \rangle \approx C_n$. Hence $\mathrm{Dih}_n/C_n \approx C_2$.

3.15. Let $C_n = \langle g \rangle$ and $C_m = \langle h \rangle$. Because both groups are cyclic, a homomorphism $f : C_n \to C_m$ is uniquely determined by where it sends a generator g, $f(g) = h^k$. But since $g^n = e$, we should have $f(g^n) = (f(g))^n = (h^k)^n = h^{kn} = e$, or $kn = 0 \bmod m$. Denote $d = \gcd(n, m)$, then n/d and m/d are mutually prime and hence k should be a multiple of m/d, i.e., take on one of the values $0, m/d, 2m/d, \ldots, (d-1)m/d$. This means there are $d = \gcd(n, m)$ distinct homomorphisms. Now let $m = q$ and $n = pq$. Since $\gcd(n, m) = q$, we obtain that $f(g) = h$ determines a valid homomorphism, obviously with $\operatorname{im} f = C_q$, and for which $\ker f = \langle g^q \rangle \approx C_p$. Hence by the theorem of homomorphism we have $C_{pq}/C_p \approx C_q$.

3.16. Let $C_{pq} = \langle g \rangle$, $N = \langle g^q \rangle$ and $H = \langle g^p \rangle$. Since g^q is of order p and g^p is of order q, we have $N \approx C_p$ and $H \approx C_q$. Then $N, H \lhd C_{pq}$ (since the group C_{pq} is abelian); $N \cap H = \{e\}$ (for if $(g^q)^k = (g^p)^l$ then, e.g., q divides l, hence $(g^p)^l = e$); and $C_{pq} = NH$ (for that due to the Bésout's lemma there exist k and l such that $\gcd(p, q) = kp + lq = 1$, hence $g = (g^q)^l (g^p)^k$). Hence $C_{pq} \approx N \otimes H$.

3.17. The group $C_2 \otimes C_2 \otimes C_2$ contains no elements of order 4 and 8 and the group $C_4 \otimes C_2$ contains no elements of order 8.

3.18. Hint: consider an element of maximal order and observe that it generates a cyclic subgroup H. Prove that the original group G is a product of H and G/H, then use induction in order of G. You may use Ex. 3.17 as collection of guiding examples.

3.19. To prove that a group G is a semidirect product, it is enough to identify two subgroups, $N \lhd G$ and $H \subset G$, such that $N \cap H = \{e\}$ and $G = NH$.

a) Recall that $N_4 = \{e, (12)(34), (13)(24), (14)(23)\}$ and let $C_3 = \{e, (123), (132)\}\rangle \approx C_3$. It is easy to observe that products of the elements of N_4 and C_3 indeed populate the whole A_4; obviously $N_4 \cap C_3 = \{e\}$ and $N_4 \lhd A_4$ (see Ex. 3.11c)).
b) For $n \geq 2$ the group Dih_n is generated by a rotation r and a reflection i, and the cycle groups C_n and C_2 that they generate have only identity in intersection (see Ex. 2.12). Since $C_n \lhd \mathrm{Dih}_n$ (see Ex. 3.14g)), we have $\mathrm{Dih}_n \approx C_n \rtimes C_2$.
c) For $n \geq 3$ the group S_n contains a normal subgroup A_n of index 2 (see Ex. 3.14f)). An order-2 representative of the non-trivial coset, say (12), generates a cycle group intersecting with A_n only at identity (see Ex. 3.4). Hence $S_n \approx A_n \rtimes C_2$.

3.20. Let $C_8 = \langle g | g^8 = e\rangle$ and $C_2 = \langle h | h^2 = e\rangle$. To construct an external semidirect product we must only specify how the conjugation hgh^{-1} is realized as an automorphism of C_8. Since C_8 is cyclic and is generated by any of the elements g, g^3, g^5 or g^7, we can set $hgh^{-1} = g^k$, where $k = 1, 3, 5, 7$. The presentation of the resulting group reads $G_k = \langle g, h | g^8 = h^2 = e, hgh^{-1} = g^k\rangle$. Note that $G_1 = C_8 \otimes C_2$ and $G_7 = \mathrm{Dih}_8$. The groups G_3 and G_5 are called quasidihedral and modular groups of order 16, respectively.

3.21. We have $Q_8 = \{1, -1, i, -i, j, -j, k, -k\}$ and $|Q_8| = 8$, as reflected in notation. The proper subgroups can be found, e.g., by direct inspection of multiplication table, there are four of them: $Z(Q_8) = \langle -1\rangle = \{1, -1\} \approx C_2$, $\langle i\rangle = \{1, i, -1, -i\} \approx C_4$, $\langle j\rangle = \{1, j, -1, -j\} \approx C_4$, and $\langle k\rangle = \{1, k, -1, -k\} \approx C_4$. The central elements $\{1, -1\}$ represent two 1-element classes and, since, e.g., $(\pm 1)i(\pm 1) = i$, $(\pm i)i(\pm i)^{-1} = (\pm i)i(\mp i) = i$, $(\pm j)i(\pm j)^{-1} = (\pm j)i(\mp j) = -i$ and $(\pm k)i(\pm k)^{-1} = (\pm k)i(\mp k) = -i$, the remaining elements are grouped into 2-element classes $\{i, -i\}$, $\{j, -j\}$ and $\{k, -k\}$. The class equation reads $8 = 1 + 1 + 2 + 2 + 2$. The subgroup $Z(Q_8)$ is normal merely by definition, and the remaining subgroups are normal since their index is 2 (see Ex. 3.4) or, alternatively, since they are assembled from the entire classes. By factorizing the presentation one can easily show that $Q_8/Z(Q_8) \approx N_4 \approx C_2 \otimes C_2$, which is obviously non-isomorphic to any subgroup. If Q_8 were a semidirect product, it would contain two subgroups with only identity in their intersection. However, in Q_8 all the proper subgroups contain $Z(Q_8)$ (alternatively: if Q_8 were a semidirect product of two subgroups, then, since all proper subgroups are normal, it would be a direct product, which is not the case).

3.22. Note first that the commutator of two elements can be represented as $ghg^{-1}h^{-1} = (ghg^{-1})h^{-1}$.
a) For S_3 this means that (see Ex. 2.4): the commutator of 3-cycles is a product of 3-cycles, thus either identity or also a 3-cycle; the commutator of transpositions is a product of two transpositions, thus either identity or a 3-cycle; finally, the commutator of a transposition and a 3-cycle is a product of two 3-cycles, thus also a 3-cycle. Thus commutators generate the group A_3 of 3-cycles, $(S_3)' = A_3$. Since the latter is abelian, $(S_3)'' = (A_3)' = \{e\}$. The subnormal series reads: $\{e\} \lhd A_3 \lhd S_3$.

b) For S_4 for brevity let us promptly observe that commutator is an even permutation and by the same argument as above we can demonstrate that each even permutation represents a commutator, thus $(S_4)' = A_4$. One can see further that commutator of 3-cycles in A_4 is either identity or a product of transpositions, thus $(S_4)'' = (A_4)' = \mathrm{N}_4$. Finally, since N_4 is abelian, $(S_4)''' = (\mathrm{N}_4)' = \{e\}$, but because $\mathrm{N}_4 \approx C_2 \otimes C_2$, it still contains three normal subgroups isomorphic to C_2. The subnormal series thus reads $\{e\} \lhd C_2 \lhd \mathrm{N}_4 \lhd A_4 \lhd S_4$.
c) For Dih_n one can observe immediately from the data of the table in solution of Ex. 3.10 that $\mathrm{Dih}_n' = \langle r^2 \rangle$, which is C_n for odd n or $C_{n/2}$ for even n. Hence $\mathrm{Dih}_n'' = \{e\}$ and the subnormal series reads: $\{e\} \lhd C_n \lhd \mathrm{Dih}_n$ for odd n or $\{e\} \lhd C_{n/2} \lhd \mathrm{Dih}_n$ for even n.

3.23. Let $\{e\} = G_0 \subset G_1 \subset G_2 \subset \cdots \subset G_r = G$ be a maximal chain of proper subgroups, and let $g_i \in G_i \backslash G_{i-1}$ for $1 \le i \le r$. Then $G_i = \langle g_1, \ldots, g_i \rangle$, in particular G can be generated by r elements. By Lagrange Theorem,

$$|G| = \prod_{i=1}^{r} |G_i : G_{i-1}| \ge 2^r$$

hence $r \le \log_2 n$. Now, by Cayley's theorem, $G \subseteq S_n$, and hence

$$\begin{aligned} N_{\text{groups}}(n) &\le (\#\text{ of subgroups of order } n \text{ in } S_n) \\ &\le (\#\text{ of } \log_2 n\text{-generator subgroups of } S_n) \\ &\le (\#\text{ of } \log_2 n\text{-element subsets of } S_n) \\ &\le (n!)^{\log_2 n} \le n^{n \log_2 n} \end{aligned}$$

Although this bound is evidently much stronger than the one discussed in Ex. 1.5, it is still far from perfect. For example, by applying more sophisticated methods one can prove [13] that $N_{\text{groups}}(n) \simeq n^{2\mu^2(n)/27}$, where $\mu(n)$ is the highest power to which any prime divides n.

3.24. In fact, these properties are crucial for many applications of group theory and will be extensively used in a number of further exercises.
a) If two orbits intersect $g_1 x_1 = g_2 x_2$ then $x_1 = g_1^{-1} g_2 x_2$, hence $G x_1 = G g_1^{-1} g_2 x_2 = G x_2$.
b) This simple observation unifies the Lagrange and centralizer theorems and is proved by the same line of arguments as each of them.
c) Using a) and b) above, we have:

$$|X| = \sum_{x \in X} 1 = \sum_{\text{orbits}} |Gx| = \sum_{\text{orbits}} |G : G_x| = \sum_{\text{orbits}} \frac{|G|}{|G_x|}$$

In particular, if $|G| = p^n$, then p divides $|G : G_x|$ unless $G_x = G$, meaning that x is a fixed point for all $g \in G$.

d) To count orbits instead of points, let us introduce a weighting factor $1/|Gx|$ and use b) from above:

$$|X/G| = \sum_{\text{orbits}} 1 = \sum_{x\in X} \frac{1}{|Gx|} = \frac{1}{|G|}\sum_{x\in X} |G_x| = \frac{1}{|G|}\sum_{x\in X}\Big(\sum_{g\in G} \delta_{x,g(x)}\Big)$$
$$= \frac{1}{|G|}\sum_{g\in G}\Big(\sum_{x\in X} \delta_{x,g(x)}\Big) = \frac{1}{|G|}\sum_{g\in G} |X_g|$$

3.25. a) Consider action of G on itself by conjugation, $gx = g^{-1}xg$. Then the fixed points are exactly those forming the group center $Z(G)$ and according to Ex. 3.24c), p divides $|Z(G)|$. But since $|Z(G)| \geq 1$, this means that $|Z(G)|$ contains at least p elements.

b) Since $Z(G)$ is non-trivial, we can consider a quotient group $G/Z(G)$ of order $p^{n'}$, $n' < n$, and proceed by induction.

c) If $n = 2$ then the two options are: (i) either $|Z(G)| = p^2$ and hence G is abelian from the beginning, or (ii) $|Z(G)| = p$, then $G/Z(G)$ is of order p and thus cyclic. In the latter case let $G/Z(G) = \langle a\rangle$, then any $g \in G$ should be written as $g = a^k z$ with $z \in Z(G)$. However, a pair of such elements evidently commutes.

3.26. Although the order $|A_4| = 24$ of A_4 is divided by 6, there are no subgroups of order 6 in A_4. For if it existed, it would be necessarily normal (see Ex. 3.4) in contradiction with the results of Ex. 3.11c). The rest of the exercise contains self-explanatory instructions.

3.27. For $G = S_p$ its order $p!$ is divided by p, and the only elements of order p are p-cycles in amount $(p-1)!$.

3.28. Since p does not divide m, the only element $g \in C_m$ satisfying $g^p = e$ is identity.

3.29. The solution reduces to direct analyses of the groups by increase of their order:

(i) Groups of prime orders $|G| = 2, 3, 5, 7, 11, \ldots$ are obviously all cyclic.

(ii) Groups of orders prime squared $|G| = 4, 9, \ldots$ are all abelian, but can have either one or two generators (see Exs 3.18, 3.25).

(iii) Group G of order $6 = 3\cdot 2$ by Cauchy theorem (see Ex. 3.26) contains two cyclic subgroups of orders 2 and 3 and is obviously spanned by their product. By Ex. 3.4 the subgroup of order 3 is normal, hence the group G is either a direct or a semidirect product of these subgroups. This results in two possibilities, $G = C_3 \otimes C_2 \approx C_6$ or $G = C_3 \rtimes C_2 \approx \mathrm{Dih}_3 \approx S_6$. The case of groups of order $10 = 5\cdot 2$ is similar.

(iv) For $|G| = 8$ there is a larger number of possibilities. By Lagrange theorem, nontrivial elements of G can have orders 8, 4 or 2. If G contains an element of order 8 then $G = C_8$. If G contains an element h of order 4 then it generates a normal subgroup $H = \langle h\rangle$ and all the remaining elements $h' \in G\backslash H$ satisfy $h'^2 \in H$. If h' has order 2 then either it commutes with h (then $G \approx C_4 \otimes C_2$) or not (then $G \approx$

Dih_4). Or if h' has order 4 then $h'^2 = h^2$ and $G \approx Q_8$. Finally, if G contains only elements of order 2 then $G \approx C_2 \otimes C_2 \otimes C_2$. The case $|G| = 12$ is still different yet it can be analyzed along similar lines yielding the results listed in the following table:

Order	Group	Remarks
1	$\{e\}$	trivial
2	C_2	cyclic
3	C_3	cyclic
4	C_4	cyclic
	$\mathrm{Dih}_2 = C_2 \otimes C_2 \approx \mathrm{N}_4$	abelian of 2 generators
5	C_5	cyclic
6	$C_6 = C_3 \otimes C_2$	cyclic
	$\mathrm{Dih}_3 = C_3 \rtimes C_2 \approx S_3$	non-abelian
7	C_7	cyclic
8	C_8	cyclic
	$C_4 \otimes C_2$	abelian of 2 generators
	$C_2 \otimes C_2 \otimes C_2$	abelian of 3 generators
	$\mathrm{Dih}_4 = C_4 \rtimes C_2$	non-abelian
	Q_8	non-abelian
9	C_9	cyclic
	$C_3 \otimes C_3$	abelian of 2 generators
10	$C_{10} = C_5 \otimes C_2$	cyclic
	$\mathrm{Dih}_5 = C_5 \rtimes C_2$	non-abelian
11	C_{11}	cyclic
12	$C_{12} = C_4 \otimes C_3$	cyclic
	$C_6 \otimes C_2$	abelian of 2 generators
	$Q_{12} \equiv \mathrm{Dic}_3 = C_3 \rtimes C_4$	non-abelian
	$A_4 = (C_2 \otimes C_2) \rtimes C_3$	non-abelian
	$\mathrm{Dih}_6 = C_6 \rtimes C_2 \approx \mathrm{Dih}_3 \otimes C_2$	non-abelian

3.30. Let H be a subgroup generated by a collection of representatives from all classes. In particular it contains representatives from each class, so that (i) any element $g_0 \in G$ can be represented as $g_0 = ghg^{-1}$, where $g \in G$ and $h \in H$. Now let us make two refinements: (ii) if $h = e$, then for all g in such way we always obtain $geg^{-1} = e$, for this reason we consider the case $h = e$ separately; (iii) if $g' \in gH$, then $g'hg'^{-1} = (gh')h(gh')^{-1} = g(h'hh'^{-1})g^{-1} = gh''g^{-1}$ for some $h'' \in H$, hence to avoid double counting when h runs through the whole H, it is enough to restrict the choice of g to representatives of cosets G/H. Taking into account (i)–(iii), we obtain: $1 + |G/H| \cdot (|H| - 1) \geq |G| = |G/H| \cdot |H|$ or $|G/H| \leq 1$, meaning that $H = G$.

3.31. Consider a regular polyhedron with n-polygonal faces and k edges meeting at each vertex. The main point is that, by definition, the *total* symmetry group acts effectively (with trivial stabilizers) and transitively (producing a single orbit) on flags, hence it is in one-to-one correspondence with them. In particular, its order can be calculated in three different ways: $|V| \cdot k \cdot 2$ (by enumerating the vertices, as for each vertex there are k adjacent edges and for each edge 2 adjacent faces); $2 \cdot |E| \cdot 2$ (by enumerating edges, as each edge has two vertices at endpoints and two adjacent faces); $2 \cdot n \cdot |F|$ (by enumerating faces, as each face should be bordered by n edges, each with 2 endpoints). Since the rotational symmetry group G is twice smaller, we have: $|G| = k|V| = 2|E| = n|F|$. However, separately on vertices, edges and faces, G acts non-effectively, with stabilizers of orders $|G|/|V|$, $|G|/|E|$ and $|G|/|F|$, respectively. Since identity stabilizes everything, it should be taken out and considered separately. To assemble G from stabilizers of each of the elements, we should thus have:

$$|G| = 1 + \frac{1}{2}\left[|V| \cdot \left(\frac{|G|}{|V|} - 1\right) + |E| \cdot \left(\frac{|G|}{|E|} - 1\right) + |F| \cdot \left(\frac{|G|}{|F|} - 1\right)\right]$$

where the correction factor 1/2 takes into account that each rotation stabilizes a pair of opposite elements rather than a single one. By substituting $|G| = 2|E|$ and using standard algebraic identities we obtain the equality $|V| - |E| + |F| = 2$ known as Euler theorem. Next, by expressing the numbers of vertices and faces through the number of edges, we obtain:

$$|E| = \frac{4}{\frac{1}{n} + \frac{1}{k} - \frac{1}{2}}$$

Since $|E| > 0$, we should have $\frac{1}{n} + \frac{1}{k} > \frac{1}{2}$. Thus all the possibilities can be listed explicitly:

k	n	$\|E\|$	$\|V\|$	$\|F\|$	$\|G\|$	Name
3	3	6	4	4	12	tetrahedron
3	4	12	8	6	24	cube
4	3	12	6	8	24	octahedron
3	5	30	20	12	60	dodecahedron
5	3	30	12	20	60	icosahedron

(A.1)

It is convenient for the reader to compare this result with the ADE classification of finite rotation subgroups discussed in Section 4.2.4. Indeed, the constraint $\frac{1}{n} + \frac{1}{k} > \frac{1}{2}$ derived in the above classification of polyhedra is equivalent to $\frac{1}{n} + \frac{1}{k} + \frac{1}{2} > 1$ which corresponds to searching the solutions of eq. (4.48) with $k_3 = 2$, $k_2 > 2$ that are those discussed in Section 4.2.3.2 yielding the three exceptional groups listed in eqs (4.61)–(4.63). Looking at the above table the reader should notice that the three exceptional solutions $(3,3,2), (4,3,2), (5,3,2)$ of the diophantine inequality just correspond to the five regular polyhedra the interchange of the first two numbers in the triplet being duality which interchanges the number of vertices with the number of faces. The tetrahedron is self-dual since $3 = 3$ and $4 = 4$.

3.32. The group acts on itself by conjugation and due to Ex. 3.24 we have:

$$|G| = \sum_{i=1}^{n} N_i = \sum_{i=1}^{n} \frac{|G|}{m_i}$$

where g_i and $C(g_i)$ are a representative of the i-th class and its centralizer, $N_i = |G|/m_i$ is the size of the i-th class, and $m_i = |C(g_i)|$. This equation is called the class equation. Let us enumerate the classes so that $N_1 \le N_2 \le \cdots \le N_n$, $m_1 \ge m_2 \ge \cdots \ge m_n$. Since identity always forms a separate 1-element class, we always have $N_1 = 1$ and $m_1 = |G|$. After reducing by common factor $|G|$ the class equation acquires the form $1 = \sum_{i=1}^{n} \frac{1}{m_i}$ and for a given n possesses only a finite number of solutions which can be easily listed. Among them we need only those dividing $m_1 = |G|$:

a) For $n = 2$ we obtain $m_2 = m_1/(m_1 - 1)$, hence $m_1 = 2$, $m_2 = 1$ and the class equation reads $2 = 1 + 1$. This only possibility uniquely describes C_2.

b) For $n = 3$ the class equation (after applying further restrictions $m_i | m_1$) has the following solutions (m_1, m_2, m_3): $(6, 3, 2)$, $(4, 4, 2)$ and $(3, 3, 3)$. They correspond to the class equations $6 = 1 + 2 + 3$, $4 = 1 + 1 + 2$ and $3 = 1 + 1 + 1$, respectively. The first one is realized by $\mathrm{Dih}_3 \approx S_3$ and the last one by C_3. As for the second one, it is ruled out by the following argument. Such group, if existed, should contain besides identity a one-element class $\{a\}$ and a 2-element class $\{g_1, g_2\}$. However, a centralizer of g_i is a subgroup of order $m_3 = 2$, thus is uniquely identified as $C(g_i) = \{e, g_i\}$. This means that $a^{-1} g_i a \ne g_i$ in contradiction with $g_i^{-1} a g_i = a$.

c) For $n = 4$ the class equation (after applying further restrictions $m_i | m_1$) has 12 solutions listed in the table below. Only three of them are identified with groups, the case $|G| = 1 + 1 + 2 + 2$ is excluded by observing that here we would have $|Z(G)| = 2$ and thus $|G/Z(G)| = 4$, necessarily resulting in commutativity of $G/Z(G)$ and hence of the whole G, in contradiction with the presence of 2-element classes. All the other spurious cases are excluded by the following general argument [15]. Consider the group of the order $|G| = 2n$ with the class equation of the form $2n = 1 + ? + n$, where the question mark stands for an unknown splitting of the remaining $n - 1$ elements into classes. Let $[n] = \{g_1, g_2, \ldots, g_n\}$, then $|C(g_i)| = 2n/n = 2$, hence $C(g_i) = \{e, g_i\}$, in particular $g_i^{-1} = g_i$, meaning that all g_i are involutions ($g_i^2 = e$), and $g_j g_i g_j \ne g_i$ for $j \ne i$. Due to the latter $(g_i g_j)^2 = (g_i g_j g_i) g_j \ne e$, hence the products $g_i g_j$ do not belong to $[n]$. In fact, by numerology the remaining $n - 1$ elements can be written as products $g_1 g_2, g_1 g_3, \ldots, g_1 g_n$ and since they are all not involutions, n can be only odd. All possible pairwise products $g_i g_j$ represent essentially the same elements, though differently enumerated. We have $g_1^{-1}(g_1 g_i) g_1 = g_i g_1 = (g_1 g_i)^{-1}$ meaning that each non-trivial element $g_i g_j$ is conjugated to its inverse. Besides,

$$(g_i g_1)^{-1}(g_1 g_j)(g_i g_1) = g_1 g_i g_1 (g_j g_i) g_1 = g_1 g_i (g_i g_j) = g_1 g_j$$

so that all the products mutually commute. Hence the only possible structure of classes in this case is

$$2n = 1 + \underbrace{2 + 2 + \cdots + 2}_{(n-1)/2} + n$$

(where each 2-element class contains mutually inverse elements). For example, the case $20 = 1 + 4 + 5 + 10$ is ruled out by that 10 is even, and $18 = 1 + 2 + 6 + 9$ is ruled out by that it includes a 6-element class instead of 2-element ones. Note that the cases $|G| = 6, 8, 24$ are ruled out also by the formula $|G| = 1 + \sum_{i=2}^{4} n_i^2$, where n_i are the dimensions of non-trivial representations.

# classes n	$\{m_i\}$	Class equation	Comment
2	$(2,2)$	$2 = 1 + 1$	C_2
3	$(6,3,2)$	$6 = 1 + 2 + 3$	$\mathrm{Dih}_3 \approx S_3$
	$(4,4,2)$	$4 = 1 + 1 + 2$	ruled out
	$(3,3,3)$	$3 = 1 + 1 + 1$	C_3
4	$(4,4,4,4)$	$4 = 1 + 1 + 1 + 1$	C_4, $C_2 \otimes C_2$
	$(6,6,3,3)$	$6 = 1 + 1 + 2 + 2$	ruled out
	$(6,6,6,2)$	$6 = 1 + 1 + 1 + 3$	ruled out
	$(8,8,4,2)$	$8 = 1 + 1 + 2 + 4$	ruled out
	$(10,5,5,2)$	$10 = 1 + 2 + 2 + 5$	Dih_5
	$(12,12,3,2)$	$12 = 1 + 1 + 4 + 6$	ruled out
	$(12,6,4,2)$	$12 = 1 + 2 + 3 + 6$	ruled out
	$(12,4,3,3)$	$12 = 1 + 3 + 4 + 4$	A_4
	$(18,9,3,2)$	$18 = 1 + 2 + 6 + 9$	ruled out
	$(20,5,4,2)$	$20 = 1 + 4 + 5 + 10$	ruled out
	$(24,8,3,2)$	$24 = 1 + 3 + 8 + 12$	ruled out
	$(42,7,3,2)$	$42 = 1 + 6 + 14 + 21$	ruled out

Hence there are only eight different groups (including a trivial one) with $n \leq 4$.

3.33. The symmetry of a square is Dih_4 (see Ex. 2.12) – not too rich to make direct counting uncontrollable. By noticing that Dih_4 is generated by four reflections, let us construct enumeration so that equivalent colorings are not double counted. Since two opposite vertices are transposed by a reflection in diagonal, there are $m(m+1)/2$ ways of their inequivalent (yet with respect to that only reflection) colorings. However, two such pairs are transposed by reflections in mid-lines. Hence the number of inequivalent colorings of two pairs of mutually opposite vertices is

$$\frac{\frac{m(m+1)}{2}\left(\frac{m(m+1)}{2} + 1\right)}{2}$$

However, there is another and much more general approach based on the application of the Burnside lemma (see Ex. 3.24d)). Indeed, action of Dih_4 splits the set of colored squares into orbits and what we are interested in is just the number of these orbits, exactly what the Burnside lemma is about. To count the number of orbits we need to know the number of *points* of that set which are fixed by each element of Dih_4. Obviously, two elements of the same conjugation class have the same number of fixed points, hence it is enough to count them for a representative of each class. This is done in the following table:

Conjugacy class	# elements	# fixed points for representative
identity	1	m^4
rotations by $\pm\pi/2$	2	m
rotation by π (inversion about origin)	1	m^2
reflections in mid-lines	2	m^2
reflections in diagonals	2	m^3

The last column is filled by observing that the only restriction imposed by invariance of a coloring is that the permuted vertices should be colored the same. For example, for identity we can paint each of the vertices independently and there are $m \cdot m \cdot m \cdot m = m^4$ such possibilities, while for rotations by $\pm\pi/2$ all vertices should be colored the same, thus giving only m possibilities. Now, by applying Burnside lemma, we obtain:

$$|X/G| = \frac{1}{8}(1 \cdot m^4 + 2 \cdot m + 1 \cdot m^2 + 2 \cdot m^2 + 2 \cdot m^3)$$

which in fact is equivalent to the result firstly obtained by means of the direct approach.

3.34. The $O_{24} \sim S_4$ group of rotational symmetries of a cube (see Ex. 2.14 and also Section 3.4.3 in the main text) acts on a set of colored cubes, thus splitting it into orbits. Calculation of the number of orbits is a standard application of Burnside lemma (see Ex. 3.24d)). Obviously, all representatives of the same conjugacy class of S_4 have the same number of fixed points.

a) For example, the fixed points of each of the order-2 rotations around the lines passing through the centers of a pair of opposite faces (which constitute a 3-element class, see Ex. 3.7) are such colorings that paint in the same color the remaining two pairs of opposite faces. Obviously, there are three possibilities for painting each of these pairs and each of the two opposite faces pierced by the rotation axis, all independent, hence there are $3 \cdot 3 \cdot 3 \cdot 3 = 81$ fixed points for each of such rotations in total. Proceeding by considering each of the remaining classes, we obtain:

Class	# elements	# fixed points for each element
identity	1	3^6
order-2 rotations, edge axis	6	3^3
order-2 rotations, face axis	3	3^4
order-3 rotations, corner axis	8	3^2
order-4 rotations, face axis	6	3^3

Hence the demanded number of colorings is

$$|X/G| = \frac{1}{24}(1\cdot 3^6 + 6\cdot 3^3 + 3\cdot 3^4 + 8\cdot 3^2 + 6\cdot 3^3) = 57$$

b) For example, the fixed points of each of the order-2 rotations around the lines passing through centers of a pair of opposite faces (which constitute a 3-element class, see Ex. 3.7) are such colorings that paint in the same colors the opposite vertices of those faces. Obviously, there are two possibilities for painting each of these four pairs of vertices, all independent, hence there are $2\cdot 2\cdot 2\cdot 2 = 16$ fixed points for each of such rotations in total. Proceeding by considering each of the remaining classes, we obtain:

Class	**# elements**	**# fixed points for each element**
identity	1	2^8
order-2 rotations, edge axis	6	2^4
order-2 rotations, face axis	3	2^4
order-3 rotations, corner axis	8	2^4
order-4 rotations, face axis	6	2^2

Hence the demanded number of colorings is

$$|X/G| = \frac{1}{24}(1\cdot 2^8 + 6\cdot 2^4 + 3\cdot 2^4 + 8\cdot 2^4 + 6\cdot 2^2) = 23$$

3.35. The symmetry of a double benzene ring is $\mathrm{Dih}_2 \approx \mathrm{N}_4$, the same as that of a rectangular: besides identity it includes reflections in vertical and horizontal symmetry axes and rotation by π. The problem is almost the same as the problem of enumerating colorings of the hydrogen sites (see, e.g., Ex. 3.33) with the only difference that now we would like to restrict the number of occurrences of each *color*. If we were not to impose such a restriction then the total number of isomers would be obtained by Burnside lemma as above, $(1\cdot 2^8 + 3\cdot 1\cdot 2^4)/4 = 76$. Imposing the restriction let us note, for a moment disregarding equivalent possibilities and focusing on combinatorics, that the coefficients of each term of the binomial expansions

$$(\mathrm{H}+\mathrm{NO}_2)^8 = \sum_{n=1}^{8} C_8^n(\mathrm{H})^n(\mathrm{NO}_2)^{8-n}, \quad ((\mathrm{H})^2+(\mathrm{NO}_2)^2)^4 = \sum_{k=1}^{4} C_4^k(\mathrm{H})^{2k}(\mathrm{NO}_2)^{8-2k}$$

obviously provide the numbers of ways to fill the vacant sites by n hydrogen atoms and $8-n$ nitrite groups, or by k pairs of hydrogen atoms and the remaining $8-2k$ nitrite groups. Hence let us come back to the Burnside lemma and replace the numbers 2^8, 2^4 of permutations with repetitions by the corresponding numbers of combinations C_8^n and $C_4^{n/2}$. In fact, following Pólya [125], we can even better construct the following polynomial:

$$P(\mathrm{H},\mathrm{NO}_2) = \frac{1}{4}[1\cdot(\mathrm{H}+\mathrm{NO}_2)^8 + 3\cdot 1\cdot((\mathrm{H})^2+(\mathrm{NO}_2)^2)^4]$$

$$= (\mathrm{H})^8 + 2(\mathrm{H})^7\mathrm{NO}_2 + 10(\mathrm{H})^6(\mathrm{NO}_2)^2 + 14(\mathrm{H})^5(\mathrm{NO}_2)^3 + 22(\mathrm{H})^4(\mathrm{NO}_2)^4 + 14(\mathrm{H})^3(\mathrm{NO}_2)^5 + 10(\mathrm{H})^2(\mathrm{NO}_2)^6 + 2\mathrm{H}(\mathrm{NO}_2)^7 + (\mathrm{NO}_2)^8$$

In particular, there is only a single stereoisomer of naphthalene, but there are two isomers of 1-nitronaphthalene and 14 isomers of 3-nitronaphthalene.

3.36. The definitions of the groups $\mathrm{GL}_n(\mathbb{F}_2)$, $\mathrm{SL}_n(\mathbb{F}_2)$ and $\mathrm{PSL}_n(\mathbb{F}_2)$ are discussed and their orders computed in Ex. 2.1.

a) For $q = 2$ we have $|\mathbb{F}_q^*| = q - 1 = 1$ and $|Z(\mathrm{SL}_n(\mathbb{F}_2)| = \gcd(n, q-1) = 1$, hence all these three groups coincide.
b) The group $\mathrm{PSL}_2(\mathbb{F}_q)$ naturally acts on a projective line $\mathbb{F}_q\mathrm{P}^1$. There are $q^2 - 1$ points in the plane $\mathbb{F}_q^2$ excluding the origin, each of them uniquely defines a line passing through origin, but each of such lines contains $|\mathbb{F}_q^*| = q - 1$ points besides the origin. Hence there are $(q^2 - 1)/(q - 1) = q + 1$ different lines passing through the origin, which are considered as points in $\mathbb{F}_q\mathrm{P}^1$. The homomorphism (*in fact, monomorphism*) $\mathrm{PSL}_2(\mathbb{F}_q) \to S_{q+1}$ arises since they are permuted by transformations of $\mathrm{PSL}_2(\mathbb{F}_q)$. Note that

$$|\mathrm{PSL}_2(\mathbb{F}_q)| = \frac{(q^2 - q)(q^2 - 1)}{(q - 1)\gcd(2, q - 1)} = \frac{(q - 1)q(q + 1)}{\gcd(2, q - 1)}$$

c) In particular, $|\mathrm{PSL}_2(\mathbb{F}_2)| = 1 \cdot 2 \cdot 3/1 = 6 = |S_3|$, hence $\mathrm{PSL}_2(\mathbb{F}_2) \approx S_3$.
d) In particular, $|\mathrm{PSL}_2(\mathbb{F}_3)| = 2 \cdot 3 \cdot 4/2 = 12$, hence the homomorphic image is a subgroup of index 2 in S_4, and thus is necessarily normal (see Ex. 3.4). But there is only a unique normal subgroup $A_4 \lhd S_4$.
e) In particular, $|\mathrm{PSL}_2(\mathbb{F}_4)| = 3 \cdot 4 \cdot 5/1 = 60$, hence the homomorphic image is a subgroup of index 2 in S_5, and thus is necessarily normal (see Ex. 3.4). But there is only a unique normal subgroup $A_5 \lhd S_5$.

3.37. The group $\mathrm{PSL}_3(\mathbb{F}_2)$ of order $(2^3 - 2^2) \cdot (2^3 - 2) \cdot (2^3 - 1) = 4 \cdot 6 \cdot 7 = 168$ (see Ex. 2.1) acts on a set of $|\mathbb{F}_2^{3*}|/|\mathbb{F}_2{}^*| = 2^2 + 2 + 1 = 7$ points of the Fano plane $\mathbb{F}_2P^2$, which can be listed by their homogeneous (not all zero) coordinates $(x \bmod 2, y \bmod 2, z \bmod 2)$.

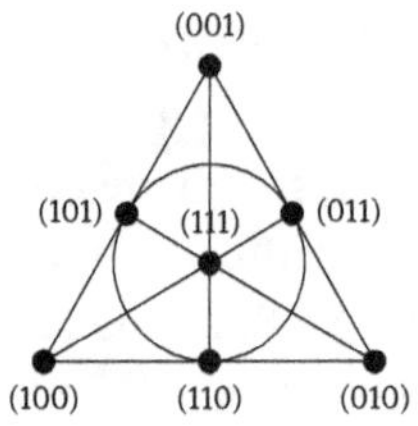

As usual, the lines are defined by a linear equation $ax + by + cz = 0 \pmod 2$, for example, the line $x + y = 0$ contains the points (001), (110) and (111). It is easy to observe that there are exactly seven pairwise intersecting lines each containing three points (see the figure above) and that a projective transformation from $PSL_3(\mathbb{F}_2)$ transforms triples of collinear points into triples of collinear points. The latter requirement is rather restrictive and by inspection one uniquely identifies the conjugacy classes:

Cycle structure	Representative	# elements
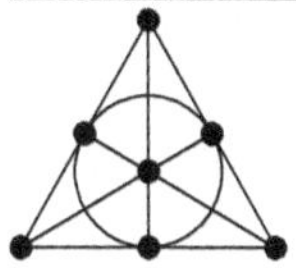	$\begin{pmatrix} 1 & 0 & 0 \\ 0 & 1 & 0 \\ 0 & 0 & 1 \end{pmatrix}$	1
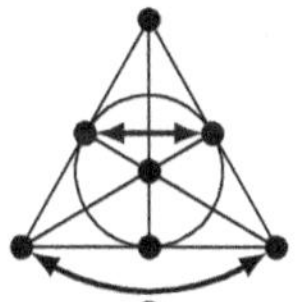	$\begin{pmatrix} 0 & 1 & 0 \\ 1 & 0 & 0 \\ 0 & 0 & 1 \end{pmatrix}$	21
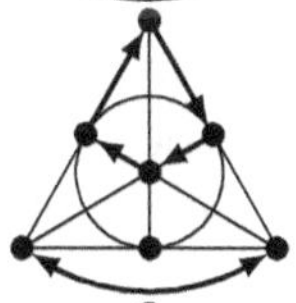	$\begin{pmatrix} 0 & 1 & 0 \\ 1 & 0 & 1 \\ 0 & 0 & 1 \end{pmatrix}$	42
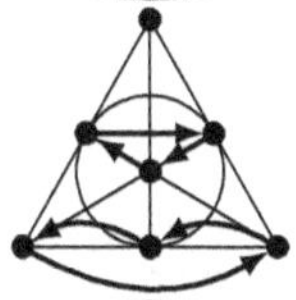	$\begin{pmatrix} 0 & 1 & 0 \\ 1 & 1 & 0 \\ 0 & 0 & 1 \end{pmatrix}$	56
	$\begin{pmatrix} 0 & 0 & 1 \\ 1 & 1 & 0 \\ 0 & 1 & 1 \end{pmatrix}$	24
	$\begin{pmatrix} 1 & 1 & 1 \\ 1 & 0 & 1 \\ 1 & 0 & 0 \end{pmatrix}$	24

The numbers of elements are determined as follows: a pair of transpositions fixes one line which can be picked up in seven ways, and an arbitrary point out of that line can be sent to any of the remaining $7 - 3 - 1 = 3$ points, thus giving $7 \cdot 3 = 21$ possibilities. For pairs of a transposition and a 4-cycle one can pick up the fixed line in seven ways, then send a point out of that line to any of the remaining $7 - 3 - 1 = 3$ points, and then again to the remaining $3 - 1 = 2$ points, thus making $7 \cdot 3 \cdot 2 = 42$ possibilities. Finally, a long cycle is uniquely determined by picking up its four first segments, the first can be chosen in $7 - 1 = 6$ ways, the second one should not go back or along the same line, thus giving two possibilities, and so on, thus making in total $6 \cdot 2^3 = 48$ representatives. However, since 48 does not divide 168, such cycles have to be split further into two classes of mutually inverse transformations, of 24 elements each. Since a proper normal subgroup, if it existed, should be combined from classes necessarily including the class of identity, its possible order can be only among $1+21=22$, $1+24=25$, $1+42=43$, $1+21+24=46$, $1+24+24=49$, $1+56=57$, $1+21+21+21=64$, $1+21+21+24=67$, $1+21+24+24=70$, $1+21+56=78$, $1+24+56=81$, $1+21+24+42=88$, $1+24+24+42=91$, $1+42+56=99$, $1+21+24+56=102$, $1+24+24+56=105$, $1+21+24+24+42=112$, $1+21+42+56=120$, $1+24+42+56=123$, $1+21+24+24+56=126$, $1+21+24+42+56=144$ or $1+24+24+42+56=147$, but none of these numbers divide 168. Finally, by using the table above and the Burnside lemma (see Ex. 3.24), for the number of inequivalent colorings in analogy with Ex. 3.33 we obtain:

$$\frac{1}{168}(1\cdot 3^7+21\cdot 3^5+2\cdot 24\cdot 3+42\cdot 3^3+56\cdot 3^3)=60$$

3.38. For abelian groups, each of the elements forms a distinct class and all the irreducible representations are 1-dimensional, meaning in particular that $\chi(e)=1$. Furthermore, the admissible values of a character on an element of order n are the n-th roots of unity. Taking into account also the orthogonality of different inequivalent irreps, we obtain:

a) For $C_2=\{e,g\}$, where $g^2=e$:

Irrep	**Class** [e]	[g]
trivial	1	1
non-trivial	1	−1

b) For $C_3=\{e,g,g^2\}$, where $g^3=e$:

Irrep	**Class** [e]	[g]	[g^2]
trivial	1	1	1
1st non-trivial	1	$\frac{-1+i\sqrt{3}}{2}$	$\frac{-1-i\sqrt{3}}{2}$
2nd non-trivial	1	$\frac{-1-i\sqrt{3}}{2}$	$\frac{-1+i\sqrt{3}}{2}$

c) For $C_4=\{e,g,g^2,g^3\}$, where $g^4=e$:

Irrep	**Class** [e]	[g]	[g^2]	[g^3]
trivial	1	1	1	1
1st non-trivial	1	−1	1	−1
2nd non-trivial	1	i	−1	$-i$
3rd non-trivial	1	$-i$	−1	i

d) For $N_4=\{e,(12)(34),(13)(24),(14)(23)\}\approx C_2\otimes C_2$, all non-identity elements are of order 2:

Irrep	Class			
	[e]	[(12)(34)]	[(13)(24)]	[(14)(23)]
trivial	1	1	1	1
1st non-trivial	1	1	−1	−1
2nd non-trivial	1	−1	1	−1
3rd non-trivial	1	−1	−1	1

3.39. The group S_3 is split into three classes (of identity, 2-cycles and 3-cycles, respectively), hence there should be three inequivalent irreps. Their dimensions $\{n_i\}$ should satisfy $|S_3| = 6 = n_1^2 + n_2^2 + n_3^2$. In our case this condition uniquely determines $n_1 = n_2 = 1$ and $n_3 = 2$. There trivial representation should be among the 1-dimensional ones. By the requirement of orthogonality or an argument relying on normality of the kernel one also uniquely identifies the second 1-dimensional irrep as the signature of a permutation. The 2-dimensional irrep is the same as constructed in Ex. 2.11. Its character can be either read off from the matrices given there, or alternatively computed from $\chi([e]) = 2$ and the orthogonality relations:

$$\chi([e]) + 3\chi([(12)]) + 2\chi([(123)]) = 0, \quad \chi([e]) - 3\chi([(12)]) + 2\chi([(123)]) = 0$$

The results are summarized in the character table:

Irrep	Class		
	[e]	[(12)]	[(123)]
1	1	1	1
1′	1	−1	1
2	2	0	−1

Since the characters of a 1-dimensional representation coincide with the representation, it only remains to construct the 2-dimensional representation D. For that, it is enough to determine the representation of the group generators. Of course, this can be done only up to unitary equivalence. The group S_3 is generated by $i = (12)$ and $r = (123)$. Let us go to a basis where $D(r)$ is diagonal. Since r is of order 3, we have $D^3(r) = D(r^3) = D(e) = 1$, hence the eigenvalues of $D(r)$ are the cubic roots of unity. Since $\chi(r) = \mathrm{Tr}(D(r)) = -1$, we can pick up $D(r) = \mathrm{diag}(\zeta, \zeta^2)$, where $\zeta^2 + \zeta + 1 = 0$. Now $D(i)$ can be determined from the requirements $D(i)D(r)D(i) = D(iri) = D(r^{-1}) = D^\dagger(r)$ and $D(i)^\dagger D(i) = 1$. Writing $D(i) = \left(\begin{smallmatrix} a & b \\ c & d \end{smallmatrix}\right)$, from the first requirement we obtain $a = d = 0$ and $bc = 1$, whereas the second one after that implies $|b| = |c| = 1$. Hence we can take $D(i) = \left(\begin{smallmatrix} 0 & 1 \\ 1 & 0 \end{smallmatrix}\right)$. The whole representation is now constructed by taking the various products of the obtained matrices $D(r)$ and $D(i)$. One can easily show that it is unitary equivalent to the representation explicitly written down in Ex. 2.11. The character of the regular representation $D^{(\mathrm{reg})}$ is given by

$$\chi^{(\mathrm{reg})}(g) = \begin{cases} |S_3| = 6, & g = e \\ 0, & g \neq e \end{cases}$$

Hence by using the orthogonality relations we have $D^{(\mathrm{reg})} = a_1 D_1 \oplus a_{1'} D_{1'} \oplus a_2 D_2$, where $a_i = \frac{1}{|G|} \sum_g \chi_i^*(g)\chi^{(\mathrm{reg})}(g)$. In particular,

$$a_1 = \frac{1}{6}(6\cdot 1 + 3\cdot 0\cdot 1 + 2\cdot 0\cdot 1) = 1$$
$$a_{1'} = \frac{1}{6}(6\cdot 1 + 3\cdot 0\cdot(-1) + 2\cdot 0\cdot 1) = 1$$
$$a_2 = \frac{1}{6}(6\cdot 2 + 3\cdot 0\cdot 0 + 2\cdot 0\cdot(-1)) = 2$$

One observes that each irrep occurs with the multiplicity equal to its dimension, which is in fact a general property of the regular representation (see eq. (3.93) of the main text).

3.40. The number of irreps is the same as the number of classes and one of them is necessarily the trivial 1-dimensional representation. Dimensions $\{n_i\}$ of the others in all the suggested cases can be uniquely guessed from the formula $|G| = \sum_i n_i^2$. Often, for groups with a small number of classes, the character table is uniquely determined by orthogonality relations.

a) For Dih_4, the equation $8 = 1^2 + n_2^2 + n_3^2 + n_4^2 + n_5^2$ has a single solution (assuming the dimensions are sorted in ascending order) $n_2 = n_3 = n_4 = 1$ and $n_5 = 2$. Taking into account that: (i) $\chi_i(e) = n_i$; (ii) on a class of elements of order n characters of 1-dimensional representations can take value only among the n-th roots of unity, (iii) which in addition should be real if that class contains their inverses (see Ex. 3.6), their possible values in our case are restricted to ± 1. Thus the character table is uniquely fixed by the orthogonality relations,

Irrep	Class				
	$\{e\}$	$\{r^2\}$	$\{r, r^3\}$	$\{i, r^2 i\}$	$\{ri, r^3 i\}$
1	1	1	1	1	1
1′	1	1	1	−1	−1
1″	1	1	−1	1	−1
1‴	1	1	−1	−1	1
2	2	−2	0	0	0

b) Since the structure of classes of Q_8 essentially repeats that of Dih_4, for which it was just shown to be uniquely singled out, the character table is just the same:

Irrep	Class				
	$\{1\}$	$\{-1\}$	$\{i, -i\}$	$\{j, -j\}$	$\{k, -k\}$
1	1	1	1	1	1
1′	1	1	1	−1	−1
1″	1	1	−1	1	−1
1‴	1	1	−1	−1	1
2	2	−2	0	0	0

Let us emphasize that the groups Q_8 and Dih_4 are not isomorphic, as can it be seen, e.g., by considering the structure of their subgroups or counting the elements of various orders.

c) For A_4, there are four classes, hence four inequivalent irreps. Their dimensions are uniquely determined by the equation $12 = 1^2 + n_2^2 + n_3^2 + n_4^2$, which admits a single solution $n_2 = n_3 = 1$, $n_4 = 3$. Hence the character table is derived along the same lines as in a) with the only difference that this time the class of 3-cycles is split into two of mutually inverse elements, hence the values of the characters of 1-dimensional representations on that classes should be complex and conjugated, thus fixed to $\zeta = (-1 + i\sqrt{3})/2$ and ζ^*:

Irrep	Class			
	$[e]$, 1	$[(ij)(kl)]$, 3	$[(ijk)]$, 4	$[(ijk)']$, 4
1	1	1	1	1
1′	1	1	ζ	ζ^*
1″	1	1	ζ^*	ζ
3	3	−1	0	0

d) For S_4, there are five classes, hence five inequivalent irreps. Their dimensions are again uniquely determined by the equation $24 = 1^2 + n_2^2 + n_3^2 + n_4^2 + n_5^2$, which admits a single solution $n_2 = 1$, $n_3 = 2$ and $n_4 = n_5 = 3$. After fixing the values $\chi_i(e)$ and determining the characters of 1-dimensional representations in the same manner as above, the remaining values can be reconstructed either mechanically by using again the orthogonality relations, or a bit more easily by using instead the completeness relations together with the expected symmetry between the two mutually adjoint 3-dimensional irreps:

Irrep	Class				
	$[e]$, 1	$[(ij)(kl)]$, 3	$[(ijk)]$, 8	$[(ij)]$, 6	$[(ijkl)]$, 6
1	1	1	1	1	1
1′	1	1	1	−1	−1
2	2	2	−1	0	0
3	3	−1	0	1	−1
3′	3	−1	0	−1	1

3.41. A rotational symmetry of a crystal lattice $\mathbb{Z}^3$ is given by a matrix of rotation by an angle φ with integer elements. But by changing the basis such matrix can be always transformed to

$$R_z(\varphi) = \begin{pmatrix} \cos\varphi & -\sin\varphi & 0 \\ \sin\varphi & \cos\varphi & 0 \\ 0 & 0 & 1 \end{pmatrix} \tag{A.2}$$

so that its trace $\chi(g) = 1 + 2\cos\varphi$ can take on integer values only for $\varphi = 0,\ \pi/3,\ \pi/2,\ 2\pi/3$ and π. These values are $3, 2, 1, 0$ and -1, respectively. Consider a (reducible) representation of G in $\otimes^n \mathbb{Z}^3$ with the character $\chi^n(g)$ (see Ex. 3.43), then the sum $\sum_{g\in G}\chi^n(g)$ equals to a product of $|G|$ and the number of instances of the trivial representation, i.e., it is divided by $|G|$. Thus $|G|$ divides $\sum_{g\in G} P(\chi(g))$ for any polynomial $P(x)$. Let us choose in particular $P(x) = (x+1)x(x-1)(x-2)$, then $P(\chi(g)) \neq 0$ only for $g = e$ and we have: $\sum_{g\in G} P(\chi(g)) = P(\chi(e)) = P(3) = 24$. This means that $|G|$ divides 24, i.e., can be only one out of $1, 2, 3, 4, 6, 8, 12, 24$. This approach can be developed further and leads to a complete classification of the Point Groups by listing all the groups of these orders (see, e.g., Ex. 3.29) and excluding groups containing elements of orders different from $1, 2, 3, 6$ (see the discussion at the beginning of Section 4.4 in the main text).

3.42. The symmetry group of a molecule is made of transformations that both permute the equilibrium sites of the atoms and rotate their shifts. Since the shifts are described by polar vectors attached to the equilibrium positions, the resulting representation is a direct product $D_G \otimes D_V$ of the representation D_G permuting the atoms and D_V transforming a polar vector. On the other hand, the group includes translations D_V and rotations D_R of a molecule as a whole, which should be excluded in order to obtain pure vibrations.

a) In the H_2O molecule the hydrogen atoms make an obtuse planer angle with the oxygen atom at the center. By directing z along the axis of order 2 (bisectrix of that angle) and y orthogonally to the plane of the molecule, the symmetry group of the molecule H_2O is $C_{2v} = \{E, C_2, \sigma_v, \sigma_v'\} \approx \mathrm{N}_4$, where E is identity, σ_v is reflection in xz plane, σ_v' is reflection in yz-plane and $C_2 = \sigma_v\sigma_v'$ is rotation by π around the z-axis. The character table for C_{2v} reads:

C_{2v}		E	C_2	σ_v	σ_v'
z	A_1	1	1	1	1
R_z	A_2	1	1	−1	−1
R_y, x	B_1	1	−1	1	−1
R_x, y	B_2	1	−1	−1	1
	D_G	3	1	3	1

where the first column identifies the actual representations of the vector components x, y, z and rotations R_x, R_y, R_z. Character of the reducible representation D_G with the values giving the number of atoms fixed by each group element is also included in the last row of the table. Obviously, we have: $D_G = 2A_1 + B_1$ and

$$\begin{aligned} D_{\mathrm{vibr}} &= D_G \otimes D_V - D_V - D_R = (2A_1 + B_1) \otimes (B_1 + B_2 + A_1) - (B_1 + B_2 + A_1) \\ &\quad - (B_2 + B_1 + A_2) = 2B_1 + 2B_2 + 2A_1 + A_1 + A_2 + B_1 - B_1 - B_2 \\ &\quad - A_1 - B_2 - B_1 - A_2 = 2A_1 + B_1 \end{aligned}$$

Hence there are $2 \cdot 1 + 1 = 3$ vibration modes, all with different frequencies, two of the symmetry A_1 (with the same eigenfrequencies) and one of the symmetry B_1, as shown in the figure:

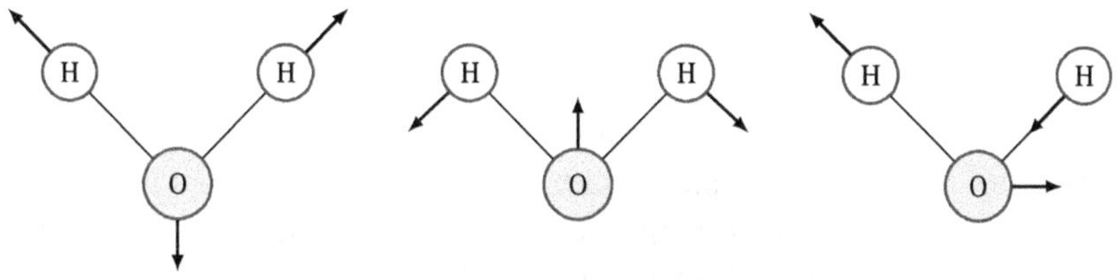

1st A_1 mode   2nd A_1 mode B_1 mode

b) By placing the $CHCl_3$ molecule to direct the CH bond vertically, the symmetry group is identified as $C_{3v} = \{E, C_3, C_3^{-1}, \sigma_v, \sigma_v', \sigma_v''\} \approx \mathrm{Dih}_3$. Its character table reads (compare to Ex. 3.39):

C_{3v}		E	$2C_3$	$3\sigma_v$
z	A_1	1	1	1
R_z	A_2	1	1	−1
$(R_x, R_y), (x, y)$	E	2	−1	0
	D_G	5	2	3

where in the last row again the character of the reducible representation D_G indicating what number of atoms is fixed by each group element is included. Obviously, we have: $D_G = 3A_1 + E$, hence the representation of the normal vibrations is computed as

$$
\begin{aligned}
D_{\text{vibr}} &= D_G \otimes D_V - D_V - D_R = (3A_1 + E) \otimes (E + A_1) - (E + A_1) - (E + A_2) \\
&= 3E + 3A_1 + (A_1 + A_2 + E) + E - E - A_1 - E - A_2 = 3A_1 + 3E
\end{aligned}
$$

This means that there are $3 \cdot 1 + 3 \cdot 2 = 9$ normal modes with 3 + 3 = 6 different frequencies.

c) The symmetry group of the CH_4 molecule is obviously $T_d \approx S_4$ (see Ex. 2.13) with the character table (compare to Ex. 3.40d)):

T_d		E	$8C_3$	$3C_2$	$6\sigma_d$	$6S_4$
	A_1	1	1	1	1	1
	A_2	1	1	1	−1	−1
	E	2	−1	2	0	0
(R_x, R_y, R_z)	T_1	3	0	−1	−1	1
(x, y, z)	T_2	3	0	−1	1	−1
	D_G	5	2	1	3	1

In this case we clearly have $D_G = 2A_1 + T_2$, hence the representation of the normal vibrations is computed as

$$\begin{aligned} D_{\text{vibr}} &= D_G \otimes D_V - D_V - D_R = (2A_1 + T_2) \otimes T_2 - T_2 - T_1 \\ &= 2T_2 + (A_1 + E + T_1 + T_2) - T_2 - T_1 = A_1 + E + 2T_2 \end{aligned}$$

This means that there are also (as in b) above) $1 + 2 \cdot 1 + 2 \cdot 3 = 9$ normal modes, but now due to extra symmetry with only $1 + 1 + 2 = 4$ different frequencies. It is suggested that the reader compares the present discussion with the detailed treatment of XY_4 molecule vibrations provided in Section 12.2.

Fascinating animations of the various vibration modes that we have just counted are provided, e.g., at the website of the chemical faculty of the Ludwig Maximilian University of Munich [146].

3.43. Let $D_{ij}(g)$ be the matrices of a representation of a group G in V with basis vectors $\{e_i\}$. Then the group G is also naturally represented in $\otimes^n V$ with basis vectors $e_{i_1} \otimes e_{i_2} \dots e_{i_n}$ by the matrices $D_{i_1 j_1}(g) D_{i_2 j_1}(g) \dots D_{i_n j_n}(g)$. The character of this representation is obviously $\chi_{\otimes^n V}(g) = \chi_V^n(g)$, where χ_V is the character of D. This representation is reducible, for example, $V \otimes V = (V \vee V) \oplus (V \wedge V)$. Note that $\otimes^n V$ carries also the regular representation of the group S_n acting by permuting the basis elements, $e_i \mapsto e_{\sigma(i)}$, hence we can define a (reducible) representation of the group $S_n \otimes G$ in $\otimes^n V$ by

$$D_{i_1 i_2 \dots i_n, j_1 j_2 \dots j_n}(\sigma, g) = D_{\sigma(i_1) j_1}(g) D_{\sigma(i_2) j_2}(g) \dots D_{\sigma(i_n) j_n}(g)$$

Its character is given by

$$\chi_{\otimes^n V}(\sigma, g) = \operatorname{Tr} D(\sigma, g) = \sum_{i_1, i_2, \dots, i_n} D_{\sigma(i_1) i_1}(g) D_{\sigma(i_2) i_2}(g) \dots D_{\sigma(i_n) i_n}(g)$$

The sum is simplified by taking into account the cycle decomposition of σ. For example, let $n = 6$ and $\sigma = (i_1 i_2 i_4)(i_3 i_5)$, then

$$\begin{aligned} &\sum_{i_1, \dots, i_6} D_{\sigma(i_1) i_1}(g) D_{\sigma(i_2) i_2}(g) \dots D_{\sigma(i_6) i_6}(g) \\ &= \Big(\sum_{i_1, i_2, i_3} D_{i_1 i_4}(g) D_{i_4 i_2}(g) D_{i_2 i_1}(g) \Big) \cdot \Big(\sum_{i_3, i_5} D_{i_3 i_5}(g) D_{i_5 i_3}(g) \Big) \cdot \Big(\sum_{i_6} D_{i_6 i_6}(g) \Big) \\ &= \Big(\sum_{i_1} D_{\sigma(i_1) i_1}(g^3) \Big) \cdot \Big(\sum_{i_3} D_{i_3 i_3}(g^2) \Big) \cdot \Big(\sum_{i_6} D_{i_6 i_6}(g) \Big) = \chi_V(g^3) \chi_V(g^2) \chi_V(g) \end{aligned}$$

Similarly, in general case we have:

$$\chi_{\otimes^n V}(\sigma, g) = \prod_i \chi_V^{r_i}(g^{l_i})$$

Now, if $\chi^{(\pm)}_{S_n}(\sigma)$ is a character of the trivial or of the alternating 1-dimensional representation of S_n, this irrep can be extracted by a projection

$$\chi^{(\pm)}_{\otimes^n V}(g) = \frac{1}{n!}\sum_{\sigma} \chi_{\otimes^n V}(\sigma, g)\chi^{(\pm)}_{S_n}(\sigma) = \frac{1}{n!}\sum_{\{r_i,l_i\}} \frac{n!}{\prod_i r_i! l_i^{r_i}} \chi_{\otimes^n V}([\{r_i,l_i\}],g)\chi^{(\pm)}_{S_n}([\{r_i,l_i\}])$$

Finally, since $\chi^{(+)}_{S_n}([\{r_i,l_i\}]) = 1$ and $\chi^{(-)}_{S_n}([\{r_i,l_i\}]) = (-1)^{\sum_i r_i(l_i-1)} = (-1)^{n-\sum_i r_i}$, we arrive at the desired formulas. Using them, for the special cases $n = 2,3$ we obtain:

$$\chi^{(\pm)}_{V\otimes V}(g) = \frac{1}{2!1^1}\chi^2_V(g) \pm \frac{1}{1!2^1}\chi_V(g^2) = \frac{1}{2}\chi^2_V(g) \pm \frac{1}{2}\chi_V(g^2)$$

$$\chi^{(\pm)}_{V\otimes V\otimes V}(g) = \frac{1}{6}\chi^3_V(g) \pm \frac{1}{2}\chi_V(g)\chi_V(g^2) + \frac{1}{3}\chi_V(g^3)$$

3.44. The number of independent components of a tensor/pseudotensor equals the multiplicity of a trivial/alternating representation in a representation by the tensors of demanded rank and symmetry.

a) Here C_{ik} is a symmetric $C_{ki} = C_{ik}$ polar tensor of rank 2, hence the corresponding representation is $\vee^2 V$. The character of the vector representation V is given in the table (see Ex. 3.42a)):

C_{2v}	E	C_2	σ_v	σ'_v
A_1	1	1	1	1
A_2	1	1	−1	−1
B_2	1	−1	−1	1
$V = A_1 + B_1 + B_2$	3	−1	1	1
$\vee^2 V$	$\frac{3^2+3}{2} = 6$	$\frac{(-1)^2+3}{2} = 2$	$\frac{1^2+3}{2} = 2$	2
$\wedge^3 V$	$\frac{3^3}{6} - \frac{3\cdot 3}{2} + \frac{3}{3} = 1$	$\frac{(-1)^3}{6} - \frac{(-1)\cdot 3}{2} + \frac{-1}{3} = 1$	$\frac{1^3}{6} - \frac{1\cdot 3}{2} + \frac{1}{3} = -1$	-1
$\vee^2(\vee^2 V)$	$\frac{6^2+6}{2} = 21$	$\frac{2^2+6}{2} = 5$	5	5

The character of the target representation $\vee^2 V$ is computed using the formula $\chi_{\vee^2 V}(g) = \frac{1}{2}(\chi^2_V(g) + \chi_V(g^2))$ (see Ex. 3.43), the computation is presented in details in the fifth row of the table. Using this data, the number of independent components is calculated as a scalar product $(\vee^2 V, A_1) = (6\cdot 1 + 2\cdot 1 + 2\cdot 1 + 2\cdot 1)/4 = 3$.

b) Here C_{ikl} is totally antisymmetric axial tensor of rank 3, hence the corresponding representation is $\wedge^3 V$. Its character is computed according to the formula

$$\chi_{\wedge^3 V}(g) = \frac{1}{6}\chi^3_V(g) - \frac{1}{2}\chi_V(g)\chi_V(g^2) + \frac{1}{3}\chi_V(g^3)$$

(see Ex. 3.43), the computation is presented in detail in the sixth row of the table in a) above. Using this data, the number of independent components is calculated as a scalar product $(\wedge^3 V, A_2) = (1\cdot 1 + 1\cdot 1 + (-1)\cdot(-1) + (-1)\cdot(-1)\cdot 1)/4 = 1$.

c) Here $C_{(ik)(lm)}$ is a rank-4 polar tensor, symmetric with respect to transposition of indices in first and second pairs and transposition of that pairs. Hence the corresponding tensor representation is $\vee^2(\vee^2 V)$. Calculation of its character is presented in the last row of the table in a) above. Using this data, the number of independent components is calculated as a scalar product $(\vee^2(\vee^2 V), A_1) = (21\cdot 1 + 5\cdot 1 + 5\cdot 1 + 5\cdot 1)/4 = 9$.

d) Here C_{ik} is a symmetric $C_{ki} = C_{ik}$ polar tensor of rank 2, hence the corresponding representation is $\vee^2 V$. The character of the vector representation V is given in the table (see Ex. 3.42b)):

C_{3v}	E	$2C_3$	$3\sigma_v$
A_1	1	1	1
A_2	1	1	−1
E	2	−1	0
$V = A_1 + E$	3	0	1
$\vee^2 V$	$\frac{3^2+3}{2} = ⑥$	$\frac{0^2+0}{2} = ⓪$	$\frac{1^2+3}{2} = ②$
$\wedge^3 V$	$\frac{3^3}{6} - \frac{3\cdot 3}{2} + \frac{3}{3} = ①$	$\frac{0^3}{6} - \frac{0\cdot 0}{2} + \frac{3}{3} = ①$	$\frac{1^3}{6} - \frac{1\cdot 3}{2} + \frac{1}{3} = (\text{-}1)$
$\vee^2(\vee^2 V)$	$\frac{6^2+6}{2} = (21)$	$\frac{0^2+0}{2} = ⓪$	$\frac{2^2+6}{2} = ⑤$

The character of the representation $\vee^2 V$ is computed in the fifth row of the table. Using this data, the number of independent components is calculated as a scalar product $(\vee^2 V, A_1) = (6\cdot 1 + 2\cdot 0\cdot 1 + 3\cdot 2\cdot 1)/6 = 2$.

e) C_{ikl} is totally antisymmetric axial tensor of rank 3, hence the corresponding representation is $\wedge^3 V$. Its character is computed in the sixth row of the table in e) above. Using this data, the number of independent components is calculated as a scalar product $(\wedge^3 V, A_2) = (1\cdot 1 + 2\cdot 1\cdot 1 + 3\cdot(-1)\cdot(-1))/6 = 1$.

f) $C_{(ik)(lm)}$ is a rank-4 polar tensor, symmetric with respect to transposition of induces in first and second pairs and transposition of that pairs. Hence the corresponding tensor representation is $\vee^2(\vee^2 V)$. Calculation of its character is presented in the last row of the table in e) above. Using this data, the number of independent components is calculated as a scalar product $(\vee^2(\vee^2 V), A_1) = (21\cdot 1 + 2\cdot 0\cdot 1 + 3\cdot 5\cdot 1)/6 = 6$.

g) Here C_{ik} is a symmetric $C_{ki} = C_{ik}$ polar tensor of rank 2, hence the corresponding representation is $\vee^2 V$. The character of the vector representation V is given in the table (see Ex. 3.42c)):

T_d	E	$8C_3$	$3C_2$	$6\sigma_d$	$6S_4$
A_1	1	1	1	1	1
A_2	1	1	1	−1	−1
$V = T_2$	3	0	−1	1	−1
$\vee^2 V$	$\frac{3^2+3}{2} = ⑥$	$\frac{0^2+0}{2} = ⓪$	$\frac{(-1)^2+3}{2} = ②$	$\frac{1^2+3}{2} = ②$	$\frac{(-1)^2+(-1)}{2} = ⓪$
$\wedge^3 V$	①	①	①	(-1)	(-1)
$\vee^2(\vee^2 V)$	$\frac{6^2+6}{2} = (21)$	$\frac{0^2+0}{2} = ⓪$	$\frac{2^2+6}{2} = ⑤$	⑤	$\frac{0^2+2}{2} = ①$

The character of the representation $\vee^2 V$ is computed in the fourth row of the table. Using this data, the number of independent components is calculated as a scalar product $(\vee^2 V, A_1) = (6 \cdot 1 + 8 \cdot 0 \cdot 1 + 3 \cdot 2 \cdot 1 + 6 \cdot 2 \cdot 1 + 6 \cdot 0 \cdot 1)/24 = 1$.

h) C_{ikl} is totally antisymmetric axial tensor of rank 3, hence the corresponding representation is $\wedge^3 V$. Its character is given in the fifth row of the table in g) above (though for brevity, the computational details for this case had to be suppressed). Using this data, the number of independent components is calculated as a scalar product $(\wedge^3 V, A_2) = (1 \cdot 1 + 8 \cdot 1 \cdot 1 + 3 \cdot 1 \cdot 1 + 6 \cdot (-1) \cdot 1 + 6 \cdot (-1) \cdot 1)/24 = 1$.

i) $C_{(ik)(lm)}$ is a rank-4 polar tensor, symmetric with respect to transposition of induces in first and second pairs and transposition of that pairs. Hence the corresponding tensor representation is $\vee^2(\vee^2 V)$. Calculation of its character is presented in the last row of the table in g) above. Using this data, the number of independent components is calculated as a scalar product $(\vee^2(\vee^2 V), A_1) = (21 \cdot 1 + 8 \cdot 0 \cdot 1 + 3 \cdot 5 \cdot 1 + 6 \cdot 5 \cdot 1 + 6 \cdot 1 \cdot 1)/24 = 3$.

As it could be expected, stronger the symmetry, smaller the amount of independent tensor components.

B MATHEMATICA NoteBooks

B.1 Introduction

Calculations of any kind in Group Theory are best suited for implementation on a computer.

The concepts behind all group theoretical constructions are clear and simple yet their actual realization typically involves the use of very large matrices and long iterative operations possible only through the running of computer codes.

As an integral part of this book we provide a certain number of MATHEMATICA NoteBooks whose conception and use is described in the following sections that enable the user to inspect explicit representations of finite groups and Lie algebras and perform manipulations on several objects of interest in this book. A particular feature of the present textbook is the explicit construction of four of the five exceptional Lie algebras[1] and the NoteBooks that make these constructions available to the reader or to the student are described in this chapter.

The MATHEMATICA NoteBooks can be downloaded by the reader at the site:

http://www.degruyter.com/books/978-3-11-055119-8

B.2 The NoteBook e88construziaNew

The structure of the $\mathbf{e}_{8(8)}$ Lie algebra was discussed in Section 9.4. This MATHEMATICA NoteBook is finalized to the explicit construction of the adjoint representation of the $\mathbf{e}_8$ Lie algebra by means of 248×248 matrices. Running this NoteBook one creates a library containing accessible objects that can be uploaded for further use. The obtained objects stored in the library are the following ones:

1. The simple roots are encoded in a file named $\boldsymbol{\alpha}$.
2. The set of all positive roots organized by height from height =1 (the simple roots) to height = 29 (the highest root). There are two files: one, named **rootaltorg** contains, organized in groups of the same height, the integer components of the roots in the simple root basis. The other, named **euclideroot** contains, organized in groups of the same height, the components of the roots in the standard Euclidean orthonormal basis of $\mathbb{R}^8$.
3. The file **Steppi** contains the 120 explicit 248×248 matrices representing the step operators E^{α} associated with each of the 120 positive roots. In this file the step operators are organized exactly in the same order as the corresponding roots in the file **rootaltorg**.

1 The only missing construction is that of the Lie algebra $\mathfrak{e}_6$ which anyhow can be obtained by truncation of the construction of $\mathfrak{e}_7$.

https://doi.org/10.1515/9783110551204-014

4. The file **Cartani** contains the eight explicit 248 × 248 matrices representing the Cartan generators $\mathscr{H}^i$ corresponding to the Euclidean components of the roots.
5. The generators $E^{-\alpha}$ associated with the negative roots are the transposed of the generators associated with the corresponding positive roots.
6. **Simples** is a file that contains the 8-step operators corresponding to the eight simple roots.
7. The file **racine** contains the enumeration of all the positive roots (in the Dynkin basis) organized in order of increasing height but not subdivided into height groups, just simply listed from 1 to 120.
8. The tensor **dimealt** contains the dimension of the sets of roots of height $h = 1, \dots, 29$.
9. The tensor **eracin** contains the roots in Euclidean form just enumerated from 1 to 120 as in the file **racine**.

B.3 The NoteBook g2construzia

This MATHEMATICA NoteBook is finalized to the explicit construction of the fundamental 7-dimensional representation of the $\mathfrak{g}_2$ Lie algebra by means of 7 × 7 matrices as it is described in Section 9.1. Running this NoteBook one creates accessible objects that can be used for other calculations. The obtained objects are the following ones:
1. The array **rutte** contains the six positive roots enumerated as in eq. (9.1).
2. The array **Steppig2** contains the step operators associated with the positive roots enumerated in the same order as in **rutte**.
3. The array **MSteppig2** contains the step operators associated with the negative roots enumerated in the same order as their opposites in **rutte**.
4. The array **Cartanig2** contains the Cartan generators in an orthonormal basis.
5. The array **TTsolv** contains the eight generators of the Borel solvable Lie algebra enumerated as follows: $\{H_1, H_2, E^{\alpha_1}, \dots, E^{\alpha_6}\}$.

B.4 The NoteBook e77construzia

This MATHEMATICA NoteBook is devoted to the construction of the $\mathfrak{e}_7$ Lie algebra in its fundamental representation which is 56-dimensional.

B.4.1 Instructions for the user

Evaluating this NoteBook one obtains a series of objects that are available for further calculations:
1. The object $\boldsymbol{\alpha}$ is an array of 7-vectors with seven components corresponding to the simple roots of $\mathfrak{e}_7$ as they are displayed in Figure 9.5 and in eq. (9.80).

2. The object $\boldsymbol{\beta}$ is an array of 7-vectors with seven components corresponding to the simple roots of $A_7 \subset \mathfrak{e}_7$ as they are displayed in eq. (9.105). A_7 is the Lie algebra of the subgroup $\mathrm{SL}(8, \mathbb{R})$ which is the so-called electric subgroup. In the fundamental 56 representation of $\mathfrak{e}_7$ the matrices representing $\mathrm{SL}(8, \mathbb{R})$ are block-diagonal with two blocks 28×28.
3. The object **rho** is an array of 28 integer-valued 7-vectors. These are the 28 positive roots of A_7 expressed as integer-valued linear combinations of the simple roots of $\mathfrak{e}_7$. Obviously these 28 roots are a subset of the 63 positive roots of $\mathfrak{e}_7$. These roots are enumerated in Table 9.7.
4. The object **EP** is an array containing the 28 positive step operators of A_7 given as 56×56 matrices, namely, within the fundamental representation of $\mathfrak{e}_7$. These step operators are enumerated in the same order as the corresponding roots are enumerated in **rho**.
5. The object **EM** is an array containing the 28 negative step operators of A_7 given as 56×56 matrices, namely, within the fundamental representation of $\mathfrak{e}_7$. These step operators are enumerated in the same order as the corresponding roots are enumerated in **rho**. They are the transposed of the corresponding **EP**.
6. The object **racine** is an array of 63 integer-valued 7-vectors. These are the 63 positive roots of $\mathfrak{e}_7$ expressed as integer-valued linear combinations of the simple roots of $\boldsymbol{\alpha}$. The roots are enumerated by height starting from the simple ones up to the highest one.
7. The object **rutte** is an array of 63 seven-vectors. These are the 63 positive roots of $\mathfrak{e}_7$ expressed in the same Euclidean basis where the simple roots are those of eq. (9.80). They are enumerated in the same order as in the object **racine**.
8. The object **E7P** is an array that contains the positive step operators of $\mathfrak{e}_7$ written as 56×56 matrices and enumerated in the same order as the corresponding roots **rutte** (or **racine**).
9. The object **E7M** is an array that contains the negative step operators of $\mathfrak{e}_7$ written as 56×56 matrices and enumerated in the same order as the corresponding roots **rutte** (or **racine**).
10. The object **Cartani** is an array that contains the seven Cartan generators of $\mathfrak{e}_7$ written as 56×56 matrices, each corresponding to an axis of the Euclidean basis in which the simple roots have been written in eq. (9.80).
11. The object **W** is an array that contains (given in the Euclidean basis) the seven fundamental weights of $\mathfrak{e}_7$.
12. The object **WW** is an array that contains (given in the Euclidean basis) the 56 weights of the fundamental representation of $\mathfrak{e}_7$.
13. The weights can be represented as the maximal weight **vmax** = $\boldsymbol{W}_1$ minus an integer-valued linear combination of simple roots. The vectors expressing these linear combinations are encoded in the array **qqq** which corresponds to the weights according to the formula: $W_i = W_{\mathrm{max}} - \sum_{j=1}^{7} \mathrm{qqq}_{[[i,j]]} \alpha_j$.

B.5 The NoteBook F44construzia.nb

The theory of the exceptional Lie Algebra $\mathfrak{f}_4$ is extensively discussed in Section 9.2 where its adjoint and fundamental representations are constructed. The reader and student of this book interested in manipulations involving the $\mathfrak{f}_4$ Lie algebra in its maximally non-compact form $\mathfrak{f}_{(4,4)}$ can easily obtain the needed objects from the MATHEMATICA NoteBook described in the present section and named **F44construzia.nb**. After evaluation of such a *math-code* a series of objects become available to the user that are named and described here below.

B.5.1 The $\mathfrak{f}_4$ Lie algebra

1. The file containing the roots of $\mathfrak{f}_4$ is named **rutteF4**.
2. The file containing the Cartan operators of $\mathfrak{f}_4$ is named **Carti**.
3. The file containing the step operators of positive roots of $\mathfrak{f}_4$ (in the same order as the root is in **rutteF4**) is named **stepsi**.
4. The file containing the step operators of negative roots of $\mathfrak{f}_4$(in the same order as the root is in **rutteF4**) is named **antistepsi**.

All the operators are in the fundamental 26×26-dimensional representation of $\mathfrak{f}_{4(4)}$. The roots of the Lie algebra are of the form:

$$\text{roots} = \begin{pmatrix} \alpha_1 \\ \alpha_2 \\ \alpha_3 \\ \alpha_4 \\ \alpha_1+\alpha_2 \\ \alpha_2+\alpha_3 \\ \alpha_3+\alpha_4 \\ \alpha_1+\alpha_2+\alpha_3 \\ \alpha_2+2\alpha_3 \\ \alpha_2+\alpha_3+\alpha_4 \\ \alpha_1+\alpha_2+2\alpha_3 \\ \alpha_1+\alpha_2+\alpha_3+\alpha_4 \\ \alpha_2+2\alpha_3+\alpha_4 \\ \alpha_1+2\alpha_2+2\alpha_3 \\ \alpha_1+\alpha_2+2\alpha_3+\alpha_4 \\ \alpha_2+2\alpha_3+2\alpha_4 \\ \alpha_1+2\alpha_2+2\alpha_3+\alpha_4 \\ \alpha_1+\alpha_2+2\alpha_3+2\alpha_4 \\ \alpha_1+2\alpha_2+3\alpha_3+\alpha_4 \\ \alpha_1+2\alpha_2+2\alpha_3+2\alpha_4 \\ \alpha_1+2\alpha_2+3\alpha_3+2\alpha_4 \\ \alpha_1+2\alpha_2+4\alpha_3+2\alpha_4 \\ \alpha_1+3\alpha_2+4\alpha_3+2\alpha_4 \\ 2\alpha_1+3\alpha_2+4\alpha_3+2\alpha_4 \end{pmatrix} = \begin{pmatrix} -1 & -1 & -1 & 1 \\ 0 & 0 & 2 & 0 \\ 0 & 1 & -1 & 0 \\ 1 & -1 & 0 & 0 \\ -1 & -1 & 1 & 1 \\ 0 & 1 & 1 & 0 \\ 1 & 0 & -1 & 0 \\ -1 & 0 & 0 & 1 \\ 0 & 2 & 0 & 0 \\ 1 & 0 & 1 & 0 \\ -1 & 1 & -1 & 1 \\ 0 & -1 & 0 & 1 \\ 1 & 1 & 0 & 0 \\ -1 & 1 & 1 & 1 \\ 0 & 0 & -1 & 1 \\ 2 & 0 & 0 & 0 \\ 0 & 0 & 1 & 1 \\ 1 & -1 & -1 & 1 \\ 0 & 1 & 0 & 1 \\ 1 & -1 & 1 & 1 \\ 1 & 0 & 0 & 1 \\ 1 & 1 & -1 & 1 \\ 1 & 1 & 1 & 1 \\ 0 & 0 & 0 & 2 \end{pmatrix} \tag{B.1}$$

B.5.2 Maximal compact subalgebra $\mathfrak{su}(2) \times \mathfrak{sp}(6,\mathbb{R}) \subset \mathrm{F}_{(4,4)}$

The generators of the maximal compact subalgebra are 24 and are of the form: $H_i = \mathrm{E}[\alpha_i] - \mathrm{E}[-\alpha_i]$.

B.5.2.1 The subalgebra $\mathfrak{su}(2)$

The $\mathfrak{su}(2)$ part of the H subalgebra is generated by:

$$\{J_x, J_y, J_x\} = \left\{\frac{H_1 - H_{14} + H_{20} - H_{22}}{4\sqrt{2}}, \frac{H_5 + H_{11} - H_{18} + H_{23}}{4\sqrt{2}}, -\frac{H_2 - H_9 + H_{16} + H_{24}}{4\sqrt{2}}\right\} \tag{B.2}$$

and it is contained in the following files:

1. Formal expression in **formSO3gen.**
2. Explicit matrix representation **SO3gen.**

All generators given in the 26 × 26 fundamental representation.

B.5.2.2 The subalgebra $\mathfrak{usp}(6)$

The other factor of the compact subalgebra is generated by 21 generators encoded in the following files:

1) Formal expression in

$$\mathbf{formUsp6gen} = \begin{pmatrix} -\frac{H_2}{2} - \frac{H_9}{2} + \frac{H_{16}}{2} - \frac{H_{24}}{2} \\ -\frac{H_2}{2} + \frac{H_9}{2} + \frac{H_{16}}{2} + \frac{H_{24}}{2} \\ \frac{H_2}{2} + \frac{H_9}{2} + \frac{H_{16}}{2} - \frac{H_{24}}{2} \\ H_{10} \\ H_7 \\ H_4 \\ -H_{13} \\ H_6 \\ -H_3 \\ -H_1 + H_{14} + H_{20} - H_{22} \\ -H_5 - H_{11} - H_{18} + H_{23} \\ H_{21} \\ -H_8 \\ H_1 + H_{14} + H_{20} + H_{22} \\ H_5 - H_{11} - H_{18} - H_{23} \\ -H_1 - H_{14} + H_{20} + H_{22} \\ H_5 - H_{11} + H_{18} + H_{23} \\ H_{17} \\ H_{15} \\ H_{12} \\ H_{19} \end{pmatrix}$$

2) Explicit matrix representation **Usp6gen** (26 × 26).

B.5.3 The $\mathfrak{sl}(2,\mathbb{R}) \times \mathfrak{sp}(6,\mathbb{R})$ subalgebra and the W-representation

B.5.3.1 The subalgebra $\mathfrak{sl}(2,\mathbb{R})$

The $\mathfrak{sl}(2,\mathbb{R})$ subalgebra is generated by

$$\{L_+, L_-, L_0\} = \left\{ \frac{E^{\alpha_{24}}}{\sqrt{2}}, \frac{E^{-\alpha_{24}}}{\sqrt{2}}, \frac{\mathcal{H}_4}{2} \right\} \tag{B.3}$$

and it is contained in the following objects:

1. Explicit matrix representation in **LLp.**
2. Explicit matrix representation **LLm.**
3. Explicit matrix representation **LL0.**

All generators are given in the 26×26 fundamental representation.

B.5.3.2 The Lie algebra $\mathfrak{sp}(6,\mathbb{R})$

1. The generators of the $\mathfrak{sp}(6,\mathbb{R})$ Lie algebra are in the file **Sp6RgenF4.**
2. Their formal expression in terms of F4 generators is contained in the file:

$$\textbf{formSp6RgenF4} = \begin{pmatrix} \mathcal{H}_1 \\ \mathcal{H}_2 \\ \mathcal{H}_3 \\ \mathrm{E}\alpha[4] \\ \mathrm{E}\alpha[3] \\ \mathrm{E}\alpha[2] \\ \mathrm{E}\alpha[7] \\ -\mathrm{E}\alpha[6] \\ \mathrm{E}\alpha[10] \\ -\mathrm{E}\alpha[9] \\ \mathrm{E}\alpha[13] \\ \mathrm{E}\alpha[16] \\ \mathrm{E}\alpha[-4] \\ \mathrm{E}\alpha[-3] \\ \mathrm{E}\alpha[-2] \\ \mathrm{E}\alpha[-7] \\ -\mathrm{E}\alpha[-6] \\ \mathrm{E}\alpha[-10] \\ -\mathrm{E}\alpha[-9] \\ \mathrm{E}\alpha[-13] \\ \mathrm{E}\alpha[-16] \end{pmatrix} \tag{B.4}$$

3. The W-generators transforming in the **14** of $\mathfrak{sp}(6,\mathbb{R})$ are encoded in the file **Wgen.**

4. The formal expression of the **W**-generators in terms of the F4 generators is encoded in the file

$$\mathbf{formWgen} = \begin{pmatrix} \mathrm{E}\alpha[5] \\ \mathrm{E}\alpha[20] \\ \mathrm{E}\alpha[14] \\ -\mathrm{E}\alpha[23] \\ \mathrm{E}\alpha[21] \\ \mathrm{E}\alpha[19] \\ -\mathrm{E}\alpha[17] \\ -\mathrm{E}\alpha[22] \\ -\mathrm{E}\alpha[11] \\ -\mathrm{E}\alpha[18] \\ -\mathrm{E}\alpha[1] \\ -\mathrm{E}\alpha[8] \\ -\mathrm{E}\alpha[12] \\ -\mathrm{E}\alpha[15] \end{pmatrix} \tag{B.5}$$

5. The matrices of the 14-dimensional representation of $\mathfrak{sp}(6,\mathbb{R})$ are encoded in the file **D14**. The generators are ordered in the following way: first the three Cartan generators, then the nine step-up operators, finally the nine step-down operators associated with the negative roots of $\mathfrak{sp}(6,\mathbb{R})$.
6. The roots of $\mathfrak{sp}(6,\mathbb{R})$ are encoded in the file

$$\mathbf{ruttasp6} = \begin{pmatrix} \alpha_1 & \{1,-1,0\} \\ \alpha_2 & \{0,1,-1\} \\ \alpha_3 & \{0,0,2\} \\ \alpha_1+\alpha_2 & \{1,0,-1\} \\ \alpha_2+\alpha_3 & \{0,1,1\} \\ \alpha_1+\alpha_2+\alpha_3 & \{1,0,1\} \\ 2\alpha_2+\alpha_3 & \{0,2,0\} \\ \alpha_1+2\alpha_2+\alpha_3 & \{1,1,0\} \\ 2\alpha_1+2\alpha_2+\alpha_3 & \{2,0,0\} \end{pmatrix} \tag{B.6}$$

7. The invariant symplectic matrix of Sp(6,R) corresponding to the 14 representation is encoded in the file **C14**.

B.6 The NoteBook L168Group

The theory of the remarkable simple group L_{168} is extensively discussed in Section 4.4. The MATHEMATICA NoteBook L168Group constructs all the objects mentioned in that section and provides an ample range of subroutines to calculate subgroups, irreducible representations, decomposition of representations into irreps and all that.

B.6.1 Description of the generated objects

B.6.1.1 Generators

The group L168 has three generators S, T, R with the relations given in eq. (4.104). Abstractly we use the Greek letter for the generators of L168 $R = \rho$, $S = \sigma$, $T = \tau$.

In the programme there are three 7-dimensional realizations of these generators:

1. **R,S,T** are 7×7 matrices in the orthonormal basis.
2. **RL,SL,TL** are 7×7 matrices in the basis of A7 simple roots where the metric is the A7 Cartan matrix (ROOT LATTICE).
3. **RW,SW,TW** are 7×7 matrices in the basis of A7 simple weights where the metric is the A7 Cartan matrix (WEIGHT LATTICE).

B.6.1.2 Conjugacy classes

The group L168 has six conjugacy classes organized as displayed in eq. (4.23). The abstract form of the group elements as **words** in the generator symbols $\boldsymbol{\rho}, \boldsymbol{\sigma}, \boldsymbol{\tau}$ is given in the file **formL168clas** which has the following appearance:

$$\begin{aligned}
\text{formL168clas}[[2]] &= \{\mathscr{E}, 1, \{\epsilon\}\} \\
\text{formL168clas}[[2]] &= \{\mathscr{R}, 21, \{\rho.\sigma.\rho.\sigma.\rho.\tau.\tau, \tau.\tau.\sigma.\rho.\tau.\tau.\tau, \sigma.\rho.\sigma.\sigma, \\
&\quad \sigma.\rho.\tau.\tau.\sigma.\rho.\tau.\tau, \sigma.\rho.\tau.\sigma.\rho.\tau, \\
&\quad \tau.\sigma, \rho.\sigma.\rho.\tau.\sigma.\rho.\tau, \tau.\tau.\sigma.\rho.\tau.\sigma.\sigma, \\
&\quad \rho.\tau.\tau.\sigma.\rho.\sigma.\rho, \tau.\sigma.\rho.\tau.\sigma, \sigma.\rho.\sigma.\rho.\tau.\sigma.\sigma, \rho.\tau.\tau.\sigma.\rho.\tau.\tau, \\
&\quad \rho.\sigma.\rho.\tau, \rho.\tau.\sigma.\rho.\tau.\sigma, \tau.\tau.\sigma.\rho.\tau.\tau.\sigma.\rho, \\
&\quad \tau.\tau.\tau.\sigma.\rho.\tau.\tau, \tau.\tau.\sigma.\rho.\sigma, \rho.\tau.\sigma.\rho, \\
&\quad \tau.\sigma.\rho.\tau.\tau.\sigma.\rho.\tau, \tau.\sigma.\rho.\sigma.\rho.\tau.\tau.\sigma, \rho\}\} \\
\cdots &= \cdots
\end{aligned} \tag{B.7}$$

This file is very important in the construction of irreducible representations in order to provide in each case the precise form of the representation of each individual group element. In the programme, after evaluation of this NoteBook the explicit matrices of the group organized by conjugacy classes are stored in two different files:

1. **L168strutclas** contains the 7×7 matrices in the simple root basis, namely, those acting on the configuration space of the root lattice.
2. **L168clasW** contains the 7×7 matrices in the simple root basis, namely, those acting on the configuration space of the root lattice. The standard realization of the seven simple roots as vectors in eight dimensions orthogonal to the vector $\{1, 1, 1, 1, 1, 1, 1, 1\}$ is encoded in the object **alp** while the simple weights are encoded in the object **lammi**; finally, the Cartan matrix is encoded in the object **CC**.

The set of all the group elements is provided in two files:
1. **grupL168** contains the 168 matrices 7×7 in the configuration space (root lattice).
2. **L168W** contains the 168 matrices 7×7 in momentum space (weight lattice).

B.6.1.3 Characters

The six irreducible representations of the group L168 are of dimensions **1,6,7,8,3** and **3**, respectively. The character table is displayed in eq. (4.107) of the main text. The character table is encoded in the object **PchiL168.** The populations of conjugacy classes are encoded in the object **PgiL168.** The names of the representations are encoded in the object **namesD168.**

B.6.1.4 Embedding into $\mathfrak{g}_2$ and τ-matrices

The embedding of the group L168 into G_2 is discussed in Section 4.4.3 and it is demonstrated by showing that there is a G_2-invariant 3-tensor that is invariant under L168. The tensor which satisfies the G_2-relations and which is L168 invariant is named $\boldsymbol{\phi}$ in the orthonormal basis and $\boldsymbol{\varphi}$ in the root basis. The basis of 8×8 gamma matrices in seven dimensions such that the G_2-invariant tensor coincides with $\boldsymbol{\phi}_{\mathbf{ijk}} = \boldsymbol{\eta}\boldsymbol{\tau}_{\mathbf{ijk}}\boldsymbol{\eta}$ are constructed and encoded in the file $\boldsymbol{\tau\tau}$ (see Section 4.4.3 of the main text).

B.6.1.5 The maximal subgroups G21 and O24A, O24B

The simple group L168 contains maximal subgroups only of index 8 and 7, namely, of order 21 and 24. The order 21 subgroup G21 is the unique non-abelian group of that order and abstractly it has the structure of the semidirect product $\mathbb{Z}_3 \propto \mathbb{Z}_7$. Up to conjugation there is only one subgroup 21 as we have explicitly verified with the computer. On the other hand, up to conjugation there are two different groups of order 24 that are both isomorphic to the octahedral group O_{24}. They are named O24A, O24B.

B.6.1.6 The group G21

The group G21 has two generators, $\mathscr{X}$ and $\mathscr{Y}$, that satisfy the relations displayed in eq. (4.134) of the main text. The conjugacy classes are displayed in eq. (4.135).

The five irreducible representations of the group G21 are of dimensions **1,1,1,3** and **3**, respectively. The character table is given in eq. (4.137). After evaluation of this NoteBook, the character table is encoded in the object **Pchi21.** The populations of conjugacy classes are encoded in the object **Pgi21.** The names of the representations are encoded in the object **names21.** Our choice of the representative for the entire conjugacy class of G21 maximal subgroups is given, abstractly, by setting the following generators:

$$\mathscr{Y} = \rho.\tau.\tau.\tau.\sigma.\rho; \quad \mathscr{X} = \sigma.\rho.\sigma.\rho.\tau.\tau \tag{B.8}$$

In the programme the entire group G21 is encoded in the following files.

1. **G21W** contains all the 21 elements as 7×7 matrices in the weight basis.
2. **G21clasW** contains all the conjugacy classes in the weight basis (momentum space).
3. **G21clasR** contains all the conjugacy classes in the root basis (configuration space).
4. **FormalG21clas** contains the group generators written symbolically as words in ρ, σ, τ organized in conjugacy classes which mention the choice of $\mathcal{Y}, \mathcal{X}$ generators, as follows:

$$\begin{aligned} \text{FormalG21clas[[1]]} &= \{\mathscr{Y}, \mathscr{X}\} \\ \text{FormalG21clas[[2]]} &= \{\{1, \{\epsilon\}\}, \\ &\quad \{3, \{\rho.\sigma.\rho.\sigma.\rho.\tau.\sigma, \rho.\tau.\tau.\tau.\sigma.\rho, \tau.\sigma.\rho.\sigma.\rho.\tau.\sigma.\sigma\}\}, \\ &\quad \{3, \{\sigma.\rho.\sigma.\rho.\tau.\tau.\sigma, \rho.\sigma.\rho.\tau.\tau.\sigma.\sigma, \tau.\sigma.\rho.\tau.\tau\}\}, \\ &\quad \dots \end{aligned} \tag{B.9}$$

On the other hand:

$$\begin{aligned} \text{FormG21YX} = \{&\{1, \{\epsilon\}\}, \{3, \{\mathscr{Y}.\mathscr{Y}, \mathscr{Y}, \mathscr{X}.\mathscr{X}.\mathscr{Y}.\mathscr{X}\}\}, \\ &\{3, \{\mathscr{Y}.\mathscr{Y}.\mathscr{Y}, \mathscr{X}.\mathscr{Y}.\mathscr{X}.\mathscr{Y}.\mathscr{X}, \mathscr{Y}.\mathscr{X}.\mathscr{X}.\mathscr{Y}.\mathscr{X}\}\}, \\ &\dots \end{aligned} \tag{B.10}$$

is a file that contains all the group elements of the G21 group organized into conjugacy classes in the standard order and expressed as formal words in the generators.

B.6.1.7 Irreducible representations of G21

The irreducible representations of G21 are provided in the form of substitution rules named **repG21D1, repG21D2, repG21D3, repG21D4, repG21D5** for the generators $\{\mathscr{Y}, \mathscr{X}\}$. On the other hand, the object:

$$\text{formRepra21} = \{\epsilon, \mathscr{Y}, \mathscr{X}.\mathscr{X}.\mathscr{Y}.\mathscr{X}.\mathscr{Y}.\mathscr{Y}, \mathscr{Y}.\mathscr{X}.\mathscr{X}, \mathscr{X}\} \tag{B.11}$$

contains a representative for each conjugacy class.

B.6.1.8 The octahedral subgroups

The octahedral group O24 has two generators, S and T, that satisfy the relations mentioned in eq. (4.139) of the main text. The conjugacy classes are displayed in eq. (4.140). The five irreducible representations of the group O24 are of dimensions **1,1,2,3** and **3**, respectively. The character table is displayed in eq. (4.141). The character table is encoded in the object **Pchi24**. The populations of conjugacy classes are encoded in the object **Pgi24**. The names of the representations are encoded in the object **names24**. In the program the entire groups O24 are encoded in the following objects: **O24A** and **O24B** that contain all the 7×7 matrices representing these maximal subgroups in the weight basis. Furthermore:

1. **O24AclasW** contains all the 24 elements as 7×7 matrices in the weight basis, organized by conjugacy classes for the group **O24A.** The file **O24AclasR** contains instead the same matrices in the root lattice basis.
2. **O24BclasW** contains all the 24 elements as 7×7 matrices in the weight basis, organized by conjugacy classes for the group **O24B.** The file **O24BclasR** contains instead the same matrices in the root lattice basis.
3. The formal encoding of the groups **O24A** and **O24B** is given by the objects **formO24Aclas** and **formO24Bclas,** where every group element in each conjugacy class is expressed in terms of the generators $\boldsymbol{\rho}, \boldsymbol{\sigma}, \boldsymbol{\tau}$ of the ambient group L168.

B.6.1.9 Irreducible representations of O24

The irreducible representations of **O24** were extensively discussed in Section 4.3.6 of the main text. They are created by evaluating this NoteBook and they are respectively named:

DD1classa, DD2classa, DD3classa, DD4classa and **DD5classa.**

They are organized by conjugacy classes in the standard order.

B.6.1.10 Generators of O24A and O24B

In the case of the group **O24A** the formal generators are:

$$\mathrm{fT} = \rho.\sigma.\rho.\tau.\tau.\sigma.\rho.\tau; \quad \mathrm{fS} = \tau.\tau.\sigma.\rho.\tau.\sigma.\sigma \tag{B.12}$$

In the case of the group **O24B** the formal generators are:

$$\mathrm{fT} = \rho.\tau.\sigma.\rho.\tau.\tau.\sigma.\rho.\tau; \quad \mathrm{fS} = \sigma.\rho.\tau.\sigma.\rho.\tau \tag{B.13}$$

The substitution rules for the embedding are **QO24A** and **QO24B.**

B.6.2 The basic commands of the present package

We list next by topic the basic commands of the present package explaining what they do and how they can be used.

B.6.2.1 Construction of the representation of the group L168 and of its maximal subgroups and decomposition into irreps

For these tasks we have the following commands:

1. **brutcaratterL168.** This routine calculates the character of any linear representation D of the group L168 provided in the following way:

$$\mathrm{Repra} = \{D[1], D[\rho], D[\sigma], D[\tau\rho\sigma], D[\tau], D[\sigma\rho]\} \tag{B.14}$$

where D[1], etc. are the matrix realization of the standard representatives of the six conjugacy classes in terms of the generators of the group, satisfying the defining relations. The routine derives the multiplicity vector encoded in the object **decompo** and it constructs the decomposition of the representation D into irreps of L168. The projectors onto the irreducible representations are named **PP168.**

2. **brutcaratter21.** This routine calculates the character of any linear representation D of the group G21 provided in the following way:

$$\text{Repra21} = \left\{ \boxed{D[e] \mid D[\mathscr{Y}] \mid D[\mathscr{X}^2\mathscr{Y}\mathscr{X}\mathscr{Y}^2] \mid D[\mathscr{Y}\mathscr{X}^2] \mid D[\mathscr{X}]} \right\} \tag{B.15}$$

where D[e], etc. are the matrix realization of the standard representatives of the five conjugacy classes in terms of the generators of the group $\mathscr{Y}$, and $\mathscr{X}$ satisfying the defining relations. The routine derives the multiplicity vector encoded in the object **decompo21** and it constructs the decomposition of the representation D into irreps of G21. The projectors onto the irreducible representations are named **PP21.**

3. **brutcaratter24.** This routine calculates the character of any linear representation D of the group O24 provided in the following way:

$$\text{Repra24} = \left\{ \boxed{D[e] \mid D[T] \mid D[\text{STST}] \mid D[\text{S}] \mid D[\text{ST}]} \right\} \tag{B.16}$$

where D[e], etc. are the matrix realization of the standard representatives of the five conjugacy classes in terms of the generators of the group T, and S satisfying the defining relations. The routine derives the multiplicity vector encoded in the object **decompo24** and it constructs the decomposition of the representation D into irreps of O24. The projectors onto the irreducible representations are named **PP24.**

4. **belcaratter21.** This routine requires a substitution rule

$$\mathbf{passarulla} = \{\mathscr{Y} \to \mathbf{D}[\mathscr{Y}], \mathscr{X} \to \mathbf{D}[\mathscr{X}]\}$$

that calculates the character of the corresponding representation and its explicit decomposition into irreps. The projection operators onto irreps are explicitly calculated.

B.6.2.2 Auxiliary group theoretical routines used by the package but available also to the user

Besides the basic commands described in the previous section, this package contains also some general group-theoretical routines that are internally utilized but also available to the user. These are:

1. **generone.** Given a set of matrices named **Allgroup** the routine generone generates the set of all their products. Repeated use of generone arrives at a set that closes under multiplication if the original matrices were elements of a finite group.

2. **generoneName**. Given a set of matrices named **AllgroupN**, associated, each of them, with a name, the routine generone generates the set of all their products keeping track of the non-commutative product of names. Repeated use of generoneName arrives at a set that closes under multiplication if the original matrices were elements of a finite group.
3. **coniugatoL** (or **coniugatoM**, they are equivalent). If you give a set of matrices forming a finite group and you name it **gruppone**, **coniugatoL** produces the set of conjugacy classes into which the finite group is organized. The output of this calculation is named **orgclas.**
4. **verifiosub**. Given a set of matrices that form a finite group, named **gruppone** and a subset named **settino**, verifiosub verifies whether settino is a subgroup and, moreover, it verifies whether it is a normal subgroup.
5. **quozientus**. Given a set of matrices forming a finite group, named **gruppone** and a normal subgroup named **gruppino**, quozientus constructs the equivalence classes G/H, namely, the quotient group. The output of this calculation is named **equaclass.**

B.6.2.3 Additional specialized routines for the analysis of subgroups and orbits in the 7-dimensional momentum lattice

1. **cercatoreorbo**. Given a subgroup G of L168 named **sottogruppo** (it should be given in the weight basis) the routine cercatoreorbo generates the general form of a 7-vector invariant under G. The output is named **invarvec** and it is displayed.
2. **stabilio**. Given a 7-vector named **veicolo** the routine stabilio constructs the subgroup of L168 (in the weight basis) that leaves it invariant. The output of this calculation is named **stab.**
3. **stabilione**. You have to call **gruppone** the set of $\mathbf{n} \times \mathbf{n}$ matrices representing a given group G. You have to name **veicolo** an **n**-vector. The routine stabilione constructs the subgroup G that leaves it invariant. The output of this calculation is named **stab.**
4. **rovnastab**. Given a 7-vector **v**, named **vectus** the routine rovnastab constructs first the L168 orbit $\mathcal{O}_v$ of vectus, named **orballo.** Next for each vector $v_i \in \mathcal{O}_v$ the routine constructs its stability subgroup Γ_i named **stab.** The program verifies whether this subgroup is contained in one of the three standard representatives of the maximal subgroups, namely O24A, O24B or G21. By definition of orbit it is obligatory that one of the Γ_i should be inside one of the three groups O24A, O24B or G21. The routine finds the standard vector named **repvec** whose stability subgroup is included in one of the three standard maximal subgroups and determines which one. The final output is a subgroup of one of the three maximal ones that represents the orbit and is named **standgroup.**
5. **dihedraleA.** (**dihedraleB**). These two routines respectively construct the dihedral subgroup $\mathrm{Dih}_3 \subset \mathrm{O}_{24}$ in the case of the maximal subgroup O24A and O24B. The out-

puts of the two routines are respectively named **DH3A** and **DH3B**. The construction is done fixing first the generator A of order 3 and then looking for a generator B of order 2 which satisfies the defining relation of the dihedral group with A. Fixing A amounts to choosing a fixed representative of the conjugacy class of Dih_3 groups inside the octahedral one.

B.6.2.4 The irreducible representation of L_{168}

The irreducible representations of the group L_{168} are defined by giving the form of the three generators ρ, σ, τ.

B.6.2.4.1 The 3-dimensional complex representation

The 3-dimensional complex representation has been constructed in two different bases by Pierre Ramond et al. and by Markusevich:

1. In Pierre Ramond basis the three standard generators are given by the matrices **R3,S3,T3**. The entire representation organized in conjugacy classes can be obtained from the file **formL168clas** performing the substitution **ramond3**. Applying the same substitution rule to the formal-presentation of a subgroup we obtain its realization inside the 3-dimensional realization of the bigger group.
2. In Markusevich basis the three standard generators are given by the matrices **RP**, **SP**, **TP**. The entire representation organized in conjugacy classes can be obtained from the file **formL168clas** performing the substitution **markus3**. Applying the same substitution rule to the formal-presentation of a subgroup we obtain its realization inside the 3-dimensional realization of the bigger group.

B.6.2.4.2 The 6-dimensional representation

In the 6-dimensional representation the three standard generators are given by the matrices **R6,S6,T6**. The entire representation organized in conjugacy classes can be obtained from the file **formL168clas** performing the substitution **ramond6**. Applying the same substitution rule to the formal-presentation of a subgroup we obtain its realization inside the 6-dimensional realization of the bigger group.

B.6.2.4.3 The 8-dimensional representation

In the 8-dimensional representation the three standard generators are given by the matrices **R8,S8,T8**. The entire representation organized in conjugacy classes can be obtained from the file **formL168clas** performing the substitution **ramond8**. Applying the same substitution rule to the formal-presentation of a subgroup we obtain its realization inside the 8-dimensional realization of the bigger group.

B.6.2.5 Prepared characters

The file **Reprano** contains the formal definition of the representatives of the 6-conjugacy classes for the group L168. Substituting the formal generators with one of the irrep-substitutions one obtains the 6-matrices whose trace provides the character of the representation. These files can be utilized with the routine brutcaratter168. The file **Reprano21** contains the formal definition of the representatives of the 6-conjugacy classes for the group G21. Substituting the formal generators with one of the irrep-substitutions one obtains the 6-matrices whose trace provides the character of the representation. These files can be utilized with the routine brutcaratter21.

B.6.2.6 Generation of orbits in the $\mathbb{C}^3$

Given a complex 3-vector that should be named **vectus**, one can generate its orbit under L168T or one of its three maximal subgroups using the following commands:
1. **orbitando168** generates the orbit of **vectus** under the full group L168. The output is named **orbita.**
2. **orbitando21** generates the orbit of **vectus** under the subgroup G21. The output is named **orbita.**
3. **orbitando24A** generates the orbit of **vectus** under the subgroup O24A. The output is named **orbita.**
4. **orbitando24B** generates the orbit of **vectus** under the subgroup O24B. The output is named **orbita.**

B.6.2.7 Constructing the group *L*168 acting on $\mathbb{C}^3$

We construct explicitly the 3×3 complex matrices that represent the simple finite group *L*168 inside SL(3,$\mathbb{C}$) in the basis utilized by Markushevich for purposes of algebraic geometry.

The generators satisfy the standard relations of the presentation mentioned in eq. (4.21).

We construct explicitly all the group elements of the group *L*168 in this 3-dimensional complex representation utilizing Markushevich basis. They are encoded in two different files:
1. **gruppo3C** contains the group elements organized in conjugacy classes exactly in the same order as in the file **formL168** which writes each element as a word in the generators. This is very important in order to identify each group element if needed in another representation.
2. **L168C3** contains the 168 elements just in one stock. This file is useful to calculate orbits of given vectors or lines and to find out stability subgroups.

We utilize the internal Mathematical command *FullSimplify* in order to write the numerical matrices representing the group elements in the simplest possible way. The matrix entries are all elementary transcendental numbers or algebraic numbers that have an algebraic representation as roots of algebraic equations with rational coefficients. For this reason *Mathematica* sometimes writes them as follows: Root[1 + 7#1^2 − 14#1^3 + 49#1^6&, 1]. This is not a problem. Originally these numbers where produced by multiplication of rational numbers extended with elementary transcendentals that are the seventh roots of unity $\sqrt[7]{1}$ or their real or imaginary parts and the expression in terms of these can be recovered by typing ToRadicals. The analytic form of the generators contained in the substitution rule **markus3** (computed in this background NoteBook) is the following one:

$$\epsilon \to \begin{pmatrix} 1 & 0 & 0 \\ 0 & 1 & 0 \\ 0 & 0 & 1 \end{pmatrix} \tag{B.17}$$

$$\rho \to \begin{pmatrix} -\frac{2\mathrm{Cos}[\frac{\pi}{14}]}{\sqrt{7}} & -\frac{2\mathrm{Cos}[\frac{3\pi}{14}]}{\sqrt{7}} & \frac{2\mathrm{Sin}[\frac{\pi}{7}]}{\sqrt{7}} \\ -\frac{2\mathrm{Cos}[\frac{3\pi}{14}]}{\sqrt{7}} & \frac{2\mathrm{Sin}[\frac{\pi}{7}]}{\sqrt{7}} & -\frac{2\mathrm{Cos}[\frac{\pi}{14}]}{\sqrt{7}} \\ \frac{2\mathrm{Sin}[\frac{\pi}{7}]}{\sqrt{7}} & -\frac{2\mathrm{Cos}[\frac{\pi}{14}]}{\sqrt{7}} & -\frac{2\mathrm{Cos}[\frac{3\pi}{14}]}{\sqrt{7}} \end{pmatrix} \tag{B.18}$$

$$\sigma \to \begin{pmatrix} 0 & 0 & -(-1)^{1/7} \\ (-1)^{2/7} & 0 & 0 \\ 0 & (-1)^{4/7} & 0 \end{pmatrix} \tag{B.19}$$

$$\tau \to \begin{pmatrix} \frac{i+(-1)^{13/14}}{\sqrt{7}} & -\frac{(-1)^{1/14}(-1+(-1)^{2/7})}{\sqrt{7}} & \frac{(-1)^{9/14}(1+(-1)^{1/7})}{\sqrt{7}} \\ \frac{(-1)^{11/14}(-1+(-1)^{2/7})}{\sqrt{7}} & \frac{i+(-1)^{5/14}}{\sqrt{7}} & \frac{(-1)^{3/14}(1+(-1)^{3/7})}{\sqrt{7}} \\ -\frac{(-1)^{11/14}(1+(-1)^{1/7})}{\sqrt{7}} & -\frac{(-1)^{9/14}(1+(-1)^{3/7})}{\sqrt{7}} & -\frac{-i+(-1)^{3/14}}{\sqrt{7}} \end{pmatrix} \tag{B.20}$$

We remind the reader that ρ, σ, τ are the abstract names for the generators of L_{168} whose 168 elements are written as words in these letters (modulo relations). Substituting explicit matrices satisfying the relations for these letters one obtains an explicit representation of the group. In the present case the irreducible 3-dimensional representation is DA_3.

For possible use in future calculation, in the present section we rewrite the generators of L_{168} in the Markusevich basis in terms of 6×6 real matrices where the imaginary unit i has been replaced by the 2×2 matrix $\left(\begin{smallmatrix} 0 & 1 \\ -1 & 0 \end{smallmatrix}\right)$ and the real part is proportional to $\left(\begin{smallmatrix} 1 & 0 \\ 0 & 1 \end{smallmatrix}\right)$. Essentially this is the canonical embedding $\mathrm{SU}(3) \mid\to \mathrm{SO}(6)$. Since the group $L168 \subset \mathrm{SU}(3)$. We can make such a construction. This representation might be useful in various applications.

The generators are named: **RP6, SP6, TP6.**

B.7 The NoteBook octagroup2

The theory of the octahedral group was developed in Sections 3.4.3 and 4.3.6. The NoteBook octagroup2 provides the explicit construction of all items described in those sections and includes also graphical routines to construct finite portions of the cubic lattice that can be displayed to visualize orbits of lattice points under the action of the group O_{24}.

Evaluating this NoteBook the following objects and routines become available to the user.

B.7.1 Objects and commands available to the user

The names of the created objects available to the user are the following ones:

1. The set of all group elements of O_h given as 3×3 orthogonal matrices and organized into conjugacy classes according to the chemical nomenclature is encoded in the object **chemclassa.**
2. The action on the three coordinates $\{x, y, z\}$ of each element of the full octahedral group organized into conjugacy classes is encoded in the object **repchemclassa.**
3. The array containing all the 48 group elements of O_h given as 3×3 orthogonal matrices but not ordered in conjugacy classes is encoded in the object **octahedroD.**
4. The array containing all the 24 group elements of O_{24} given as 3×3 orthogonal matrices but not ordered in conjugacy classes is encoded in the object **octahedroP.**
5. The array containing all the 24 group elements of O_{24} given as 3×3 orthogonal matrices and labeled by their name n_r where $n = 1, 2, 3, 4, 5$ enumerates the conjugacy classes C_n and $r = 1, \dots, |C_n|$ is encoded in the object **GroupO.**
6. The multiplication table of the O_{24} group is encoded in the objects **tabloid** or **multab** (with or without column and row heads).
7. The 10 irreducible representations of the full octahedral group O_h are encoded in the object **DHclassa** whose elements have the following structure:

$$\textbf{DHclassa[[irrep,conj class]]} = \{\textbf{name of class, \# of el. in the class}, \{D[g_1], D[g_2], \dots, D[g_p]\}\} \quad \text{(B.21)}$$

   The characters of the 10 irreducible representations of the full octahedral group O_h are encoded in the object **chi.** The characters are organized into a table encoded in the object named **Tabella.**
8. The five irreducible representations of the proper octahedral group O_{24} are encoded in the object **PDclassa** whose elements have the same structure as those of DHclassa.
9. The characters of the five irreducible representations of the proper octahedral group O_{24} are encoded in the object **Pchi.** The characters are organized into a table encoded in the object named **PTabella.**

B.7.2 Auxiliary group theoretical routines used by the package but available also to the user

Besides the basic commands described in the previous section, this package contains also some general group-theoretical routines that are internally utilized but available to the user. These are:

1. **generone**. Given a set of matrices named **Allgroup** the routine generone generates the set of all their products. Repeated use of generone arrives at a set that closes under multiplication if the original matrices were elements of a finite group.
2. **generoneName**. Given a set of matrices named **AllgroupN**, associated, each of them, with a name, the routine generone generates the set of all their products keeping track of the non-commutative product of names. Repeated use of generoneName arrives at a set that closes under multiplication if the original matrices were elements of a finite group.
3. **coniugatoL**. If you give a set of matrices forming a finite group and you name it **gruppone**, **coniugatoL** produces the set of conjugacy classes into which the finite group is organized. The output of this calculation is named **orgclas**.
4. **verifiosub**. Given a set of matrices that form a finite group, named **gruppone** and a subset, named **settino**, verifiosub verifies whether settino is a subgroup and, moreover, it verifies whether it is a normal subgroup.
5. **quozientus**. Given a set of matrices forming a finite group, named gruppone and a normal subgroup, named gruppino, quozientus constructs the equivalence classes G/H, namely, the quotient group. The output of this calculation is named **equaclass**.
6. The routines **genorb** and **genorbP**. Given any 3-vector $\{v_1, v_2, v_3\}$ named **vec**, the routines create its orbit under the action of the full octahedral group (genorb) or under the action of the proper octahedral group (genorbP).
7. **latticePS**. This is a routine that generates finite portions of the space lattice and of the momentum lattice.
   In order to activate this routine you have to give two data:
   - **spaz** = ? is the lattice spacing $\boldsymbol{a}$ that you want.
   - **nplan** = ? is the number of lattice points that you want to place on the cubic rays.
     Then you type **latticePS**.
   - Running the routine one obtains a lattice with a certain finite number **N** of points that depends on the initial user chosen number of points **nplan.** The computer calculates N and also calculates how many points of the lattice intersect the spherical surface centered at the origin $\{0, 0, 0\}$ and of radius n a where n ∈ ℕ. We name such collection of points *spherical arrays* and they are encoded into an object named **strata.** The various radii and the number of lattice points intersected by the sphere of the corresponding radius are contained in a file named **multip.**

8. When you run **latticePS** you generate graphical objects that can be displayed by typing **Show[...]**. The generated graphical objects are **skeletonPS** and **vertexPS**, namely, the links between points in the cubic lattice (skeletonPS) and the points of the cubic lattice (vertexPS). These objects can be combined with other graphical objects as polyhedra, spheres and special points in the lattice.
9. **orbitandus**. The routine orbitandus organizes the lattice points contained in a spherical layer into orbits of the proper octahedral group. In order to utilize this routine you have first to generate a lattice by means of the routine **latticePS**. Next if you type **orbitandus** you obtain a file named **filtro** that is an array of arrays. Each entry of the array has three entries and it looks like following:

$$\{25, \quad 30, \quad \{\{O_1^{25}, 6\}, \{O_2^{25}, 24\}\}\} \tag{B.22}$$

The first entry 25 is the radius squared of the spherical orbit, the second entry 30 is the number of lattice points that intersect this sphere, the third entry tells you that in this case the 30 points organize into an orbit of 6 points under the proper octahedral group ⊕ an orbit of 24 points under the action of the same group.
10. Routines for the analysis of group representations. Suppose that you have constructed a group representation. The matrices have to be organized into an array of five entries corresponding to the five conjugacy classes each of which has the form **Repra**[[**r**]] = {**p**, $\{\boldsymbol{D}[\boldsymbol{g}_1], \boldsymbol{D}[\boldsymbol{g}_2], \dots, \boldsymbol{D}[\boldsymbol{g}_p]\}$} where p is the population of the rth class and $\{\boldsymbol{D}[\boldsymbol{g}_1], \boldsymbol{D}[\boldsymbol{g}_2], \dots, \boldsymbol{D}[\boldsymbol{g}_p]\}$ are the matrices representing the p elements in that class. Then you can utilize the following routines:
    - **homverifioinvers** verifies that the representation is indeed a homomorphism where the product is from right to left D[A]D[B] == D[BA]. Next it utilizes the routine ortogverifio that verifies if it is an orthogonal representation. Finally it utilizes the routine brutcaratter to decompose the given representation into irreducible ones.
    - **homverifiodirec** verifies that the representation is indeed a homomorphism where the product is from left to right D[A]D[B] == D[AB]. Next it utilizes the routine ortogverifio that verifies if it is an orthogonal representation. Finally it utilizes the routine brutcaratter to decompose the given representation into irreducible ones.
    - **brutcaratter**. It can be called directly to decompose the given representation into irreducible ones.

B.8 The NoteBook metricgrav

In this NoteBook we provide a package to calculate Einstein equations for any given metric in arbitrary dimensions using the metric formalism.

B.8.1 Description of the programme

Let us describe the content of the NoteBook.

B.8.1.1 What the NoteBook does

Given are an n-dimensional manifold $\mathcal{M}$ whose coordinates we denote x_i and a metric defined over it and provided in the form

$$\mathrm{ds}^2 = g_{\mathrm{ij}}(x)\,\mathrm{dx}^i \otimes \mathrm{dx}^j \tag{B.23}$$

The programme extracts the metric tensor $g_{\mathrm{ij}}(x)$, calculates its inverse $g^{\mathrm{ij}}(x)$, calculates the Christoffel symbols $\Gamma_{\mathrm{ij}}{}^{k}(\mathrm{x})$, then the Riemann tensor, the Ricci tensor and the Einstein tensor.

B.8.2 Initialization and inputs to be supplied

After reading the NoteBook, calculations are initialized in the following way:

1. First the user types **nn** = positive integer number (which is going to be the dimension **n** of the considered manifold).
2. Then the user types **mainmetric**. The computer will ask the user to supply three inputs in the following form:
   a) the set of coordinates as n-vector. That vector must be named **coordi** = $\{x_1, \ldots, x_n\}$;
   b) the set of coordinates differentials as n-vector. That vector must be named **diffe** = $\{\mathrm{dx}_1, \ldots, \mathrm{dx}_n\}$;
   c) the metric given as a quadratic differential that must be named ds2. The user will type **ds2** = $g_{[[i,j]]}\,\mathrm{dx}^i\,\mathrm{dx}^j$.
3. After providing these inputs the user will type the command **metricresume.**

B.8.3 Produced outputs

1. The Christoffel symbols $\Gamma^{\lambda}_{\mu\nu}$ are encoded in an array **Gam**$[[\lambda,\mu,\nu]]$.
2. The Riemann tensor $\mathcal{R}^{\lambda}_{\mu\nu\rho}$ is encoded in an array **Rie**$[[\lambda,\mu,\nu,\rho]]$.
3. The curvature 2-form $\mathfrak{R} = d\Gamma + \Gamma \wedge \Gamma$ is encoded in an array named **RR**$[[\lambda,\mu]]$.
4. The Ricci tensor $\mathcal{R}_{\mu\rho} \equiv R^{\lambda}_{\mu\lambda\rho}$ is encoded in an array named **ricten**$[[\mu,\rho]]$.
5. The Einstein tensor $G_{\mu\rho} \equiv \mathcal{R}_{\mu\rho} - \frac{1}{2} g_{\mu\rho}\mathcal{R}$ is encoded in an array **einst**$[[\mu,\rho]]$.

B.9 The NoteBook Vielbgrav

In this NoteBook we provide a package to calculate the geometry of an arbitrary manifold in arbitrary dimension and with an arbitrary signature using the vielbein formalism and the intrinsic components of all tensors.

B.9.1 Instructions for the user

B.9.1.1 The inputs

In order to initialize the calculation, the user has to type five lines of inputs providing the following information:

1. the dimension $n =$ **dimse**;
2. the set of coordinates as an n-vector = **coordi**;
3. the set of differentials, as an n-vector = **diffe**;
4. the set of vielbein 1-forms as an n-vector = **fform**;
5. the signature of the space as a n-vector = **signat** of $+/-1$.

B.9.1.2 Activating the calculation

After providing the above information the user will start the calculations by typing **mainstart**.

B.9.1.3 The obtained outputs

The MATHEMATICA NoteBook calculates the following objects:

1. The contorsion $\mathbf{c}^{i}_{\mathbf{jk}}$ defined by the equation $dV^i = c^i_{jk} V^i \wedge V^k$ and encoded in an array $\mathbf{contens}_{[[i,j,k]]}$.
2. The spin connection 1-form $\boldsymbol{\omega}^{ij}$ defined by the equation $dV^i - \omega^{ij} \wedge V^k \eta_{jk} = 0$ and encoded in an array $\boldsymbol{\omega}_{[[i,j]]}$.
3. The intrinsic components of the spin connection defined by the equation $\omega^{ij} = \omega^{ij}_k V^k$ and encoded in a tensor $\mathbf{ometen}_{[[i,j,k]]}$.
4. The curvature 2-form $\mathfrak{R}^{ij}$ defined by the equation $\mathfrak{R} = d\omega - \omega \wedge \omega$ and encoded in a tensor $\mathbf{RF}_{[[i,j]]}$.
5. The Riemann tensor with flat indices, defined by $\mathfrak{R}^{ij} = \mathrm{Rie}^{ij}{}_{pq} V^p \wedge V^q$ and encoded in an array $\mathbf{Rie}_{[[i,j,a,b]]}$.
6. The Ricci tensor with flat indices defined by $\mathrm{Ric}^i_p = \mathrm{Rie}^{iq}{}_{pq}$ and encoded in an array **ricten[[e,b]]**.

B.10 Roots of $\mathfrak{e}_8$

In this appendix we display the list of all positive roots of the $\mathfrak{e}_8$ Lie algebra explicitly calculated by the NoteBook described in Section B.2.

B.10.1 Listing of the $\mathfrak{e}_8$ roots

===

# of roots of height = 1 is 8

———————————————————-

$\alpha_1 : \{1,0,0,0,0,0,0,0\} \to \{0,1,-1,0,0,0,0,0\}$
$\alpha_2 : \{0,1,0,0,0,0,0,0\} \to \{0,0,1,-1,0,0,0,0\}$
$\alpha_3 : \{0,0,1,0,0,0,0,0\} \to \{0,0,0,1,-1,0,0,0\}$
$\alpha_4 : \{0,0,0,1,0,0,0,0\} \to \{0,0,0,0,1,-1,0,0\}$
$\alpha_5 : \{0,0,0,0,1,0,0,0\} \to \{0,0,0,0,0,1,-1,0\}$
$\alpha_6 : \{0,0,0,0,0,1,0,0\} \to \{0,0,0,0,0,1,1,0\}$
$\alpha_7 : \{0,0,0,0,0,0,1,0\} \to \{-\frac{1}{2},-\frac{1}{2},-\frac{1}{2},-\frac{1}{2},-\frac{1}{2},-\frac{1}{2},-\frac{1}{2},-\frac{1}{2}\}$
$\alpha_8 : \{0,0,0,0,0,0,0,1\} \to \{1,-1,0,0,0,0,0,0\}$

===

# of roots of height = 2 is 7

———————————————————-

$\alpha_9 : \{1,1,0,0,0,0,0,0\} \to \{0,1,0,-1,0,0,0,0\}$
$\alpha_{10} : \{1,0,0,0,0,0,0,1\} \to \{1,0,-1,0,0,0,0,0\}$
$\alpha_{11} : \{0,1,1,0,0,0,0,0\} \to \{0,0,1,0,-1,0,0,0\}$
$\alpha_{12} : \{0,0,1,1,0,0,0,0\} \to \{0,0,0,1,0,-1,0,0\}$
$\alpha_{13} : \{0,0,0,1,1,0,0,0\} \to \{0,0,0,0,1,0,-1,0\}$
$\alpha_{14} : \{0,0,0,1,0,1,0,0\} \to \{0,0,0,0,1,0,1,0\}$
$\alpha_{15} : \{0,0,0,0,0,1,1,0\} \to \{-\frac{1}{2},-\frac{1}{2},-\frac{1}{2},-\frac{1}{2},-\frac{1}{2},\frac{1}{2},\frac{1}{2},-\frac{1}{2}\}$

===

# of roots of height = 3 is 7

———————————————————-

$\alpha_{16} : \{1,1,1,0,0,0,0,0\} \to \{0,1,0,0,-1,0,0,0\}$
$\alpha_{17} : \{1,1,0,0,0,0,0,1\} \to \{1,0,0,-1,0,0,0,0\}$
$\alpha_{18} : \{0,1,1,1,0,0,0,0\} \to \{0,0,1,0,0,-1,0,0\}$
$\alpha_{19} : \{0,0,1,1,1,0,0,0\} \to \{0,0,0,1,0,0,-1,0\}$
$\alpha_{20} : \{0,0,1,1,0,1,0,0\} \to \{0,0,0,1,0,0,1,0\}$
$\alpha_{21} : \{0,0,0,1,0,1,1,0\} \to \{-\frac{1}{2},-\frac{1}{2},-\frac{1}{2},-\frac{1}{2},\frac{1}{2},-\frac{1}{2},\frac{1}{2},-\frac{1}{2}\}$
$\alpha_{22} : \{0,0,0,1,1,1,0,0\} \to \{0,0,0,0,1,1,0,0\}$

===

# of roots of height = 4 is 7

———————————————————-

$\alpha_{23} : \{1,1,1,1,0,0,0,0\} \to \{0,1,0,0,0,-1,0,0\}$
$\alpha_{24} : \{0,1,1,1,1,0,0,0\} \to \{0,0,1,0,0,0,-1,0\}$
$\alpha_{25} : \{0,1,1,1,0,1,0,0\} \to \{0,0,1,0,0,0,1,0\}$
$\alpha_{26} : \{1,1,1,0,0,0,0,1\} \to \{1,0,0,0,-1,0,0,0\}$

$\alpha_{27} : \{0,0,1,1,0,1,1,0\} \rightarrow \{-\frac{1}{2},-\frac{1}{2},-\frac{1}{2},\frac{1}{2},-\frac{1}{2},-\frac{1}{2},\frac{1}{2},-\frac{1}{2}\}$
$\alpha_{28} : \{0,0,1,1,1,1,0,0\} \rightarrow \{0,0,0,1,0,1,0,0\}$
$\alpha_{29} : \{0,0,0,1,1,1,1,0\} \rightarrow \{-\frac{1}{2},-\frac{1}{2},-\frac{1}{2},-\frac{1}{2},\frac{1}{2},\frac{1}{2},-\frac{1}{2},-\frac{1}{2}\}$

===

# of roots of height = 5 is 7

$\alpha_{30} : \{1,1,1,1,1,0,0,0\} \rightarrow \{0,1,0,0,0,0,-1,0\}$
$\alpha_{31} : \{1,1,1,1,0,1,0,0\} \rightarrow \{0,1,0,0,0,0,1,0\}$
$\alpha_{32} : \{0,1,1,1,0,1,1,0\} \rightarrow \{-\frac{1}{2},-\frac{1}{2},\frac{1}{2},-\frac{1}{2},-\frac{1}{2},-\frac{1}{2},\frac{1}{2},-\frac{1}{2}\}$
$\alpha_{33} : \{0,1,1,1,1,1,0,0\} \rightarrow \{0,0,1,0,0,1,0,0\}$
$\alpha_{34} : \{0,0,1,1,1,1,1,0\} \rightarrow \{-\frac{1}{2},-\frac{1}{2},-\frac{1}{2},\frac{1}{2},-\frac{1}{2},\frac{1}{2},-\frac{1}{2},-\frac{1}{2}\}$
$\alpha_{35} : \{1,1,1,1,0,0,0,1\} \rightarrow \{1,0,0,0,0,-1,0,0\}$
$\alpha_{36} : \{0,0,1,2,1,1,0,0\} \rightarrow \{0,0,0,1,1,0,0,0\}$

===

# of roots of height = 6 is 7

$\alpha_{37} : \{1,1,1,1,0,1,1,0\} \rightarrow \{-\frac{1}{2},\frac{1}{2},-\frac{1}{2},-\frac{1}{2},-\frac{1}{2},-\frac{1}{2},\frac{1}{2},-\frac{1}{2}\}$
$\alpha_{38} : \{1,1,1,1,1,1,0,0\} \rightarrow \{0,1,0,0,0,1,0,0\}$
$\alpha_{39} : \{0,1,1,1,1,1,1,0\} \rightarrow \{-\frac{1}{2},-\frac{1}{2},\frac{1}{2},-\frac{1}{2},-\frac{1}{2},\frac{1}{2},-\frac{1}{2},-\frac{1}{2}\}$
$\alpha_{40} : \{0,1,1,2,1,1,0,0\} \rightarrow \{0,0,1,0,1,0,0,0\}$
$\alpha_{41} : \{0,0,1,2,1,1,1,0\} \rightarrow \{-\frac{1}{2},-\frac{1}{2},-\frac{1}{2},\frac{1}{2},\frac{1}{2},-\frac{1}{2},-\frac{1}{2},-\frac{1}{2}\}$
$\alpha_{42} : \{1,1,1,1,1,0,0,1\} \rightarrow \{1,0,0,0,0,0,-1,0\}$
$\alpha_{43} : \{1,1,1,1,0,1,0,1\} \rightarrow \{1,0,0,0,0,0,1,0\}$

===

# of roots of height = 7 is 7

$\alpha_{44} : \{1,1,1,1,1,1,1,0\} \rightarrow \{-\frac{1}{2},\frac{1}{2},-\frac{1}{2},-\frac{1}{2},-\frac{1}{2},\frac{1}{2},-\frac{1}{2},-\frac{1}{2}\}$
$\alpha_{45} : \{1,1,1,2,1,1,0,0\} \rightarrow \{0,1,0,0,1,0,0,0\}$
$\alpha_{46} : \{0,1,1,2,1,1,1,0\} \rightarrow \{-\frac{1}{2},-\frac{1}{2},\frac{1}{2},-\frac{1}{2},\frac{1}{2},-\frac{1}{2},-\frac{1}{2},-\frac{1}{2}\}$
$\alpha_{47} : \{0,1,2,2,1,1,0,0\} \rightarrow \{0,0,1,1,0,0,0,0\}$
$\alpha_{48} : \{1,1,1,1,1,1,0,1\} \rightarrow \{1,0,0,0,0,1,0,0\}$
$\alpha_{49} : \{0,0,1,2,1,2,1,0\} \rightarrow \{-\frac{1}{2},-\frac{1}{2},-\frac{1}{2},\frac{1}{2},\frac{1}{2},\frac{1}{2},\frac{1}{2},-\frac{1}{2}\}$
$\alpha_{50} : \{1,1,1,1,0,1,1,1\} \rightarrow \{\frac{1}{2},-\frac{1}{2},-\frac{1}{2},-\frac{1}{2},-\frac{1}{2},-\frac{1}{2},\frac{1}{2},-\frac{1}{2}\}$

===

# of roots of height = 8 is 6

$\alpha_{51} : \{1,1,1,2,1,1,1,0\} \rightarrow \{-\frac{1}{2},\frac{1}{2},-\frac{1}{2},-\frac{1}{2},\frac{1}{2},-\frac{1}{2},-\frac{1}{2},-\frac{1}{2}\}$
$\alpha_{52} : \{1,1,2,2,1,1,0,0\} \rightarrow \{0,1,0,1,0,0,0,0\}$
$\alpha_{53} : \{0,1,1,2,1,2,1,0\} \rightarrow \{-\frac{1}{2},-\frac{1}{2},\frac{1}{2},-\frac{1}{2},\frac{1}{2},\frac{1}{2},\frac{1}{2},-\frac{1}{2}\}$
$\alpha_{54} : \{0,1,2,2,1,1,1,0\} \rightarrow \{-\frac{1}{2},-\frac{1}{2},\frac{1}{2},\frac{1}{2},-\frac{1}{2},-\frac{1}{2},-\frac{1}{2},-\frac{1}{2}\}$

$\alpha_{55}:\{1,1,1,2,1,1,0,1\}\to\{1,0,0,0,1,0,0,0\}$
$\alpha_{56}:\{1,1,1,1,1,1,1,1\}\to\{\frac{1}{2},-\frac{1}{2},-\frac{1}{2},-\frac{1}{2},-\frac{1}{2},\frac{1}{2},-\frac{1}{2},-\frac{1}{2}\}$

==

# of roots of height = 9 is 6

$\alpha_{57}:\{1,1,1,2,1,2,1,0\}\to\{-\frac{1}{2},\frac{1}{2},-\frac{1}{2},-\frac{1}{2},\frac{1}{2},\frac{1}{2},\frac{1}{2},-\frac{1}{2}\}$
$\alpha_{58}:\{1,1,2,2,1,1,1,0\}\to\{-\frac{1}{2},\frac{1}{2},-\frac{1}{2},\frac{1}{2},-\frac{1}{2},-\frac{1}{2},-\frac{1}{2},-\frac{1}{2}\}$
$\alpha_{59}:\{1,2,2,2,1,1,0,0\}\to\{0,1,1,0,0,0,0,0\}$
$\alpha_{60}:\{0,1,2,2,1,2,1,0\}\to\{-\frac{1}{2},-\frac{1}{2},\frac{1}{2},\frac{1}{2},-\frac{1}{2},\frac{1}{2},\frac{1}{2},-\frac{1}{2}\}$
$\alpha_{61}:\{1,1,2,2,1,1,0,1\}\to\{1,0,0,1,0,0,0,0\}$
$\alpha_{62}:\{1,1,1,2,1,1,1,1\}\to\{\frac{1}{2},-\frac{1}{2},-\frac{1}{2},-\frac{1}{2},\frac{1}{2},-\frac{1}{2},-\frac{1}{2},-\frac{1}{2}\}$

==

# of roots of height = 10 is 6

$\alpha_{63}:\{1,1,2,2,1,2,1,0\}\to\{-\frac{1}{2},\frac{1}{2},-\frac{1}{2},\frac{1}{2},-\frac{1}{2},\frac{1}{2},\frac{1}{2},-\frac{1}{2}\}$
$\alpha_{64}:\{1,2,2,2,1,1,1,0\}\to\{-\frac{1}{2},\frac{1}{2},\frac{1}{2},-\frac{1}{2},-\frac{1}{2},-\frac{1}{2},-\frac{1}{2},-\frac{1}{2}\}$
$\alpha_{65}:\{1,2,2,2,1,1,0,1\}\to\{1,0,1,0,0,0,0,0\}$
$\alpha_{66}:\{1,1,2,2,1,1,1,1\}\to\{\frac{1}{2},-\frac{1}{2},-\frac{1}{2},\frac{1}{2},-\frac{1}{2},-\frac{1}{2},-\frac{1}{2},-\frac{1}{2}\}$
$\alpha_{67}:\{0,1,2,3,1,2,1,0\}\to\{-\frac{1}{2},-\frac{1}{2},\frac{1}{2},\frac{1}{2},\frac{1}{2},-\frac{1}{2},\frac{1}{2},-\frac{1}{2}\}$
$\alpha_{68}:\{1,1,1,2,1,2,1,1\}\to\{\frac{1}{2},-\frac{1}{2},-\frac{1}{2},-\frac{1}{2},\frac{1}{2},\frac{1}{2},\frac{1}{2},-\frac{1}{2}\}$

==

# of roots of height = 11 is 6

$\alpha_{69}:\{2,2,2,2,1,1,0,1\}\to\{1,1,0,0,0,0,0,0\}$
$\alpha_{70}:\{1,1,2,3,1,2,1,0\}\to\{-\frac{1}{2},\frac{1}{2},-\frac{1}{2},\frac{1}{2},\frac{1}{2},-\frac{1}{2},\frac{1}{2},-\frac{1}{2}\}$
$\alpha_{71}:\{1,2,2,2,1,2,1,0\}\to\{-\frac{1}{2},\frac{1}{2},\frac{1}{2},-\frac{1}{2},-\frac{1}{2},\frac{1}{2},\frac{1}{2},-\frac{1}{2}\}$
$\alpha_{72}:\{1,2,2,2,1,1,1,1\}\to\{\frac{1}{2},-\frac{1}{2},\frac{1}{2},-\frac{1}{2},-\frac{1}{2},-\frac{1}{2},-\frac{1}{2},-\frac{1}{2}\}$
$\alpha_{73}:\{1,1,2,2,1,2,1,1\}\to\{\frac{1}{2},-\frac{1}{2},-\frac{1}{2},\frac{1}{2},-\frac{1}{2},\frac{1}{2},\frac{1}{2},-\frac{1}{2}\}$
$\alpha_{74}:\{0,1,2,3,2,2,1,0\}\to\{-\frac{1}{2},-\frac{1}{2},\frac{1}{2},\frac{1}{2},\frac{1}{2},\frac{1}{2},-\frac{1}{2},-\frac{1}{2}\}$

==

# of roots of height = 12 is 5

$\alpha_{75}:\{2,2,2,2,1,1,1,1\}\to\{\frac{1}{2},\frac{1}{2},-\frac{1}{2},-\frac{1}{2},-\frac{1}{2},-\frac{1}{2},-\frac{1}{2},-\frac{1}{2}\}$
$\alpha_{76}:\{1,1,2,3,2,2,1,0\}\to\{-\frac{1}{2},\frac{1}{2},-\frac{1}{2},\frac{1}{2},\frac{1}{2},\frac{1}{2},-\frac{1}{2},-\frac{1}{2}\}$
$\alpha_{77}:\{1,2,2,3,1,2,1,0\}\to\{-\frac{1}{2},\frac{1}{2},\frac{1}{2},-\frac{1}{2},\frac{1}{2},-\frac{1}{2},\frac{1}{2},-\frac{1}{2}\}$
$\alpha_{78}:\{1,2,2,2,1,2,1,1\}\to\{\frac{1}{2},-\frac{1}{2},\frac{1}{2},-\frac{1}{2},-\frac{1}{2},\frac{1}{2},\frac{1}{2},-\frac{1}{2}\}$
$\alpha_{79}:\{1,1,2,3,1,2,1,1\}\to\{\frac{1}{2},-\frac{1}{2},-\frac{1}{2},\frac{1}{2},\frac{1}{2},-\frac{1}{2},\frac{1}{2},-\frac{1}{2}\}$

==

# of roots of height = 13 is 5

$\alpha_{80}:\{2,2,2,2,1,2,1,1\}\to\{\frac{1}{2},\frac{1}{2},-\frac{1}{2},-\frac{1}{2},-\frac{1}{2},\frac{1}{2},\frac{1}{2},-\frac{1}{2}\}$
$\alpha_{81}:\{1,2,2,3,2,2,1,0\}\to\{-\frac{1}{2},\frac{1}{2},\frac{1}{2},-\frac{1}{2},\frac{1}{2},\frac{1}{2},-\frac{1}{2},-\frac{1}{2}\}$

$\alpha_{82}: \{1,2,2,3,1,2,1,1\} \rightarrow \{\frac{1}{2},-\frac{1}{2},\frac{1}{2},-\frac{1}{2},\frac{1}{2},-\frac{1}{2},\frac{1}{2},-\frac{1}{2}\}$
$\alpha_{83}: \{1,2,3,3,1,2,1,0\} \rightarrow \{-\frac{1}{2},\frac{1}{2},\frac{1}{2},\frac{1}{2},-\frac{1}{2},-\frac{1}{2},\frac{1}{2},-\frac{1}{2}\}$
$\alpha_{84}: \{1,1,2,3,2,2,1,1\} \rightarrow \{\frac{1}{2},-\frac{1}{2},-\frac{1}{2},\frac{1}{2},\frac{1}{2},\frac{1}{2},-\frac{1}{2},-\frac{1}{2}\}$

===

# of roots of height = 14 is 4

$\alpha_{85}: \{2,2,2,3,1,2,1,1\} \rightarrow \{\frac{1}{2},\frac{1}{2},-\frac{1}{2},-\frac{1}{2},\frac{1}{2},-\frac{1}{2},\frac{1}{2},-\frac{1}{2}\}$
$\alpha_{86}: \{1,2,2,3,2,2,1,1\} \rightarrow \{\frac{1}{2},-\frac{1}{2},\frac{1}{2},-\frac{1}{2},\frac{1}{2},\frac{1}{2},-\frac{1}{2},-\frac{1}{2}\}$
$\alpha_{87}: \{1,2,3,3,2,2,1,0\} \rightarrow \{-\frac{1}{2},\frac{1}{2},\frac{1}{2},\frac{1}{2},-\frac{1}{2},\frac{1}{2},-\frac{1}{2},-\frac{1}{2}\}$
$\alpha_{88}: \{1,2,3,3,1,2,1,1\} \rightarrow \{\frac{1}{2},-\frac{1}{2},\frac{1}{2},\frac{1}{2},-\frac{1}{2},-\frac{1}{2},\frac{1}{2},-\frac{1}{2}\}$

===

# of roots of height = 15 is 4

$\alpha_{89}: \{2,2,2,3,2,2,1,1\} \rightarrow \{\frac{1}{2},\frac{1}{2},-\frac{1}{2},-\frac{1}{2},\frac{1}{2},\frac{1}{2},-\frac{1}{2},-\frac{1}{2}\}$
$\alpha_{90}: \{2,2,3,3,1,2,1,1\} \rightarrow \{\frac{1}{2},\frac{1}{2},-\frac{1}{2},\frac{1}{2},-\frac{1}{2},-\frac{1}{2},\frac{1}{2},-\frac{1}{2}\}$
$\alpha_{91}: \{1,2,3,3,2,2,1,1\} \rightarrow \{\frac{1}{2},-\frac{1}{2},\frac{1}{2},\frac{1}{2},-\frac{1}{2},\frac{1}{2},-\frac{1}{2},-\frac{1}{2}\}$
$\alpha_{92}: \{1,2,3,4,2,2,1,0\} \rightarrow \{-\frac{1}{2},\frac{1}{2},\frac{1}{2},\frac{1}{2},\frac{1}{2},-\frac{1}{2},-\frac{1}{2},-\frac{1}{2}\}$

===

# of roots of height = 16 is 4

$\alpha_{93}: \{2,2,3,3,2,2,1,1\} \rightarrow \{\frac{1}{2},\frac{1}{2},-\frac{1}{2},\frac{1}{2},-\frac{1}{2},\frac{1}{2},-\frac{1}{2},-\frac{1}{2}\}$
$\alpha_{94}: \{2,3,3,3,1,2,1,1\} \rightarrow \{\frac{1}{2},\frac{1}{2},\frac{1}{2},-\frac{1}{2},-\frac{1}{2},-\frac{1}{2},\frac{1}{2},-\frac{1}{2}\}$
$\alpha_{95}: \{1,2,3,4,2,2,1,1\} \rightarrow \{\frac{1}{2},-\frac{1}{2},\frac{1}{2},\frac{1}{2},\frac{1}{2},-\frac{1}{2},-\frac{1}{2},-\frac{1}{2}\}$
$\alpha_{96}: \{1,2,3,4,2,3,1,0\} \rightarrow \{-\frac{1}{2},\frac{1}{2},\frac{1}{2},\frac{1}{2},\frac{1}{2},\frac{1}{2},\frac{1}{2},-\frac{1}{2}\}$

===

# of roots of height = 17 is 4

$\alpha_{97}: \{2,2,3,4,2,2,1,1\} \rightarrow \{\frac{1}{2},\frac{1}{2},-\frac{1}{2},\frac{1}{2},\frac{1}{2},-\frac{1}{2},-\frac{1}{2},-\frac{1}{2}\}$
$\alpha_{98}: \{2,3,3,3,2,2,1,1\} \rightarrow \{\frac{1}{2},\frac{1}{2},\frac{1}{2},-\frac{1}{2},-\frac{1}{2},\frac{1}{2},-\frac{1}{2},-\frac{1}{2}\}$
$\alpha_{99}: \{1,2,3,4,2,3,1,1\} \rightarrow \{\frac{1}{2},-\frac{1}{2},\frac{1}{2},\frac{1}{2},\frac{1}{2},\frac{1}{2},\frac{1}{2},-\frac{1}{2}\}$
$\alpha_{100}: \{1,2,3,4,2,3,2,0\} \rightarrow \{-1,0,0,0,0,0,0,-1\}$

===

# of roots of height = 18 is 3

$\alpha_{101}: \{2,2,3,4,2,3,1,1\} \rightarrow \{\frac{1}{2},\frac{1}{2},-\frac{1}{2},\frac{1}{2},\frac{1}{2},\frac{1}{2},\frac{1}{2},-\frac{1}{2}\}$
$\alpha_{102}: \{2,3,3,4,2,2,1,1\} \rightarrow \{\frac{1}{2},\frac{1}{2},\frac{1}{2},-\frac{1}{2},\frac{1}{2},-\frac{1}{2},-\frac{1}{2},-\frac{1}{2}\}$
$\alpha_{103}: \{1,2,3,4,2,3,2,1\} \rightarrow \{0,-1,0,0,0,0,0,-1\}$

===

# of roots of height = 19 is 3

$\alpha_{104}: \{2,2,3,4,2,3,2,1\} \rightarrow \{0,0,-1,0,0,0,0,-1\}$
$\alpha_{105}: \{2,3,3,4,2,3,1,1\} \rightarrow \{\frac{1}{2},\frac{1}{2},\frac{1}{2},-\frac{1}{2},\frac{1}{2},\frac{1}{2},\frac{1}{2},-\frac{1}{2}\}$

α_{106} : {2, 3, 4, 4, 2, 2, 1, 1}→$\{\frac{1}{2}, \frac{1}{2}, \frac{1}{2}, \frac{1}{2}, -\frac{1}{2}, -\frac{1}{2}, -\frac{1}{2}, -\frac{1}{2}\}$
==

# of roots of height = 20 is 2

α_{107} : {2, 3, 3, 4, 2, 3, 2, 1}→{0, 0, 0, −1, 0, 0, 0, −1}
α_{108} : {2, 3, 4, 4, 2, 3, 1, 1}→$\{\frac{1}{2}, \frac{1}{2}, \frac{1}{2}, \frac{1}{2}, -\frac{1}{2}, \frac{1}{2}, \frac{1}{2}, -\frac{1}{2}\}$
==

# of roots of height = 21 is 2

α_{109} : {2, 3, 4, 4, 2, 3, 2, 1}→{0, 0, 0, 0, −1, 0, 0, −1}
α_{110} : {2, 3, 4, 5, 2, 3, 1, 1}→$\{\frac{1}{2}, \frac{1}{2}, \frac{1}{2}, \frac{1}{2}, \frac{1}{2}, -\frac{1}{2}, \frac{1}{2}, -\frac{1}{2}\}$
==

# of roots of height = 22 is 2

α_{111} : {2, 3, 4, 5, 2, 3, 2, 1}→{0, 0, 0, 0, 0, −1, 0, −1}
α_{112} : {2, 3, 4, 5, 3, 3, 1, 1}→$\{\frac{1}{2}, \frac{1}{2}, \frac{1}{2}, \frac{1}{2}, \frac{1}{2}, \frac{1}{2}, -\frac{1}{2}, -\frac{1}{2}\}$
==

# of roots of height = 23 is 2

α_{113} : {2, 3, 4, 5, 3, 3, 2, 1}→{0, 0, 0, 0, 0, 0, −1, −1}
α_{114} : {2, 3, 4, 5, 2, 4, 2, 1}→{0, 0, 0, 0, 0, 0, 1, −1}
==

# of roots of height = 24 is 1

α_{115} : {2, 3, 4, 5, 3, 4, 2, 1}→{0, 0, 0, 0, 0, 1, 0, −1}
==

# of roots of height = 25 is 1

α_{116} : {2, 3, 4, 6, 3, 4, 2, 1}→{0, 0, 0, 0, 1, 0, 0, −1}
==

# of roots of height = 26 is 1

α_{117} : {2, 3, 5, 6, 3, 4, 2, 1}→{0, 0, 0, 1, 0, 0, 0, −1}
==

# of roots of height = 27 is 1

α_{118} : {2, 4, 5, 6, 3, 4, 2, 1}→{0, 0, 1, 0, 0, 0, 0, −1}
==

# of roots of height = 28 is 1

α_{119} : {3, 4, 5, 6, 3, 4, 2, 1}→{0, 1, 0, 0, 0, 0, 0, −1}

==

# of roots of height = 29 is 1

—————————————————.

α_{120} : $\{3, 4, 5, 6, 3, 4, 2, 2\} \rightarrow \{1, 0, 0, 0, 0, 0, 0, -1\}$

Bibliography

[1] O. Aharony, O. Bergman, D. L. Jafferis, and J. Maldacena, "$N = 6$ superconformal Chern-Simons-matter theories, M2-branes and their gravity duals," *J. High Energy Phys.*, vol. 2008, no. 10, p. 091, 2008. doi:10.1088/1126-6708/2008/10/091, arXiv:0806.1218 [hep-th].

[2] L. Andrianopoli, R. D'Auria, S. Ferrara, P. Fré, R. Minasian, and M. Trigiante, "Solvable Lie algebras in type IIA, type IIB and M-theories," *Nucl. Phys. B*, vol. 493, no. 1–2, pp. 249–277, 1997. hep-th/9612202.

[3] L. Andrianopoli, R. D'Auria, S. Ferrara, P. Fré, and M. Trigiante, "R-R scalars, U-duality and solvable Lie algebras," *Nucl. Phys. B*, vol. 496, no. 3, pp. 617–629, 1997. hep-th/9611014.

[4] L. Andrianopoli, R. D'Auria, and S. Ferrara, "U-duality and central charges in various dimensions revisited," *Int. J. Mod. Phys. A*, vol. 13, no. 3, pp. 431–492, 1998.

[5] M. A. Awada, M. J. Duff, and C. N. Pope, "$\mathcal{N} = 8$ supergravity breaks down to $\mathcal{N} = 1$," *Phys. Rev. Lett.*, vol. 50, no. 5, p. 294, 1983.

[6] J. Bagger and N. Lambert, "Comments on multiple M2-branes," *J. High Energy Phys.*, vol. 2008, no. 02, p. 105, 2008. doi:10.1088/1126-6708/2008/02/105, arXiv:0712.3738 [hep-th].

[7] V. Bargmann, "Zur Theorie des Wasserstoffatoms," *Z. Phys. A Hadrons Nucl.*, vol. 99, no. 7, pp. 576–582, 1936.

[8] A. Barut and R. Raczka, *Theory of Group Representations and Applications*. World Scientific Publishing Co Inc., Singapore, 1986.

[9] A. O. Barut, P. Budini, and C. Fronsdal, "Two examples of covariant theories with internal symmetries involving spin," in *Proceedings of the Royal Society of London A: Mathematical, Physical and Engineering Sciences*, vol. 291, pp. 106–112, The Royal Society, London, 1966.

[10] M. Billó, D. Fabbri, P. Fré, P. Merlatti, and A. Zaffaroni, "Shadow multiplets in AdS_4/CFT_3 and the super-Higgs mechanism: hints of new shadow supergravities," *Nucl. Phys. B*, vol. 591, no. 1–2, pp. 139–194, 2000. hep-th/0005220.

[11] M. Billó, D. Fabbri, P. Fré, P. Merlatti, and A. Zaffaroni, "Rings of short $\mathcal{N} = 3$ superfields in three dimensions and M-theory on $AdS_4 \times N^{010}$," *Class. Quantum Gravity*, vol. 18, no. 7, p. 1269, 2001. doi:10.1088/0264-9381/18/7/310, hep-th/0005219.

[12] B. Biran, F. Englert, B. de Wit, and H. Nicolai, "Gauged N=8 supergravity and its breaking from spontaneous compactification," *Phys. Lett. B*, vol. 124, no. 1–2, pp. 45–50, 1983.

[13] S. R. Blackburn, P. M. Neumann, and G. Venkataraman, *Enumeration of Finite Groups*. Cambridge Univ. Press, Cambridge, 2007.

[14] E. Brown and N. Loehr, "Why is $PSL(2,7) \approx GL(3,2)$?," *Am. Math. Mon.*, vol. 116, no. 8, pp. 727–732, 2009.

[15] W. Burnside, *Theory of Groups of Finite Order*. Cambridge University Press, Cambridge, 1911.

[16] É. Cartan, Über die einfachen Transformationsgruppen, Berichte über die Verhandlungen der Königlich-Sächsischen Gesellschaft der Wissenschaften zu Leipzig, Mathematisch-Physische Klasse, 1893, pp. 395–420 (online archve: https://archive.org/stream/berichteberdiev02klasgoog#page/n446/mode/2up); reprint, *Oeuvres completes, Vol. I, no. 1*, Gauthier-Villars, Paris, 1952, pp. 107–132.

[17] É. Cartan, *Sur la structure des groupes de transformations finis et continus*, Thése, vol. 826. Nony, Paris, 1894.

[18] N. Carter, *Visual Group Theory*. Mathematical Association of America, Washington, 2009.

[19] L. Castellani, R. D'Auria, and P. Fré, "$SU(3) \otimes SU(2) \otimes U(1)$ from D=11 supergravity," *Nucl. Phys. B*, vol. 239, no. 2, pp. 610–652, 1984.

https://doi.org/10.1515/9783110551204-015

[20] L. Castellani, R. D'Auria, P. Fre, K. Pilch, and P. van Nieuwenhuizen, "The bosonic mass formula for Freund-Rubin solutions of D=11 supergravity on general coset manifolds," *Class. Quantum Gravity*, vol. 1, no. 4, p. 339, 1984.

[21] L. Castellani, L. J. Romans, and N. P. Warner, "A classification of compactifying solutions for D=11 supergravity," *Nucl. Phys. B*, vol. 241, no. 2, pp. 429–462, 1984.

[22] L. Castellani, R. D'Auria, and P. Fre, *Supergravity and Superstrings: A Geometric Perspective. Vol. 1: Mathematical Foundations*. World Scientific, Singapore, 1991.

[23] A. Cayley, "On the theory of groups, as depending on the symbolic equation $\theta^n = 1$," *Philos. Mag.*, vol. 7, pp. 40–47, 1854, appears in *The Collected Mathematical Papers of Arthur Cayley*, Cambridge University Press, Cambridge, 1889, vol. 2, pp. 123–130.

[24] A. Cayley, "A memoir on the theory of matrices," *Philos. Trans. R. Soc. Lond.*, vol. 148, pp. 17–37, 1858.

[25] A. Ceresole, P. Fré, and H. Nicolai, "Multiplet structure and spectra of $\mathcal{N} = 2$ supersymmetric compactifications," *Class. Quantum Gravity*, vol. 2, no. 2, p. 133, 1985.

[26] A. Ceresole, G. Dall'Agata, R. D'Auria, and S. Ferrara, "Spectrum of type IIB supergravity on $AdS_5 \times T^{11}$: predictions on $\mathcal{N} = 1$ SCFT's," *Phys. Rev. D*, vol. 61, no. 6, p. 066001, 2000. hep-th/9905226.

[27] S. Coleman, *Aspects of Symmetry: Selected Erice Lectures*. Cambridge University Press, Cambridge, 1988.

[28] F. Cordaro, P. Fré, L. Gualtieri, P. Termonia, and M. Trigiante, "$\mathcal{N} = 8$ gaugings revisited: an exhaustive classification," *Nucl. Phys. B*, vol. 532, no. 1–2, pp. 245–279, 1998. doi:10.1016/S0550-3213(98)00449-0, hep-th/9804056.

[29] J. F. Cornwell and J. Cornwell, *Group Theory in Physics, vol, 1*. Academic Press, London, 1984. Part A.

[30] A. P. Cracknell, *Group Theory in Solid-State Physics*. Halsted Press, London, 1975.

[31] A. Das and S. Okubo, *Lie Groups and Lie Algebras for Physicists*. World Scientific, Singapore, 2014.

[32] R. D'Auaria, P. Fré, and P. Van Nieuwenhuizen, "$\mathcal{N} = 2$ matter coupled supergravity from compactification on a coset G/H possessing an additional killing vector," *Phys. Lett. B*, vol. 136, no. 5-6, pp. 347–353, 1984.

[33] R. D'Auria and P. Fré, "Spontaneous generation of Osp(4|8) symmetry in the spontaneous compactification of D=11 supergravity," *Phys. Lett. B*, vol. 121, no. 2–3, pp. 141–146, 1983.

[34] R. D'Auria and P. Fré, "On the spectrum of the $\mathcal{N} = 2$ SU(3) ⊗ SU(2) ⊗ U(1) gauge theory from D=11 supergravity," *Class. Quantum Gravity*, vol. 1, no. 5, p. 447, 1984.

[35] R. D'Auria and P. Fré, "Universal Bose–Fermi mass-relations in Kaluza–Klein supergravity and harmonic analysis on coset manifolds with Killing spinors," *Ann. Phys.*, vol. 162, no. 2, pp. 372–412, 1985.

[36] P. A. M. Dirac, "Quantised singularities in the electromagnetic field," in *Proceedings of the Royal Society of London A: Mathematical, Physical and Engineering Sciences*, vol. 133, pp. 60–72, The Royal Society, London, 1931.

[37] M. S. Dresselhaus, G. Dresselhaus, and A. Jorio, *Group Theory: Application to the Physics of Condensed Matter*. Springer Science & Business Media, Berlin, 2007.

[38] M. J. Duff and C. N. Pope, "Kaluza-Klein supergravity and the seven sphere," *Supersymmetry and supergravity*, p. 183, 1983. ICTP/82/83-7, Lectures given at September School on Supergravity and Supersymmetry, Trieste, Italy, Sep. 6–18, 1982. Published in Trieste Workshop 1982:0183 (QC178:T7:1982).

[39] J. Elliott and P. Dawber, *Symmetry in Physics, Volume 1 and 2*. Macmillan, London, 1979.

[40] L. Euler, *Methodus inveniendi lineas curvas maximi minimive proprietate gaudentes sive solutio problematis isoperimetrici latissimo sensu accepti*. Springer Science &

Business Media, Berlin, 1952. (Ed. Caratheodory Constantin, reprint of 1744 edition), ISBN 3-76431-424-9.

[41] D. Fabbri, P. Fré, L. Gualtieri, and P. Termonia, "M-theory on $AdS_4 \times M^{1,1,1}$: the complete $Osp(2|4) \times SU(3) \times SU(2)$ spectrum from harmonic analysis," *Nucl. Phys. B*, vol. 560, no. 1–3, pp. 617–682, 1999. hep-th/9903036.

[42] D. Fabbri, P. Fré, L. Gualtieri, C. Reina, A. Tomasiello, A. Zaffaroni, and A. Zampa, "3D superconformal theories from Sasakian seven-manifolds: new non-trivial evidences for AdS_4/CFT_3," *Nucl. Phys. B*, vol. 577, no. 3, pp. 547–608, 2000. hep-th/9907219.

[43] D. Fabbri, P. Fre, L. Gualtieri, and P. Termonia, "Osp(N|4) supermultiplets as conformal superfields on ∂AdS_4 and the generic form of $\mathcal{N} = 2$, D=3 gauge theories," *Class. Quantum Gravity*, vol. 17, no. 1, p. 55, 2000. hep-th/9905134.

[44] G. Fano, "Sui postulati fondamentali della geometria proiettiva in uno spazio lineare a un numero qualunque di dimensioni," *Giornale di Matematiche*, vol. 30, pp. 106–132, 1892. http://www.bdim.eu/item?id=GM_Fano_1892_1.

[45] G. Fano, *Metodi Matematici della Meccanica Quantistica*. Zanichelli, Bologna, 1967.

[46] S. Ferrara and C. Fronsdal, "Conformal Maxwell theory as a singleton field theory on AdS_5, IIB 3-branes and duality," *Class. Quantum Gravity*, vol. 15, no. 8, p. 2153, 1998. doi:10.1088/0264-9381/15/8/004, hep-th/9712239.

[47] S. Ferrara and C. Fronsdal, "Gauge fields as composite boundary excitations," *Phys. Lett. B*, vol. 433, no. 1, pp. 19–28, 1998. doi:10.1016/S0370-2693(98)00664-9, hep-th/9802126.

[48] S. Ferrara, C. Fronsdal, and A. Zaffaroni, "On $\mathcal{N} = 8$ supergravity in AdS_5 and $\mathcal{N} = 4$ superconformal Yang-Mills theory," *Nucl. Phys. B*, vol. 532, no. 1–2, pp. 153–162, 1998. doi:10.1016/S0550-3213(98)00444-1, hep-th/9802203.

[49] V. Fock, "Zur Theorie des Wasserstoffatoms," *Z. Phys. A Hadrons Nucl.*, vol. 98, no. 3, pp. 145–154, 1935.

[50] P. Fré, "Gaugings and other supergravity tools of p-brane physics," *arXiv preprint hep-th/0102114*, 2001. Proceedings of the Workshop on Latest Development in M-Theory, Paris, France, 1–9 Feb. 2001.

[51] P. G. Fré, *Gravity, a Geometrical Course, vol. 1, 2*. Springer Science & Business Media, Berlin, 2012.

[52] P. Fré, "Supersymmetric M2-branes with Englert fluxes, and the simple group PSL(2,7)," *Fortschr. Phys.*, vol. 64, no. 6-7, pp. 425–462, 2016. arXiv:1601.02253 [hep-th].

[53] P. G. Fré, *A Conceptual History of Symmetry from Plato to Special geometries*. To be published (Editorial process infieri with Springer).

[54] P. G. Fré, *Advances in Lie Algebra and Geometry from Supergravity*. To be published (Editorial process infieri with Springer).

[55] P. Fre and A. S. Sorin, "The Weyl group and asymptotics: all supergravity billiards have a closed form general integral," *Nucl. Phys.*, vol. B815, pp. 430–494, 2009.

[56] P. Fre and A. S. Sorin, "Classification of Arnold-Beltrami flows and their hidden symmetries," *Phys. Part. Nucl.*, vol. 46, no. 4, pp. 497–632, 2015.

[57] P. Fre and M. Trigiante, "Twisted tori and fluxes: a no go theorem for Lie groups of weak G2 holonomy," *Nucl. Phys. B*, vol. 751, no. 3, pp. 343–375, 2006. (doi:10.1016/j.nuclphysb.2006.06.006, hep-th/0603011).

[58] P. Fré, L. Gualtieri, and P. Termonia, "The structure of $\mathcal{N} = 3$ multiplets in AdS_4 and the complete $Osp(3|4) \times SU(3)$ spectrum of M-theory on $AdS_4 \times N^{0,1,0}$," *Phys. Lett. B*, vol. 471, no. 1, pp. 27–38, 1999. hep-th/9909188.

[59] P. Fre, V. Gili, F. Gargiulo, A. S. Sorin, K. Rulik, and M. Trigiante, "Cosmological backgrounds of superstring theory and solvable algebras: oxidation and branes," *Nucl. Phys.*, vol. B685, pp. 3–64, 2004.

[60] P. Fré, F. Gargiulo, J. Rosseel, K. Rulik, M. Trigiante, and A. Van Proeyen, "Tits–Satake projections of homogeneous special geometries," *Class. Quantum Gravity*, vol. 24, no. 1, pp. 27–78, 2006. doi:10.1088/0264-9381/24/1/003, hep-th/0606173.

[61] P. Fré, F. Gargiulo, and K. Rulik, "Cosmic billiards with painted walls in non-maximal supergravities: a worked out example," *Nucl. Phys. B*, vol. 737, no. 1, pp. 1–48, 2006. doi:10.1016/j.nuclphysb.2005.10.023, hep-th/0507256.

[62] P. Fre, A. S. Sorin, and M. Trigiante, "Black hole nilpotent orbits and Tits Satake universality classes," 2011, arXiv:1107.5986 [hep-th].

[63] P. Fre, A. S. Sorin, and M. Trigiante, "Integrability of supergravity black holes and new tensor classifiers of regular and nilpotent orbits," *J. High Energy Phys.*, vol. 04, p. 015, 2012.

[64] P. Fre, A. S. Sorin, and M. Trigiante, "The c-map, Tits Satake subalgebras and the search for $\mathcal{N} = 2$ inflaton potentials," *Fortschr. Phys.*, vol. 63, pp. 198–258, 2015.

[65] D. Z. Freedman and H. Nicolai, "Multiplet shortening in Osp(N|4)," *Nucl. Phys. B*, vol. 237, no. 2, pp. 342–366, 1984.

[66] P. G. O. Freund and M. A. Rubin, "Dynamics of dimensional reduction," *Phys. Lett. B*, vol. 97, no. 2, pp. 233–235, 1980.

[67] F. G. Frobenius, "Uber lineare Substitutionen und bilineare Formen," *J. Reine Angew. Math.*, vol. 84, pp. 1–63, 1878.

[68] C. Fronsdal, "Infinite multiplets and the hydrogen atom," *Phys. Rev.*, vol. 156, no. 5, pp. 1665–1677, 1967.

[69] D. Gaiotto and X. Yin, "Notes on superconformal Chern-Simons-Matter theories," *J. High Energy Phys.*, vol. 2007, no. 08, p. 056, 2007. doi:10.1088/1126-6708/2007/08/056, arXiv:0704.3740 [hep-th].

[70] H. Georgi, *Lie Algebras in Particle Physics: From Isospin to Unified Theories*, vol. 54. Westview Press, Boulder, 1999.

[71] R. Gilmore, *Lie Groups, Physics, and Geometry: An Introduction for Physicists, Engineers and Chemists*. Cambridge University Press, Cambridge, 2008.

[72] R. Gilmore, *Lie Groups, Lie Algebras, and Some of Their Applications*. Courier Corporation, North Chelmsford, 2012.

[73] H. Goldstein, "Prehistory of the "Runge–Lenz" vector," *Am. J. Phys.*, vol. 43, no. 8, pp. 737–738, 1975.

[74] H. Goldstein, "More on the prehistory of the Laplace or Runge–Lenz vector," *Am. J. Phys.*, vol. 44, no. 11, pp. 1123–1124, 1976.

[75] H. Goldstein, *Classical Mechanics*. Pearson Education, India, 2011.

[76] H. Grassmann, *Die Lineale Ausdehnungslehre, ein neuer Zweig der Mathematik*. Otto Wigand, Leipzig, 1844.

[77] M. J. Greenberg and J. R. Harper, *Algebraic Topology, a First Course*. The Benjamin/Cummings Publishing Company, London, 1981.

[78] M. Günaydin and N. P. Warner, "Unitary supermultiplets of Osp(8|4,R) and the spectrum of the S^7 compactification of 11-dimensional supergravity," *Nucl. Phys. B*, vol. 272, no. 1, pp. 99–124, 1986.

[79] M. Günaydin, D. Minic, and M. Zagermann, "4D doubleton conformal theories, CPT and IIB strings on $AdS_5 \times S^5$," *Nucl. Phys. B*, vol. 534, no. 1–2, pp. 96–120, 1998. hep-th/9806042.

[80] M. Günaydin, D. Minis, and M. Zagermann, "Novel supermultiplets of SU(2,2|4) and the AdS_5/CFT_4 duality," *Nucl. Phys. B*, vol. 544, no. 3, pp. 737–758, 1999. hep-th/9810226.

[81] M. Hamermesh, *Group Theory and Its Application to Physical Problems*. Courier Corporation, North Chelmsford, 1962.

[82] W. Heidenreich, "All linear unitary irreducible representations of de Sitter supersymmetry with positive energy," *Phys. Lett. B*, vol. 110, no. 6, pp. 461–464, 1982.

[83] V. Heine, *Group Theory in Quantum Mechanics: An Introduction to Its Present Usage*. Courier Corporation, North Chelmsford, 2007.

[84] S. Helgason, *Differential Geometry and Symmetric Spaces*. Academic Press, San Diego, 1962.

[85] M. Henneaux, D. Persson, and P. Spindel, "Spacelike singularities and hidden symmetries of gravity," *Living Rev. Relativ.*, vol. 11, no. 1, 2008. doi:10.12942/lrr-2008-1, arXiv:0710.1818 [hep-th].

[86] R. Hermann, *Lie Groups for Physicists, vol. 5*. WA Benjamin, New York, 1966.

[87] L. Hulthén, "Über die quantenmechanische Herleitung der Balmerterme," *Z. Phys. A Hadrons Nucl.*, vol. 86, no. 1, pp. 21–23, 1933.

[88] J. E. Humphreys, *Introduction to Lie Algebras and Representation Theory*. Springer, Berlin, 1972.

[89] J. E. Humphreys, *Reflection Groups and Coxeter Groups*. Cambridge University Press, Cambridge, 1992.

[90] A. Hurwitz, "Über algebraische Gebilde mit Eindeutigen Transformationen in sich," *Math. Ann.*, vol. 41, no. 3, pp. 403–442, 1892.

[91] T. Inui, Y. Tanabe, and Y. Onodera, *Group Theory and Its Applications in Physics*, vol. 78. Springer Science & Business Media, Berlin, 2012.

[92] N. Jacobson, *Lie Algebras*. Courier Corporation, North Chelmsford, 1979.

[93] G. James and M. W. Liebeck, *Representations and Characters of Groups*. Cambridge University Press, Cambridge, 2001.

[94] H. F. Jones, *Groups, Representations and Physics*. Taylor & Francis Group, LLC, New York, 1998.

[95] C. Jordan, "*Traité des substitutions et des équations algébriques*," Gauthiers-Villars, Parıs, 1870, Reedición: Gabay, Parıs, 1989.

[96] R. Kallosh and A. Van Proeyen, "Conformal symmetry of supergravities in AdS spaces," *Phys. Rev. D*, vol. 60, no. 2, p. 026001, 1999. doi:10.1103/PhysRevD.60.026001, hep- th/9804099.

[97] W. Killing, "Die Zusammensetzung der stetigen endlichen Transformationsgruppen," *Math. Ann.*, vol. 33, no. 1, pp. 1–48, 1888.

[98] R. C. King, F. Toumazet, and B. G. Wybourne, "A finite subgroup of the exceptional Lie group G2," *J. Phys. A, Math. Gen.*, vol. 32, no. 48, pp. 8527–8537, 1999.

[99] F. Klein, "Ueber die Transformation siebenter Ordnung der elliptischen Functionen," *Math. Ann.*, vol. 14, no. 3, pp. 428–471, 1878.

[100] F. Klein, "Vergleichende Betrachtungen über neuere geometrische Forschungen," *Math. Ann.*, vol. 43, no. 1, pp. 63–100, 1893. (Also: Gesammelte Abh. Vol. 1, Springer, Berlin, 1921, pp. 460-497).

[101] A. W. Knapp, *Lie Groups Beyond an Introduction*, vol. 140 of *Progress in Mathematics*. Springer Science & Business Media, Berlin, 2013.

[102] R. S. Knox and A. Gold, *Symmetry in the Solid State*. WA Benjamin, New York, 1964.

[103] S. Lie, "Zur theorie des Integrabilitetsfaktors," *Christiana Forh*, pp. 242–254, 1874.

[104] S. Lie, *Theorie der Transformationsgruppen, vol. I*, Written with the help of Friedrich Engel. B. G. Teubner, Leipzig. 1888.

[105] S. Lie, *Theorie der Transformationsgruppen, vol. II*, Written with the help of Friedrich Engel. B. G. Teubner, Leipzig. 1890.

[106] S. Lie, *Theorie der Transformationsgruppen, vol. III*, Written with the help of Friedrich Engel. B. G. Teubner, Leipzig. 1893.

[107] S. Lie, *Vorlesungen über continuierliche Gruppen mit geometrischen und anderen Anwendungen*. Written with the help of Georg Scheffers, B. G. Teubner, Leipzig. 1893.

[108] H. J. Lipkin, *Lie Groups for Pedestrians*. Courier Corporation, North Chelmsford, 2002.

[109] C. Luhn, S. Nasri, and P. Ramond, "Simple finite non-Abelian flavor groups," *J. Math. Phys.*, vol. 48, no. 12, p. 123519, 2007. arXiv:0709.1447 [hep-th].

[110] J. Maldacena, "The large-N limit of superconformal field theories and supergravity," *Int. J. Theor. Phys.*, vol. 4, no. 38, pp. 1113–1133, 1999. doi:10.1023/A:1026654312961, hep-th/9711200.

[111] I. A. Malkin and V. I. Man'ko, "Symmetry of the hydrogen atom," *Sov. J. Nucl. Phys.*, vol. 3, pp. 267–274, 1966.

[112] A. Miller, "Application of group representation theory to symmetric structures," *Appl. Math. Model.*, vol. 5, no. 4, pp. 290–294, 1981.

[113] J. Milnor, "On the geometry of the Kepler problem," *Am. Math. Mon.*, vol. 90, no. 6, pp. 353–365, 1983.

[114] J. Moser, "Regularization of Kepler's problem and the averaging method on a manifold," *Commun. Pure Appl. Math.*, vol. 23, no. 4, pp. 609–636, 1970.

[115] S. Mukhi and N. Mukunda, *Introduction to Topology, Differential Geometry and Group Theory for Physicists*, Wiley Eastern, New Delhi 1990.

[116] M. Nakahara, *Geometry, Topology and Physics*. CRC Press, Boca Raton 2003.

[117] C. Nash and S. Sen, *Topology and Geometry for Physicists*. Academic Press, London, 1983.

[118] H. Nicolai, "Representations of supersymmetry in anti de Sitter space," In *Supersymmetry and Supergravity'84*, World Scientific, Singapore, p. 368, 1985.

[119] V. I. Ogievetskii and I. V. Polubarinov, "Wave equations with zero and nonzero rest masses," *Sov. Phys. JETP*, vol. 37, no. 2, pp. 335–338, 1960.

[120] Y. S. Osipov, "The Kepler problem and geodesic flows in spaces of constant curvature," *Celest. Mech. Dyn. Astron.*, vol. 16, no. 2, pp. 191–208, 1977.

[121] D. N. Page and C. N. Pope, "Stability analysis of compactifications of D=11 supergravity with $SU(3) \times SU(2) \times U(1)$ symmetry," *Phys. Lett. B*, vol. 145, no. 5-6, pp. 337–341, 1984.

[122] D. N. Page and C. N. Pope, "Which compactifications of D=11 supergravity are stable?," *Phys. Lett. B*, vol. 144, no. 5–6, pp. 346–350, 1984.

[123] W. Pauli, "Über das Wasserstoffspektrum vom Standpunkt der neuen Quantenmechanik," *Z. Phys. A Hadrons Nucl.*, vol. 36, no. 5, pp. 336–363, 1926.

[124] G. Peano, *Calcolo Geometrico secondo l'Ausdehnungslehre di H. Grassmann, preceduto dalle operazioni della logica deduttiva*. Fratelli Bocca Editori, Torino, 1888.

[125] G. Pólya, *Combinatorial Enumeration of Groups, Graphs, and Chemical Compounds*, Springer, New York, 1987.

[126] H. Rademacher, "On the expansion of the partition function in a series," *Ann. Math.*, vol. 44, no. 3, pp. 416–422, 1943.

[127] P. Ramond, *Group Theory: A Physicist'S Survey*. Cambridge University Press, Cambridge, 2010.

[128] F. C. Santos, V. Soares, and A. C. Tort, "An English translation of Bertrand's theorem," *Lat. Am. J. Physics Educ.*, vol. 5, no. 4, pp. 694–696, 2011. arXiv:0704.2396.

[129] E. Schmutzer, *Symmetrien und Erhaltungssätze der Physik*, Akademie Verlag, Berlin, 1972.

[130] R. Slansky, "Group theory for unified model building," *Phys. Rep.*, vol. 79, no. 1, pp. 1–128, 1981.

[131] S. Sternberg, *Group Theory and Physics*. Cambridge University Press, Cambridge, 1995.

[132] J. Stickforth, "On the complementary Lagrange formalism of classical mechanics," *Am. J. Phys.*, vol. 46, no. 1, pp. 71–73, 1978.

[133] J. Stickforth, "The classical Kepler problem in momentum space," *Am. J. Phys.*, vol. 46, no. 1, pp. 74–75, 1978.

[134] J. Stillwell, *Naive Lie theory*. Springer Science & Business Media, New York, 2008.

[135] H.-W. Streitwolf, *Group Theory in Solid-State Physics*. Macdonald and Co., London, 1971.

[136] J. J. Sylvester, "A demonstration of the theorem that every homogeneous quadratic polynomial is reducible by real orthogonal substitutions to the form of a sum of positive and negative squares," *Philos. Mag. Ser. 4*, vol. 4, no. 23, pp. 138–142, 1852. (http://www.maths.ed.ac.uk/~aar/sylv/inertia.pdf).

[137] M. Tinkham, *Group Theory and Quantum Mechanics*. Courier Corporation, North Chelmsford, 2003.

[138] V. S. Varadarajan, *Lie Groups, Lie Algebras, and Their Representations*. Springer-Verlag New York Inc., New York, 2013.

[139] H. Weyl, *The Classical Groups: Their Invariants and Representations*. Princeton University Press, Princeton, 1939.

[140] H. Weyl, *The Theory of Groups and Quantum Mechanics*. Courier Corporation, North Chelmsford, 1950.

[141] E. Wigner, *Group Theory: and Its Application to the Quantum Mechanics of Atomic Spectra*, vol. 5. Academic Press, New York, 1959.

[142] E. Witten, "Search for a realistic Kaluza–Klein theory," *Nucl. Phys. B*, vol. 186, no. 3, pp. 412–428, 1981.

[143] P. Woit, "Quantum theory, groups and representations: An introduction," *Department of Mathematics, Columbia University*, 2015, https://www.math.columbia.edu/~woit/QM/qmbook.pdf.

[144] B. G. Wybourne, *Classical Groups for Physicists*. Wiley, New York, 1974.

[145] A. Zee, *Group Theory in a Nutshell for Physicists*. Princeton University Press, Princeton, 2016.

[146] "Chemie im Computer." http://wchem.cup.uni-muenchen.de/wvib/. Accessed: 2017-06-30.

[147] "Keith Conrad personal webpage." http://www.math.uconn.edu/~kconrad/. Accessed: 2017-06-30.

[148] "Mathematics Stack Exchange forum." https://math.stackexchange.com/. Accessed: 2017-06-30.

www.ingramcontent.com/pod-product-compliance
Lightning Source LLC
Chambersburg PA
CBHW080122220326
41598CB00032B/4926
* 9 7 8 3 1 1 0 5 5 1 1 9 8 *